Friction in textile materials

The Textile Institute and Woodhead Publishing

The Textile Institute is a unique organisation in textiles, clothing and footwear. Incorporated in England by a Royal Charter granted in 1925, the Institute has individual and corporate members in over 90 countries. The aim of the Institute is to facilitate learning, recognise achievement, reward excellence and disseminate information within the global textiles, clothing and footwear industries.

Historically, The Textile Institute has published books of interest to its members and the textile industry. To maintain this policy, the Institute has entered into partnership with Woodhead Publishing Limited to ensure that Institute members and the textile industry continue to have access to high calibre titles on textile science and technology.

Most Woodhead titles on textiles are now published in collaboration with The Textile Institute. Through this arrangement, the Institute provides an Editorial Board which advises Woodhead on appropriate titles for future publication and suggests possible editors and authors for these books. Each book published under this arrangement carries the Institute's logo.

Woodhead books published in collaboration with The Textile Institute are offered to Textile Institute members at a substantial discount. These books, together with those published by The Textile Institute that are still in print, are offered on the Woodhead web site at: www.woodheadpublishing.com. Textile Institute books still in print are also available directly from the Institute's web site at: www.textileinstitutebooks.com.

A list of Woodhead books on textile science and technology, most of which have been published in collaboration with The Textile Institute, can be found at the end of the contents pages.

Woodhead Publishing in Textiles: Number 78

Friction in textile materials

Edited by
B. S. Gupta

The Textile Institute

CRC Press
Boca Raton Boston New York Washington, DC

WOODHEAD PUBLISHING LIMITED
Cambridge, England

Published by Woodhead Publishing Limited in association with The Textile Institute
Woodhead Publishing Limited, Abington Hall, Granta Park,
Great Abington
Cambridge CB21 6AH, England
www.woodheadpublishing.com

Published in North America by CRC Press LLC, 6000 Broken Sound Parkway, NW,
Suite 300, Boca Raton, FL 33487, USA

First published 2008, Woodhead Publishing Limited and CRC Press LLC

British Library Cataloguing in Publication Data
A catalogue record for this book is available from the British Library.

Library of Congress Cataloging in Publication Data
A catalog record for this book is available from the Library of Congress.

Woodhead Publishing ISBN 978-1-85573-920-8 (book)
Woodhead Publishing ISBN 978-1-84569-472-2 (e-book)
CRC Press ISBN 978-0-8493-3478-8
CRC Press order number WP3478

The publishers' policy is to use permanent paper from mills that operate a sustainable forestry policy, and which has been manufactured from pulp which is processed using acid-free and elementary chlorine-free practices. Furthermore, the publishers ensure that the text paper and cover board used have met acceptable environmental accreditation standards.

Project managed by Macfarlane Book Production Services, Dunstable, Bedfordshire, England (e-mail: macfarl@aol.com)
Typeset by Replika Press Pvt Ltd, India
Printed by T J International Limited, Padstow, Cornwall, England

Dedication

To Frank Bowden and David Tabor, their associates, and other scientists, who have pioneered the work in the field of the tribology of fibers.

Contents

Contributor contact details

(* = main contact)

Editor, Chapters 1, 2 and 3

Professor B. S. Gupta
College of Textiles
North Carolina State University
2401 Research Drive
Box 8301
Raleigh, NC 27695
USA

E-mail: bgupta@ncsu.edu

Chapter 4

Professor R. Kovar* and Professor Z. Kus
Faculty of Textile Engineering
Technical University of Liberec
Studenska 2
461 17 Liberec
Czech Republic

E-mail: radko.kovar@tul.cz

Professor B. S. Gupta
College of Textiles
North Carolina State University
2401 Research Drive
Box 8301
Raleigh, NC 27695
USA

Chapter 5

Professor B. S. Gupta*
College of Textiles
North Carolina State University
2401 Research Drive
Box 8301
Raleigh, NC 27695
USA

E-mail: bgupta@ncsu.edu

Professor J. O. Ajayi
School of Science
Abubakar Tafawa Balewa University
740004 Bauchi
Nigeria

Dr M. Kutsenko
Goulston Technologies Inc.
700 North Johnson Street
Monroe, NC 28110
USA

Chapters 6 and 8

Professor Y. E. El Mogahzy
Department of Textiles
Auburn University
101 Textile Building
Auburn, AL 36849
USA

E-mail: yehiae@eng.auburn.edu

Chapter 7

Dr J. A. Rippon
Centre for Material and Fibre Innovation
Faculty of Science and Technology
Deakin University
Pigdons Road
Geelong
Victoria 3217
Australia

E-mail: john.rippon@deakin.edu.au

Chapter 9

Professor V. K. Kothari* and
Dr A. Das
Department of Textile Technology
Indian Institute of Technology
Hauz Khas
New Delhi 110016
India

E-mail: kothari@textile.iitd.ernet.in

Chapter 10

Professor J. O. Ajayi
School of Science
Abubakar Tafawa Balewa University
740004 Bauchi
Nigeria

E-mail: joajayi@atbunet.org

Chapter 11

Dr M. Kutsenko* and
Dr T. W. Theyson
Goulston Technologies Inc.
700 North Johnson Street
Monroe, NC 28110
USA

E-mail: mkutsenko@goulston.com

Chapter 12

Y. Li and Dr J. Hinestroza*
Department of Fiber Science
Cornell University
213 Martha Van Rensselaer Hall
Ithaca, NY 14853
USA

E-mail: jh433@cornell.edu

Woodhead Publishing in Textiles

1 **Watson's textile design and colour Seventh edition**
Edited by Z. Grosicki

2 **Watson's advanced textile design**
Edited by Z. Grosicki

3 **Weaving Second edition**
P. R. Lord and M. H. Mohamed

4 **Handbook of textile fibres Vol 1: Natural fibres**
J. Gordon Cook

5 **Handbook of textile fibres Vol 2: Man-made fibres**
J. Gordon Cook

6 **Recycling textile and plastic waste**
Edited by A. R. Horrocks

7 **New fibers Second edition**
T. Hongu and G. O. Phillips

8 **Atlas of fibre fracture and damage to textiles Second edition**
J. W. S. Hearle, B. Lomas and W. D. Cooke

9 **Ecotextile '98**
Edited by A. R. Horrocks

10 **Physical testing of textiles**
B. P. Saville

11 **Geometric symmetry in patterns and tilings**
C. E. Horne

12 **Handbook of technical textiles**
Edited by A. R. Horrocks and S. C. Anand

13 **Textiles in automotive engineering**
W. Fung and J. M. Hardcastle

14 **Handbook of textile design**
J. Wilson

15 **High-performance fibres**
Edited by J. W. S. Hearle

16 **Knitting technology Third edition**
D. J. Spencer

17 **Medical textiles**
Edited by S. C. Anand

18 **Regenerated cellulose fibres**
Edited by C. Woodings

19 **Silk, mohair, cashmere and other luxury fibres**
Edited by R. R. Franck

20 **Smart fibres, fabrics and clothing**
Edited by X. M. Tao

21 **Yarn texturing technology**
J. W. S. Hearle, L. Hollick and D. K. Wilson

22 **Encyclopedia of textile finishing**
H-K. Rouette

23 **Coated and laminated textiles**
W. Fung

24 **Fancy yarns**
R. H. Gong and R. M. Wright

25 **Wool: Science and technology**
Edited by W. S. Simpson and G. Crawshaw

26 **Dictionary of textile finishing**
H-K. Rouette

27 **Environmental impact of textiles**
K. Slater

28 **Handbook of yarn production**
P. R. Lord

29 **Textile processing with enzymes**
Edited by A. Cavaco-Paulo and G. Gübitz

30 **The China and Hong Kong denim industry**
Y. Li, L. Yao and K. W. Yeung

31 **The World Trade Organization and international denim trading**
Y. Li, Y. Shen, L. Yao and E. Newton

32 **Chemical finishing of textiles**
W. D. Schindler and P. J. Hauser

33 **Clothing appearance and fit**
J. Fan, W. Yu and L. Hunter

34 **Handbook of fibre rope technology**
H. A. McKenna, J. W. S. Hearle and N. O'Hear

35 **Structure and mechanics of woven fabrics**
J. Hu

36 **Synthetic fibres: nylon, polyester, acrylic, polyolefin**
Edited by J. E. McIntyre

37 **Woollen and worsted woven fabric design**
E. G. Gilligan

38 **Analytical electrochemistry in textiles**
P. Westbroek, G. Priniotakis and P. Kiekens

39 **Bast and other plant fibres**
R. R. Franck

40 **Chemical testing of textiles**
Edited by Q. Fan

41 **Design and manufacture of textile composites**
Edited by A. C. Long

42 **Effect of mechanical and physical properties on fabric hand**
Edited by Hassan M. Behery

43 **New millennium fibers**
T. Hongu, M. Takigami and G. O. Phillips

44 **Textiles for protection**
Edited by R. A. Scott

45 **Textiles in sport**
Edited by R. Shishoo

46 **Wearable electronics and photonics**
Edited by X. M. Tao

47 **Biodegradable and sustainable fibres**
Edited by R. S. Blackburn

48 **Medical textiles and biomaterials for healthcare**
Edited by S. C. Anand, M. Miraftab, S. Rajendran and J. F. Kennedy

49 **Total colour management in textiles**
Edited by J. Xin

50 **Recycling in textiles**
Edited by Y. Wang

51 **Clothing biosensory engineering**
Y. Li and A. S. W. Wong

52 **Biomechanical engineering of textiles and clothing**
Edited by Y. Li and D. X-Q. Dai

53 **Digital printing of textiles**
Edited by H. Ujiie

54 **Intelligent textiles and clothing**
Edited by H. Mattila

55 **Innovation and technology of women's intimate apparel**
W. Yu, J. Fan, S. C. Harlock and S. P. Ng

56 **Thermal and moisture transport in fibrous materials**
Edited by N. Pan and P. Gibson

57 **Geosynthetics in civil engineering**
Edited by R. W. Sarsby

58 **Handbook of nonwovens**
Edited by S. Russell

59 **Cotton: Science and technology**
Edited by S. Gordon and Y-L. Hsieh

60 **Ecotextiles**
Edited by M. Miraftab and A. Horrocks

61 **Composite forming technologies**
Edited by A. C. Long

62 **Plasma technology for textiles**
Edited by R. Shishoo

63 **Smart textiles for medicine and healthcare**
Edited by L. Van Langenhove

64 **Sizing in clothing**
Edited by S. Ashdown

65 **Shape memory polymers and textiles**
J. Hu

66 **Environmental aspects of textile dyeing**
Edited by R. Christie

67 **Nanofibers and nanotechnology in textiles**
Edited by P. Brown and K. Stevens

68 **Physical properties of textile fibres Fourth edition**
W. E. Morton and J. W. S. Hearle

69 **Advances in apparel production**
Edited by C. Fairhurst

70 **Advances in fire retardant materials**
Edited by A. R. Horrocks and D. Price (forthcoming)

71 **Polyesters and polyamides**
Edited by B. L. Deopora, R. Alagirusamy, M. Joshi and B. Gupta

72 **Advances in wool**
Edited by N. A. G. Johnson and I. Russell (forthcoming)

73 **Military textiles**
Edited by E. Wilusz

74 **3D fibrous assemblies**
J. Hu (forthcoming)

75 **Medical textiles 2007**
Edited by J. Kennedy, A. Anand, M. Miraftab and S. Rajendran (forthcoming)

76 **Fabric testing**
Edited by J. Hu (forthcoming)

77 **Biomimetics and textiles**
Edited by A. Abbott and M. Ellison (forthcoming)

78 **Friction in textile materials**
Edited by B. S. Gupta

Preface

The first major effort in summarizing and comprehensively presenting a work on friction in textile materials was made by Howell, Mieszkis and Tabor in 1959 when they wrote the book, *Friction in Textiles* (Butterworths). Since then much has been published in the field, both at the theoretical and the practical levels, in understanding the mechanisms that apply to fibrous materials and evaluating, modeling and understanding the effects of material, structural and operational variables. Although a number of smaller reviews have appeared, a more comprehensive discussion on the subject has not been published. The project of writing the current monograph was undertaken to put together a more thorough and current treatise on friction in textiles.

The book contains 12 chapters, addressing different aspects of the subject matter (see the 'Introduction,' that follows); they are arranged in a manner that the reader can get a coherent picture of the inter-related concepts of friction in fibers. Except for the first two chapters, which are meant to provide the necessary background information about the structure and properties of textile fibers, and about friction classical laws and concepts known in the 1950s, other chapters provide the current status of the work in the respective areas and future prospects that will stimulate further research. The addition of this monograph to the literature is hoped to provide fundamental scientific and engineering knowledge for those in academia, involved in the teaching of fiber science and in the research on the behavior of assemblies, and in industry, involved in developing automated/high speed processes and in designing fibrous assemblies for new and novel products. The book should also serve as a suitable text for a course on the tribology of textiles at the graduate and senior undergraduate levels.

The chapters have undergone an extensive peer-review process in that each chapter has been reviewed and edited by two to three other authors and, in several cases, also by external experts in the field. Accordingly, the authors feel that the book is an up-to-date treatise on the subject area and will be a useful addition to the literature on textile science and engineering.

With great pleasure I extend my gratitude to the contributing authors for agreeing to develop their chapters and working with me in revising the contents and serving as peer-reviewers of other contributions. Without their co-operation and support this major undertaking could not have been completed. Several external textile scientists also participated in the review process, notably Professor David Buchanan, North Carolina State University, USA, Professor John Hearle, UMIST, UK, and Professor Stephen Michielsen, North Carolina State University, USA. I thank these and other professional colleagues for reviewing the material and making many useful suggestions. Thanks are also due to the secretarial staff of the Department of Textiles Chemistry, Engineering and Science, for their help with word processing and miscellaneous matters, and the administration of the Department for encouragement and support throughout the project. Two of my graduate students, Mr. Ajit Moghe, Ph.D. in Fiber and Polymer Science, and Mr. Venugopal Boppa, M.S. in Textile Engineering, deserve special recognition and thanks. They spent countless hours in conducting computer searches of the literature, developing soft copies of the figures for many chapters, and formatting and developing the chapters for final submission. Finally, I thank Woodhead Publishing Limited for approaching me to consider this publication. All individuals with whom I have had the opportunity to interact, in particular, Emma Starr, Kathryn Picking, Francis Dodds and Amanda Macfarlane, were most understanding, co-operative, and helpful.

Bhupender S. Gupta
Raleigh, North Carolina

Introduction

B S GUPTA, North Carolina
State University, USA

The great role friction plays in our life was expressed fully by Bowden and Tabor in their book, *Introduction to Tribology* (Anchor Press, 1973): "from the ticking of clock in the morning, throughout our work and leisure until we clean our teeth at night, we are involved in tribology." Indeed it is difficult to imagine a physical phenomenon in which friction does not play a significant role. We note its works in many different forms: the force we exert in performing mechanical work, the electrical power and fuel we use in running equipment, the wear damage we find on surfaces, the temperature rise we observe in materials and devices during use, and the sound we hear in most industrial and non-industrial operations. Friction is required to be controlled for improving the efficiency of industrial operations, the quality of our products, the performance of transport equipment and devices, and the quality, durability and comfort of materials we wear or use.

Specific to textiles, friction governs the quality of the product made from and the efficiency of staple processing, twisting and winding of yarns, weaving, knitting and braiding of fabrics, and forming and finishing of final products from textiles. Surface physical properties, such as smoothness, roughness or hand, and mechanical properties, such as drape, shear, stiffness, strength and elastic recovery, are all influenced by this property. Friction also affects wear of textiles and guides. The introduction of synthetic fibers with potential for new apparel and non-apparel applications, and increasing interest in high speed automated operations, required that the tribology of fibers be understood.

It is well known that the requirement for friction is not constant; it varies from process to process and application to application, and also within a process from point-to-point and within an application from function-to-function. Usually, different values are desired for optimum performance. As an example, in the use of braided textiles such as sutures for closing wounds, a relatively lower value will be preferred for knot tying and sliding it to the wound, but a higher value will be desired for insuring knot security. Likewise, in use of textiles as apparel, this duality of requirement prevails. A lower value of friction will enhance drape and comfort properties whereas a higher value will contribute to strength and dimensional stability. While one cannot instantly and exactly vary the frictional resistance during a process or application, except by varying pressure which is not always feasible, understanding this property and the effects the material, the processing and the finishing factors produced on it can provide a useful tool for selecting parameters that are optimum for the process or the application.

The project of writing the book was undertaken to put together a current and comprehensive monograph on friction in textiles. Because the field is broad, the treatise is given as a collection of works developed by different authors. Collectively, the chapters include a useful and thorough review of the concepts and models and of the experimental procedures and results. The entire subject area is presented in two parts. The first (Chapters 1 to 5) is devoted to the understanding of the structure of materials, the laws and mechanisms applying to friction in both metals and fibers, the nature of the friction profile generally obtained and the factors affecting it, and the test methods that have been and can be used in examining the property in different types of materials and for different purposes. The second part (Chapters 6 to 12) then deals with the behavior found in specific fiber types, processes and structures.

Included in Chapter 1 is an introduction to fibers, and their structures and properties. The premise being that although friction is a surface phenomenon, it is governed by both the surface and the bulk properties. In order to understand and control friction, an understanding of a fiber's morphology, chemical and physical structures and properties is necessary. Numerous fibers exist and, therefore, the details of the above knowledge will provide the needed foundation for understanding the differences in the friction behavior found in different materials and under different sets of conditions.

The early work in friction was necessarily performed on metals, which were the materials of interest of the time. Studies on these led to the development of the classical laws. A deeper understanding of friction, however, evolved in the twentieth century with the work on non-metals, which were the elastomers, plastics, Teflon®, and fibers. The century referred to can in fact be called the renaissance of the science of tribology. The studies led to the refinement of the concept of friction by recognizing the role actual adhesion between contacting surfaces played in determining friction force. The departures from the classical laws were observed when one or both of the surfaces rubbed were either a rubber or a plastic. The role of the bulk properties of the underlying layers in the sliding or rolling action, when one of the materials was viscoelastic, was recognized. The use of a lubricant in reducing friction, heating and wear in metals, in both the low pressure-high speed and high pressure-low speed operations, was well understood. Chapter 2 gives details of this early work, in large part as discovered on and applied to metals, but also, to smaller degree, of what has been known about the behavior of other materials, i.e. rubbers, plastics and fibers.

Extensive studies conducted on fibrous materials have shown that the relationship between the frictional force and the normal force is not linear but given by the power law, $F = aN^n$, in which n is usually less than 1. The values of the constants a and n are found experimentally and have been shown to vary from one material to another, as well as, on a given material, from one set of conditions to another. The coefficient of friction, μ, given by the ratio of the frictional force to the normal force is not a constant in these materials but a function of the normal force and the geometrical area of contact. Assuming a non-linear general model for the pressure-area relationship, i.e. $P = KA^\alpha$, in which K and α are the constants of the material, a model for the true area of contact between the bodies was developed. This treatment gave detailed structures to a and n, which, heretofore, were strictly empirical constants. The structures given allow one to understand the effects the bulk properties, the morphological features and the procedure related factors produce on the values of the constants and through them on the value of μ. Chapter 3 includes a detailed discussion of the theory and shows the usefulness of the model in understanding the results found in a number of experimental investigations.

It is generally recognized that it takes greater force to initiate sliding than to maintain it. This fact combined with the elasticity of the measuring system, under certain conditions, leads to a profile that is characterized by the presence of peaks and troughs. The fluctuations may be regular, as expected in a smooth and uniform material, or irregular, as expected in assemblies characterized by variations in structure. Usually the pattern expected in textile product is complex, due to the superimposition of the effects from the dynamics of the measuring system and from variations at each level of organization in the hierarchy of the assembly structure. A careful analysis of the friction profile can provide useful information about the nature of the material and the assembly structure. An insight into the phenomenon of stick-slip, along with the identification of the range of factors influencing the profile is given in Chapter 4. Studies conducted using model yarns proved highly instructional and showed the effects of each of the many factors that have a bearing on the pattern. Also included in this chapter is a model that accounts for the stick-slip phenomenon and illustrates the effects of the material and the procedure related factors on the profile.

A large number of test methods have been described in the literature for measuring friction. They differ in terms of the form in which the textile material is used, the type of the contact maintained between the surfaces, the test environment that surrounds the contact region, and the methods by which the relative motion is actuated, the normal force is imposed and the friction resistance is measured. Irrespective of the nature of the textile material used, the contact still occurs between two fibers or between a fiber and another surface. It has been of fundamental interest to the fiber scientists and engineers, therefore, to understand the frictional behavior of single fibers. Accordingly, much effort noted in the literature is on being able to effectively measure friction at that level. In the textile industry, however, with increased emphasis on high speed processing, especially in filament yarn manufacturing, the control of friction with the use of a topically applied finish has been of great interest. Thus, in some of the tests of friction used, a finish has been incorporated as a primary variable, along with the speed of sliding and the nature of the second surface as associate variables. Finally, with textiles used as apparel, one of the interests at the fabric level has been on understanding the role the material and the construction factors produce on the subjective feeling of hand and comfort. Correlating the objective measures of surface friction with the subjective measures of hand has been the aim in some studies on fabrics. Chapter 5 includes a review of the measuring techniques that have been and can be used and the models that characterize friction based on them. The presentation is made in three parts. In the first, the general methods along with the associated equations are included. In the second, the testing and characterization of friction in fabrics, along with the examination of the role the topographical features of a fabric play in influencing friction and hand, are considered. In the last part, a description of the standard test methods and commercially available friction devices is given.

Covered in Part II, is the examination of the frictional behavior found in specific fibers, structures and processes. The first three chapters of this part discuss friction and associated issues, challenges and scopes, in three types of fibers, i.e. cotton, wool and synthetic. These fibers differ greatly in their structure and morphology and, therefore, also in their frictional properties. Chapter 6 focuses on cotton, highlights its characteristics and structural features, both of which influence the fiber's surface forces, and examines its frictional behavior. A special test method is discussed that is particularly suited for use on cotton and is said to be capable of separating the role of fiber dimensions from that of fiber surface in determining the frictional resistance. The method presents the potential for standardization of the technique for use on the fiber and inclusion of the parameters obtained from it in the data base for more effectively classing/grading cotton.

Chapter 7 discusses the structure, properties, and friction related issues of the second most popular natural textile fiber, the wool. The fiber has unique chemical, physical and morphological structures. The surface has scales, oriented root to tip, which gives the fiber directional frictional properties and, due to it, the tendency to felt or shrink when subjected to aqueous washing. Alleviating this tendency requires that the scales be either significantly destroyed or masked, the latter by deposition of a layer of suitable polymer. Much research has gone into finding efficient and economical methods that not only impart shrink-proofing properties to the fiber but also add value to the product. These and other aspects of wool are covered in Chapter 7. It starts with a discussion of the fiber's chemical and physical structure. Following this is a review of wool's frictional properties and the mechanisms that cause felting. The final section includes comments on some of the problems associated with the current shrink-proofing methods and highlights some of the directions that are being investigated for future industrial applications.

The surface characteristics of synthetic fibers are dealt with in Chapter 8. The emphasis is placed on bringing out the important design parameters that can assist a manufacturer in modifying the surface structure, and through it, the fiber's frictional behavior. The parameters that affect friction and can be conveniently changed during the extrusion process include: fiber size, shape, crimp, molecular orientation, and surface texture. The impressive developments taking place in nano-fiber technology makes it possible to alter a fiber's surface at the microscopic scale and, therefore, friction, without significantly affecting bulk properties. The effect of wettability on friction and that of surface texture on wettability are also discussed in this chapter.

A vast majority of apparel and home furnishing textiles and significant fraction of nonwovens are based on staple fiber processing. The staple fibers are usually received at processing plants in highly compacted bales, which must be opened, (in case of cotton also cleaned), carded/combed, drafted and converted to products. The fibers interact with each other and with other surfaces during the process. The efficiency of the operation and the quality of the end product are greatly influenced by fiber frictional properties. The magnitude of friction required for optimum effect is not constant but varies from point-to-point in the process. If friction cannot be varied as needed then the process parameters must be adjusted for smoother operation. A detailed discussion on, and the requirements for, the role of friction in the staple fiber processing is included in Chapter 9.

Fabrics make contact with other fabrics as well as with other surfaces during use. Drape, style, hand, abrasion resistance and elastic recovery are some of the characteristics affected by fabric friction. The latter is governed by the structure of the fabric and the frictional properties of the constituent elements. Any protrusion on the surface can be expected to alter the resistance to motion detected when a fabric is stroked tactually by hand or rubbed by another surface. The number of stick-slip pulses and their amplitudes can be expected to be related to the fabric sett and cord/crown height. A finishing treatment given, mechanical or chemical, produces a change in the frictional response. Friction being a surface phenomenon, fabric topography, in addition to the contents, can have a large influence on the behavior. A detailed discussion of these and other aspects of fabric friction are covered in Chapter 10.

The necessity of modifying friction in man-made fibers was identified soon after the development of the first commercially successful synthetic fiber in 1939. Since then, spin finishes or, more correctly, fiber lubricating systems, have been used to control friction and facilitate spinning of all man-made fibers and yarns. It is well known that without such control on surface behavior, the efficient production of man-made fiber based products, especially those involving high speed operations, would be impossible. Today, the contemporary fiber lubricating systems are designed not only to provide high efficiency in fiber spinning, drawing and texturing processes, but also to impart a range of secondary properties appreciated by the consumer, as, for example, acceptable hand (tactile feel) and antistatic characteristics. Most surprisingly, the modern lubricating systems work at minute add-ons, normally not exceeding 1% by weight of fiber. Expectedly, the development of the spin finishes has always started with an analysis of their impact on the frictional behavior of the textiles for which they are formulated. Chapter 11 explores the mechanisms that underlie lubrication and the role lubrication plays in textiles, examines the factors that modify friction and enhance processing, and highlights some of the key trends found in experimental investigations.

It is to be expected that a significant part of friction in textiles, given a finish, is governed by the boundary lubrication phenomena. This includes fiber-to-fiber friction in most situations, and fiber to other surface frictions during processing of staple fibers into yarns and fabrics, conversion of fabrics into products, and day-to-day handling of textiles by consumers. In boundary lubrication, the finish, ideally a monolayer-thick, is not considered to totally mask the morphology and the chemical potential of a fiber surface, but to modify them. At such microscopic or nanometer level, the lubricant tends to form self assembled molecular layers due to interactions with the surface of the fiber. These interactions, which can be investigated using the technique of lateral force microscopy, appear to determine the alignment modes of the molecules of a lubricant and hence its lubrication efficiency. Chapter 12 gives details of the lateral force microscopy technique, as it applies to lubrication, reviews the chemical structures and properties of lubricants frequently used in textile processing, and explores in detail the boundary lubrication behavior of a number of finishing agents. It is visualized that a fuller understanding of the behavior of surfaces, treated with finishes and lubricants, at the molecular level will allow the industry to develop optimum treatment compositions for each different material and application.

The 12 chapters, introduced above, are arranged in a sequence that facilitates a reader to advance progressively from the fundamental concepts of friction to more practical aspects of the property in influencing behavior during processing and use. Although a list of the contents of the chapters is given at the beginning of the book, a more detailed index presented at the end will allow the reader to identify the specific area of interest and find it quickly in the text.

Although the subject matter of the book has a single theme and the chapters, written by different authors and covering different aspects of the subject, address it, some redundancy exists which is unavoidable. The most noteworthy of these is that each different author has taken the liberty of including a small section that describes the basic concept of friction, the classical laws addressing it and the departure from these noted in the behavior of fibrous materials. The readers will also note some apparent difference in the way the kinetic frictional force is taken from the friction force profile. Some authors have taken strictly the conceptual view of the parameter by considering only the material properties affecting it and described it as the average value corresponding the troughs of the friction trace. Others have taken into consideration both the material properties and the dynamics of the measuring system in defining the parameter. They assumed the value to lie half way between the troughs and the peaks. This difference in the manner of defining the kinetic frictional force parameter in the book has caused no difficulty in interpreting results and drawing conclusions.

Part I

Understanding friction in textiles

1
Textile fiber morphology, structure and properties in relation to friction

B S GUPTA, North Carolina State University, USA

1.1 Introduction

Friction is a surface phenomenon; however, it is governed by both the surface and the bulk properties of the material. This entails a comprehensive understanding of a material's morphology and its chemical and physical structures and properties. Morphology, i.e. surface architecture, should influence the nature of contact between two bodies brought against each other. Chemical structure should affect the bonding between molecular layers within the body as well as the forces on the free surface that interact with other surfaces. The physical structure combined with the chemical affect the bulk properties, especially the specific shear strength and the compressional modulus (hardness) of material, both of which are known to affect the frictional force.

Another structure-dependent bulk mechanical property, i.e. the viscoelastic behavior, affects the nature of the friction profile obtained as a function of time. Surface energetics, a function of morphology and chemical structure, play an important role in governing the tribological behavior of materials. For example, the application of any finish, such as a lubricant, to modify friction, requires that the chemical used is compatible with the surface. Compatibility between two materials is reflected by the contact angle, which in turn is a function of the magnitude and the nature of the surface energies of the substrate and the agent.

There are other properties, as well, that one should consider in fully understanding and controlling friction. Textiles are exposed to both dry and wet conditions during processing and use. Fibers vary in their interaction with moisture; those which are hydrophilic absorb aqueous fluids into their internal structure, swell and undergo a change in their frictional behavior. The greater the absorption, the greater is the change. Those that are hydrophobic will tend to accept oils and oil-based treatments on the surface more readily. Another property of interest is the response of fibers to heat. It is well known that frictional energy is converted into thermal, which can lead to significant rises in temperature. Fibers being particularly non-conducting, the changes can be enormous, as the temperature increase will tend to be localized. If a lubricant is present, its viscosity can change which will affect friction.

In this chapter, the information on structure and properties are provided that may serve as useful background knowledge for fibers and textiles whose frictional behaviors are discussed in the book.

1.2 Fiber classification

Numerous fibers exist that are used in producing textile products. They vary in chemical and physical structure and, therefore, also in chemical, physical, and mechanical properties. Some fibers are natural, others are man made or manufactured. A general classification is given in Table 1.1. The natural cellulose fibers include many more than listed, for example ramie, hemp, abaca, pina, coir, kapok, and more, which are obtained from plant stems, leaves or seeds, but their volume is small and applications limited to non-apparel, non-conventional products. Chemically, these fibers are similar to cotton but their physical structures, dimensions, mechanical properties and production volume, restrict them to limited uses. Among the protein fibers, the two most significant are silk and wool, the latter obtained primarily from sheep. Wool is a hair fiber and is also obtained, although in much smaller volume, from other animals, such as goat, llama, angora and camel. All hair fibers tend to have similar chemical structures but they differ in dimensions, volume produced and availability. Among the manufactured fibers, many more names have existed in the past, but either they were never produced as fibers or their production has been discontinued for economic or other reasons. With the current interest in ultra high performance materials, which have higher modulus, strength or thermal stability than available in existing synthetic fibers, several other fibers have been introduced and more are no doubt

Table 1.1 A general classification of fibers*

Natural fibers	
Cellulosics:	Cotton, Flax, Jute
Proteins:	Wool, Silk
Man-made or manufactured fibers	
Cellulosics:	Rayon, Acetate, Lyocell
Synthetic:	Nylon, Polyester, Polypropylene, High modulus Polyethylene, Acrylic, Modacrylic, Spandex, Rubber, Saran, Aramid, Sulfar, PBI, Polytetrafluoroethylene
Inorganic:	Glass, Metallic, Carbon, Graphite, Ceramic

*The list is not all-inclusive, the fibers named are those that are better known and continue to be of commercial significance.

under development. Two noteworthy are PBO, or poly (p-phenylene benzobisoxazole) and M5, also known as PIPD, or poly [2,6-diimidazo (4,5-b:4′,5′-e) pyridinylene – 1,4-(2,5-dihydroxy) phenylene] [1]. These fibers have 40 to 60% higher strength and 20 to 40% higher modulus than found in the high strength carbon fibers. Accordingly, the list of fibers in the manufactured category will grow as new polymers become available to meet man's ever-increasing need for yet higher performing materials.

There is a special category of fibers not mentioned yet. These are the fibers specially developed for use in medicine. Some of these are bioabsorbable. The well known among these are polyglycolic acid, polylactic acid, polyglactin, polydiaxanone, polybutester and polyglyconate [2]. Since these fibers are more or less restricted for use as sutures, ligatures and implants, wherein largely mechanical and biological properties are considered important, little has been published in the technical literature about their physical structure and other properties.

1.3 Formation of man-made fibers

The first man-made fiber ever produced was rayon, which was derived from an existing cellulose polymer in the form of wood. The fiber is known as regenerated because the cellulose chains already existed in the form of wood. A modified cellulose fiber, cellulose acetate, also regenerated, followed, which was produced by modifying the chemical nature of the cellulose chains. Over the years, an effort has been made to similarly regenerate a protein fiber from naturally occurring proteins, but a commercially successful fiber has not resulted. A vast majority of known man-made and commercially successful fibers are, however, developed from small and basic chemical compounds based on petroleum, by first synthesizing chains and then extruding the resulting polymer into fiber [3].

Innovative structures covering different cross-sectional shapes, including hollow, and sizes, including microdenier or micrometer or less in diameter, could be developed. Chemistries of two different monomer compounds could be combined by co-polymerization in forming a polymer chain that has new properties. Likewise, two different polymers could be co-extruded through the same spinneret hole to form an alloy or a bicomponent fiber. These are, thus, the truly synthetic fibers that can cover a broad range of sizes and shapes and a very wide range of mechanical and physical properties. Availability of these materials has resulted in the improved performance of textiles in many existing products and in the emergence of many new and novel products, especially technical.

The natural polymers, mentioned above, are synthesized and grown into fibers by nature. The synthetic materials, on the other hand, must be first polymerized into chains, by linking small molecules together end to end, and then extruded into fibers. Chains are built by either a condensation or an addition process. Nylon and polyester are examples of polymers synthesized by the condensation process whereas polyethylene, polypropylene, acrylic and polytetrafluoroethylene (Teflon®) are some examples of polymers prepared by the addition process.

The polymers so synthesized are often in the form of powder or chips. These are either reacted with suitable chemicals or melted to cause chains to disentangle and flow freely. The fluid, i.e. the solution or the melt, is then extruded into fine streams by forcing the polymer through one to thousands of holes of required diameter in a spinneret. The streams must then solidify to yield a fiber. Three methods are conventionally used to extrude a fiber: melt spinning, dry spinning and wet spinning. Some fibers, in particular nylon and polyester, are frequently made by a continuous process in which the intermediate step of first making the chips from the synthesis of polymer and then melting them to form fiber is bypassed.

The fiber as extruded, whether regenerated or synthetic, is unoriented and inelastic. It must be stretched to some degree so that the molecules are aligned parallel to the axis and bonds developed between them. This assists the fiber to develop a memory of the new state and have the desired mechanical properties and elasticity. This change in structure is achieved by drawing the as spun fiber, usually under heat, and then annealing. High speed spinning assists in the orientation process.

The natural fibers are endowed with crimp or curls that provide the desired bulk, hand, softness and stretchability. The normal man-made fibers as extruded and oriented are usually straight and lack these properties. Most man-made fibers are, therefore, texturized or crimped and set in this configuration before they are used in forming yarns and from them the fabrics.

1.4 Concepts of semi-crystalline two-phase structures of fibers

At places in the fiber, whether the latter is natural or oriented manufactured, the chains are packed optimally like the atoms in an inorganic crystal. In these crystalline regions, the molecules are firmly bonded. At other places, the molecules are irregularly arranged with relatively large distances between them. In these, the amorphous or disordered regions, the bonds also form but they are weak and range in strength. While the crystalline regions anchor the chains and contribute to the thermal and mechanical stability, the amorphous regions contribute to stretch and recovery and to physiological comfort properties that are important in many applications of textiles. Most versatile fibers are, therefore, only partially crystalline, i.e. they have a two-phase morphology in which the crystalline and amorphous regions are scattered throughout the structure.

The polymer properties and the manufacturing conditions govern the nature of the so-called fine or macromolecular structure of the fiber. Concentrated melt or solution, normal speed of spinning and fast quenching or regeneration lead to the most prevalent of the structures, i.e. the fringed micelle. It consists of numerous small brick-like crystallites scattered in a matrix of amorphous structure (Fig. 1.1(a)) [4, 5]. Somewhat thinner melt or solution, high shear stresses or elongational flow during spinning and relatively slower rate of quenching or regeneration tend to yield a fibrillar morphology in which the

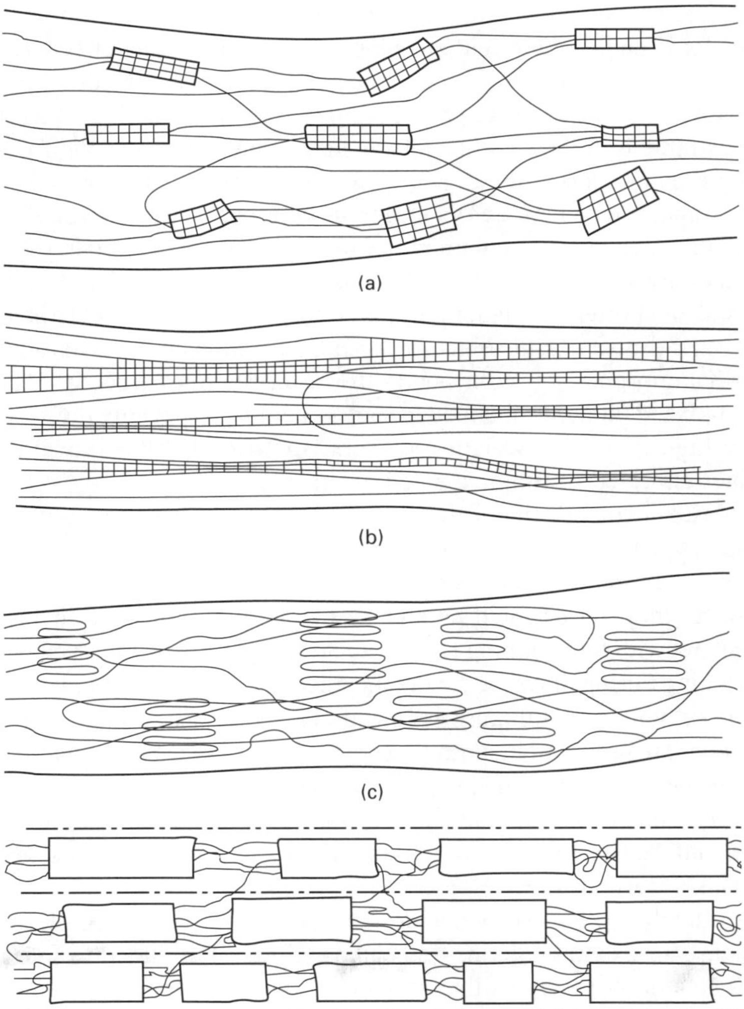

(a) (b) (c) (d)

1.1 Concepts of two-phase structures found in fibers. (a) fringed micelle, (b) fringed fibril, (c) fringed lamella, (d) modified fringed fibril with micelles lined up in columns.

crystals are long and thin rod-like units, oriented more or less parallel to the fiber axis. In between the fibrils are disordered regions. This structure is referred to as fringed fibrillar (Fig. 1.1 (b)) [5]. Most natural fibers tend to conform to this morphology, although the fibrils may not lie exactly parallel to the axis but may spiral around or be inclined at some angle to it. In cotton, for example, the fibrils spiral around the axis at an angle of 25–30° and in flax or jute the angle is less, about 6°. Another structure is found in polymers, primarily synthetic, it being the folded chain fringed lamella (Fig. 1.1(c)) [4, 6]. If the chains are flexible, forces between them weak, the solution or melt thin, the quenching rate low and the external stresses low, the chains may then fold roughly between two planes and lead to platelet-type crystalline units called lamellas. Since polymer chains must be extended for optimum translation of their tensile properties into properties of fiber, such folds are undesirable in fibers. Fortunately, a significant fraction of the folds tend to get pulled out during the drawing and setting process [6], but it is expected that some will remain and continue to be a part of the morphology of the flexible polymer based commodity fibers. One of the objectives of a spinning technology known as gel spinning is to draw out the folds into fibrillar crystals and orient the latter parallel to the fiber axis. Such method, applied to even the most flexible of the polymers, polyethylene, leads to fully oriented fibrillar morphology, with nearly 100% crystalline structure. The tensile strength and modulus obtained in such structures, especially those containing ultra high molecular weight polymers, approach the theoretical maximum values expected of fully extended chains [7].

Most general use fibers after extrusion are quenched or solidified rapidly. This leads to a structure having numerous small crystallites, resembling micelles. In the as spun fiber, the micelles are oriented randomly but upon drawing they get oriented parallel to the axis. The flow of chains through spinneret with force pulling the extrudate along the axis, followed by uniaxial stretching during drawing, favors the micelles lining up behind each other in columns with fringes linking these within and between columns. Accordingly, the structure produced in most fibers resembles an assembly of inter-connected fibrils with each fibril made up of alternating non-crystalline and crystalline regions, much like the one shown in Fig. 1.1 (d). Fracturing or peeling a fiber and then examining the structure under a high power microscope usually reveals a fibrillar morphology.

In flexible polymers, which do not have significant intermolecular interactions, for example olefins, the chains will normally tend to fold and lead to a lamellar-type crystalline structure. Upon drawing, however, the folds get pulled out to different degrees and the structure obtained is again fibrillar, presumably with some folds still remaining as a part of the structure. The extreme example of a fiber based on flexible polymer with little inter-chain interaction is gel spun high molecular weight polyethylene, alluded to earlier.

Finally, there is a class of fibrous materials known as liquid crystalline polymers that have inherently stiff chains, for example aramid, PBO and PIPD (Fig. 1.2). The chains of these polymers remain extended even in the liquid form. When extruded, they line up parallel to each other, register and bond into extended chain fibrillar crystals oriented, parallel to fiber axis.

Accordingly, in contrast to the unique morphologies that each natural fiber ends up having, man-made fibers, in spite of vast differences in their chemical structures, can be engineered to have more or less similar morphologies. Most present day fibers, manufactured at increasingly higher speeds and finished using optimum drawing and annealing conditions, favor a form of fibrillar structure. The fibrils themselves, however, are expected to have structures that reflect differences in the chemical nature of the individual materials and the intended end use. In Section 1.5, that follows, therefore, mostly the other noteworthy structural characteristics of selected individual fibers are pointed out.

1.5 Structure and morphology of textile fibers

1.5.1 Natural fibers

Cellulose fibers

The fiber of greatest commercial interest in this group is cotton, which still accounts for the largest single fiber volume used in the world. The chemical structure of one repeat of cellulose chain is shown in Fig. 1.2. Each repeat has three hydroxyl groups that are capable of linking with neighboring chains by hydrogen bonds. These groups, when free or weakly bonded, also attract and bond with water. The oxygen bridges between the repeat units allow chains to bend and twist, making the polymer flexible. The chains are quite long, and the fiber has approximately 66% crystalline structure. Although cotton fiber has distribution of lengths and diameters, and these vary from type to type, the most widely used types have the average values of 2.5 ± 0.5 cm for length and 18 ± 2 μm for diameter. The fiber has kidney bean shape cross-section which when swollen becomes round. Longitudinally, the fiber appears convoluted or twisted. Under ambient conditions, the material contains about 8% moisture. Accordingly, the fiber is strong, extensible, and absorbent, and, unlike most other fibers, not only retains its strength but becomes stronger when wet. Applications requiring an absorbent strong and soft material, and frequent laundering, find cotton as essentially an irreplaceable fiber.

Protein fibers

These include keratin fibers (animal hair, wool) and fibroin fibers (silk). The basic chemical structure is shown in Fig. 1.2. The pendant groups (R) are the

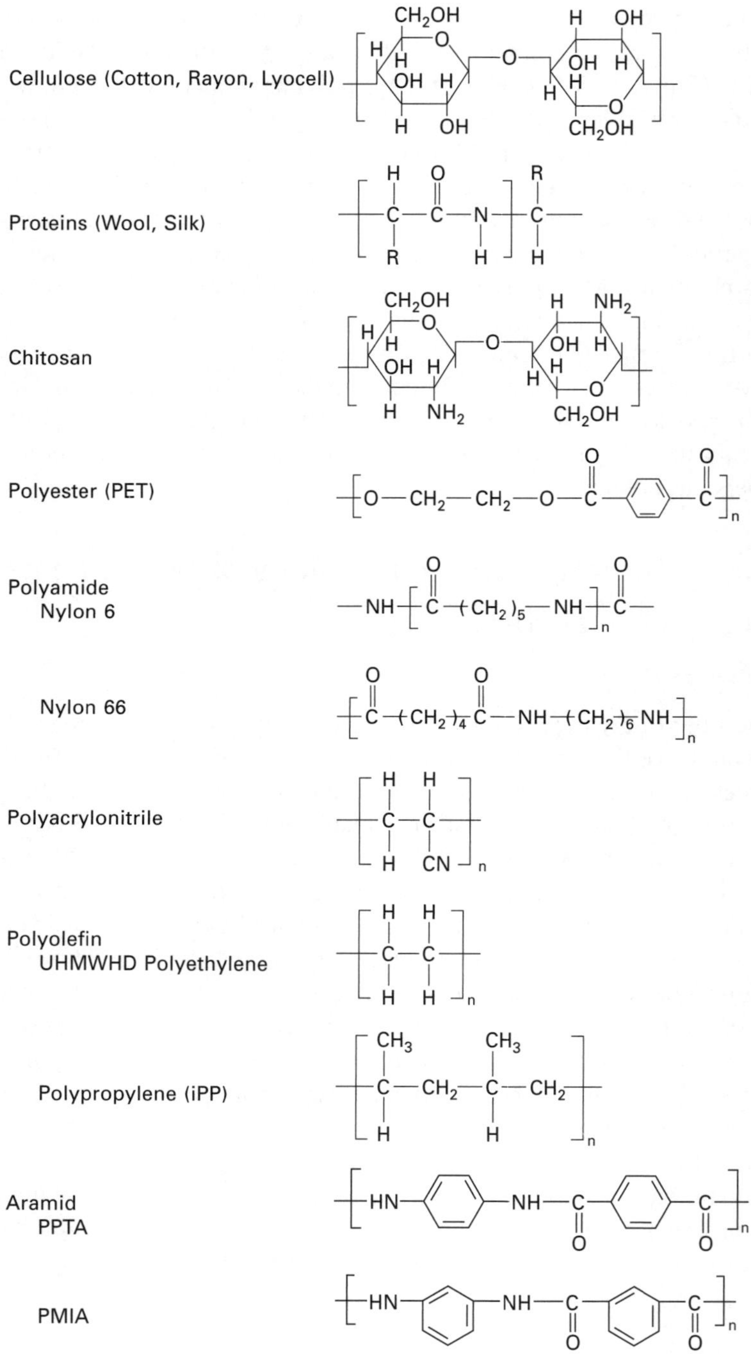

1.2 Chemical constitutions of the primary repeat units of fibers.

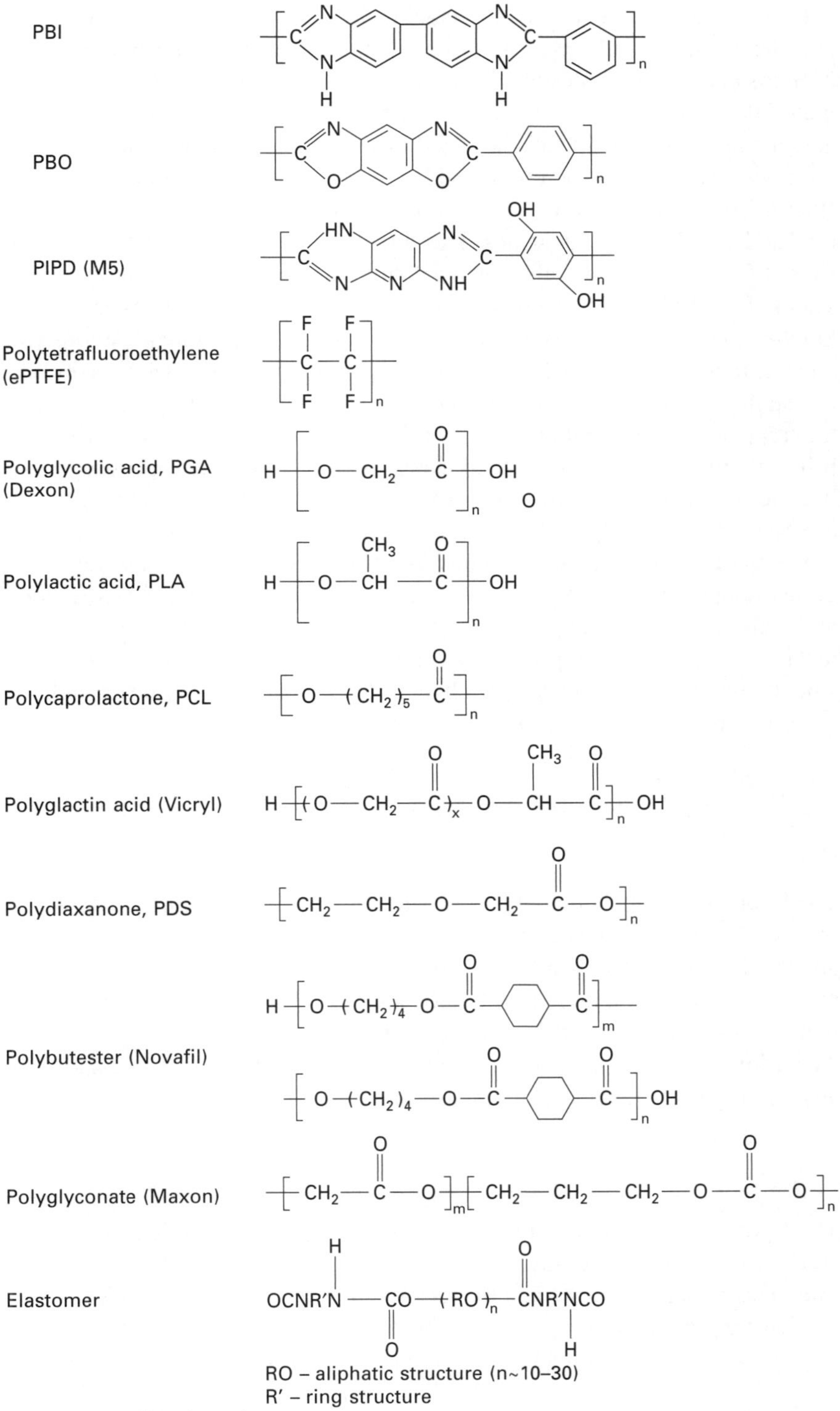

1.2 Continued

side groups that make up the amino acids. It is the variation in the sequence and the amount of different amino acids that determines the properties of proteins and differentiates one from the other. In keratins, which are the hair materials, about 20 different side groups are found [8, 9]. A unique amino acid in this fiber is cystine that is the sulfur-containing compound and amounts to about 12% of the structure. Abundance of hydrophilic groups allows the fiber to absorb high amount of moisture. The fiber has a cross-section that is oval and a surface that has scales. The latter, pointed from root to tip, give the fiber different frictional values when it slides against the scales and with the scales. This difference is made use of in developing felted nonwoven fabrics. On the other hand, it is this difference that leads to dimensional instability in apparel fabrics and is the target of finishing treatments designed to reduce it. A detailed discussion of morphology and frictional properties of wool and the treatments given to it is covered in Chapter 7. The silk fibroin, produced primarily by the silkworm for textile application, has the same primary chain but the pendant groups are fewer (4) and much simpler. For example, the amino acid accounting for the largest fraction (~ 44%) of protein is glycine in which the pendant group is simply H. In contrast, hair fiber has only 6.5% of this compound. The simple structure allows the fibroin chains to be extended and better packed, leading to a fiber that has an exceptional combination of mechanical and physical properties for many applications. Its surface is also smooth and the cross-section, which tends to be triangular, is fine enough for braiding and weaving into high quality textiles.

1.5.2 Regenerated fibers

Rayon

The fiber, although chemically similar to cotton (Fig. 1.2), differs from it in physical structure: rayon's molecular weight is about 20% and crystallinity about 50% of the values applying to cotton. These differences make rayon relatively weaker and more extensible, but more absorbent (about two times) than cotton. The regular fiber is also much weaker when wet than dry. Accordingly, one of the major applications of the fiber has been in disposable absorbent pads, sponges, and sanitary napkins.

Lyocell

The rayon fiber has undergone extensive modifications in chemical processing and extrusion and a number of types have become available for specific uses. The most noteworthy and the recent of these is Lyocell that has dry and wet tensile properties comparable to those of cotton and is, therefore, suitable for use in regular apparel and other durable products. Additionally, the fiber is made by an environmentally friendly process in which the contents of spinning bath are recovered and reused [10]. The final fiber has a smooth surface and round cross-section, and is said to be stronger than cotton and regular rayon in both the dry and wet states [11].

Cellulose acetate

This is the chemically modified cellulose material in which the acetyl groups, CH_3COO, replace some of the hydroxyl groups, OH, on the chain. Usually, on an average, about two of the three hydroxyl groups present on a repeat unit are modified. The acetyl groups being bulky and lacking functionality, the fiber produced has lower density, lower strength and lower ability to absorb moisture. The cross-sectional shape of the fiber tends to be serrated or bulbous.

1.5.3 Synthetic fibers

Polyester

This is one of the most successful and versatile of the manufactured fibers that finds applications in most categories of products, i.e. apparel, industrial and home furnishings. The chemical constitution of the most common material, poly(ethylene terephthalate) or PET, is shown in Fig. 1.2. The average molecular weight lies in the 15000 to 25000 g/mole range. This will be equivalent to the degree of polymerization of about 80 to 130. The fiber has an aromatic component and an aliphatic sequence. Although the polymer lacks strong functional groups, the molecules when drawn pack closely and lead to a semi-crystalline mechanically and thermally strong fiber. Lacking polar groups, the fiber has very small attraction for water, which makes the material hydrophobic. A number of variations of the basic repeat are available but they vary primarily in terms of the proportion of the aromatic and the aliphatic components of the repeat and, as a result, vary in their physical and mechanical properties. A number of aliphatic polyesters are made but they are used either as low melt adhesives for binder applications or as bioabsorbable polymers for manufacture of absorbable suture and other medical products.

Nylon

Although many different types of nylons have been produced, the two most common and widely used are nylon 66 and nylon 6 (Fig. 1.2). The chains, being void of aromatic compounds, are flexible and prone to fold. As is the case with other flexible polymeric materials, after extrusion, drawing at high temperature and annealing orients and extends the chains parallel to fiber axis. It is expected that some chain folds will remain in the structure. The fiber tends to have low modulus and high extensibility. The presence of amide groups allows hydrogen bonding between NH and CO groups of adjacent chains, which gives the fiber excellent mechanical and thermal stability. As compared to polyester, the fiber interacts better with moisture and wets more easily. Being one of the most resilient and elastic of the common materials, nylon is a fiber of choice for floor covering into which about 75% of the total production goes. The rest of the consumption is split between industrial (primarily tire cord) and apparel (largely intimate and sports wear).

Polyolefins

The two fibers of interest are polyethylene and polypropylene (Fig. 1.2). The fiber based on the simplest of the hydrocarbon polymers, polyethylene, is useful only when produced in the extended form using the gel spinning process. This leads to the extended chain polyethylene with ultra high modulus and tenacity fibers (HMPE) suitable for use in many high performance technical textiles. Molecular weight of the order of 1 million g/mole has been used in producing these fibers. The emergence of the second, more versatile, olefin fiber, polypropylene, took place after the polymer chain in the isotactic form could be synthesized. In this, the pendant group CH_3 on alternating carbon atoms appears in the same direction on the chain and allows neighboring molecules to register, pack and bond with stronger dispersion forces. In polypropylene also, the chains tend to fold but they open up during drawing and annealing and develop fibrillar structure with crystallinity approaching 50%. The fiber develops excellent mechanical properties, which are comparable to those of polyester and nylon.

The olefin fibers have the lowest density (0.91–0.96 g/cc) of all fibrous materials. Both fibers being strictly hydrocarbons are hydrophobic and, therefore, do not wet. Accordingly, they do not lend to apparel applications except in a very limited sense. The fibers also have relatively low melting points, which are 150°C for the gel spun polyethylene and 175°C for the isotactic polypropylene. Polypropylene has mechanical properties that favor many, non-apparel, textile applications, some important being carpet, upholstery, civil engineering fabrics, surgical ligatures and implants, and facings for sanitary products. The gel spun high molecular weight high modulus polyethylene, on the other hand, has a stretch of the order of only 2% which, when combined with its even lower melting point, limits its applications largely to products such as body armor, moorings, cables and composites.

Aramids

The two most widely manufactured aramid fibers are the poly (para-phenylene terephthalamide) or PPTA, and poly (meta-phenylene isophthalamide) or

PMIA. Although the chemical constitution is the same, a difference exists in the configuration of the two chains (Fig. 1.2), which leads to fibers with vastly different mechanical properties, discussed later. PPTA is an ultra-high strength, high modulus and low extension fiber whereas PMIA is a fiber with more textile-like properties. Both have high thermal resistance and flame resistance characteristics; however, since PMIA has more textile-like or wearable characteristics, this is the fiber selected for making of fire protective garments. The PPTA fiber, because of its high tenacity and high modulus properties, which are maintained to quite high temperatures (> 300°C), is used in technical products where not only high strength and stiffness are required but also high thermal resistance is needed.

Polybenzimidazole

Polybenzimidazole (PBI) is a fiber based on aromatic structure and has recurring imidazole groups in the chain (Fig. 1.2). The fiber has unusual thermal resistance, does not burn in air, and although it may char after prolonged exposure to high heat flux, the charred fabric tends to retain its integrity and flexibility and undergoes little shrinkage. In addition to having low modulus, acceptable strength and high extension (~25%), the fiber has four times the moisture absorption of Nomex® or twice that of cotton! Accordingly, the fiber is not only flexible but also highly comfortable to wear when used in protective clothing, which is its primary application.

Polyacrylonitrile

Two fibers that fall in this group are acrylic and modacrylic, with the former used more widely in textile applications. The chemical structure of the base polymer is illustrated in Fig. 1.2. The pendant acrylonitrile groups (C≡N) are highly polar and lead to extensive bonding between chains. Because of this, a second component, which is another vinyl monomer, e.g. methyl acrylate or vinyl acetate, is inserted in the chain to improve polymer solubility for extrusion into fiber. The proportion of the second monomer determines the type of fiber: 15% or less for acrylic, and greater than 15% but less than 65% for modacrylic. Among several useful properties is the fiber's very high resistance to UV light, which is much higher than those of olefin, cotton, wool and nylon fibers. This remarkable resistance against UV degradation makes the fiber particularly useful for a number of outdoor applications, including awning, lawn furniture, tents, and automobile upholstery. About 80% of the total production used for textiles, however, goes into the making up of knitted apparel, including sweaters and socks. The cross-section of the fiber is non-circular; usually it is either kidney bean or dog-bone in shape.

Elastomeric fibers

The most common fiber is rubber, either natural or synthetic. It can be stretched out many times its original length and yet it recovers almost completely from it. The traditional rubber has been used in textile applications but with limited success, due to the presence of unreacted double bonds that make them susceptible to oxidation. Also, the material has low resistance to fluids used in laundry and dry cleaning. Accordingly, new thermoplastic elastomers have been developed that are linear block copolymers. These contain soft amorphous sections, which impart stretchability, and hard crystalline components, which act as tie points and hold the structure together in a memory-endowed, mechanically stable, material. These are the spandex fibers, which are usually known as the segmented polyurethanes. General chemical structure of the polymer is given in Fig. 1.2. The hard segments of neighboring chains tend to associate with each other and crystallize with the soft segments remaining largely coiled and unassociated (Fig. 1.3). Although the elastomeric materials are expected to stretch extensively and recover from it almost completely, it should be noted that the recovery is not instantaneous but usually time dependent. This means that some mechanical energy is lost in each cycle of loading and unloading. Such loss of energy in rubber tires, as explained in Chapter 2, plays a positive role in controlling the behavior of an automobile when brakes are abruptly applied.

Polytetrafluoroethylene

The chemical structure of polytetrafluoroethylene or PTFE (trade name Teflon®) (Fig. 1.2) consists of a core chain of carbon atoms with a fluorine sheath, which essentially shields the core. The inter-chain forces are low, which allow the individual molecules to be able to slide past each other with relative ease. Compared with many other polymers, the creep, therefore,

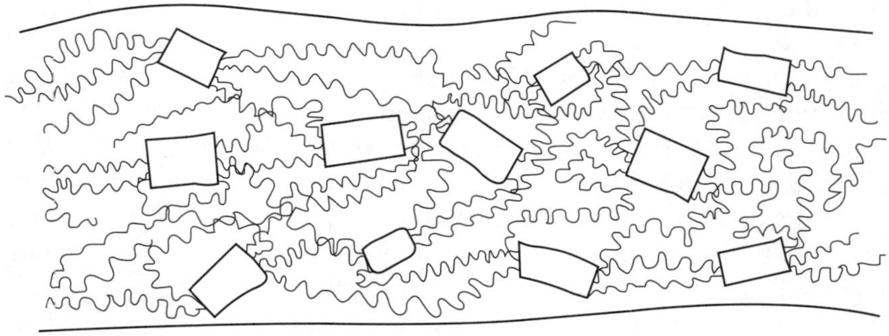

1.3 Illustration of fine structure in elastomeric fibers, compound of blocks of crystallizable hard segments and non-crystallizable coiled soft segments.

tends to be large. The polymer chains can pack closely and, therefore, the degree of order can be high. The material has a high melting point (~ 330°C) and is virtually inert to all known solvents, acids and bases. The polymer has no affinity for water. The smooth surface combined with low surface energy makes the surface so inert that it resists sticking to any material. This is also supported by the fact that the coefficient of friction of PTFE is the lowest of all materials (~ 0.007), with essentially no stick-slip character. This property has allowed the material to be used as a non-stick, low friction, coating for metals and work surfaces.

In most textile related applications, the polymer is used in the expanded form, known as ePTFE [12]. The structure of the product, usually a film, consists of "nodes" and "fibrils", the former found perpendicular to the direction of stretch and interconnected by fibrils (Fig. 1.4). The pores created are unique in size in that they make ePTFE film or fabric breathe by allowing vapor to pass through but act as a barrier against passage of water (rain), chemicals and microbes.

1.6 Fiber properties

1.6.1 Fiber cross-sectional shape

Typical cross-sectional shapes of some of the fibers, natural as well as manufactured, are illustrated in Fig. 1.5. The textile fibers have a range of shapes, including circular, oval, multi-lobal, triangular, serrated, kidney bean,

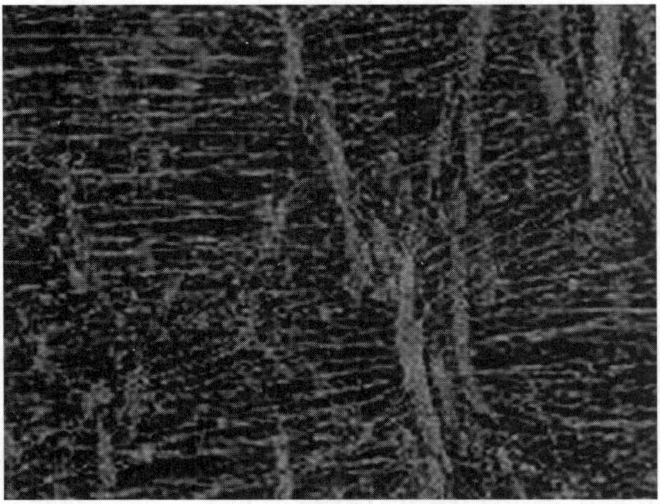

1.4 Micrograph of ePTFE, showing a porous structure containing nodes interconnected by fibrils (Courtesy Menardi–MikroPul, LLC; www.micropul.com).

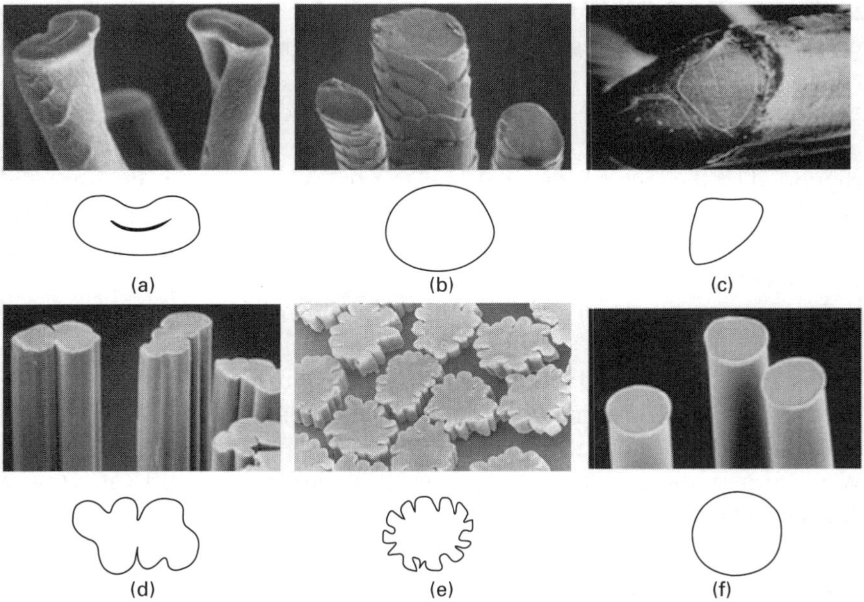

1.5 Typical cross-sectional shapes of textile fibers. (a) cotton, (b) wool, (c) silk, (d) acetate, (e) Enka rayon (Courtesy Sulky of America, Kennejaw GA) and (f) melt spun synthetic fibers, i.e. nylon, polyester and olefin. (Figures (a), (b), (c), (d), (f), courtesy Japan Chemical Fibers Industry Association, Tokyo).

dog-bone, hollow and square. Since in melt spinning only heat is lost during the solidification process, the cross-sectional shape of the extrudate closely resembles that of the spinneret hole. Accordingly, the melt-spun fibers can be engineered to have almost any desired cross-sectional shape, including hollow and shapes with multiple voids. Some examples of melt-spun fibers with unusual shapes are given in Fig. 1.6. These are used in order to enhance one or more of a fiber's desirable physical characteristics, such as the hand, resiliency, translucency, wick-ability, insulation or cohesion. The cross-sectional shapes of the dry and the wet spun fibers, on the other hand, could be expected to deviate from those of the spinneret holes since mass is lost during the solidification process. Still, however, through a control on the regeneration process, a range of shapes has been possible in these materials as well. (Figure 1.7 illustrates the cross-sectional morphologies that have been possible in the solution-spun rayon [3, 13].) Clearly, the lobes of the multi-lobal and the ridges and corners noted in many others will run along the length of the fiber (Figs 1.5–1.7). Since friction is a surface phenomenon, it is expected that the shape of the cross-section will have a profound effect on the coefficient of friction. No major controlled study on the effect of fiber cross-section shape on friction has appeared in the literature in the knowledge of the author.

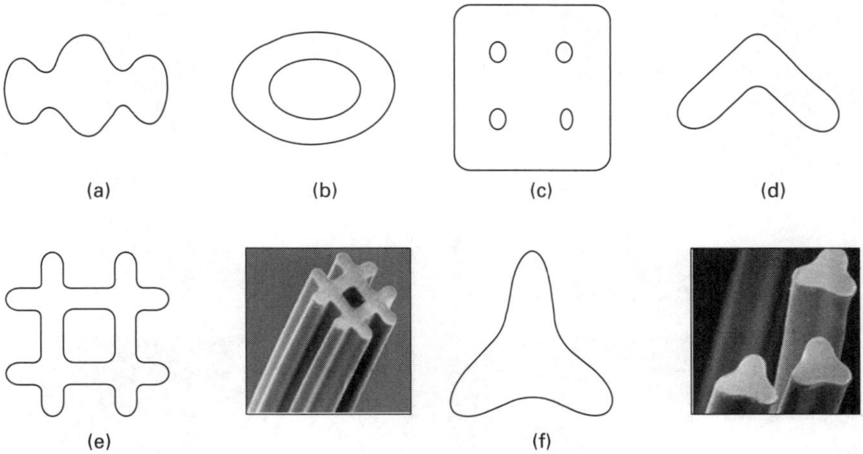

1.6 Example of unusual shapes with which the melt spun polyester and nylon fibers have been produced (a) Du pont's Coolmax®, (b) Du pont's Thermax®, (c) Antron® nylon, (d) nylon L letter shape, (e) nylon hollow fiber-Microart by Unitika (Courtesy Japan Industrial Design Promotion Organization (Tokyo) and (f) triangular polyester (Courtesy Japan Industrial Chemical Fibers Industry Association, Tokyo).

1.6.2 Fiber cross-sectional size

The cross-sectional size of materials has been commonly expressed by their diameters. This also continues to be one of the practices in the field of textiles. However, because fibers are small and their shapes not always circular, a direct measurement of size has been difficult. Accordingly, the size of fibers has usually been expressed in terms of their mass per unit length, or linear density. Two primary units of linear density used are denier and tex. Denier represents the mass in gram of 9 km and tex the mass in gram of 1 km. Linear densities are directly related to cross-section size of fibers, as follows:

$$\text{Linear density (denier)} = \text{cross-section area (cm}^2\text{)} \times \text{density (g/cc)} \times 9 \times 10^5$$

$$\text{Linear density (tex)} = \text{cross-section area (cm}^2\text{)} \times \text{density (g/cc)} \times 1 \times 10^5$$

Tex being a more metric of the two units for linear density but denier resulting in a more appropriate value for fibers (1–5), frequently the unit decitex (dtex) is used to specify the size of single fibers:

$$\text{Linear density (dtex)} = \text{cross-section area (cm}^2\text{)} \times \text{density (g/cc)} \times 10 \times 10^5$$

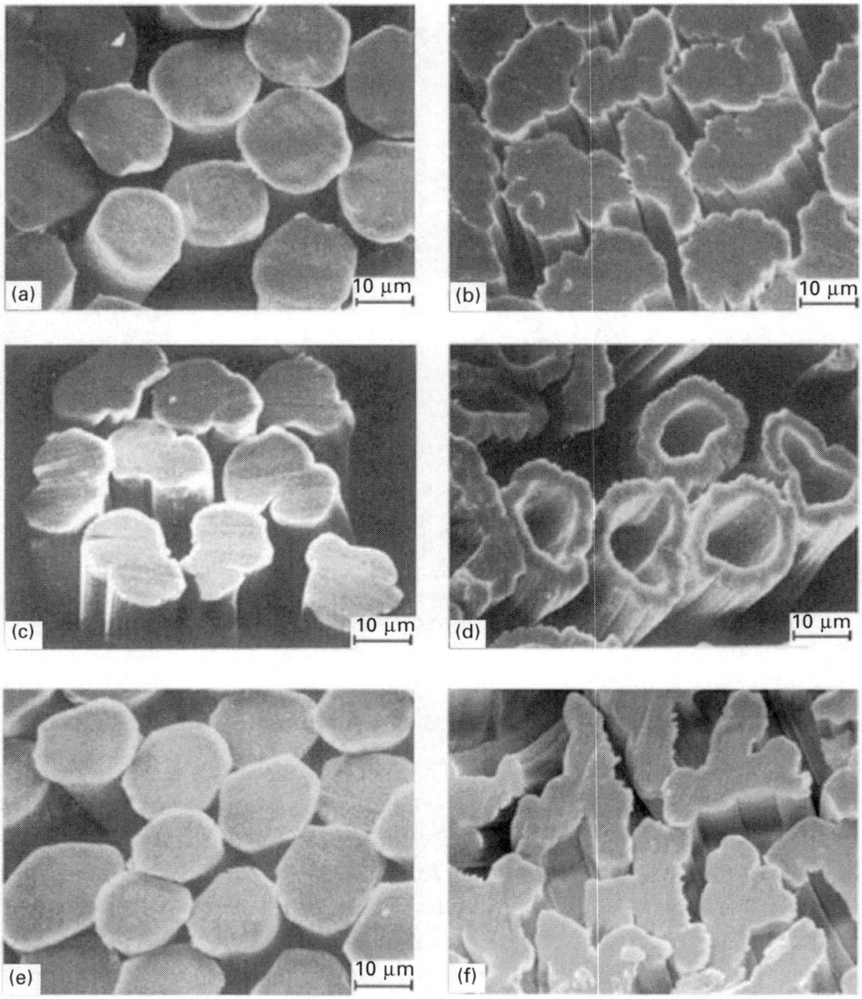

1.7 Cross-sectional shapes of some of the regenerated cellulose rayon fibers. (a) high wet modulus, (b) regular rayon, (c) crimped high wet modulus, (d) hollow, (e) cuprammonium, (f) trilobal. [3, 13]

These lead to the following conversion factors:

$$1 \text{ (denier)} = 0.111 \text{ (tex)}$$
$$= 1.11 \text{ (dtex)}$$

Natural fibers vary in size from type to type, and from fiber to fiber in the same type. The typical values are: cotton 15–20 μm (2–4 denier), wool 15–45 μm (2–20 denier) and silk 12–30 μm (1.3–8 denier). The man-made fibers can be manufactured over a broad range of sizes (10–400 μm diameter, or 1–1000 denier), using the conventional spinning technologies. This range can be extended greatly on the lower end, for example to 1 μm by the melt blowing process or even down to 0.1 μm by the electrospinning technology.

Since the strength and stiffness of a fiber are a direct function of size, fibers for different applications are made in different linear densities (deniers), which are 1–3 for apparel, 6–15 for carpets, and 3–1000 for industrial products, including medical. One of the latter is suture, used in closing wound after surgery. The linear density (denier) of a monofilament ophthalmic suture can be as low as 4 and that of one used in bringing chest muscles together after open heart procedure can be as much as 1000. Very high linear density filaments are also used in making high strength cables and cords.

1.6.3 Moisture absorption and swelling

Fibers absorb moisture to varying degrees depending on their chemical and physical structures. Some fibers have hydrophilic groups (OH and NH) that can form hydrogen bonds with water molecules and, therefore, attract the latter from air or from aqueous media if in contact with it. Many other fibers have largely inert groups and do not have significant attraction for water. How much is absorbed per unit mass will depend on the density of water absorbing groups in the fiber, the fraction of total mass that is amorphous and, if absorption is from atmosphere, the relative humidity and temperature of the surrounding air. Another factor affecting absorption is whether or not the fiber can swell and make room for water to penetrate and bond. With swelling, the structure undergoes changes in dimensions. Absorption of moisture being exothermic, heat is generated. The rise in temperature, however, is small, i.e. of the order of only a few degrees. Since absorption involves breakage of intermolecular bonds, most mechanical and physical properties, including friction, are affected.

The amount of water absorbed is usually given by the parameter moisture regain, which is the mass absorbed per unit dry mass of fiber and expressed as percentage. Two quantities are of interest: regain from ambient air under standard conditions (65% relative humidity and 20°C) and regain from liquid water under conditions of saturation. These are given for selected fibers in Table 1.2. With absorption, dimensions change [16]. With saturation, fibers may increase in cross-section area or volume by 25–75%. Fiber length also increases, but by a small amount (~ 1%).

Values in Table 1.2 show that the cellulose fibers (cotton, rayon), the modified cellulose fiber (acetate) and the protein fibers (wool, silk) absorb what is considered a high amount of moisture under the standard and the saturation conditions. Among the synthetic fibers, the most absorbent is PBI, with a standard regain of 15%. The aramids and nylons also absorb, from the physiological comfort standpoint, reasonable amount of water. The fiber M5

Table 1.2 Values of standard (at 65% RH and 70°F) and saturation moisture regains of some fibers

Fiber	Standard regain (%)*	Saturation regain (%)*
Wool	15	23
Silk	11	22
Rayon	14	68
Cotton	7.5	28
Acetate	6	17
PBI	15	
Aramid	5	
Nylon	4.5	11
Polyester	0.4	2.5
Polypropylene	0	
M5	2	
PBO	0.6	
HMPE	0	
Oxide glass	0	
Carbon T300	0	
Steel cord	0	

*The standard values are generally as frequently published [1, 14, 15]. The saturation values are estimated from published swelling values [16] and a model due to Gupta and Chatterjee [17]. Blanks represent data not available.

also has hydrophilic groups and manages to absorb 2% or so of moisture. The remaining fibers absorb either little or none and may be termed hydrophobic. These fibers do not wet with water but may, nevertheless, wet readily with oils. The section on surface energetics, included later, discusses conditions required for wetting with fluids having different characteristics.

1.6.4 Thermal behavior

As fibers rub against each other or against another surface, heat is generated. Since fibers are poor conductors of heat, the temperature rise could be rapid, and because they also have generally low softening, melting and/or decomposition points, the temperature rise could be high enough to adversely affect structure and properties. The problem could be severe for systems in which threads or filaments pass continuously over guides, which occurs during formation of fibers, winding of yarns, and weaving, knitting and braiding of fabrics. In these processes, the heat generated by friction can build up on the contact surface and affect the properties of the textile material sliding against it. In high-speed sewing, for example, the rise in needle temperature could be extensive and cause melting and fusing of thermoplastic fibers or charring of non-thermoplastic materials.

Table 1.3 Thermal properties of selected fibers

Fiber	T_g (°C)	T_m (°C)	Specific heat (J/kg. K)	Thermal conductivity (W/m. K)
Steel		1600	434	60.5
Alumina			765	46
Carbon fiber		3700	710	15–500
Cellulose			1250	0.07
Wool			1340	
PTFE	130	347	938	0.35
Polyester	125	255	1103	0.14
Nylon 6 6	50	260	1419	0.25
Nylon 6	90	215	1419	0.25
PE	– 80	130	1855	0.24
PP	– 18	165	1789	0.12
Acrylic	90		1286	
PBI			1000	0.038
PBO		650*		
M5		530*	1000	0.038
Aramid (Kevlar® 149)		550		
Aramid (Nomex®)		420		

T_g – glass transition temperature; T_m – observed melting or decomposition point. Some fibers do not have observable values of these parameters. Thermal conductivity, being difficult to measure on fibers, is given for materials that are available as films.
Sources: Steven B. Warner, *Fiber Science,* Prentice Hall, Englewood cliffs, 1995 [20]. Morton W E and Hearle J W S, *Physical Properties of Textile Fibers,* 3rd Edition, The Textile Institute, Manchester, 1993 [4]. Hearle J W S, *High performance fibers,* Woodhead Publishing Ltd., Cambridge, 2001 [1, 18, 19].
*On-set of decomposition in air.

Some of the thermal quantities of interest for selected fibers are given in Table 1.3. These are given only for general appreciation of the magnitudes and the ranges involved as, in many cases, exact conditions and the method used in measuring them are not available. It is noted that the fibers soften at relatively low temperatures (50–150°C for most commodity fibers). Some fibers, for example thermo-sets, do not melt, as much before they could reach their melting temperatures, they start to decompose and degrade. Although as compared to metals, the polymeric materials have high heat capacity (2–4 times), their ability to conduct away heat is 2 to 3 orders of magnitude lower. This, combined with their relatively much lower softening/degradation temperature, makes them vulnerable to damage by heat generated due to friction during processing and use, especially industrial. Fiber-to-fiber differences in thermal properties are small but they differ greatly together from metals. Carbon fiber is unique in that it has some of the mechanical properties comparable to those of many of the high performance organic fibers but also the thermal stability that exceeds those of metals.

1.6.5 Expression of stress values in fibers

The breaking stress of fiber is traditionally expressed in terms of force divided by linear density. This leads to the units as g_f/denier, g_f/tex, mN/dtex, etc. Using the conversion factors given in Section 1.6.2, and 1 N = 102 g_f, one can convert the value from one set of textile units to another. In engineering discipline, however, the stress is given in the units of Pa (N/m^2). The following equation shows the relation between the stress in engineering unit (σ_e) and that in textile unit (σ_t):

$$\sigma_e \text{ (Pa)} = \sigma_t \text{ (g}_f\text{/denier)} \times \text{density (g/cc)} \times 88.29 \times 10^6$$

A value of 2 g_f/denier (18 g_f/tex or 1.8 g_f/dtex) is considered minimally desirable for a fiber to be able to go through the rigors of mechanical processing and be converted into a yarn or a fabric. Commodity fibers used in apparel manufacturing range in strengths from 2 to 5 g_f/denier. With the densities varying from 0.9 to about 1.5 g/cc, this will result in engineering values to range from about 160 to 660 MPa. Industrial fibers can be expected to have much higher values.

1.6.6 Tensile properties

One of the great attributes of textile fibers is that they are significantly stronger than metals (on a given mass basis) and at the same time significantly lighter (on a given volume basis). It is well recognized that the tensile properties of a given polymer can be varied widely by changing the molecular weight and the degree of orientation. However, for reasons of economics and the fact that a desirable change in one property frequently results in an adverse effect in another, each type of fiber is usually manufactured with a narrower range of properties. The typical average values for selected fibers are given in Table 1.4.

The results show that on a given mass basis, i.e. in textile units, fibers are very strong; some, such as aramid (Kevlar®) and HMPE, are an order of magnitude stronger than steel. In a number of applications, for example body armor and structural composites, high stiffness is a critical requirement. Clearly, several fibers stand out in this regard, whose initial modulus are half an order or more greater than that of steel. A limitation of polymeric materials is in their having a low melting or decomposition temperature. In this regard, carbon fiber that structurally bridges the gap between organic and inorganic materials excels over organic fibers; it is, accordingly, a preferred material for products that must be strong, stiff and light weight and yet be able to endure very high temperatures.

The tensile properties of fibers are affected by a rise in temperature. Fiber becomes weaker and more extensible with increase in temperature [23]. This

Table 1.4 Tensile properties of fibers

Fiber	Tenacity cN/Tex	(MPa)	Breaking strain (%)	Initial modulus N/Tex	(GPa)
Cotton	40	(600)	7	5	(7.5)
Rayon	22	(331)	18	6	(9)
Wool	14	(180)	40	2.2	(2.9)
Silk	38	(498)	23	7	(9)
Nylon (MT)	48	(547)	20	3	(3.5)
Polyester (MT)	47	(648)	15	11	(16)
Polypropylene	66	(605)	17	7	(9)
PBI	24	(341)	25	2.8	(4)
Aramid (Nomex®)	49	(669)	25	8	(11)
Aramid (Kevlar®149)	213	(3090)	1.3	128	(186)
HMPE (Spectra® 1000)	320	(3104)	3.3	110	(107)
PBO	353	(5500)	2.5	179	(280)
M5	294	(5000)	1.5	194	(330)
Carbon T 300	196	(3530)	1.5	128	(230)
Steel cord	27	(2100)	8	24	(185)

Sources: R Meredith [21], S Kumar [22], *Manufactured Fiber Fact Book* [15], J W S Hearle, *High Performance Fibers* [18]. MT: medium tenacity.

is relevant as during a friction process, heat is generated, which could cause an adverse change in fiber's mechanical properties. Another factor affecting these properties is the amount of moisture absorbed. The greater the absorption, the greater the disruption of intermolecular bonds and, therefore, the greater is the effect on mechanical properties. All absorbing fibers undergo an increase in breaking extension and decrease in initial modulus and, except the natural cellulose fibers, a decrease in strength. The exception noted, cotton, flax, etc., may have as much as 20% increases in strength with wetting.

1.6.7 Bending, shear and transverse compression properties

According to our best understanding of the mechanism of friction, bending, shear and hardness properties of a fiber affect friction. Bending rigidity should influence conformability and intimacy of contact between two bodies, shear modulus should influence lateral deformability and strength of junctions, and transverse compressional modulus or hardness should influence and correlate with the true area of contact. Resistance to bending is determined by the tensile modulus (E) and the second power of the linear density (d) directly and the density (ρ) indirectly. The shape of fiber cross-section (ε) also has an effect, with the resistance increasing as the shape changes from elliptical to circular to trilobal to tubular [4]. The model is as follows:

$$\text{Bending rigidity} = \left[\frac{1}{4\pi\beta_0}\right]\frac{\varepsilon E d^2}{\rho}$$

In this, ε, the shape factor, has a value of 1 for circular, less than 1 for flat or elliptical and greater than 1 for hollow fiber. Although no definitive data is available, its value should also be more than 1 for trilobal and multi-lobal fibers with circular symmetry. β_0 is a constant whose value will depend on the units in which E, d and ρ are expressed. Applying this model and assuming ε to be 1 and linear densities comparable, one can rank fibers based on their bending stiffness. This is done in Table 1.5 for selected fibers. For convenience, the results are expressed in dimensionless form by dividing the value of a fiber with that of a hypothetical material whose value in the same units is 1. The results show that on an equal mass basis, Spectra®, PBO and M5 are, among the fibers listed, the stiffest materials to bend. Kevlar® 149 and Carbon T300 follow these. One notes that steel cord ranks with the most flexible of the fibers, i.e. nylon, cotton and wool. Just as is the case with tensile modulus, bending stiffness of fibers decreases with increase in temperature and, in hydrophilic materials, with increase in moisture regain.

Table 1.6 gives the values of the shear rupture stress for a few fibers and compares these with those of the tensile rupture values. The shear value is expectedly less, between 2 to 3 times smaller than the tensile value. If one is concerned with the resistance to deformation at only low strains, then the

Table 1.5 Relative bending rigidity of fibers

Fiber	Density g/cc	Initial modulus N/Tex	Relative bending rigidity*
Cotton	1.52	5	3.3
Rayon	1.50	6	4.0
Wool	1.30	2.2	1.7
Silk	1.34	7	5.2
Nylon	1.14	2.5	2.2
Polypropylene	0.93	7	7.5
Polyester	1.38	11	8.0
PBI	1.43	2.8	2.0
Aramid (Nomex®)	1.38	8	5.8
Aramid (Kevlar®149)	1.45	128	88.3
HMPE (Spectra® 1000)	0.96	110	114.6
PBO	1.56	179	114.7
M5	1.70	194	114.1
Carbon T300	1.80	128	71.1
Steel cord	7.80	24	3.1

*Based on the equation given in Section 1.6.7 with $\varepsilon = 1$ and $d = 1$; the result is made dimensionless by dividing the value of a fiber with that of a hypothetical fiber whose value in the same units is 1.

Table 1.6 Tensile, bending and shear properties of textile fibers

Fiber	Tensile modulus/shear modulus	Tensile strength (cN/Tex)	Shear strength (cN/Tex)	Shear modulus (GPa)
Cotton	3.7	235	84	2.2
Wool	3.2			1.3
Silk	3.9	314	116	2.4
Nylon	5.8	392	112	0.4
Polyester	13.8			0.8
Aramid (Kevlar® 149)	154			1.2
HMPE (Spectra® 1000)	594			0.18
PBO	330			0.97
Carbon T 300	49			4.7
Steel	2.8			66
E Glass	2.0			36
Polypropylene	12			0.75

Sources: V Mehta and S Kumar [30], W E Morton and J W S Hearle, *Physical Properties of Textile Fibers*, 3rd Edition, The Textile Institute, Manchester, 1993 [4], H Bohringer and W Schieber [31], and J D Owen [32].

difference between the shear and the tensile moduli may be considered. For isotropic materials, the relationship between the tensile (E) and the shear (G) modulus is given by the following model in which v is the Poisson's ratio:

$$E = 2\,G\,(1 + v)$$

Only limited measurements of v have been made on fibers [4]. A value of about 0.39 has been reported for nylon and between 0.42 and 0.63 for wool [24–28]. Assuming that the values will range between 0.3 and 0.6, irrespective of polymer type and structure, one can conclude that for isotropic polymers the shear modulus will be 2.6 to 3.2 times smaller than the tensile modulus. The values of the ratio of tensile to shear modulus in Table 1.6 show that most commodity fibers tend to support this range. The more extended the chains and the oriented the fiber, the greater the ratio. This is largely because with increase in orientation E increases greatly but G decreases some, usually by a factor of less than 2 [29]. The extreme values are given by the high performance fibers. One notes that with very few exceptions (inorganic fibers on the higher end and Spectra® and nylon on the lower end), the fiber-to-fiber differences in the shear properties (modulus) are quite small. On this account, one could expect that the differences in the frictional properties of different fibers would also tend to be small.

When an oriented cylindrical fiber is compressed transversely (Fig. 1.8), only a thin vertical segment in the center will get compressed but all others lying off the direct pressure point will get sheared sideways. The material forced to compress between the two pressure points will offer very high resistance.

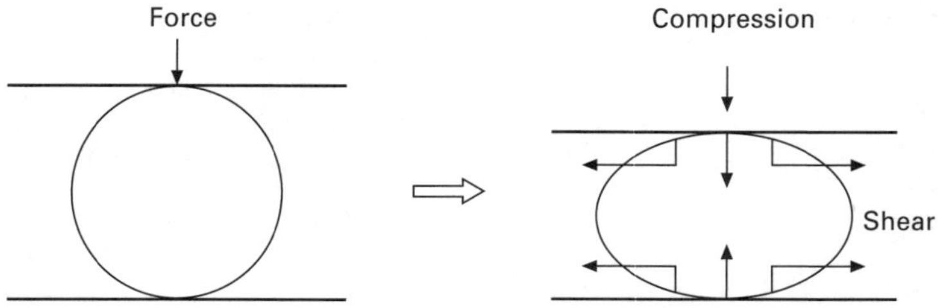

1.8 Behavior during transverse compression of a fiber.

Table 1.7 Comparison of tensile and compressive properties of fibers [33]

Fiber	Axial modulus (GPa)	Transverse compression modulus (GPa)
Nylon	4.3	1.2
Polypropylene	5	1.0
Polyester	16	1.2
Spectra® 1000	89	1.2
Kevlar® 149	179	2.5
Carbon T 300	168	6.8
E Glass	77	68

This will be reduced as the structural units lying off the center move sideways and create room for the compressed units to slip laterally and reposition. Such combined compression-cum-shear will not occur if instead of a cylindrical fiber, a flat fiber or film was compressed. The resistance to compression in this case will be very high, as the action will generally tend to densify the structure. Accordingly, in general, one can expect that in fibers and fiber-based products the transverse compressive modulus will be greater than the shear modulus but, since only a small fraction of the total volume undergoes compression and the rest shear, the difference will not be large. The values of the transverse compression modulus for a number of fibers are given in Table 1.7. These values are greater than the values of the shear modulus (Table 1.6), but in most cases, the difference is marked by a factor of less than 2. Also, ignoring the inorganic fibers (carbon included), the values of most fibers are quite close.

1.6.8 Viscoelastic properties

Textile fibers, being two-phase structures, have both elastic and time-dependent recovery behaviors. Even from small strains, very few materials show perfect instantaneous recovery, which will be possible only if the strain imposed fell within the elastic limit of the weakest of the bonds supporting the load. When load is applied, a disproportionately high amount of stress will be transferred to the weakest of the elements and cause them to exceed their elongation limits and break. This will cause the material to flow. Unless the stresses are large, the flow will be limited and when the external load is removed, the elastically deformed but unbroken bonds will return to their original states and assist even some of the broken elements to return to their initial positions. Accordingly, even though the macroscopic dimensions may be regained, time is taken for it to happen and some energy is lost in the process. The loading and unloading curve usually follows a loop, reflecting time dependent behavior and loss in energy, known as hysteresis. The elastic recovery test involves several cycles of loading and unloading with each successive cycle taken to higher load over the previous cycle [4]. A plot of elastic recovery (fraction of elongation or strain recovered) against strain for a number of textile fibers is shown in Fig. 1.9 [34]. Fiber-to-fiber differences in elastic recovery are obvious. It is generally found and accepted that recovery from strain even as small as 2–5% in most fibers is less than 100%.

Because fibers are viscoelastic, they increase in length, or creep, if a constant load is imposed and maintained for a period of time, or decrease in stress, i.e. undergo stress relaxation, when a given deformation is imposed and maintained over a period of time. It has been shown that these are the non-Newtonian flow-related behaviors in polymers and they occur due to kinetic processes that take place in disordered regions at temperatures that are sufficiently high [35]. Since most polymers have relatively low softening points, the ambient conditions are sufficient to invoke kinetic processes in amorphous regions. The creep and relaxation phenomena are intimately related

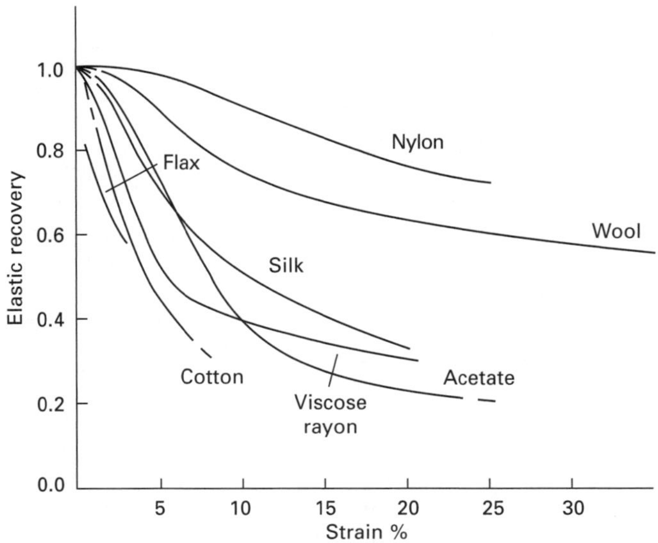

1.9 Elastic recovery properties of textile fibers [34].

so that a fiber that creeps at a high rate also relaxes similarly. Creep behaviors of four textile fibers are shown in Fig. 1.10 [36]. Some fibers creep faster than others but all undergo time-dependent growth in length. The magnitude of creep is affected by the load imposed and that of relaxation by the amount of the deformation maintained, and the rates of both increase with temperature in all materials and with increase in moisture in hydrophilic materials. One expects that the creep and relaxation behaviors, generally noted in tensile tests, will also be found in bending, shear and transverse compression, although little information is available on these in the published literature.

In friction tests, one body slides in contact with another under an imposed normal force. The viscoelastic nature of fibers indicates that the speed of sliding, the magnitude of normal force and the length of time during which the latter exists before sliding takes place, will all affect the area of contact and, therefore, the magnitude of frictional force. As noted above, temperature and moisture that affect creep will also affect the area of contact and, therefore, the frictional force.

Many factors affect the difference between the static and the kinetic values of friction; one among these will clearly be the time dependence of the mechanical properties, in particular the shear, properties.

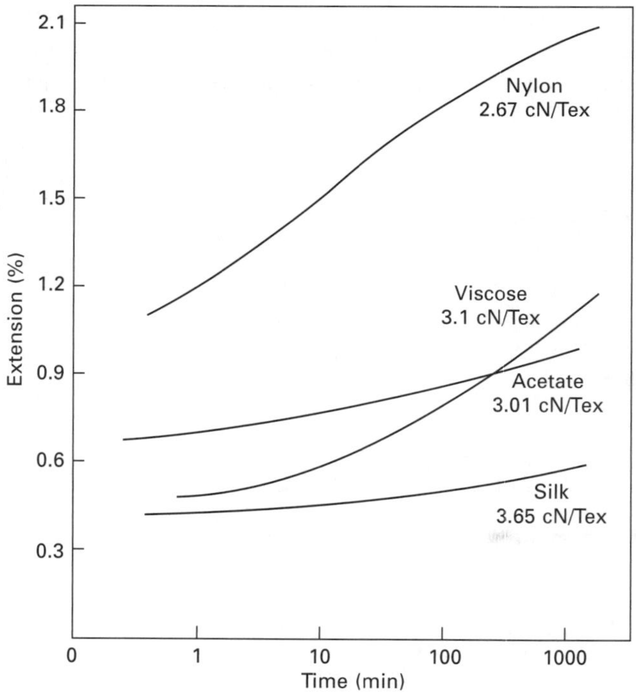

1.10 Creep behavior of selected textile fibers [36].

1.6.9 Surface energetics of fibers

Lubricants and finishes are applied to surfaces to control friction, facilitate processing and/or improve hand and functionalities. A chemical applied must spread and adhere to the surface so that it can withstand the shear forces encountered during processing and use. The surface wettability characteristics of fibers and films for aqueous and non-aqueous fluids are, therefore, important. Several parameters determine interaction between a solid and fluid, which together form a part of the discipline known as surface energetics. These concepts are inter-related and include, among others, contact angle, solid surface energy and critical surface tension.

Contact angle (θ) relates to the wettability of a solid with a liquid (Fig. 1.11). An angle less than 90° means that the fluid will spread, whereas a value equal to or greater than 90° indicates that the fluid will bead up and not spread. The Wilhelmy technique, in which the force of advancing a solid into a fluid or receding it out of the fluid is measured (Fig. 1.12), is a useful tool for characterizing the contact angle as it yields the values of both the advancing and the receding angles. For wettability, the advancing value is more directly applicable and is given for several materials against water in Table 1.8. The lower the value the more rapidly wettable the surface. Fibers such as polypropylene and polyethylene do not wet with aqueous fluids but do so easily with oils for which they have lower value of θ. Measuring the surface energy of the solid and determining its composition in terms of the polar and the dispersion fractions allows one to understand whether a surface is hydrophilic or hydrophobic and is compatible with polar or non-polar fluids. The approximate values of surface energy γ_s^{p+d} and its division into the polar γ_s^p and the dispersion γ_s^d values, assessed with a geometric mean model used by several workers [37–39], are given in Table 1.8 for a number of fibers. Clearly, the fibers such as cotton and rayon are hydrophilic and should have high attraction for polar liquids, whereas the fibers such as olefin and polyester are predominantly hydrophobic and should have greater compatibility with the non-polar liquids. These materials are often known as

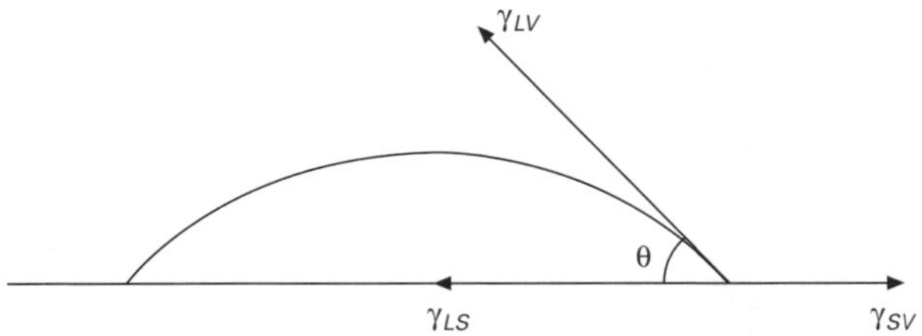

1.11 Contact angle θ formed by a drop of liquid on a solid surface at equilibrium between the three interfacial surface tensions, liquid-vapor, liquid-solid and solid-vapor.

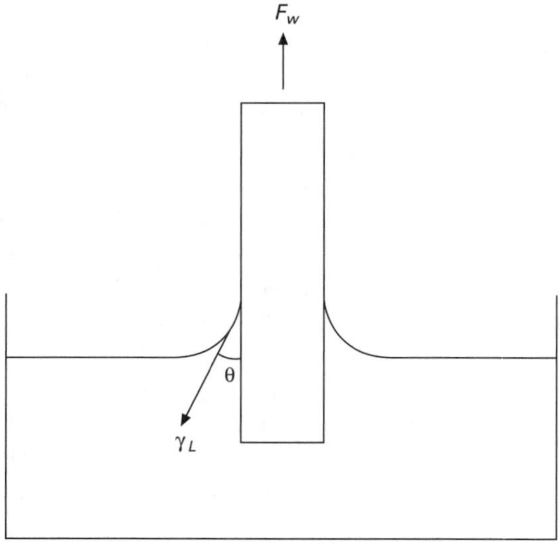

1.12 The Wilhelmy force technique for measuring contact angle. $F_w = \gamma_L P \cos\theta$, in which F_w is the force of advancing a fiber into fluid (for advancing contact angle) or receding the fiber out of fluid (for receding contact angle), P is the fiber perimeter, and θ is the contact angle.

Table 1.8 Surface contact angle and solid surface energy values of fibers

Fiber	θ_A	Surface energy (dyne/cm)		
		γ_s^p	γ_s^d	γ_s^{p+d}
Cotton	34	41.6	19.3	60.9
Rayon				
Regular	55	24.4	21.7	46.1
Trilobal	22	51.7	16.4	68.1
Wool	73	8.8	31.3	40.1
Nylon	68	15.8	35.0	50.8
Polyester	71	4.1	37.7	41.8
Polypropylene	98	1.8	21.2	23.0
Aramid (PPTA)	48.9	28.3	18.9	47.2
PBO	51.7	25.8	17.5	43.3
Carbon	37.3	4.5	35.6	40.1
S_2 Glass	62	14.4	32.0	46.4
E Glass	84	7.8	18.9	26.7
PTFE	112	1.7	17.3	19.0
PE	95	3.4	34.2	37.6

Sources: B S Gupta and H S Whang [40–42]; F Hoecker and J Karger-Kocsis [43]; S Wu [44]; A Bismarck *et al.* [45]; G M Wu *et al.* [46].

oleophilic, as oil will spread and adhere to the surface more readily than will water.

A concept of critical surface tension γ_c was introduced by Zisman [47] for the spreading of fluids and finishes on low energy surfaces. The latter are characteristic of most polymers and fibers, especially those known to have little or no intermolecular hydrogen bonding. Wetting of such a surface by a series of homologous fluids having different surface tensions, γ_L, yields a plot of the cosine of the contact angle against fluid surface tension. Extrapolation of the curve to $cos\ \theta$ of 1 corresponds to γ_c of the solid surface (Fig. 1.13). The concept of the critical surface tension γ_c implies that a fluid of the series having a surface tension $\gamma_L < \gamma_c$ will instantly spread and wet the surface. The tests for critical surface tension for a material can be conducted with sets of

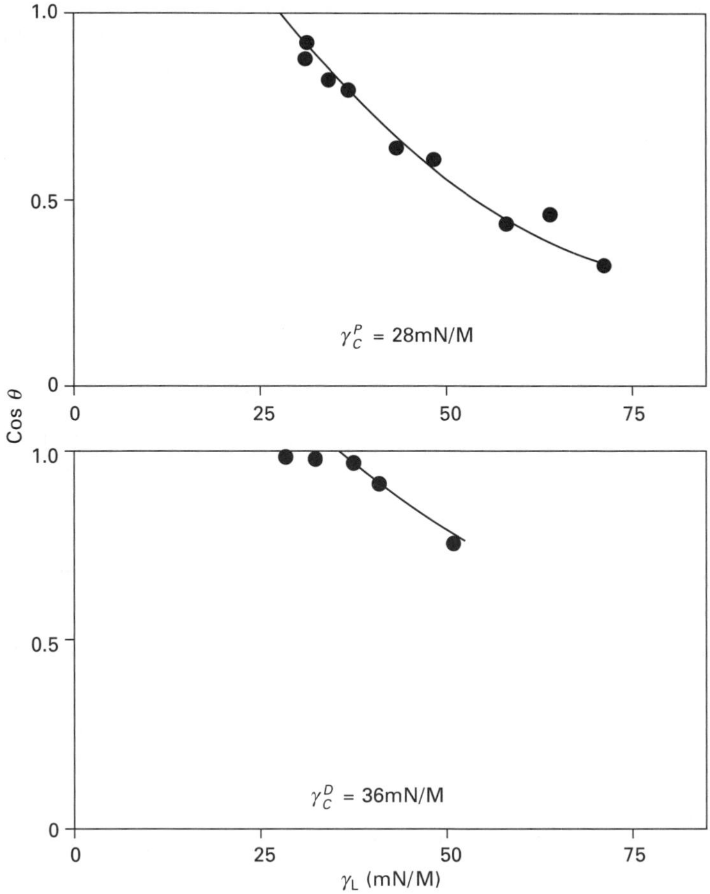

1.13 Critical surface tension (γ_c) of the sheet molding compound (SMC) determined by the Zisman's method with polar and non-polar fluids [48].

Table 1.9 Critical surface energy of different substrates by Zisman's method

Substrate	γ_c (mN/m) Measured with liquids those are:	
	Polar	Non-polar
Paraffin	18	25
Poly (propylene)	24	30
Poly (ethylene terepthalate)	30	42
SMC (Sheet Molding Compound)	28	36
Poly (methyl methacrylate)	28	39
Poly (ethylene)	29	33

Sources: K C Sehgal and M A Sherwin [48]

both polar (P) and non-polar (D) fluids and γ_c determined for each type. Knowing the value of γ_c for a material provides a convenient criterion for selecting or formulating a treatment for topical application. An appropriate surfactant that lowers the surface tension of the fluid may be used if the criteria $\gamma_L < \gamma_c$ is not met. Some example results are given in Table 1.9 [48]. A higher value of γ_c for non-polar fluid reflects compatibility between the non-polar surface and the non-polar fluid. If a polar fluid is used, its surface tension will, most likely, have to be decreased with the use of an additive in order to make the surface wettable.

1.7 Summary

A large number of fibers are used in making traditional textile and specialty technical products. They differ from each other in chemical and physical structures and, as a consequence, in their surface and bulk properties. The surface morphology and the surface and bulk physical properties, including mechanical, play important roles in influencing a fiber's frictional properties and through it the latter's performance in processing, handling and end use. The details given in this chapter should therefore, serve as useful background information for understanding the frictional behaviour of fibers discussed in this book.

1.8 References

1. Beers D, Young R J, So C L, Sikkema D, Perepelkin K E and Weedon G, 'Other High Modulus – High Tenacity (HM–HT) fibres from linear polymers', in Hearle J W S, Ed., *High-performance Fibres*, Woodhead Publishing Limited, Cambridge, p 93, 2001.
2. Chu C C, 'Chemical Structure and Manufacturing Processes', in Chu C C, von Fraunhofer J A, and Greisler H P, Eds, *Wound Closure Biomaterials and Devices*, CRC Press, New York, 1997.

3. Gupta B S, 'Manufactured Textile Fibers', in Kent J A, Ed., *Riegel's Handbook of Industrial Chemistry and Biotechnology*, 11th edition, Volume 1, Springer, Secacus, NJ, p 431, 2007.
4. Morton W E and Hearle J W S, *Physical Properties of Textile Fibers*, The Textile Institute, Manchester, 1993.
5. Hearle J W S, 'The structural mechanics of fibers', *J Polymer Sci.*, C, 1967, **20**(1): 215.
6. Peterlin A, 'Molecular Aspects of Oriented Polymers', in Ward I M, Ed., *Structure and Properties of Oriented Polymers*, Applied Science Publishers, London, p 36, 1975.
7. Van Dingenen J L J, 'Gel spun high-performance polyethylene fibers', in Hearle J W S, Ed., *High-performance fibers*, Woodhead Publishing Limited, Cambridge, p 62, 2001.
8. Harris M (Editor), *Handbook of Textile Fibers*, Harris Research Laboratories, Washington D C, USA, p 60, 1954.
9. Lindley H, 'The chemical composition and structure of wool', in Asquith R S, Ed., *Chemistry of Natural Protein Fibres*, Plenum Press, London, p 147, 1977.
10. Cole D J, 'A new cellulosic fiber- Tencel®', in *Advances in Fiber Science*, The Textile Institute, Manchester, 1992.
11. Albrecht W, Reintjes M and Wulfhorst B, 'Lyocell fibers (Alternative regenerated cellulose fibers)', *Chem. Fibers Int.*, 1997, **47**, p 298.
12. Gore R W, 'Process for producing porous products', *US Patent 3,953,566 to W L Gore & Associates Inc.*, April 27, 1973.
13. Turbak A, 'Rayon', in Klingsberg A and Baldwin T, Eds., *Encyclopedia of Polymer Science and Engineering*, 2nd edition, John Wiley, New York, Vol 14, p 55, 1985.
14. Ford J E, *Fiber Data Summaries*, Shirley Institute, Manchester, 1966.
15. *Manufactured Fiber Fact Book*, American Fiber Manufacturers Association Inc., Washington D C, p 13, 1988.
16. Preston J M and Nimkar M V, 'Measuring the swelling of fibres in water', *J. Text. Inst.*, 1949, **40**, p 674.
17. Gupta B S and Chatterjee P K, 'Measurement techniques for absorbent materials and products', in Chatterjee and Gupta Eds., *Absorbent Technology*, Elsevier, Amsterdam, p 410, 2002.
18. Hearle J W S, *High-performance Fibers*, Woodhead Publishing Ltd, Cambridge, 2001.
19. Horrocks A R, Eichhorn H, Schwaenke H, Saville N and Thomas C, 'Thermally Resistant Fibers', in Hearle, Ed., *High-performance Fibers*, Woodhead Publishing Ltd., p 281, Cambridge 2001.
20. Warner S B, *Fiber Science*, Prentice Hall, Englewood Cliffs, 1995.
21. Meredith R, 'The tensile behaviour of raw cotton and other textile fibres', *J. Text. Inst.*, 1945, **36**, T107.
22. Kumar S, 'Advances in high performance fibers', *Indian J. of Fiber and Textile Research*, 1991, **16**, 52.
23. *Technical Bulletin X-82, E I du Pont de Nemours & Co., Inc.*, Wilmington DC, USA, 1958.
24. Morris S, 'The determination of the lateral-compression modulus of fibres', *J. Text. Inst.*, 1968, **59**, 536.
25. Davis V V, 'A diffraction method for the measurement of diameter changes in strained fibres', *J. Text. Inst.*, 1959, **50**, T 688.

26. Frank F I and Ruoff A L, 'A method of measuring Poisson's ratio of fibers', *Textile Res. J.*, 1958, **28**, 213.
27. Banky E C and Slen S B, 'Dimensional changes and related phenomena in wool fibers under stress', *Textile Res. J.*, 1956, **26**, 204.
28. Hadley D W, Pinnock P R and Ward I M, 'Anisotropy in oriented fibres from synthetic polymers,' *J. Mater. Sci.*, 1969, **4**, 152.
29. Cumberbirch R J E and Owen J D, 'The mechanical properties and birefringence of various monofils', *J. Text. Inst.*, 1965, **56**, T389–T408.
30. Mehta V and Kumar S, '*Temperature dependent torsional and compression behavior of high-performance fibers'*, American Physical Society, Seattle, March 1993.
31. Bohringer H and Schieber W, cited by Schiefer H F, Fourt L and Kropf R T, 'Serviceability evaluation of Textiles', *Textile Res. J.*, 1948, **18**, p 18.
32. Owen J D, 'The application of Searle's single and double pendulum methods to single fibre rigidity measurements', *J. Text. Inst.*, 1965, **56**, T329.
33. Kawabata S, 'Measurement of the transverse mechanical properties of high-performance fibres', *J. Text. Inst.*, 1991, **81**, 432.
34. Meredith R, 'A comparison of the tensile elasticity of some textile fibres', *J. Text. Inst.*, 1945, **36**, T 147.
35. Halsey G, White H J and Eyring H, 'Mechanical properties of Textiles, I', *Textile Res. J.*, 1945, **15**, 295.
36. Leaderman H, '*Elastic and Creep Properties of Filamentous Materials and Other High Polymers'*, The Textile Foundation, Washington D C, USA, 1943.
37. Owens D K, and Wendt R C, 'Estimation of surface free energy of polymers', *J. Appl. Polymer Sci.*, 1969, **13**, 1741.
38. Kaelble D H, Dynes P J and Cirlin E H, 'Interfacial bonding and environmental stability of polymer matrix composites', *Journal of Adhesion*, 1974, **6**(1), 23.
39. Wu S, *Polymer Interface and Adhesion*, Marcel Dekker, New York, 1982.
40. Gupta B S and Whang H S, 'Surface wetting and energy properties of cellulose acetate, polyester and polypropylene fibers', *International Nonwovens Journal (INJ)*, 1999, **8**, 36.
41. Whang H S and Gupta B S, 'Surface wetting characteristics of cellulosic fibers', *Textile Res. J.*, 2000, **70**(4), 351.
42. Whang H S, *Wetting characteristics and surface energy properties of cellulosic and non-cellulosic fibers*, Ph. D. Thesis, N C State Univ., USA, 1997.
43. Hoecker F and Karger-Kocsis J, 'Surface energetics of carbon fibers and its effects on the mechanical performance of CF/EP composites', *J. Appl. Polymer Sci.*, 1996, **59**, 139–153.
44. Wu S, 'Polar and nonpolar interactions in adhesion', *J. Adhesion*, 1973, **5**(1), 39.
45. Bismarck A, Boccaccini A R, Egia Ajuriagojeaskoa E, Hulsenberg D and Leutbecher T, 'Surface characterization of glass fibers made from silicate waste: Zeta-potential and contact angle measurements', *J. Mater. Sci.*, 2004, **39**, 401–412.
46. Wu G M, Hung C H, You J H and Liu S J, 'Surface modification of reinforcement fibers for composites by acid treatments', *J. Polym. Res.*, 2004, **11**, 31–36.
47. Zisman WA, 'Surface energetics of wetting, spreading, and adhesion', *J. of Paint Technology*, 1972, **44** (564), 41.
48. Sehgal K C and Sherwin M A, 'New waterborne adhesives for low and high energy surfaces', *The Adhesives and Sealant Council Proceedings*, 1989, **18**(1), 137.

2

Fundamental concepts of friction and lubrication affecting textile fibers

B S GUPTA, North Carolina State University, USA

2.1 Introduction

Although many of the basic ideas about friction were developed through work with metals, it is the result of the investigations with non-metals, in particular textiles and rubbers, which led to a deeper understanding of the field. Work with metals led to the creation of classical laws that were simple to understand and convenient to apply to problems of great technological value. The reader is directed to excellent works published in this area by Bowden, Tabor and their associates [1–3]. The general observations of the past five centuries, starting with those of Leonardo de Vinci (1452–1519) in the fifteenth, through those of Amonton (1663–1705), and Coulomb (1736–1806) in the seventeenth and eighteenth centuries, respectively, were given a rational base in the middle of the twentieth century by the works of Bowden and Tabor. The primary concept that adhesive junctions develop at points of contact and that these must be sheared in order for sliding to occur, along with the corollary concept that this process leads to wear and to reduce which lubrication may be used, led to the introduction of a new discipline, "Tribology," in engineering colleges of the institutions of higher learning. It implies a study of friction, wear and lubrication in materials. The seemingly simple laws that apply to metals, the materials that deform plastically, show significant departures when applied to polymeric materials, the textiles, which deform visco-elastically. However, a general adhesion-based theory developed for materials deforming visco-elastically [4], discussed in Chapter 3, can also apply to those that deform plastically and elastically; the behavior of the latter two essentially becoming special cases of that of the former.

In this chapter, we shall discuss the many useful concepts of friction that have been developed over the years and continue to be of value in our understanding of the tribological behavior of the materials in general, starting with the behavior observed in metals, and then with those observed in other materials, i.e. elastomers and polymers and fibers. Considered first will be the behavior observed in unlubricated surfaces and then that observed on surfaces on which a lubricant is applied.

2.2 The basic laws of friction

Friction is the tangential force that opposes relative motion when one body slides over another. The two basic laws of friction deduced by Leonardo de Vinci in the middle of the fifteenth century, rediscovered by Amonton in 1699 [5], and investigated and substantiated by Coulomb in 1788 [6] may be stated as follows:

- The frictional force, F, is proportional to the normal force, N. This implies that the ratio, F/N, called the coefficient of friction and commonly denoted by the symbol μ, is a constant for a given pair of bodies.
- The frictional force is independent of the geometric area of contact.

Coulomb also pointed out that the force needed to initiate sliding, i.e. the static frictional force, F_s, is greater than the force needed to maintain sliding, i.e. the kinetic frictional force, F_k. He also stated that the kinetic frictional force was generally independent of the speed of sliding, which is often referred to as the third law of friction. These may be expressed as follows:

1. $F \propto N$; $\mu = F/N$
2. F independent of the geometric area of contact
3. $F_s > F_k$; $\mu_s > \mu_k$
4. F_k independent of the speed of sliding

Most of the studies until about the end of the nineteenth century were concerned with friction as it was encountered in practical problems. No serious doubts had been cast on the validity of the laws and, for that reason, on the causes of friction. Several schools of thought explaining the phenomena had emerged; among these being the surface roughness, the adhesion and the deformation theories. Amonton, the first to carry out a scientific study of the friction, had suggested that friction was due to the force needed to lift one surface over the asperities of the other; others thought it was due to the attractive forces between the atoms on the two surfaces, or to electrostatic forces. These theories all assumed that the surfaces remained separate and, while they may sometimes play a role, the work of Bowden and his associates [1–3] in the first half of the twentieth century showed that the predominant effects are usually an actual union or welding of the two surfaces at the points of contact. The frictional force is the result of the effort needed to shear these junctions in order to initiate sliding. This led to the introduction of the most successful of the concepts of friction, i.e. the adhesion-shearing theory, applying to metals and many other materials.

2.3 Adhesion-shearing theory

On a molecular scale, most surfaces are generally rough and the contact only occurs at the tips of asperities (Fig. 2.1). The normal force acting on these points exerts a pressure that is initially high, i.e. higher than the elastic limit or the yield value. This causes deformation of the junctions and lateral flow in them until the area has increased and the pressure decreased to the point that the force can be supported elastically. The mechanical properties of the material play a role in determining the nature and the extent of the deformation. Most of the early work was conducted on metals that deformed plastically at pressures exceeding the yield value. The behavior is illustrated in Fig. 2.2.

If A is the total area of contact, then:

$$A = \Sigma A_i = \Sigma \frac{N_i}{p_y} = \frac{N}{p_y} \qquad 2.1$$

In this, N is the applied normal force and p_y is the yield pressure of the material. Thus, the area of real contact is proportional to the applied normal force. A rationale for the formation of the junctions is that under the pressure that causes viscous or plastic flow the work is dissipated as heat in the local regions. This aids in forming the junctions that are welded. In metals, the temperature in minute localized regions may reach several hundred degrees [7]. With non-metals, since they are poor thermal conductors, the effect could be even more pronounced [8]. These are the so-called cold junctions, apparently because there is no detectable change in the average temperature of the contacting bodies.

In order to initiate sliding, these junctions must be ruptured by shearing. If S is the bulk specific shear strength of the junctions (N/m^2), then the frictional force needed to rupture them will be given as follows:

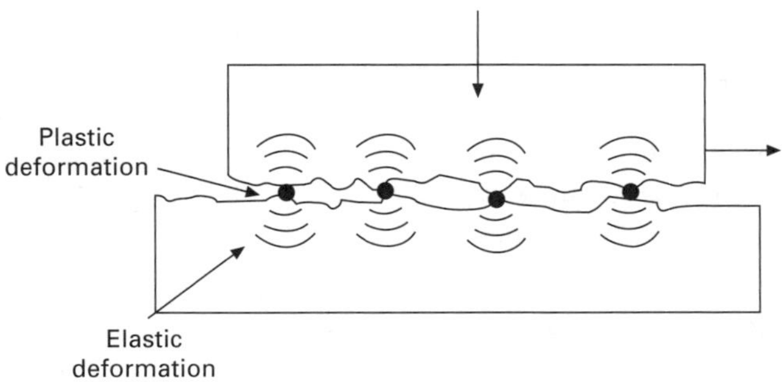

2.1 Contact between two surfaces at tips of asperities. Plastic deformation occurs at these points, leading to formation of junctions. The surrounding areas are elastically deformed, which snap back when the junctions are broken with the removal of load.

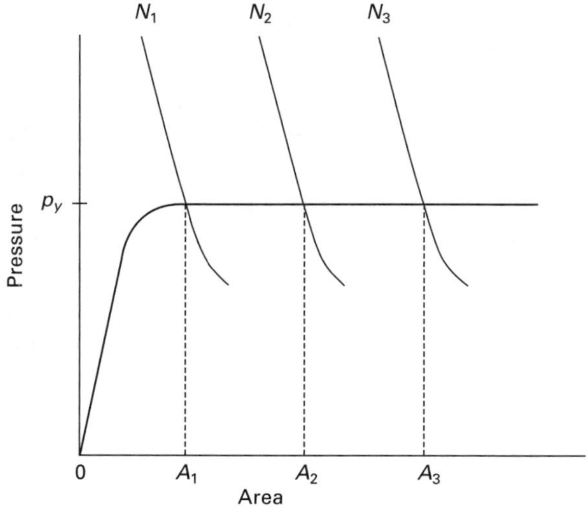

2.2 Pressure–area behavior of metals showing plastic deformation with a relatively sharp yield point. Lines 1, 2, 3 represent iso-load curves of asperities supporting normal forces, N_1, N_2, N_3, respectively.

$$F = SA = S\frac{N}{p_y} = \left(\frac{S}{p_y}\right)N \qquad 2.2$$

With p_y and S both being constants for a given material, μ should be a material property. If N is doubled, so should be F, observed Leonardo de Vinci. This supports the first law. Since, for general surfaces and/or reasonably high loads (greater than yield value), A is independent of the number of points of contact and dependent only on the normal force, F is independent of the geometric area of contact. This supports the second law. Often a ploughing term P is added to the adhesion term in equation 2.2 to describe frictional force. This is the force needed for the asperities of the harder surface to plough or groove through the material of the softer surface. However, for most materials, the ploughing term is considered to be small as compared to the adhesive term and is dropped [2]. The concept of ploughing and deformation is discussed in Section 2.6, and also later in Section 2.14 when the frictional behavior of rubber is discussed.

2.4 Slip-stick patterns in sliding

As clear from the stated laws and general observations, it takes a greater force to initiate the sliding of one object against another than to maintain sliding after motion has commenced. Growth in junctions, i.e. area of contact, due to creep and/or application of a tangential stress, or strengthening of junctions, due to inter-diffusion of surface atoms across the interface, are two of the possible reasons responsible for such a difference [3]. Because of the difference in friction values, the friction trace is intermittent, or stick-slip, in character, although the magnitude of it is determined by many factors. These include the visco-elastic properties of the two materials, the speed, and the visco-elastic nature of the electro-mechanical system used in running the friction device and in measuring the force. A detailed discussion of stick-slip phenomena is contained in Chapter 4. In a classical illustration of the phenomena, a body resting on a traversing platform is attached to a deformable spring that is held fixed at the other end (Fig. 2.3(a)). The frictional force is measured by the extension of the spring or by the load cell connected to it at the fixed end. As the platform traverses, the resting body moves with the platform until the tension in the spring equals and just exceeds F_s, the static frictional force. At this point, due to the kinetic value being lower than the static ($F_k < F_s$), the rider slips and accelerates back until the tension has dropped to F_k and, due to inertia, continues to move but decelerates until it comes to rest at some value of the force F, where $F < F_k$. It then starts to move forward until it reaches F_s. A plot of tension against time yields a characteristic stick-slip pattern (Fig. 2.3(b)). The slope during slipping is steeper than during the sticking phase. Although the classical pattern given is usually a saw tooth type, with linear rise and rapid fall in tension, in actual practice the changes in tension can be quite complex, the topography of a material can influence the fluctuations. It is important to note that the first slip point may be at a higher force value (not shown) than the successive slip points. This is due to the extra time of contact involved before the junctions break for the first time. Also, it should be noted that certain types of variations in tension might give a stick-slip appearance while in reality these are caused by other factors. For example, a localized irregularity on a surface caused by transfer of material in a journal bearing can give rise to periodic variation, even though there is no significant change in sliding velocity with time.

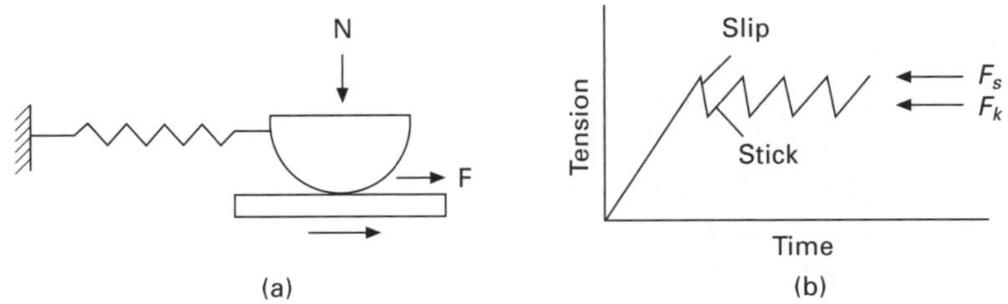

2.3 (a) Basic apparatus showing method of measuring friction between a plane surface and a rider. (b) Record of tension profile.

2.5 The role of surface roughness

Coulomb originally suggested that the frictional energy expended in sliding represents the work necessary to lift one surface over the asperities on the other [9]. As noted by Tabor [2, p 4; and 3, p 20], there were two difficulties in accepting this concept. First, it was not easy to see how this could be a dissipative process. If adhesion between surfaces was absent, then the energy lost in going up an inclined asperity would be recovered in sliding down. There would be no net change if the vertical positions of the starting and the ending points were the same. This difficulty is avoided by assuming that the work is done in sliding the surface up the asperity and then the upper body falls sharply losing its gained energy on impact as heat. Thus, more work will be needed in order to continue to slide the surface further. This will partly lend support to the deformation theory. The second difficulty is that two surfaces in contact have multiple asperities. Under normal circumstances, therefore, some asperities will be sliding up and some down; the net effect, thus, should be little loss.

In special cases, however, surface roughness will be very important. An example will be two similar files slid against each other. This will not be a true Coulomb effect, but one of mechanical interlocking.

2.6 Deformation and ploughing

If a rough hard surface slides over a soft surface, the roughness of the harder surface will groove or plough out the surface of the softer body. In the adhesion model, a term representing the ploughing force is usually added to that representing the force of adhesion:

$$F = SA + P \tag{2.3}$$

This would suggest that if a roughness factor was required, it would be most marked with materials for which the adhesion term is small. Thus, with polytetrafluoroethylene (PTFE), for which the adhesion term is small, the ploughing term may be considerable [10]. In textiles, surface roughness may play a role when one woven or knitted fabric slides over another. The friction may depend markedly on the relative directions and tightnesses of the construction. This will clearly not be a ploughing effect in the sense discussed, but may well be due to the way in which the individual yarns are repeatedly stretched and relaxed during sliding. The yarns are not perfectly elastic bodies and although they may revert to their original shape, energy will still be lost in the process. A similar dissipative process occurs to some extent in rubber. Deformation losses of this type occur in sliding and in rolling friction [11]. A brief account of this is given later in Sections 2.14 and 2.15.

2.7 The area of contact between metals

If we could obtain molecularly smooth surfaces and place them one on top of the other, one would think that the real area of contact would closely match the geometric area. With metals, this is not usually the case. Examination of metal surfaces shows that howsoever carefully a surface is prepared, it is rough on an atomic scale. Tabor notes that even electrolytic polishing leaves hills and valleys of the order of several hundred Angstrom units [2, p7]. When two such surfaces are placed in contact, they first touch only at the tips (Fig. 2.1). Since the surface fields of force only extend over a few Angstrom units, the rest of the surfaces can play no appreciable part in supporting the load, electrically or otherwise. The contact between the surfaces is, therefore, primarily determined by the deformation of asperities. The yield pressure p_y, at which plastic flow occurs, is essentially the same as the hardness of the metal. This property is generally determined on a macroscopic scale by pressing a large indenter onto the surface and measuring the mean pressure over the indentation [12]. The orders of the values for some materials are given in Table 2.1.

One of the striking properties of metals is that the plastic yield pressure p_y is a material constant. This means that the area of contact can be indirectly given as:

$$A = \frac{N}{p_y} \qquad 2.4$$

It is, however, clearly desirable to find more direct methods of assessing the true area of contact and validating the calculated value. Tabor has reviewed a number of methods for such direct measurement; these are based on the measurement of the electrical resistance, the use of phase contrast microscopy and optical interference microscopy. In general, the values obtained for most materials agreed well with those calculated by the plastic theory model [13–15].

2.8 The formation and growth of junctions

The junctions formed when two bodies are pressed together are sheared when sliding occurs. Pieces of one surface are plucked out and transferred

Table 2.1 Values of yield pressure of selected metals [3]

Metal	Yield pressure (kg/mm^2)
Lead	5
Copper	80
Mild steel	120
Hard steel	800
Tungsten carbide	2000

on to the other. A similar phenomenon occurs when a metal slides over a polymer film. As discussed earlier, deformation energy at points of contact is dissipated as heat. This can cause an enormous rise in temperature in tiny regions surrounding the contact points and lead to the development of junctions.

In the theory considered, it is assumed that the true area of contact, i.e. of the junctions formed and sheared, is the same as that existed prior to sliding. This is not so; the junctions grow before noticeable sliding occurs. When two clean metals are placed one on top of the other, the plastic yielding of the tips of asperities determines the area of contact. When a tangential force is applied, these junctions, having a plastic character, flow further and increase the true area of contact before actual sliding occurs on an observable scale [16]. This is illustrated in Fig. 2.4. Wherever the new regions touch, they form junctions. On clean metal surfaces, such growth can continue microscopically until the whole geometric area has become one enormous friction junction. Coefficients of friction of 10 or more may be observed and sliding results in heavy damages to the surfaces [3, p 96]. This is, however, for clean surfaces, produced by heating under vacuum and cooling. If a small amount of water vapor or oxygen is present, the friction drops from a very large value to the usual value of around 1. This is accompanied by a corresponding decrease in surface damage. An oxide film only a few molecules thick, produced on the surface, prevents surfaces sticking strongly together. This allows only a small growth in the junctions. Fortunately, the majority of metals used in industry readily form oxide films on their surfaces.

Junction growth is a major factor in metals, but for polymeric materials the increase in junctions is limited, probably to 20–40% [17–19]. In spite of the growth in the area of contact, the two basic laws of friction hold. Since the contact involves a large number of junctions, the size of the surface has little effect on the fundamental behavior of a given junction; consequently

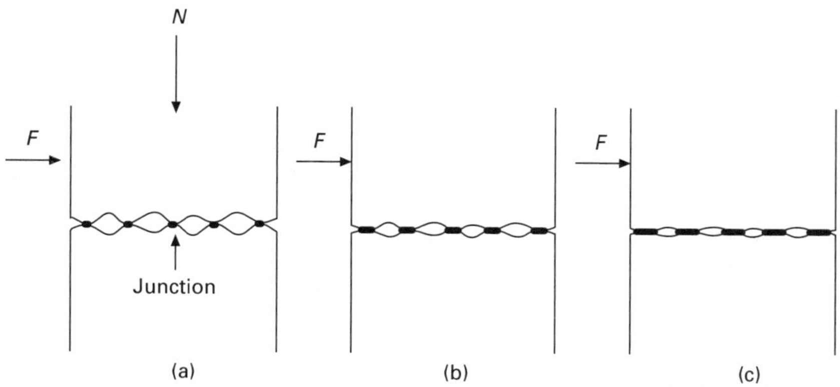

2.4 Contact between metals showing growth of junctions prior to sliding. (a) Force $F = 0$; (b) $0 < F < F_s$; (c) $F \approx F_s$.

the geometric area has no appreciable effect on μ. Also, for a given pair of surfaces, if junction growth is constant, both the initial area and the final or steady state area being sheared will still be proportional to the load.

2.9 Normal adhesion

An observation is often made that although junctions develop at points of contact, strong normal adhesion between the surfaces, i.e. the force needed to lift one surface vertically off the other, is generally not observed. Two reasons are given to account for this. First the metallic surfaces are always covered with oxide films and these are more readily penetrated in sliding than in normal loading [2]. A second reason is that when friction is measured in sliding, normal load is maintained, but in assessing normal adhesion, the normal load must be removed. In the process, the elastic stresses around the individual contact regions are released and the interface changes its shape. The relatively brittle junctions are thus stretched and easily stripped apart one by one so that by the time the load is completely removed, practically no junctions are left. In shear when the compressive stress is maintained, all junctions respond together and a large amount of work is expended in shearing them. That this is so is validated by the work with soft ductile metals, such as lead, tin and indium, which allow junctions to deform and adjust, not snap, when elastic stresses are released during lifting. In these cases, high normal forces are required to lift off one surface from the other. Using an ingenious technique involving two crossed polyester monofilaments in which one monofilament acted as a cantilever (see Section 5.5, Chapter 5), Briscoe, *et al.* [20] measured the force due to interfacial adhesion between the surfaces. In one experiment, these forces were of the order of 1.6 μN that augmented the normal force, which was of the order of 1 μN imposed by the cantilever action of the bent fiber.

2.10 The effect of hardness on friction

One may recall that the coefficient of friction is given by the ratio of the bulk specific shear strength of the junctions to the yield pressure of the asperities. Intuitively, therefore, one may assume that a soft metal will give a higher friction than would a hard metal. However, it is found that the influence of hardness is, in fact, rather small. The reason for this is that a metal that has high value of p_y also tends to have a high value of S, and vice-versa [3, p 72]. One way to obtain low friction from a metal is to deposit a thin layer of a relatively soft material, e.g. indium, on its surface. The area of contact will then be small because the harder material supports the load and S will also be small as the shearing will mostly take place in the film (Fig. 2.5). The values given in Table 2.2, illustrate the concept.

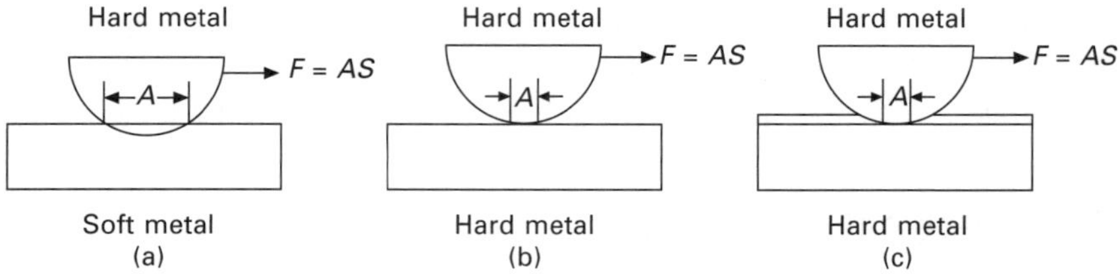

2.5 Effects of hardness and oxide film on friction in metals. Friction is not greatly affected by the hardness of metal, as in (a) the area *A* is large but the strength of junctions *S* is small, whereas in (b) *A* is small but *S* is large. In (c) an oxide film forming on the surface can lead to small *A* and small *S* and, therefore, to small *F* [3].

Table 2.2 Effect of a film of soft metal deposited on a hard surface on the value of the coefficient of friction [3]

Material	μ
Copper	1.5
– with oxide film	0.5
Steel	0.8
– with film of Indium	0.1

2.11 Heating at junctions during sliding

During sliding, the greater part of the work expended is liberated as heat, although often it is very small as the area of contact is minute. However, since the heat liberated is localized, the temperature rise at junctions may be substantial (several hundred degrees) [7]. Bulk heating is generally negligible. With non-metals, such as polymers, which are usually poor conductors of heat, high localized temperatures may be reached even more readily. Compensating for this effect, however, will be the fact that non-metals, as compared to clean metals, usually have lower friction. The high surface temperature developed plays an important role in many physical processes. In the sliding of skis or skates on ice, it leads to the development of a thin film of water and, therefore, to the characteristic low friction of ice. It aids in the development of smooth surfaces in polishing [21] and even plays a fundamental role in initiating chemical reactions [22].

2.12 The effect of speed on friction

It is observed that friction decreases somewhat with increase in speed of sliding. The reason for this is not clear as many changes take place and the effect related to each has not been worked out. For example, both p_y and S

are expected to change with speed and with the increased temperature resulting from sliding. It is generally thought that the two quantities should vary in a similar way and, therefore, the effect should not be large. It is, however, generally accepted that there is a finite time required for a junction to reach its full strength. Accordingly, the less the time of contact or the higher the speed, the smaller is the adhesion force per unit area. This should lead to a decrease in friction with increase in speed. Two other possible effects should be considered. On the unlubricated surfaces considered thus far, an increase in temperature may soften the surface, giving rise to an increase in friction. On lubricated surfaces, an increase in speed may lead to an increase in lubricant viscosity and, therefore, to an increase in friction. At the same time, an increase in temperature with speed may also cause a decrease in viscosity and, thus, in a decrease in frictional resistance. The classical concepts of lubrication are discussed later in this chapter and the effect of lubrication on frictional properties of fibers is discussed in detail in Chapter 11.

2.13 Wear and abrasion

In general, if the interfacial adhesion is weaker than the strength of the bulk, i.e. the material on either side of the interface (Fig. 2.6), then shearing will occur in the interface and the amount of plucking and wear will be minimal. This is the case when a strong oxide layer covers the surface, and shearing occurs within this film. This is one reason why hard steels resist wear. If, on the other hand, the surface film is weak or broken, junctions may form between the actual metal layers. Because of work hardening, they may become stronger than the bulk. In this situation, shearing may not occur within the interface, but at a short distance inside the bulk. This may result in relatively large fragments being plucked during sliding. The wear in either case will be proportional to the normal load, and also to the speed and the time of traverse. In addition to the above, another mechanism is operative if one of the two surfaces is relatively harder and rougher. The asperities of the harder surface will penetrate the softer one and scratch or abrade it. Shape of the asperities will play a role on the nature of the damage. Smooth asperities are likely to

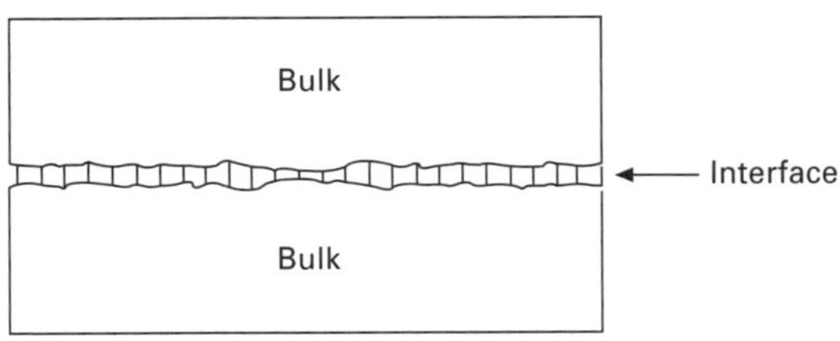

2.6 Contact between two materials with the shear strength of the interfacial bonds weaker than the strength of the surrounding bulk.

plough, whereas sharp ones may cut and remove material. Work by Tabor suggests that a minimum difference in hardness or yield pressure of 20% is necessary before asperities can cause abrasion [23].

2.14 Frictional behavior of elastomers

If we were to classify materials in terms of their deformation behavior, they will fall into three distinct groups: those deforming plastically (after a short elastic region), elastically over substantial change in length, and visco-elastically. Whereas behaviorally the ductile metals are identified with the first group, the elastomers or rubbers are identified with the second group. The behaviors of the polymeric materials and fibers bridge the gap between the two and are identified with the third group.

Rubber is a material that can be stretched to very large extension (over 100%) but when released it returns to very nearly its original shape and size. If one contrasts this with the behavior of metals that flow plastically, it should not be surprising that the frictional behavior of rubber is marked by an appreciable deviation from that supported by Amonton's laws.

The model developed by Hertz [24] for elastic materials gave the area of contact as:

$$A \propto N^{2/3} \tag{2.5}$$

Measurements of both A and F on vulcanized rubber sliding on flat glass by Schallamach [25] indicated that both properties varied nearly as $N^{2/3}$. These results, i.e. $F \propto N^n$, imply that the classical adhesion-shearing mechanism also operates in rubbers. However, the work with rubber in which the tangential force was held constant, but the speed of sliding was measured with changing temperature showed that the log of the former varied linearly as a function of the reciprocal of the latter [26]. An activation energy can be obtained from this, which could be confirmed by independent measurements. This lent support to a viscous flow process in the bulk in which a dissipative mechanism was operating. Keeping this process in mind, one can understand what may happen when an object such as a hard steel hemisphere slides over a flat rubber surface. Assume that the surfaces are lubricated such that interfacial adhesion is minimal. One will find that the work is still required to produce sliding. As the hemisphere is pulled forward, it causes elastic deformation in the rubber in front of it, but also similar recovery in the back (Fig. 2.7). However, if one examines the stress-strain and recovery curve of rubber, it is found to be not ideally elastic. Accordingly, finite energy loss occurs and this must correlate with the work of sliding. The elastic hysteresis, sometimes referred to as internal friction, is analogous to viscous loss within the bulk of rubber. Referring to Fig. 2.8, as one moves from "a" (ideally elastic), through "b" (low hysteresis), to "c" (high hysteresis), energy loss and, thus, frictional

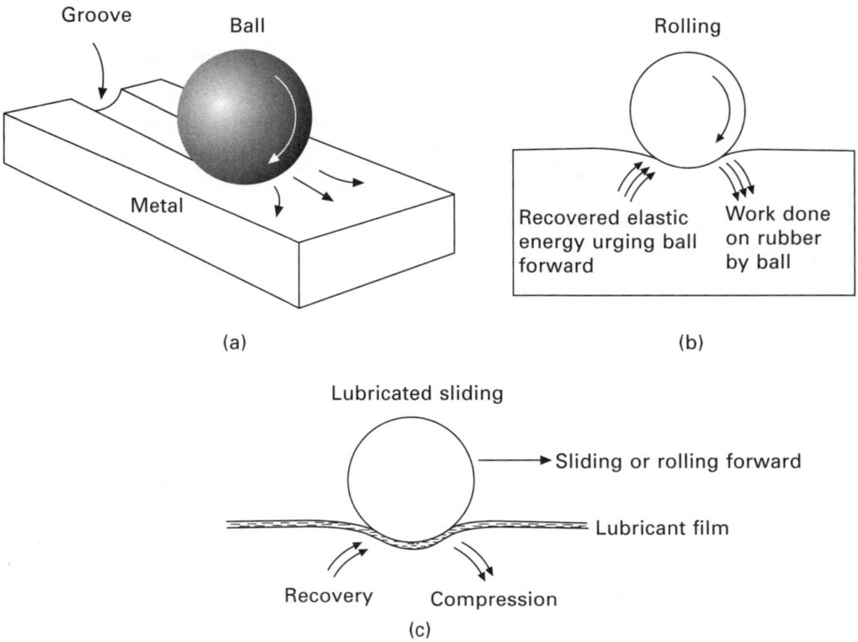

2.7 Rolling and sliding behaviors of a hard steel ball or a rod over metal and rubber. (a) Rolling or sliding on a soft metal leaves permanent deformation. (b) When the same happens on elastically deforming rubber, no permanent deformation occurs, but energy is lost due to hysteresis. In sliding on unlubricated rubber, the behavior will be dominated by both the adhesion between the slider and the rubber and the deformation properties of rubber. In a thoroughly lubricated system (c) friction will be governed only by the deformational behavior of rubber, as in rolling [3].

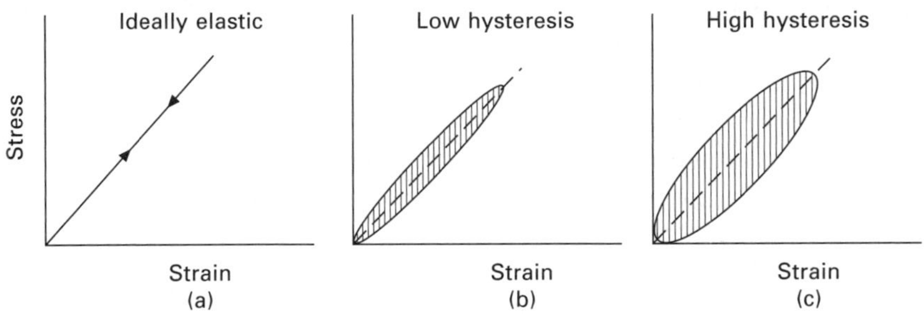

2.8 Stress–strain behavior of ideally or instantaneously elastic and visco-elastic materials.

resistance, will increase. For a hard sphere sliding over the surface of rubber characterized by high hysteresis loss (Fig. 2.8(c)) in the presence of an effective lubricant, the greater part of the sliding friction will be found to arise from hysteresis loss in the rubber itself [27].

In automobile tire applications of rubber, these losses play a crucial role in governing comfort and control of vehicles. Whether the latter is rolling under normal road conditions or skidding on a slippery surface, much of the friction and, therefore, control, arises from the hysteresis loss. Adhesion continues to play a role, especially in slipping, when, for example, the vehicle is given acceleration, deceleration, or abrupt braking; however, the potential danger is greatly reduced by the friction produced from hysteresis losses in the tires. A brief discussion on rolling friction follows.

2.15 Rolling friction

In addition to sliding, one can also move materials by rolling. The great advantage of rolling a body over sliding is that the friction in rolling may be two or more orders of magnitude lower than that in sliding. The greatest application of this concept is made in the use of ball and roller bearings, in which the balls or rollers are free to roll in a groove called a race without any shaft or trunnion to support them.

It was thought that low friction in rolling was due to minute slips between ball and surface; however, studies have shown that although slip did occur, it contributed only a small part to the rolling resistance. This is supported by the fact that lubricants that reduce sliding resistance had little effect on total rolling resistance [3, p 131]. The underlying mechanism governing rolling friction is again the deformation of the bulk discussed in the previous section. If a hard ball is rolled over a soft metal plate (Fig. 2.7(a)) it produces a groove by displacing metal plastically. If one measures the force required to displace metal, it is found to equal the observed rolling friction. This also illustrates why lubricants have so little effect on the latter. Consider now the set up shown in Fig. 2.7(b) in which a hard steel ball is rolled over a flat rubber sheet. Although no permanent groove is formed, work is still lost. In rolling, the ball deforms rubber ahead of it and does work; but much of this deformation is recovered at the rear as the ball is urged forward. As alluded to earlier, no material is ideally elastic, some energy (due to elastic hysteresis caused by internal friction between rubber molecules as they rub over each other) is lost. The amount lost will vary, a bouncy rubber may recover 95% (Fig. 2.8(b)), whereas a "soggy" rubber may recover only a small amount (Fig. 2.8(c)) and lose most of the work expended. Ball bearings are made of hard steel and the devices are designed so that the stresses are small, i.e. they lie within the elastic region. This gives very small rolling resistance ($\mu \sim 0.001$). Theoretically, the total frictional resistance is given by the sum of the forces due to adhesion and deformation (equation 2.6):

$$F = F_{adh} + F_{def} \qquad 2.6$$

As mentioned above, although adhesion does play some role, it is generally very small. This is primarily because junctions are greatly limited and peeling imposes a lower force. Therefore, the overwhelming effect in rolling friction comes from the deformation losses, or $F_{def} > F_{adh}$, the difference depending upon the hysteresis properties of the softer material. For example, if there is a layer of lubricant (Fig. 2.7(c)) between the surfaces, then $F \cong F_{def}$.

2.16 The frictional behavior of polymers and fibers

The experimental methods for investigating polymer and fiber friction fall into two groups: the point contact, which involves rubbing at a single point or at a single cluster of points, and extended line contact in which contact is maintained over a substantial length of the materials. For the point contact tests, it has been customary to measure friction in polymers by using a hemispherical probe sliding on a flat surface or to use one fiber sliding on another orthogonally for measuring friction in fibers. The point contact method has been considered well suited for a fundamental study of the mechanism of friction and most of our understanding of fiber friction is obtained this way. Two line contact methods have been used to measure friction between the fibers. In Gralen's twist method [28] two fibers are twisted together and the force to slide one fiber relative to the other is measured (see Section 5.5.5, Chapter 5, for details). In the case of the capstan method [29] a fiber is passed over a cylindrical surface and the force measured to pull the fiber against a pretension (see Section 5.5.4, Chapter 5, for details). The cylindrical surface can be a metal, ceramic, horn, or another fiber. The line contact methods, in particular the second, are valuable technological tools for assessing friction in textiles. Although they are very useful for measuring the effects of speed and lubrication on the frictional value, they are less well suited for a basic study of fiber friction. These and other methods are discussed in detail in Chapter 5.

Some of the early work, which is still relevant, showed that on clean surfaces the friction of a plastic material sliding on itself was approximately the same as for steel sliding on the plastic [10]. Over the load range investigated (1–10 Kg_f), μ was found to be approximately constant. However, for a steel hemispherical slider riding on plastic, there was a tendency for μ to be higher at heavy loads. This was attributed to a large value of the ploughing term resulting from the production of grooved track in the surface of the plastic. For a plastic hemisphere of polyvinyl chloride sliding on a clean metal, large blobs of plastic were observed that were strongly attached to the metal surface. For a metal hemisphere sliding on flat polyvinyl chloride, autoradiography showed that fragments of metal were transferred to the plastic [2, p 29]. This clearly suggested that an adhesion mechanism could be applied to the friction of plastics as well as to the friction of metals.

In one experiment [10], a steel hemisphere slid over a clean flat plastic surface left a grooved track of width *d*. It was assumed that intimate contact occurred over the front half of the slider giving an area $A = \pi d^2/8$. A plot of d^2 against load *N* gave an approximately straight line, thus showing that the area of contact was roughly proportional to the load. If we assume that junctions of bulk specific shear strength *S* formed over this area, then, for a small ploughing force, the frictional force *F* is simply given by $F = SA$. This is the same as found for metals.

Measurements of the bulk specific shear strength produced by shearing cylinders of plastic showed reasonable agreement between the values of *S* so obtained and those derived from frictional measurements (Table 2.3) [10]. The only marked discrepancy was observed with polytetrafluoroethylene (PTFE) where the calculated frictional value of *S* was much smaller than the observed bulk value. The work of King and Tabor [30] over a wide range of temperatures provided support for the theory.

The quantities measured were the static compression hardness p_y (which to a first approximation determines the area of contact *A*, given by N/p_y), the bulk shear strength *S*, and the friction coefficient μ. From the concept presented earlier, μ should equal S/p_y. The results showed that the variation of S/p_y with temperature was nearly the same as that of μ with temperature, although the quantitative agreement between μ and S/p_y was moderate to good [30]. PTFE in general has been seen to have a very low coefficient of friction (~ 0.1). Only at temperatures near the softening point was an increase in μ observed. According to Tabor, the surface energy of PTFE is very small, being of the order of 12 dynes/cm, which suggests very weak surface interactions between the molecules of two PTFE bodies [2, p 30]. The molecular chains within each are, however, firmly intertwined, which leads to a high value of yield pressure. Consequently, the interaction that is weak to start with takes place over a small area, giving rise to the observed very low μ. At high temperatures,

Table 2.3 The values of the Vicker's Hardness, specific shear strength from friction measurements ($S = F/A$), and bulk specific shear strength from direct measurements, for selected polymers [10]

Polymer	Vicker's Hardness (Kg/mm²)	Specific shear strength (Kg/mm²)	
		Friction tests	Direct tests
PTFE		0.4	1.2–2.4
Polystyrene	20	8	4
Polymethylmethacrylate	20	10	6
Nylon	10	5	6
Polyethyleneterephthalate	20	5	5
Polyvinyl Chloride (PVC)	15	7	5
Polyethylene	2	0.8	1.4

thermal motion weakens the molecular network and causes increase in friction, presumably due to an increase in the area of contact.

The frictional behavior of PTFE suggests that for long-chain molecules that can become entangled, a large difference between interfacial and bulk strength properties may have an important effect on the friction of polymers. A similar type of anisotropy has been seen to occur in materials with laminar or plate-like structures such as graphite or molybdenum sulphide. The plates themselves are very strong but the linkage to neighboring plates is weak, resulting in very low values of friction ($\mu \sim 0.1$). Such an issue does not arise with small molecules or with metals. In the latter case, weak molecular forces at the surface are generally accompanied by low bulk strength, so that S and p_y vary together.

A major problem in studying friction in fibers occurs when they have a finishing treatment on their surfaces. In these cases the results obtained are difficult to interpret. An exact understanding can be obtained only when fibers and polymers with clean surfaces are used. On polymers this is achieved by abrasion under running water. This is, however, not possible with fibers. Most workers have necessarily used some type of solvent extraction, which in some cases can modify the physical, chemical and structural properties and, through these, frictional properties. One solution is to collect fibers directly from the spinneret before a finish has been applied. This is, however, not always possible.

2.17 The relationship describing friction in polymers and fibers

One of the earliest models describing relationship between F and N was given by Makinson [31]:

$$F = a_0 + aN^n \tag{2.7}$$

In this, the first is an adhesion term and the second is a load dependent term. In examining the friction between crossed fibers in a vacuum, Pascoe [32] found no detectable normal adhesion between surfaces. In the presence of water vapor, however, a small but finite adhesion was observed with hydrophilic fibers; but even here the contribution to friction was extremely small. In both a vacuum and a humid atmosphere, the variation of F with N was substantially the same. It seemed, therefore, more satisfactory to use the following simpler relationship:

$$F = aN^n \tag{2.8}$$

where n is less than unity. This relationship has been used by several workers and well substantiated experimentally [33–35]. All workers adopting this relationship assumed that friction arose from adhesion at the points of real contact and the junctions had a constant shear strength, S. Consequently, they all explained the variation of F with N in terms of the variation of the true area of contact A with N. Clearly, if S is a constant, it must follow from the above equation that:

$$A = k_1 N^n \qquad 2.9$$

Lincoln [34] examined friction between a sphere and a flat nylon surface at low loads (1 to 100g_f) and found that F varied as $N^{2/3}$, in agreement with Hertz's solution for the elastic deformation of spherical surfaces. Howell and Mazur [36] measured friction between fibers at right angles under loads varying from 0.3 to 400 mg and found the same relationship with–equaling 0.8 for drawn and 0.9 for undrawn nylon and 0.96 for cellulose acetate. Finch [37] studied the imprint formed by curved cylinders of nylon on a sooted glass plate and found the area of the imprint to be proportional to $N^{0.89}$, in excellent agreement with frictional dependence of nylon on N. These methods of determining area of contact may not be precise but, nevertheless, they clearly support the validity of the adhesion-shearing mechanism of friction in fibers and polymers.

2.18 Lubrication

Lubricants are the materials that reduce friction and, through it, noise, vibration and wear. In the processing and use of textiles, metal/metal, fiber/metal, and fiber/fiber contacts occur. In many of the contacts involving these surfaces, a low friction is required to reduce wear, noise and also energy use. Most concepts of lubrication were developed to minimize and control friction when a metal slides or rotates against another metal. In this section, we shall first briefly review the various concepts of lubrication as they were evolved and largely applied to metals running against metals. The next two sections discuss the application of these concepts to the lubrication of textiles.

2.18.1 Fluid or hydrodynamic lubrication

If a shaft or journal rotates in a dry bearing, it crawls up against the bearing as shown in Fig. 2.9(a). If, however, the bearing contains oil and the system is aligned properly, a converging wedge of oil is dragged between the shaft and the bearing and keeps the two surfaces separated. Petroff in 1883 first analyzed this behavior, assuming that the shaft ran concentrically in the bearing and showed that the friction resistance was proportional to the viscosity of the oil and the speed of rotation and inversely proportional to the clearance [38].

Reynolds in 1886 [39] showed that for a loaded shaft, the equilibrium position of the shaft was not concentric with the bearing, but displaced

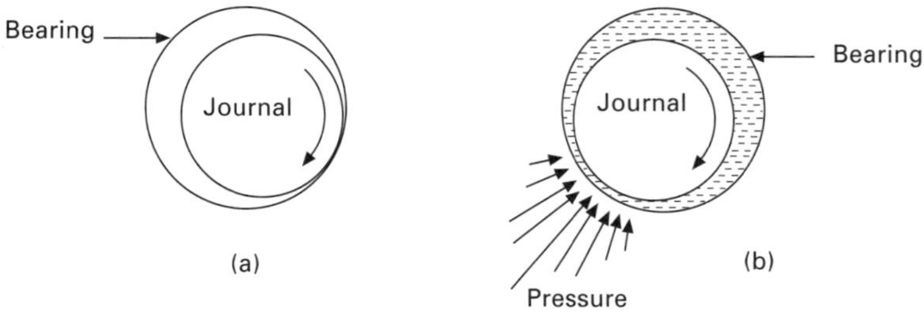

2.9 Journal rotating in a bearing. (a) unlubricated bearing; (b) bearing containing a lubricant.

downwards and sideways (Fig. 2.9(b)). This led to a converging film of lubricant being dragged between the surfaces so that the mean velocity of the oil increased as it approached the thinner part of the 'wedge'. By applying the classical laws of viscosity, he showed that the increase in velocity produced a large counteracting pressure in the oil film and if the pressure was sufficiently great, it would support the load on the shaft. The pressure distribution for a 'half-bearing' is also illustrated in this figure.

In this type of lubrication, termed hydrodynamic, in the ideal case, the resistance to motion arises only from the viscosity of the oil film. If Z is the viscosity, V is the number of revolutions per minute and N is the load on the bearing, the torque is a simple function of these parameters. If the torque (or couple) is divided by the bearing diameter, it gives the force that must be applied to the shaft to produce rotation and, if this force is divided by the load N, we have an effective coefficient of friction (μ) for the bearing. Since μ is dimensionless it is best expressed as a function of a dimensionless parameter in the journal-bearing system. The most appropriate parameter is the quantity ZV/P, where P is the nominal pressure of the bearing (N divided by projected area). For bearings in which the clearance ratios are fixed, Reynolds showed that μ is a single-valued function of ZV/P (Fig. 2.10). It is noted that over much of the practical range, the simple Petroff result is adequate. The above relationship shows that increasing the load causes μ to decrease. Accordingly, Amonton's laws do not hold. Also, the coefficient of friction increases linearly with speed, which is another deviation from the law.

The absolute value of μ for a journal bearing depends on the size of the bearing, the nominal clearance of the bearing, the speed, the load, and the viscosity, but for the range in which most bearings operate, μ is of the order 0.001 to 0.01. Because of the great advantage of hydrodynamic lubrication, engineers strive to achieve the needed conditions to have minimum friction wear. Clearly, the lower the viscosity, the lower the friction. There is, however, a lower limit; the viscosity cannot be too low, otherwise the separation distance may become low enough for asperities to penetrate and make metallic

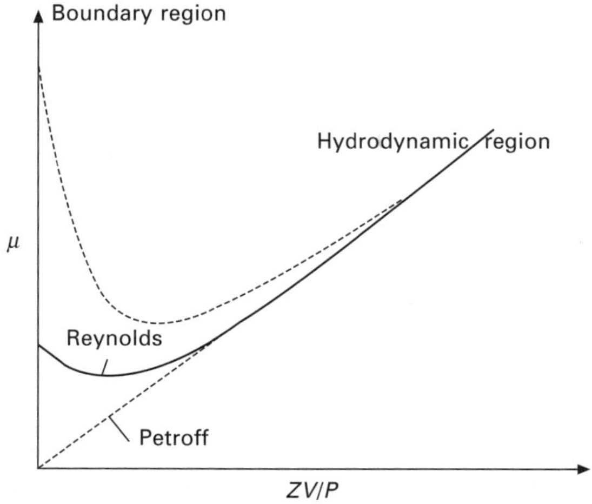

2.10 Coefficient of friction plotted against *ZV/P* for a journal bearing.

contact. It is also understood that the generated heat can increase the risk of contact by decreasing the viscosity.

2.18.2 Aerodynamic lubrication

If the system uses highly polished and well-aligned surfaces, accurate clearances, high speeds, and low loads, one can use air or some other gas as the lubricating agent. This, the so-called aerodynamic lubrication, can be considered as an extreme example of hydrodynamic lubrication [3]. With a viscosity that is three or more orders of magnitude smaller than that of thin mineral oil, this will lead to lower viscous resistance, lower heating and lower costs. In addition to providing a low coefficient of friction, the aerodynamic system can also operate at high temperatures because chemical degradation of the lubricant does not occur.

2.18.3 Elasto-hydrodynamic lubrication

In the above two types of lubrication, the shafts and bearings are perfectly rigid and they retain their shape during running. Let us consider, however, a situation where gear teeth are engaged in running one shaft from another and where pressure on the bearing could be very high; e.g. two or more orders of magnitude higher than that encountered in journal bearings. It is found that with careful alignment of the engaging teeth and a high surface finish, gears can be run successfully with minimal wear. In this system, although gear teeth deform elastically, the amount of deformation is very small because of the stiffness of the metal. A very thin film of oil gets in between the engaging teeth, preventing the contact. Although the distances at such pressures could be of the order of a molecular diameter, the film prevents contact between surfaces as its viscosity increases appropriately with the increase in pressure [3]. For the system to operate, the surfaces must be very smooth and carefully aligned. Systems such as gears or cams can operate effectively at very high contact pressures without any metallic contact occurring.

2.18.4 Boundary lubrication

If the existing conditions are such that the pressures are high, the running speeds are low and/or the surface roughness is too great, then penetration of the lubricant film will occur and the lubrication obtained will be the so-called boundary. Sir William Bate Hardy was the first to undertake a systematic study of boundary lubrication in the 1920s using a series of pure long-chain organic molecules as the lubricating agents [40]. He found that, to a first approximation, both of Amonton's laws were obeyed and this has been generally supported by later work. The role of a boundary lubricant is to reduce the contact between asperities by interposing a film that attaches to the surface. The best boundary lubricants are long chain molecules with an active end group, typically an alcohol (OH), an amine (NH_2) or a fatty acid (COOH) (Fig. 2.11(a)). When such a material, dissolved in an ordinary mineral oil, meets a metal or other solid surface, the active end group attaches to the solid and builds up a thin layer (Fig. 2.11(b)). Not only is the attachment to the surface strong, but the chains also attract each other so that it is difficult for a rubbing surface to penetrate the film. Surfaces covered with such solid

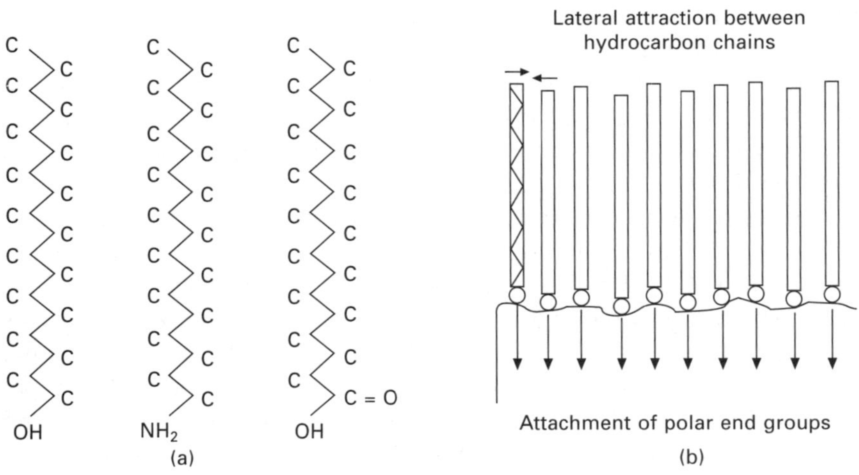

2.11 (a) Boundary lubricant molecules with polar end groups, (b) adsorption of these molecules onto a metal surface with strong lateral attraction [3].

films tend to slide over their outermost faces. Some penetration may occur, but it is far less than would occur if only a mineral oil was present on the surface. With good boundary lubricants, the friction may be reduced to a value of μ of about 0.05. The motion is generally smooth and the friction decreases only slightly with increasing speed. In general a single molecular film is sufficient to produce a large reduction in friction and wear. However, it is necessary to keep in mind that because penetration may occur if the loads are high a thicker film will be preferred.

Since temperatures can rise during sliding, it is also necessary that the film has a high melting point. Among the materials mentioned, fatty acids have the advantage that if the metal is reactive, they will react and form soaps. These have desirable shear properties, as well as melting points considerably higher than that of the original fatty acid. As an example, the melting point of stearic acid is 69°C, but of copper stearate it is about 120°C [3]. The best boundary lubricants cease to be effective above about 200°C because of oxidation of the lubricant. To extend the effectiveness to higher operating temperatures, lubricants containing a small quantity of a high temperature additive, dissolved in lubricating base oil, may be used. The most common additives used contain phosphorous, chlorine, or sulfur. These react with the surface to form a film that prevents metal-to-metal contact at higher temperatures.

The protective properties of a lubricant can be evaluated by making one of the rubbing surfaces radioactive and determining the transfer of material from one surface to the other. In one such experiment [41], the results showed that while friction was reduced by a factor of about 20 (μ from 1 for clean to 0.05 for lubricated surface), the transfer decreased by a factor of about 20,000. The enormous change in transfer was considered primarily due to a decrease in the size of the particles. Thus, in boundary lubrication, contact occurred but the size of it and the strength of the weld greatly decreased. The effectiveness of the lubrication of film did not change much until around the melting point of the film, when both the friction and the transfer increased very significantly.

Another type of boundary lubricant, known as grease, is formed when a metal soap, such as calcium stearate, is mixed with an ordinary mineral oil. The soap acts as a gelling agent and the material turns into a soft deformable paste. In most soap-based greases, the soaps form long chain fibrous structures that interlock to form a three-dimensional network, with the oil held within the cavities. Such lubricants are ideal in cases where it should stay put and not drip away. Thus, these are appropriate lubricants for ball and roller bearings and also slideways. Two other attractive properties of these lubricants are that they can be used as a means of reducing the corrosion of metal surfaces and as seals to prevent access of dust and dirt into running parts.

2.18.5 Mixed lubrication

Conditions conducive for boundary lubrication are high loads and low sliding speeds, whereas for hydrodynamic lubrication, speeds should be high enough to cause separation of surfaces with a wedge of lubricant. In practice however, there will be a wide range of experimental conditions where a part of the load is borne by a hydrodynamic film and part by a boundary film. This is often referred to as a regime of "mixed" lubrication.

2.19 Lubrication of polymers and fibers

Polymer materials may be lubricated with either fluids or solids. These substances are introduced into the contact zone either by external deposition or by emergence from within the matrix of internal reservoirs [19]. For example, a paraffin, built as part of the internal structure, can migrate to the free surface. Commercially, however, lubricating fluids that represent a broad class are the most widely used. Common examples are silicone and hydrocarbon oils as well as aqueous based agents. It is, however, generally accepted that, with the exception of elastomers, polymers are not very effectively lubricated by fluids [19, 32]. One of the reasons given for this was that the shear strength of the lubricating material, polymer itself, was almost indistinguishable from that of the underlying material. With a fatty acid as a lubricant, a greater reduction in friction is observed with plastics sliding on metals than for plastics sliding on plastics. Presumably, the soap formed by chemical reaction between the fatty acid and the metal has a lower shear strength than the fatty acid itself. Here again, the decrease in friction is far less than for metals sliding on metals lubricated with fatty acids [42]. An alternative explanation given for the lower effect on polymer friction is that the polar lubricant chains find relatively fewer polar sites for attachment on a fiber surface than on a metal surface [43].

2.20 Friction between lubricated fibers

In studies on lubricated fibers, the capstan method is usually used. In this technique a yarn is passed over a pin and either the speed of sliding or the viscosity of the lubricant, or both, are systematically varied. Two of the earliest studies are due to Roder [44] and Lyne [45]. Their results showed that as the speed V or viscosity Z increased, the final tension T (with the initial T_0 kept constant), or the coefficient of friction μ calculated by the capstan method, increased. The plots of T against the product VZ, with either V varying or Z varying, were nominally the same, indicating that a good part of friction in fibers or yarns sliding on guides or pins came from the hydrodynamic mechanism [46].

Analysis by Hansen and Tabor [46] of both Lyne's data and of their own results showed that what they termed the "effective coefficient of friction," given by F/N as $(T - T_0)/(T + T_0)$ (Fig. 2.12(a)), was a monotonic function of ZV/N (Figure 2.12(b)) in fibrous systems just as it was of ZV'/P in journal bearings. In these, V is the linear speed and N is the normal force (given by

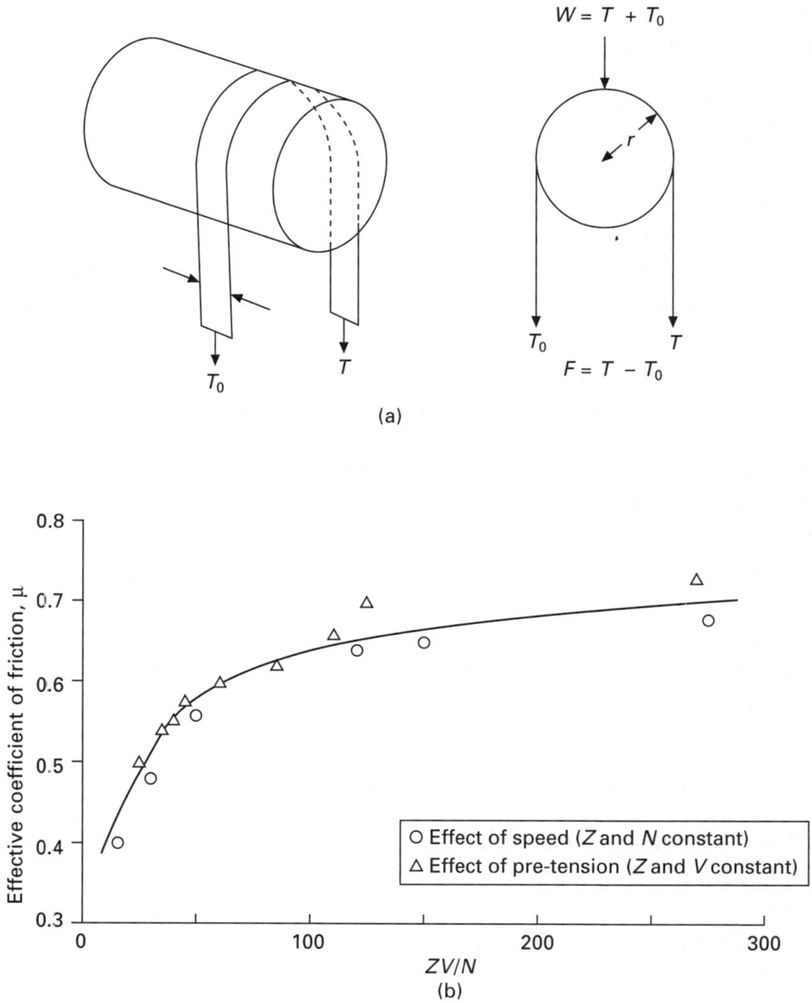

2.12 (a) Illustration of model used in characterizing hydrodynamic lubrication for a yarn sliding over a cylindrical surface.
(b) Hydrodynamic behavior of an acetate yarn sliding over a chromium pin [2].

$T + T_0$) in fiber system, whereas V' is the angular velocity or rpm and P is the normal pressure applicable to systems involving journal bearings[1].

The roughness of the cylinder was found to have an effect on the coefficient of friction: as the roughness increased, μ decreased, presumably due to decrease in the hydrodynamic area of contact [45, 48]. It is understood that if the roughness is too marked, the lubricant film may be penetrated and the friction and surface damage could increase. One of the reasons for different yarns having different frictions for the same nominal values of *ZV/N* may be attributed to different "hydrodynamic areas" for the yarns.

Tabor found that with monofilaments of large diameter (125–250 microns), the viscosity of the oil was not as important, presumably because the fiber was too rigid for the hydrodynamic effect to be appreciable [2]. With multifilament yarns, however, the viscosity of oil was found to be important. Additionally, there was evidence that wettability of the fiber surface by the lubricant also played some role. The general behavior of a yarn passing over a pin as the speed is increased is illustrated in Fig. 2.13. At very low speeds, friction will be boundary as the conditions do not favor a hydrodynamic state. Friction is not low as most organic lubricants have relatively poor boundary properties when applied to polymeric materials. With increase in speed, however, the contribution from the boundary film diminishes whilst that from the hydrodynamic film increases, and, at reasonably high speed, the hydrodynamic film supports a greater fraction of the load. It is seen that the resultant frictional behavior shows a minimum.

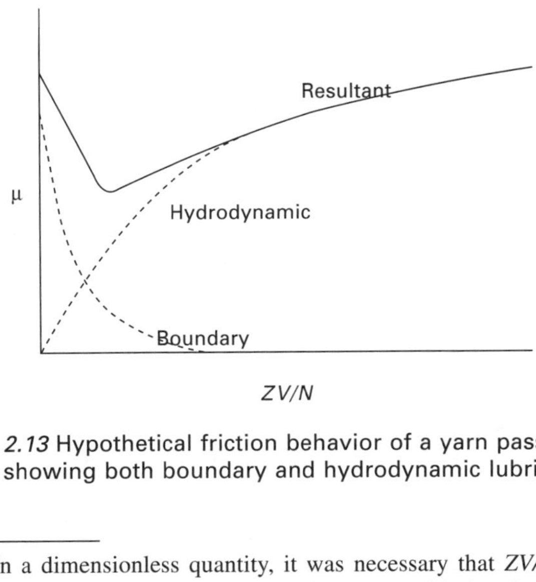

2.13 Hypothetical friction behavior of a yarn passing over a pin showing both boundary and hydrodynamic lubrication effects.

[1]To obtain a dimensionless quantity, it was necessary that *ZV/N* be multiplied by $2L$, where L is the width of the yarn ribbon in contact with the pin [47].

As alluded to above, for effectiveness, the lubricating substance should have a strong affinity for the contacting surfaces and not be ruptured readily by the asperities. This in essence is equivalent to specifying a large work of adhesion between the solid and the lubricant [19]. Unfortunately, the same affinity may adversely affect the mechanical integrity of the polymer as the fluid may plasticize the latter, or may promote brittle failure or craze formation. Thus, where fluids are widely used to lubricate polymers, for example elastomeric seals in hydraulic systems, these requirements must be carefully balanced. This explains the dilemma one faces when selecting a lubricant for a polymeric material.

Briscoe and his associates have examined the influence of such adhesion on the lubrication efficiency for a few fluid systems [19, 49]. A general result is shown in Fig. 2.14 in which the frictional force for orthogonal contacts between PET monofilaments (see Section 5.5, Chapter 5) are plotted against the surface tension of various isopropanol – water mixtures for three normal loads. The results indicate that solutions whose surface tensions are less than about 40 mJ/m^2, which approximately equals the surface energy of polyester [50], wet the fibers (contact angle is zero) [49]. Above this energy the coefficient of friction is high and not a marked function of the surface tension of the fluid. For the two lowest loads, below about 40 mJ/m^2, friction decreases with decreasing surface tension of the fluid. At the higher loads, little discontinuity is observed as the wetting condition is reached. These data indicate that, below a certain critical contact stress, a wetting solution will effectively lubricate a polymeric contact. Also, below this stress, the quality of the lubrication improves as the surface tension of the solution decreases [19].

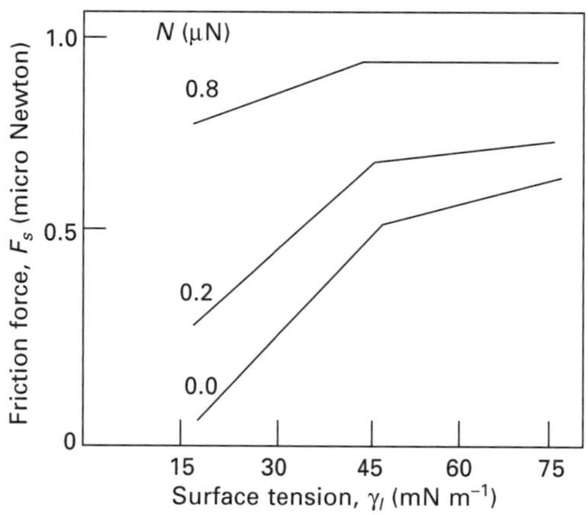

2.14 Friction force, F_s, of polyester monofilaments lubricated with aqueous solutions of isopropanol of surface tension γ_l for three different values of contact load [19].

The above results clearly indicate that a decrease in lubricant surface tension below the solid surface energy will lead to wetting of the surface and a decrease in solid-solid contact. An increase in load above a certain maximum value, however, should cause displacement of the film by the asperities and lead to an increase in direct contact. An alternative explanation for the decrease in friction produced by lubrication, suggested by Briscoe, is that the fluid reduces the strength of the junctions by interacting with the material. Measuring the peel strength of an adhesive junction in the presence of a lubricant substantiates this concept. It is, however, generally agreed that it is the lubricant's ability to reduce the extent of solid/solid contact that is important, although the idea that the lubricant influences the rupture strength of the interface is not discarded.

2.21 Summary

The tangential force needed to slide one body against another is composed of an adhesion and a deformation term. The behavior of a metal sliding on a metal is predominantly governed by the adhesion mechanism. Transfer of material from one surface to another supports this. The deformation term assumes a significant role when two materials differ greatly in their hardness values, the normal force is high, and/or the harder surface has a significant degree of roughness. Even in these circumstances, the major part of the friction force arises from adhesion between junctions. As a tangential force is applied to initiate sliding, the local contact regions flow laterally and grow in area of contact, leading to an increase in friction, especially with smooth and clean metals. The presence of an oxide or a contaminant film greatly reduces the potential for such occurrence. Kinetic friction is smaller than static. A major reason for this is a decreased opportunity for junctions to develop and grow.

In elastomeric materials, friction mostly arises from deformation and hysteresis losses. In these products, friction energy is dissipated largely in deforming the bulk material surrounding the interface and its lack of full recovery after sliding. This implies a viscous flow process. Frictional force is found to be proportional to a power (~ 2/3) of the normal force.

For plastics and fibers sliding on themselves or on other surfaces, the transfer of material from one to the other that is observed, suggests the operation of the adhesion mechanism. The coefficient of friction is affected by the normal load and the geometry of the surfaces and this is explained in terms of the deformation characteristics of the polymeric materials. A behavior intermediate to the pure plastic flow ($F \propto N$) and elastic deformation ($F \propto N^{2/3}$) is observed. Consequently, friction is affected by the time of loading and the geometry of contact. Junction growth is suggested to be low, which contrasts with the behavior found in metals. This indicates that the simple theory of friction based on ($F = SA$) is even more directly applicable to polymers.

Lubricants are materials that control friction and, through it, wear and noise. Two different types of lubrication behaviors, hydrodynamic and boundary, can be obtained. The conditions in the former lead to a complete separation of the surfaces and in the latter to a decrease in but not elimination of contact between asperities. The best boundary lubricants are long chain molecules with end groups that will bond to the surface and resist penetration by asperities. Friction of the order of $\mu \sim 0.05$ can be obtained. Observed transfer of material from one surface to the other indicates that contacts do occur but they are fewer than would be the case if no lubricant was used. In the operation of journal bearings (hydrodynamic), friction is essentially determined by the viscous resistance of the lubricant film to shear. A value of the coefficient of friction in the range 0.01–0.001 can be obtained.

In general, polymers and fibers are not effectively lubricated by fluids. The efficiency of lubrication improves if the surface energy of the lubricant favors its spreading on the solid surface. Controlling friction, static generation and wear will be relatively more important goals than reducing friction in lubricating fibers. The selection of a lubricant for fibers requires care as the lubricant could interact and adversely affect the fiber surface. The passing of yarns over guides with a lubricant, during processing, favors hydrodynamic lubrication.

2.22 Acknowledgements

This chapter is primarily a summary of the results and views of F. P. Bowden and D. Tabor and of many of their colleagues, in particular Dr. B. J. Briscoe, and of other scientists of the times who pioneered the work in the field of friction, wear and lubrication of materials, including textiles.

2.23 References

1. Bowden F P and Tabor D, *Friction and Lubrication of Solids,* Oxford, Clarendon Press, 1954.
2. Howell H G, Mieszkis K W and Tabor D, *Friction in Textiles*, Butterworths Scientific Publications, London, 1959.
3. Bowden F P and Tabor D, *Friction: An introduction to tribology*, Anchor Press, New York, 1973.
4. Gupta B S and El-Mogahzy Y E, 'Friction in fibrous materials, Part-I: Structural model', *Textile Res. J.*, 1991, **61**(9), 547–555.
5. Amontons G, *Histoire de l'Academie Royale des Sciences avec Memoires, de Mathematique et de Physique,* 1699, Page 206.
6. Coulomb C A, *Memoires de Mathematique et de Physique Royale des Sciences,* Paris, 1788, Page 161.

7. Bowden F P and Ridler K E W, 'Physical properties of surfaces. III. The surface temperature of sliding metals. The temperature of lubricated surfaces', *Proc. Roy. Soc. A,* 1936, **154**, 640.
8. Bowden F P and Thomas P H, 'The surface temperature of sliding solids', *Proc. Roy. Soc. A*, 1954, **223**, 29.
9. Bikerman J J, 'Surface roughness and sliding friction', *Rev. Mod. Phys.*, 1944, **16**, 53.
10. Shooter K V and Tabor D, 'The frictional properties of plastics', *Proc. Phys. Soc. Lond. B*, 1952, **65**, 661.
11. Tabor D, 'The mechanism of rolling friction. II. The elastic range', *Proc. Roy. Soc. A,* 1955, **229**, 198.
12. Tabor D, *The Hardness of Metals,* Oxford; Clarendon Press, 1951.
13. Holm R, *Electrical Contacts,* Springer, Berlin, 1958.
14. Dyson J and Hirst W, 'The true contact area between solids', *Proc. Phys. Soc. Lond. B*, 1954, **67**, 309.
15. Bailey A L and Courtney-Pratt J S, 'The area of real contact and the shear strength of monomolecular layers of a boundary lubricant', *Proc. Roy. Soc. A*, 1955, **227**, 500.
16. Bowden F P and Young J E, 'Friction of clean metals and the influence of absorbed films', *Proc. Roy. Soc. A,* 1951, **208**, 311.
17. Courtney-Pratt J S and Eisner E, 'The effect of a tangential force on the contact of metallic bodies', *Proc. Roy. Soc. A*, 1957, **238**, 529.
18. Tabor D, 'Friction, lubrication, and wear', in Matijevic E, Ed., *Surface and colloid science,* 1972, Vol **5**, John Wiley, New York.
19. Briscoe B J, 'The role of adhesion in the friction, wear and lubrication of polymers', in Allen K W, Ed., *Adhesion-5*, Applied Science Publishers Ltd. London, UK, ISBN-0-85334-929-0, 49–80, 1981.
20. Briscoe BJ and Kremnitzer SL, 'Study of the friction and adhesion of polyethylene terephthalate monofilaments', *J. Phys. D: Appl. Phys.*, 1979, **12**(4): 505–516.
21. Bowden F P, 'Friction on snow and ice', *Proc. Roy. Soc. A*, 1953, **217**, 462.
22. Bowden F P and Yoffe A D, *Initiation and Growth of Explosion in Liquids and Solids.* Cambridge University Press, 1952.
23. Tabor D, 'Mohs's hardness scale – A physical interpretation', *Proc. Phys. Soc. Lond. B,* 1954, **67**, 249.
24. Hertz H, 'Ueber die Berührung fester elastischer Körper', *J. reine angew. Math.*, 1882, **92**, 156.
25. Schallamach A, 'The load dependence of rubber friction', *Proc. Phys. Soc. Lond. B*, 1952, **65**, 657.
26. Schallamach A, 'The velocity and temperature dependence of rubber friction', *Proc. Phys. Soc. Lond. B*, 1953, **66**, 386.
27. Greenwood J A and Tabor D, 'The friction of hard sliders on lubricated rubber: The importance of deformation losses', *Proc. Phys. Soc. Lond.*, 1958, **71**, 989.
28. Lindberg J and Gralen N, 'Measurement of friction between single fibers: II. Frictional properties of wool fibers measured by the fiber-twist method', *Text. Res. J.,* 1948, **18**, 287.
29. Roder H L, 'The relation between fibre friction and the behaviour of fibres and yarns during processing', *J. Text. Inst.*, 1955, **46**, 84.
30. King R F and Tabor D, 'The effect of temperature on the mechanical properties and the friction of plastics', *Proc. Phys. Soc. Lond. B*, 1953, **66**, 728.

31. Makinson K R, 'Some comments on the paper by Dr N Gralen (Friction between single fibres, *Proc. Roy. Soc. A*, 1952, **212**, 491)', *Proc. Roy. Soc. A*, 1952, **212**, 495.
32. Pascoe M W, *Ph.D. Dissertation*, 1955, Cambridge.
33. Howell H G, 'The general case of friction of a string round a cylinder', *J. Text. Inst.*, 1953, **44**, T359.
34. Lincoln B, 'Frictional and elastic properties of high polymeric materials', *Brit. J. Appl. Phys.*, 1952, **3**, 260.
35. Viswanathan A, 'Frictional forces in cotton and regenerated cellulosic fibres', *J. Text. Inst.,* 1966, **57**, T30.
36. Howell H G and Mazur, 'Amonton's law and fiber friction', *J. Text. Inst.*, 1953, **44**, T59.
37. Finch R B, 'Interfiber stress and its transmission. Part II. Theory of the pressure distribution and contact area between fibers', *Text. Res. J.*, 1951, **21**, 383.
38. Petroff N, *Eng. J. St. Petersburg,* 1883, **71**, 228, 377, 535 (Reference given by Howell H G, Mieszkis K W and Tabor D, Butterworths Scientific Publications, London, 1959, Page 55).
39. Reynolds O, 'On the theory of lubrication and its application to Mr. Beauchamp Tower's experiments, including an experimental determination of the viscosity of Olive oil', *Phil. Trans.*, 1886, **177**, 157.
40. Hardy W B, *Collected Works*, Cambridge University Press, 1936 (Reference given by Howell H G, Mieszkis K W and Tabor D, Butterworths Scientific Publications, London, 1959, Page 57).
41. Rabinowicz E and Tabor D, 'Metallic transfer between sliding metals: An autoradiographic study', *Proc. Roy. Soc. A*, 1951, **208**, 455.
42. Tabor D, 'Friction, lubrication and wear of synthetic fibers', *Wear*, 1957, **1**, 5.
43. Bowers R C, Clinton W C and Zisman W A, 'Friction and lubrication of nylon', *Indust. Eng. Chem.*, 1954, **46**, 2416.
44. Roder H L, 'Measurements of the influence of finishing agents on the friction of fibres', *J. Text. Inst.,* 1953, **44**, T247.
45. Lyne D G, 'The dynamic friction between cellulose acetate yarn and a cylindrical metal surface', *J. Text. Inst.,* 1955, **46**, 112.
46. Hansen W W and Tabor D, 'Role of hydrodynamic lubrication in the friction of fibers and yarns', J. A*ppl. Phys.,* 1956, **27**, 1558.
47. Baird M E and Mieszkis K W, 'Frictional properties of nylon yarn and their relation to the function of textile guides', *J. Text. Inst.,* 1955, **46**, 101.
48. Adams M J, Briscoe B J and Kremnitzer S L, 'Survey of the adhesion, friction and lubrication of PET monofilaments', in Mittal K L, Ed., *Physicochem Aspects Polym. Surface, Proc-Int'l-Symp.*, Plenum, NY, 1, 425–450, 1981 (Publ. 1983).
49. Kremnitzer S L, *Ph.D. Thesis,* University of Cambridge, 1979.
50. Gupta B S and Whang H S, 'Surface wetting and energy properties of cellulose acetate, polyester and polypropylene fibers', *International Nonwovens Journal (INJ)*, 1999, **8**, 36.

3

Friction behavior of fibrous materials used in textiles

B S GUPTA, North Carolina State University, USA

3.1 Introduction

The fundamental laws of friction were developed from the observations made about the behavior of metals. These laws, proposed prior to the nineteenth century, were validated through studies conducted on different metals that were the materials of great engineering importance of the times. A number of interesting phenomena were observed. Junctions that developed at points of contact needed to be sheared in order for sliding to occur. On clean surfaces, the value of friction was found to be high. As sliding took place, junctions grew, increasing the area of contact. The expected extreme case was for the entire geometrical area to become one enormous junction that results in such a high level of friction that seizure can occur. Such a phenomenon was possible, however, only if the metals were cleaned immediately before conducting the frictional tests. Fortunately, metals readily develop an oxide film that prevents such increase in junctions and, thus, in area of contact. The fundamental laws of friction seemed, however, to apply in either case. Understanding the mechanism governing friction has led tribologists to design systems that result in low friction and thus in low wear.

Another mechanism of friction that plays a role with metals is deformation of surfaces involving hysteresis losses. This mechanism has been shown to play a predominant role in governing resistance to rolling of a spherical or a cylindrical body over a flat surface. By using hard metals and keeping stresses low enough to lie within the elastic regions, deformation losses can be reduced to lead to very low friction ($\mu \sim 0.01$). This is the basis on which ball and roller bearings are designed.

Deformation losses during sliding and rolling have been shown to be the major mechanism governing the behavior of elastomers, e.g. the behavior of rubber tires in normal running as well as slipping during abrupt braking.

Friction and wear being caused by contact and formation of junctions between surfaces, a major role of lubricants has been to introduce a resilient but weak layer at the interface, thereby reducing or eliminating solid to solid contact. This allows the frictional work to mainly dissipate within this interposed layer. Properties and thickness of lubricant, along with the normal force and the speed under which the bodies slide against each other determine the frictional forces.

Clearly, tribological studies have produced a large amount of information on the behavior of metals and elastomers, in particular on the former.

How far the above laws of friction apply to fibrous materials is our next consideration. Friction in fibers and textiles affects the efficiency of processing, mechanical integrity and dimensional stability of assemblies, wear of textiles and guides, and tactile comfort and drape of apparel. The introduction of synthetic fibers, with promising properties and potential for new apparel and non-apparel applications, and an increased interest in high speed automated operations, required that the tribology of fibers be understood. In this chapter, the frictional behavior of fibrous materials is discussed. Emphasis is placed on understanding the way in which behavior departs from that supported by the classical laws.

3.2 Differences in structure and properties of fibers, metals and elastomers

The properties of fibers fall between those of metals and rubbers. While metals deform plastically and rubbers elastically, fibers deform visco-elastically. Metals are crystalline, rubbers are non-crystalline, and fibers are semi-crystalline. Molecules in metals are essentially single atoms that are bonded by "metallic" bonds in a 3-dimensional network. Although these bonds involve sharing of electrons, as do co-valent bonds, there is a major difference. In a metallic bond, a given electron is not shared between two specific atoms but between many atoms of the crystal lattice. Accordingly, when a metal is subjected to stress, once this inter-atomic resistance is overcome, atoms are able to shear apart and flow plastically. In contrast, in a rubber, the molecules are made up of long chains with several thousand atoms linked end-to-end by strong co-valent bonds. The chains are not bonded with each other but are loose and randomly tangled or folded. When subjected to compressive force, the folds press against each other but bounce back when the force is withdrawn. When tensile stress is applied, the molecules stretch and extend against a retractile force and, when released, recover back to very nearly their original high entropy configuration. While the deformational behavior of a metal is driven by inter-molecular bond energy, that of a rubber is driven by entropy.

Fibers fall somewhere in between these two extremes. Like rubbers, they are polymers, but their molecular weights are usually not as large. Unlike rubbers, they have functional groups that allow them to bond with each other, leading, under favorable conditions, to the development of crystalline regions. Owing, however, to the restrictions imposed by their length, the chains are usually not able to untangle sufficiently to fully register with each other to produce a fully crystalline material. Thus, fibers are usually semi-crystalline, having both highly ordered, crystalline, and non-ordered, or less ordered, amorphous regions. The degree of crystallinity in man-made fibers varies from material to material, according to the chemical nature and the architecture of the chain. Crystallinity also varies between manufacturing processes and depends on the melt/solution viscosity, elongational stresses, quenching conditions and the amount of drawing involved. A fiber usually has a large number of small crystallites, ranging in size and perfection, scattered throughout the structure. Non-ordered regions of various sizes lie between the crystallites. These also differ in terms of packing, bonding and orientation of molecules. At ordinary temperatures, the bonds in amorphous regions are strong enough to cause molecules to resist deformation and, therefore, the behavior is dominated by the internal energy of the material. Only when the temperature is significantly higher than the glass transition temperature (Tg) does sufficient bond disruption occur in the amorphous regions for them to start to exhibit rubbery properties. At such temperatures, the deformation of the fiber becomes largely controlled by entropy [1]. Under these conditions, the crystalline regions act as tie points and provide stability to the structure. In other words, the structure and behavior of fibers and fibrous polymers above their Tg is somewhat equivalent to those of elastomers at ordinary, or room temperatures.

At ambient temperatures, the friction mechanism governing the behavior of fibers is closer to that applicable to metals, i.e. adhesion–shearing. Deformation losses due to ploughing or hysteresis are expected to be present, but they are usually small and certainly much less than that expected above the Tg. The presence in a fiber of a two-phase structure, the functional groups, a broad distribution of the strength of the bonds, and the kinetic processes in the disordered regions, together result in a visco-elastic stress-strain curve [2]. This is the case both in transverse compression and in tensile extension. In other words, the stress-strain curve is neither linear, as would be expected for an ideally elastic material, nor flat (after small deformation), as found in metals. Over a practical range of stresses it is intermediate between the two. This is the nature of the pressure area curve that should serve as the base for determining the frictional behavior of fibrous materials.

3.3 Classical friction behavior

To recapitulate, junctions form at points of contact and these must be sheared in order for sliding to occur (Fig. 2.1, Chapter 2). The frictional force (F) is, therefore, given by the product of the true area of contact (A) and the bulk specific shear strength of the junctions (S).

$$F = AS \tag{3.1}$$

For materials that deform plastically (Fig. 2.2, Chapter 2), the area (A) is given by the ratio of the normal force (N) to the yield pressure (P_y) of the material. This, when substituted for A in equation 3.1, gives the classical equation (equation 3.2) in which μ is a property of the material.

$$A = N/P_y$$

$$F = (S/P_y)\, N$$

$$F = \mu\, N \tag{3.2}$$

3.4 Frictional behavior of fibers in experimental investigations

Studies conducted during the latter half of the twentieth century, on natural and synthetic fibers, have shown that the coefficient of friction is not a material property, but is a function of the normal force and the geometric area of contact. An increase in normal force causes a decrease in the coefficient of friction. The size of the fiber, the smoothness of the surface and the mode of contact (point, line or area) also affect the value of the coefficient of friction. In the classical materials (metals), frictional force (F) is directly proportional to the normal force (N). However, a non-linear relation occurs with most polymeric materials.

An alternative relationship, giving the observed dependence of F on N, has been the subject of many investigations [3–11]. Several empirical equations giving a fit of experimental data have been suggested, but assuming that the ploughing term is small and the normal force is not excessively high (see Section 3.9 for behavior under high value of N), the simplest and the most widely accepted of these is the following:

$$F = aN^n \tag{3.3}$$

In this equation, a and n are empirical constants whose values are found by fitting the data on the model and conducting a least square analysis. Equation 3.3 was proposed by Bowden and Young [3] for application to certain nonmetals; it was applied to fibers first by Lincoln [5] and then by Howell and Mazur [8] and by most other authors who followed and worked on friction in fibrous materials. The value of n was found to be less than but close to unity (0.7–0.9). The value of a was found to be similar to that of the classical parameter, μ. The values of the constants varied with the fiber material examined, but no structure-based explanation had been proposed for the differences. A structural model developed by Gupta and El Mogahzy [12] gave theoretical meanings to these otherwise empirical constants and showed the factors, structural as well as procedural, which affected their values. The theory proposed by the authors was general and could account for the behaviors of a broad range of materials, including the plastically deforming metals and the elastically deforming rubbers.

3.5 A structural model of friction behaviour

In developing the model, it was assumed that the mechanism governing friction was the adhesion shearing. Also, it was assumed, that the specific shear strength of the junctions, *S*, was a material property; i.e. it was not affected by either the magnitude of the normal force or the area of contact.

3.5.1 The pressure–area relationship

An equation describing the pressure (P)–true area (A) relationship that covers the behavior of a range of materials, is as follows:

$$P = KA^{\alpha} \qquad 3.4$$

In this, K and α are the constants that, together, define the nature of the P–A relationship. Among the latter, K should be taken to represent the stiffness or hardness of the material in lateral compression and α, a factor that should characterize the shape of the curve. Figure 3.1 shows the range of behaviors that can occur with various materials. These include plastic, as found in metals ($\alpha = 0$), elastic ($\alpha = 1$), and visco-elastic ($0 < \alpha < 1$). The behavior of materials that undergo strain hardening can also be represented by equation 3.4 ($\alpha > 1$, Fig. 3.1).

According to the adhesion-shearing concept, the load is distributed on the points of contact (Fig. 3.2), the asperities deform and the areas of contact increase until the pressure on each decreases to the value that can be supported by the inherent ability of the material (solid line, Fig. 3.3). The latter is represented by the curve given by equation 3.4. The behavior of an asperity i, supporting a load w_i, is given by the iso-load curve I, which is a plot of the equation:

$$w_i = \text{Constant} = P_i A_i$$

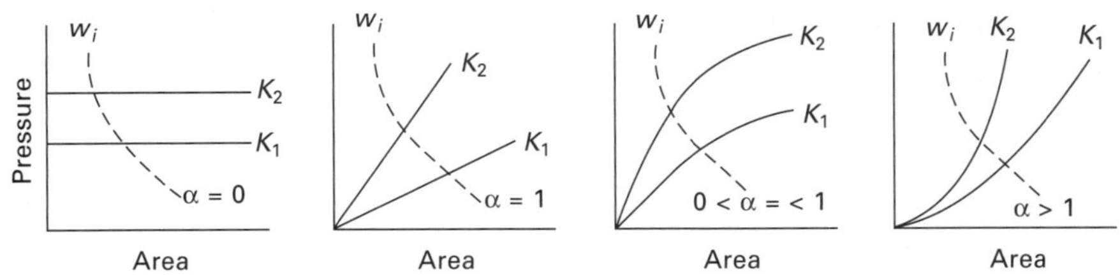

3.1 Pressure-area behaviors supported by equation 3.4 ($K_2 > K_1$) [12].

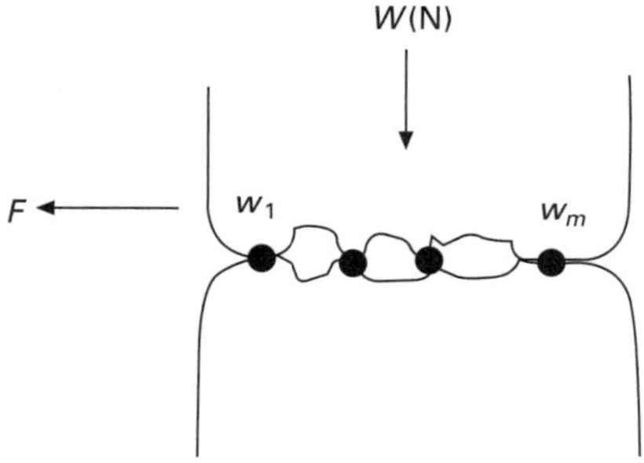

3.2 Distribution of load over contact regions.

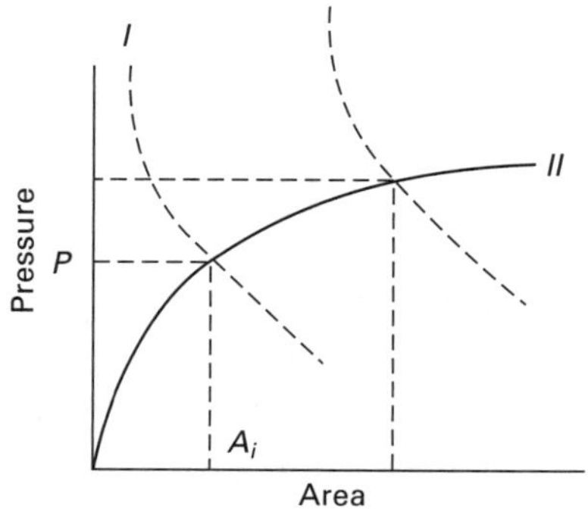

3.3 Area of contact for asperities at equilibrium.

The interaction of this curve with the *P–A* curve of the material determines the true area of contact, A_i, for asperity, *i*. Area A_i is determined by the intersection of I with II:

$$w_i = P_i A_i = K(A_i)^{\alpha} A_i = K(A_i)^{\alpha+1}$$

Or,

$$A_i = K^{-\gamma} w_i^{\gamma} \tag{3.5}$$

where,

$$\gamma = (\alpha + 1)^{-1}$$

The total true area of contact is given by the summation of the individual areas, or:

$$A = \sum^{m} A_i = K^{-\gamma} \sum^{m} w_i^{\gamma} \tag{3.6}$$

In this, *m* is the number of asperities making contact between the bodies. In order to assess the total area *A*, the summation term $\sum w_i^{\gamma}$ must be determined over the contacts *m*. For this, it is necessary to know how the stress is distributed over the contact region at equilibrium. This distribution, however, is not constant and could vary broadly in practice. The approach Gupta and El Mogahzy used was to obtain solutions for three widely different but well-defined distributions, namely uniform, spherical and conical (Fig. 3.4), and then to generalize the result [12].

3.5.2 Uniform stress distribution

All asperities are identical, i.e., they have the same shape and size at equilibrium. The load is distributed uniformly over the junctions (Fig. 3.4(a)).

$$w = N/m$$

$$\sum^{m} w_i^{\gamma} = mw^{\gamma}$$

Substitution in equation 3.6 yields:

$$A = K^{-\gamma} m^{1-\gamma} N^{\gamma} \qquad 3.7$$

3.5.3 Spherical stress distribution

The pressure existing on an annulus region of contact at radius x is given by:

$$P = P_0[1 - (x^2/x_c^2)]^{1/2} \qquad 3.8$$

In this, P_0 is the maximum pressure and x_c is the radius of the circle over which the spherical contact is constructed (Fig. 3.4(b)). The load δN supported by the annulus of width dx at x is given by:

$$\delta N = P_0[1 - (x^2/x_c^2)]^{1/2}\, 2\pi x dx \qquad 3.9$$

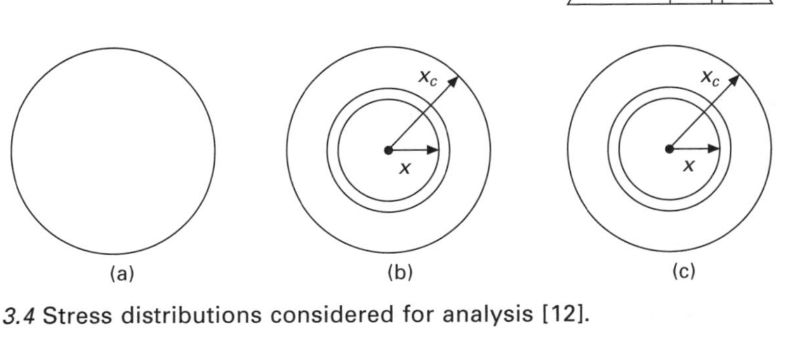

3.4 Stress distributions considered for analysis [12].

This leads to the total load N as follows:

$$N = \int_0^{x_c} P_0 \, [1 - (x^2 / x_c^2)]^{1/2} \, 2\pi \, x dx$$

which yields:

$$N = (2/3)\pi \, P_o x_c^2 \qquad 3.10$$

If M represents the number of asperities per unit area, the load δN on the annulus at radius x will be supported by $M2\pi xdx$ asperities. The average load per asperity, w_x, in the annulus will be given by:

$$w_x = \delta N/(M2\pi xdx)$$

By substituting for δN from (3.9) and for P_0 from (3.10), one gets:

$$w_x = (3N/(2\pi \, x_c^2 M))[1 - x^2/x_c^2]^{1/2} \qquad 3.11$$

Substituting w_x from (3.11) for w_i in (3.5), an expression for the true average area of contact of an asperity in the annulus is obtained:

$$K^{-\gamma} (3/2\pi)^{\gamma} (N/Mx_c^2)^{\gamma} [1 - x^2/x_c^2]^{\gamma/2} \qquad 3.12$$

The total area of contact within the annulus is obtained by multiplying the expression above with the number of asperities $M2\pi xdx$ in this region. Summation of this over the entire area yields the expression for the total true area of contact:

$$A = \int_0^{x_c} (K)^{-\gamma} (3/2\pi)^{\gamma} (N/Mx_c^2)^{\gamma} \times [1 - x^2/x_c^2]^{\gamma/2} M2\pi x dx \qquad 3.13$$

Integration yields:

$$A = [3^{\gamma} 2^{1-\gamma}/(\gamma + 2)] \, K^{-\gamma} [M\pi x_c^2]^{1-\gamma} N^{\gamma}$$

Substituting m back for $M\pi x_c^2$ gives the following expression for the true area of contact for the spherical stress distribution:

$$A = [3^{\gamma} 2^{1-\gamma}/(\gamma + 2)] \, K^{-\gamma} m^{1-\gamma} N^{\gamma} \qquad 3.14$$

3.5.4 Conical stress distribution

The pressure P (Fig. 3.4(c)) supported by an asperity at distance x from the center is given by equation 3.15:

$$P = P_0[1 - x/x_c] \qquad 3.15$$

Using equation 3.15 instead of 3.8 and following the procedure used for spherical stress distribution, the summation obtained, representing the total area of contact, is given by (3.16) and its solution by (3.17).

$$A = \int_0^{x_c} (K)^{-\gamma} (3/\pi)^{\gamma} (N/Mx_c^2)^{\gamma} \times [1 - x/x_c]^{\gamma} M 2\pi x dx \qquad 3.16$$

$$A = [(2)(3)^{\gamma}/(\gamma + 1)(\gamma + 2)] \times (K)^{-\gamma} (m)^{1-\gamma} (N)^{\gamma} \qquad 3.17$$

3.6 Generalized equations and the structure of the friction constants *a* and *n*

The area of contact equations, 3.7, 3.14 and 3.17, obtained on three very different stress distributions can be generalized and represented by equation 3.18:

$$A = C_M K^{-\gamma} m^{1-\gamma} N^{\gamma} \qquad 3.18$$

In this, C_M is a model constant whose value depends on the value of γ and the nature of the stress distribution. The values of this constant for different values of γ and stress distributions are given in Table 3.1; they range from 0.924 to 1.0. If $\gamma = 1$, i.e., $\alpha = 0$, the model constant will be unity. All equations derived from various models then reduce to the following simple equation:

$$A = N/K$$

This is the case for materials that deform plastically (equation 3.2), with K representing the yield pressure, P_y (Fig. 3.1 (a)).

Substitution for A from equation 3.18 into equation 3.1 yields equation 3.19, the equation for the frictional force F:

$$F = SC_M K^{-\gamma} m^{1-\gamma} N^{\gamma} \qquad 3.19$$

Table 3.1 Values of the model constant, C_M, for different values of γ

	C_M		
γ	Uniform	Spherical	Conical
0	1	1.0	1.0
0.1	1	0.992	0.966
0.3	1	0.982	0.930
0.5	1	0.980	0.924
0.7	1	0.984	0.940
0.9	1	0.993	0.976
1.0	1	1.0	1.0

By comparing equations 3.19 and 3.3, one can now provide structure to the empirical constants n and a. The following relations are immediately established:

$$n = \gamma = (1 + \alpha)^{-1} \quad 3.20$$

$$a = SC_M K^{-n} m^{1-n} \quad 3.21$$

Defining the coefficient of friction as the ratio of F to N (equation 3.2), it now has the form as given by equation 3.22:

$$\mu = SC_M K^{-n} m^{1-n} N^{n-1} \quad 3.22$$

3.7 Factors playing a role in friction

The models presented above illustrate the factors that affect the values of the frictional force, the coefficient of friction and the empirically determined indices, a and n. Some of these factors are related to the material and the others to the morphology of contact and the nature of stress distribution. A brief discussion follows which highlights the general effects of some of the important factors.

3.7.1 Normal force

If a material deforms plastically, i.e., $n = 1$ ($\alpha = 0$), then the normal force (N) has no effect on μ and equation 3.22 reduces to the dimensionless quantity (S/K). For all other materials, an increase in N causes a decrease in μ.

3.7.2 Mechanical behavior of junctions

The three factors that govern the mechanical behavior of junctions are the bulk specific shear strength, S, the hardness factor, K, and the shape of the pressure–area curve, given by the value of the constant n (or α). The value of S can be expected to vary with the chemical nature of the material and, for a given material, with physical factors that modify molecular interaction between surfaces, such as molecular weight, molecular orientation and crystallinity.

The parameters K and n affect the area A and, through it, F and μ. In the special case, when n is unity, K equals the yield pressure, μ and a become equivalent and are given by S/K. For all other cases, n is less than 1, which means that μ and a are different and their values are affected not only by S and K, but also by m (the number of points of contact) and by n. In the case of the parameter μ, the normal force N also has an effect.

3.7.3 Number of asperities in contact

With the normal force maintained constant, an increase in the number of points of contact, m, causes an increase in the area, A, and, through it, an increase in the values of F and μ. With an increase in m, the value of the index a also increases.

3.7.4 Other factors

Some other factors that affect friction are the mode of contact (line, point or area) during tests, the morphology of the surfaces (degree of roughness or smoothness), the testing environment (temperature and relative humidity), and the time of contact (time before sliding and speed of sliding). The mode of contact used will affect the number of points of contact, m, and through it the values of a, A, F and μ. For example, one can expect that line contact, with larger m, will lead to a larger value of μ than will point contact [13]. Area contact takes place when two flat surfaces are pressed together. One would generally expect that the larger the area the larger would be the value of μ.

Differences in surface morphology can affect friction. A rougher surface, with fewer contacts per unit apparent area, would mean a smaller value of m and, therefore, should result in lower friction. If surfaces are highly polished the contact area should be very large and so should the friction force. One could expect, however, that a small amount of debris between the surfaces could break the contacts and drastically reduce friction.

The changes in environmental conditions, i.e. relative humidity and temperature, can affect friction by altering the values of S and K. If the variations in environment also affect the visco-elastic properties of the contacting surfaces, then the value of n could also be affected.

Finally, since fibers are visco-elastic materials, their properties can change with the time of loading. The parameter that might be primarily affected is m. It is expected that the contact area will increase as the sliding speed is decreased, thus resulting in an increase in the coefficient of friction.

3.8 Experimental results from selected studies

3.8.1 General methods

The above model can be used to examine the results from experimental investigations, some of which, conducted at the North Carolina State University, are considered here. These involved studies on (1) two different synthetic fibers, varying in terms of size, shape and molecular orientation, (2) human hair, subjected to a number of cosmetic and recreation related treatments, and (3) several different surgical suture materials, varying in terms of construction, polymer type and surface finish. The mode of contact, the testing environment and the normal force were varied in selected cases.

Line contact For this, the twist method of Lindberg and Gralen [4], discussed in detail in Chapter 5 was used. The value of the coefficient of friction was calculated using equation 3.23, shown below:

$$T/T_0 = e^{\pi \tau \beta \mu} \tag{3.23}$$

In this equation, T_0 is the initial tension, T is the withdrawing tension, τ is the amount of twist inserted, and β is the twist angle, i.e. the angle at which the axes of the two fibers are inclined during the test. For materials that do not follow classical behavior, the value of μ is related to the tension T_0 and, therefore, also to the constants a and n. The model showing the relationship between these, is given by El Mogahzy and Gupta as follows [13]:

$$\mu = a(T_0 \beta^2/4\ r)^{n-1} \tag{3.24}$$

By conducting tests at different values of the initial tension T_0, calculating the corresponding values of μ (equation 3.23), and fitting the results to equation 3.24, the values of the indices a and n could be estimated. For this, it was necessary that the value of the radius r of the test specimen was known, i.e. either measured directly or estimated with a suitable model.

Point contact For point contact between two filament yarns, the method adopted was the one based on capstan geometry (Section 5.5.4, Chapter 5). In this, one fiber is held fixed horizontally in a U-frame. One end of the second fiber is attached to a load cell and the other end is led over the first fiber and tensioned with a weight [13]. The angle of wrap and the diameters of the fibers being small, the contact area was also small and assumed to be a point. The value of μ was calculated with the well-known capstan equation, given below, in which θ is the angle of wrap:

$$T = T_0\ e^{\mu\theta} \tag{3.25}$$

For the capstan geometry, the equation relating μ, a and n is given by (3.26):

$$\mu = a\ (T_0 / r)^{n-1} \tag{3.26}$$

As done for the twist method, the values of the constants a and n were determined by conducting tests at a number of values of T_0. The devices used for measuring friction by the point and the line methods had provisions for measuring friction while the contact regions were submerged in a fluid [13].

3.8.2 Effect of fiber cross-section shape

Friction was measured on monofilaments of polypropylene of 18 denier and three different cross-sectional shapes: circular, triangular, and trilobal. In both the line and the point contact tests, the circular fiber gave a significantly higher value of μ than did the triangular or the trilobal materials, the average difference being about 31% [13]. The values of a and n for the circular fibers were also higher, i.e. 28% and 3.6%, respectively, higher than the corresponding average values for the non-circular materials. The differences in the values of μ and a were considered significant. It was visualized that the trilobal or triangular fibers were transversely stiffer and, therefore, less deformable than the circular. This would lead to lower values for K and, therefore, to higher values of μ in the circular fibers, as compared to the values in the two non-circular ones.

3.8.3 Effect of testing conditions

Tests were conducted with the contact region submerged in water. The materials used were acrylic and polypropylene yarns. The tests involved the line contact given by the twist method. The results obtained are summarized in Table 3.2. While changing the environment from dry to wet did not affect the value of n (i.e. the shape of the P–A curve did not change), it caused appreciable increases in the values of μ and a. These fibers were hydrophobic and, therefore, did not swell as a result of absorbing moisture. A possible explanation, therefore, is that water acted as an anti-lubricant, i.e. it removed the topically applied finish present on the surfaces and brought the fibers more intimately in contact with each other.

3.8.4 Effect of molecular orientation

Polypropylene and acrylic multifilament yarns, drawn to different levels, were used in this study [13, 14]. For polypropylene, the stretch caused the denier and, therefore, the diameter, to decrease. In the case of the acrylic yarns, the jet stretch (which does not significantly affect orientation) and

Table 3.2 Values of μ, a, and n obtained in dry and wet tests using line contact method [13]

Fiber	Parameter	Dry	Wet
Polypropylene	$\mu(T_0/r = 328$ gf/mm$)$*	0.29	0.34
	a	0.77	0.94
	n	0.83	0.84
Acrylic	$\mu(T_0/r = 87$ gf/mm$)$	0.19	0.23
	a	0.32	0.44
	n	0.68	0.67

*Since μ is a function of the factor T_0/r (equation 3.26), the values of μ between the two fiber types cannot be compared.

cascade stretch (which affects molecular orientation) were varied such that orientation changed but the final denier remained constant. The amounts by which the acrylic yarns were stretched during processing, the values of the orientation factor (assessed by a sonic modulus procedure) and the values of μ (determined by the line and the point contact methods) are shown in Table 3.3. The values of μ for the polypropylene yarn are given in Table 3.4. The coefficient of friction increased consistently with orientation. For both yarns, higher values were obtained by line contact than by point contact, as one may expect from the relatively larger value of m in the case of the former. SEM analyses of the surfaces of the acrylic yarns (Fig. 3.5) showed that the morphology changed gradually with increase in cascade stretch. This resulted in progressive alignment and closer packing of the macrofibrils, and an increasingly smoother surface [14]. These changes could be expected to lead to more intimate contact (higher m) and thus to larger μ. Although no measurements were made, it is also feasible that an increase in orientation led to a decrease in hardness (i.e. a decrease in the value of K), which would also support an increase in μ.

3.8.5 Effect of fiber type

Two materials on which controlled experiments were conducted and whose frictional properties could be compared under equivalent conditions were

Table 3.3 Specifications and values of sonic modulus orientation factor (f) and coefficient of friction μ for acrylic yarns used in the study [14]

Sample	Stretch				μ	
	Jet	Cascade	Total	f	Point	Line
1	2.50	2	5.00	0.69	0.134	0.186
2	1.68	3	5.03	0.73	0.135	0.221
3	1.25	4	5.02	0.76	0.136	0.230
4	1.00	5	5.02	0.77	0.138	0.235
5	0.84	6	5.04	0.78	0.138	0.238
6	0.72	7	5.02	0.79	0.141	0.243

Table 3.4 Coefficient of friction of polypropylene yarns at draw ratios (DR) of 1X and 2X using line contact method [13]

T_0/r (gf/mm)	μ	
	DR: 1X	DR: 2X
100	0.32	0.38
200	0.31	0.37
300	0.30	0.36

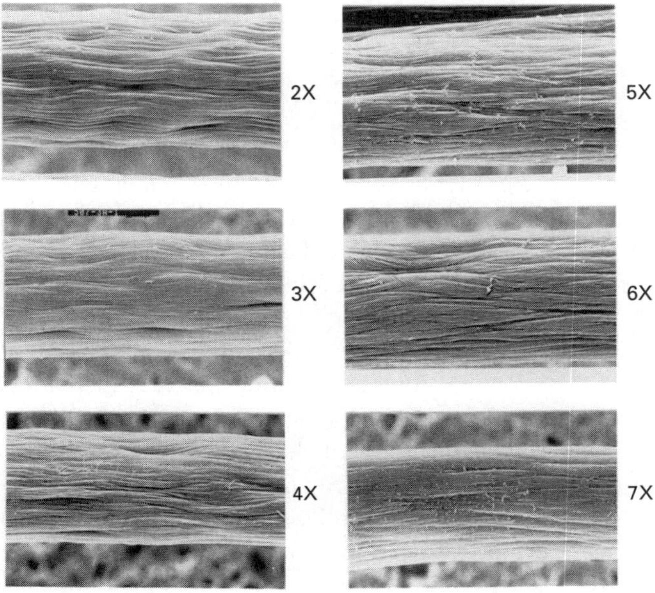

3.5 SEM micrographs of acrylic fibers given different levels of cascade stretch [14].

Table 3.5 Values of frictional parameters of acrylic and polypropylene yarns at equal values of T_o/r (line contact method) [13]

Yarns	μ (T_0/r = 200 gf/mm)	a (gf/mm)$^{1-n}$	n
Polypropylene	0.37	0.59	0.85
Acrylic	0.16	0.32	0.68

polypropylene and acrylic yarns. The results obtained on these are given in Table 3.5 [13].

The values of all three parameters, i.e. μ, a and n were higher for polypropylene than for acrylic. These differences would be expected to arise from differences in the chemical nature as well as in the physical structures of the two materials. A higher value of n indicates that polypropylene exhibits a more plastic behavior in compression than does the acrylic. Several factors could account for the difference in the values of a and, through a and n, in the values of μ (equations 3.21 and 3.24). These are the strength of the junctions (S), the deformational behavior of the junctions (K, n), and the nature and the size of the contacts during friction tests (C_M, m). Which one or more of these possibilities accounted for the differences in the values of a and μ could not be established as independent measurements of these parameters were not available. This illustrates the difficulty one faces when rationalizing differences in the frictional properties of two different visco-elastic materials.

3.8.6 Friction in human hair

Introduction Inter-fiber friction is perhaps the most important of the physical properties that influence several of the assembly characteristics of hair, such as combability, manageability, hand, and appearance. Most hair aids tend to influence assembly properties by modifying the fiber surface and, through it, friction.

Hair is a keratin fiber, with a chemical and physical structure similar to wool (Chapter 7). It contains long polypeptide chains, which are composed of about 18 amino acids. The most important amino acid is the sulfur-containing cystine that is present in the form of disulfide bonds that cross-link the polypeptide chains. These bonds contribute greatly to the chemical, physical and mechanical stability of hair. Morphologically, the fiber consists of two major components: (1) the cuticle, which is the outermost protective covering, composed of 6 to 8 layers of overlapping scales, which contain a highly cross-linked amorphous matrix; and (2) the cortex, which constitutes the bulk of the fiber. The cortex contains spindle-shaped, crystalline, microfibrils, which provides much of the mechanical stability and strength. Obviously, the component of most interest from the standpoint of hair performance and treatments is the cuticle, in particular its frictional characteristics. The scales, which are about 0.5 micrometer thick and 5 micrometer wide, cover about three quarters of the circumference, and point from root to tip. This causes the frictional properties to be directional, i.e., friction is higher in the tip to root direction, when the scales engage and oppose each other ("against scale" friction) and lower in the root to tip direction, when the scales slip over each other ("with scale" friction).

The disulfide bonds present in the cuticle are susceptible to damage from chemicals used in cosmetic treatments and from sunlight, environmental pollutants, and chlorine found in swimming pools. Exposure to these agents can affect the hair characteristics mentioned earlier by modifying the surface through chemical changes. As swimming and the use of cosmetic treatments are highly popular, a major concern is the effects of exposure of hair to chlorine in pool water and to chemical compounds present in treatment lotions.

Materials and methods A controlled study was performed by Fair and Gupta [15–17], which explored the effects of chlorine treatment on the friction and morphology of hair. Parameters investigated included, chlorine concentration, pH of the solution, and combinations of exposure to chlorine with treatments involving permanent waving, dyeing and bleaching. The hair types used were dark brown Caucasian and natural blond. Friction measurements were made by the twist method using 2 turns of twist and 3 g_f initial tension (T_0 in equation 3.23). The tests involved both "with scale" and "against scale" traverses, which provided values of μ_w and μ_a, respectively. The pH of the solution was adjusted with hydrochloric acid. Changes in the morphology of the hair fibers, after various treatments, were followed by examination in the SEM. Control samples, which had been taken through the same number of treatment cycles at neutral pH and zero reagent concentration, were also examined under the SEM. The regions viewed and photographed were those where the rubbing during friction testing had actually taken place.

Results The selected results in Fig. 3.6 and Tables 3.6 to 3.8 show that the values of the coefficient of friction increased and that of the DFE "$\mu_a - \mu_w$" decreased with the number of cycles of chlorine treatment, the chlorine concentration, and the acidity of the solution. The effects were not linear with relatively greater changes in friction values being produced following the treatments with shorter times and lower chemical concentrations. Some examples of the changes in morphology with treatments are given in Fig. 3.7. The 60 cycles control shows distinct scale structure and minimal damage due to rubbing (Fig. 3.7(a)). Treatment with chlorine concentration of 10 parts per million (PPM) and for 20 one-hour cycles led to a significant loss of scale definition (Fig. 3.7(b)). The increase in the concentration of chlorine to 50 PPM resulted in a greatly softened surface with bulk deformation produced by rubbing (Fig. 3.7(c)).

It will be instructive to examine the friction results given in Fig. 3.6 and Tables 3.6 to 3.8 in light of the changes in morphology noted in Fig. 3.7. For the control sample (0 PPM chlorine), friction and DFE increased with the

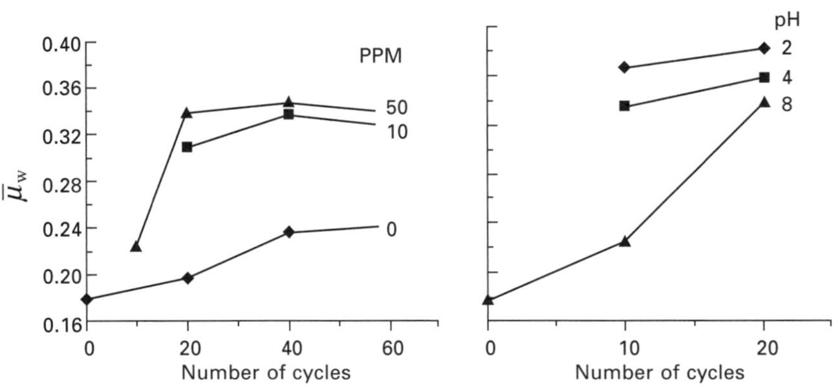

3.6 Effect of the number cycles of chlorine treatment on the average with-scale coefficient of friction: (a) at pH 8 but different concentrations, and (b) at 50 PPM but different pH [15].

Table 3.6 Effect of cycles of treatment and concentration of chlorine on DFE (x10^2) at pH of 8 [15]

Cycles	0	20			40			60		
PPM	0	0	10	50	0	10	50	0	10	50
DFE	6.5	8.8	4.3	1.6	9.4	3.9	3.0	9.5	3.2	1.9

Table 3.7 The effect of cycles and pH on DFE (x10^2) at 50 PPM [15]

Cycles	0	10			20		
pH	8	8	4	2	8	4	2
DFE	6.5	5.7	3.9	2.7	1.6	3.0	1.0

Table 3.8 Effect of cosmetic treatment combined with cycles of chlorination on average μ in the with-scale rubbing [17]. Cosmetic treatment was given prior to chlorination. Chlorine concentration 10 PPM

	Cycles				
	0	5	10	15	30
Control	0.13	0.18	0.22	0.22	0.30
Bleach	0.19	0.27	0.27	0.31	0.31
Dye	0.18	0.22	0.28	0.25	0.36
Perm	0.20	0.19	0.21	0.25	0.31

number of cycles. The increase in friction is believed to be due to swelling of the structure and, therefore, to a decrease in *K*; whereas the increase in the DFE is caused by the scales becoming more prominent. The increase in concentration from 0 to 10 PPM led to a significant decrease in scale definition (Fig. 3.7(b)) and, therefore, also to a decrease in DFE. The increase in the coefficient of friction (Fig. 3.6) was probably due to a decrease in *K*, produced by softening of the surface, and increase in *m*, resulting from the surface becoming smoother. Increasing the chlorine concentration from 10 to 50 PPM greatly increased surface softening as illustrated by the occurrence of extensive bulk deformation (Fig. 3.7(c)). The coefficient of friction, however, changed very little (Fig. 3.6). Clearly, the three parameters, *K*, *S* and *m*, and perhaps also the index *n*, which represents the shape of the pressure-area curve, all changed, but the changes were such, however, that the net effect on friction was small. Increasing the acidity of the solution heightened the damage (not shown). Under extreme acidic conditions (pH 2), dissolution of part of the surface, combined with frictional damage was enough in some cases to reveal the underlying cortex [15].

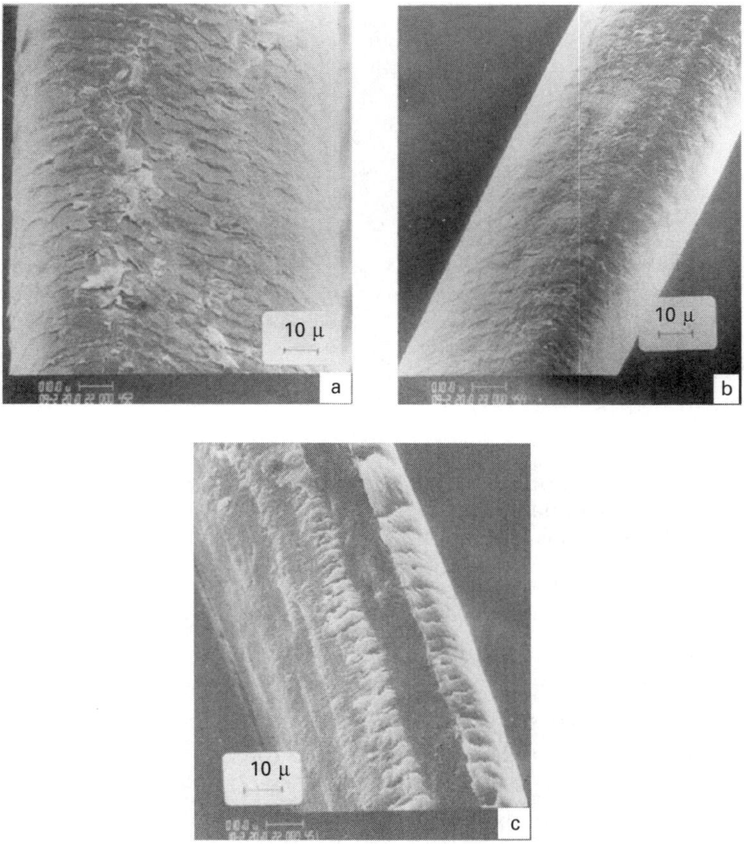

3.7 Scanning electron micrographs of friction tested brown hair fibers given chlorine treatment: (a) control, 60 cycles at 0 PPM and 8 pH, "against" scale; (b) 20 cycles at 10 PPM and 8 pH, "with" scale; and (c) 60 cycles at 50 PPM and 8 pH "against" scale [15].

The effects of the cosmetic treatments combined with the number of one-hour cycles at a concentration of 10 PPM chlorine on the average value of the coefficient of friction are shown in Table 3.8. The results considered are for samples in which the cosmetic treatment was applied first, followed by exposure to chlorine. The treatment and cycle effects are evident. Cosmetic treatments alone significantly increased the coefficient of friction beyond that of the untreated control (0 cycles). When chlorination was added to the cosmetically treated specimens, the coefficient of friction increased, indicating that the effects of the two were additive.

3.8.7 Friction in surgical sutures

Following surgery, a wound is usually closed with a suture thread, which is looped around the cut or repaired vessels and held in place with a knot.

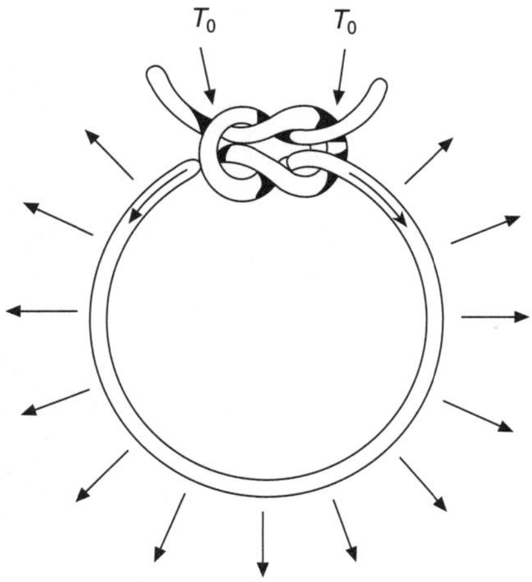

3.8 Model of a surgical knot with two throws [19]. The tissues enclosed, being compressed, exert a tension in an effort to relax back to their original state.

Friction is the primary mechanism by which the latter holds. The frictional force between the elements of the knot depends upon the coefficient of friction, the number of crossing points, the angle of contact between the threads at each crossing point, and the normal force pressing the threads against each other (Fig. 3.8). The normal force is initially governed by the tension used in tying the knot but later by the compressed tissues in an effort to expand and regain their original sizes. Clearly, if the total frictional force is high, the knot may hold firmly; if not, the components of it may slide and result in failure.

Materials and methods A knot consists of a number of throws pressed against each other and held by friction contacts (Fig. 3.8). The method that closely simulates the construction of a surgical knot and used in conducting friction tests is the twist method. Surgeons tie knots by hand and, therefore, there is no control on the tension used. Accordingly, in one detailed study, the tests of friction were conducted using many different values of T_0. The number of turns used was three [18]. The suture materials used were: Dexon®, a polyglycolic acid braided multifilament yarn; Mersilene®, a polyester braided multifilament yarn; Tevdek II®, a polyester braided and Teflon® impregnated multifilament yarn; Polydek®, a polyester braided multifilament yarn with a lighter coating of Teflon®, Ticron®, a polyester braided and silicon treated multifilament yarn; silk, a braided yarn; Ethilon®, a nylon monofilament yarn; Surgilon®, a nylon braided silicon treated yarn; and Prolene®, a polypropylene monofilament yarn. Knots were also tied under controlled conditions with different number of throws and tested for the knot holding force (KHF), i.e. the force at which a knot slipped or broke [19]. The greater the value of KHF the more secure or stronger is the knot.

Results Figure 3.9 shows the results of friction. Differences in the frictional properties of sutures are obvious. All materials were approximately of the same diameter and, therefore, had the same T_0/r value at any given value of T_0 (equation 3.24). For any given material, the value of μ decreased, as expected, with an increase in the applied tension. The two monofilament materials (polypropylene and nylon) had the highest value of μ at the lowest tension (0.56 N) and ended up having the lowest value of μ at the highest tension. The effect of geometric configuration could be seen by comparing the results for Surgilon® and Ethilon®, both nylon but one braided and the other monofilament. Although the former was silicon-treated, its μ was higher than that of the latter. The effect of surface modification could be examined by comparing the results for Mersilene®, Ticron® and Tevdek®, all polyester materials and braided but differing in terms of the treatment given to the surface. The ones with special finishes gave consistently lower values. The values of KHF, the knot holding force, are shown in Fig. 3.10 [19]. Clearly, the special coatings present on Ticron® and Polydek® led them to have much lower values of KHF at two throws, and, in the case of Ticron®, also at three throws, than obtained for the uncoated Mersilene®. These three sutures, having the same chemical constitution and physical size and constructions

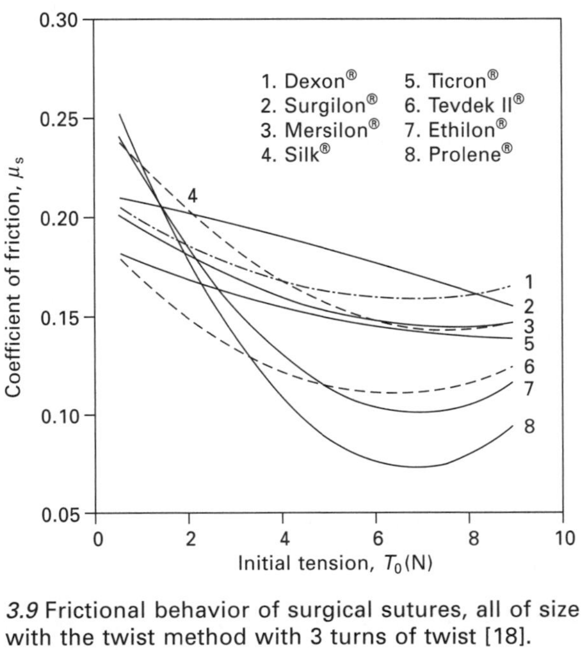

3.9 Frictional behavior of surgical sutures, all of size 2-0, measured with the twist method with 3 turns of twist [18].

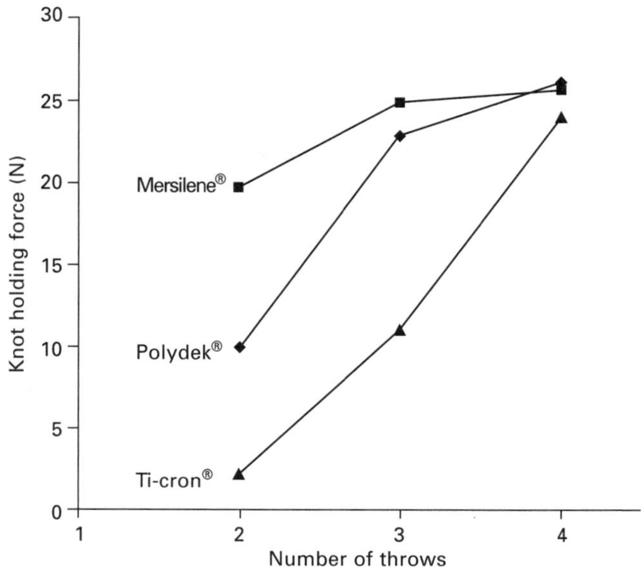

3.10 Effect of throws on knot holding force in dry tests (suture size 2-0) [19].

gave, as expected, approximately the same values when the knot was made fully secure by using four throws.

Knot security is a major concern when sutures are used in closing wounds after surgery. A surgeon usually lowers the risk of a knot slipping or untying by adding more throws in constructing the knot. This practice, although apparently necessary, is undesirable as it leaves more foreign material in the body and prolongs surgery. In an effort to make the smallest knot, i.e. the 2-throw, secure, these knots were tied in a number of sutures and exposed to different levels of energy from a CO_2 laser [20–22]. Figure 3.11 shows the effect of the energy of exposure on knot strength of the polyester (Mersilene®) suture of size 3-0. The strength increased with exposure until about 1.2 J; after this the value decreased. The increase in strength is considered to be due to an increase in friction caused by the melting and bonding of a few surface fibers within the structure of the knot. The decrease in strength is attributed to excessive melting and destruction of the fiber structure. An SEM of the surface of an exposed 2-throw knot (Fig. 3.12) supports this mechanism [23].

3.9 Friction in textiles under extreme normal forces

The behavior discussed in the sections above applies to textiles subjected to forces that lie in what one may call a nominal or practical range. The latter

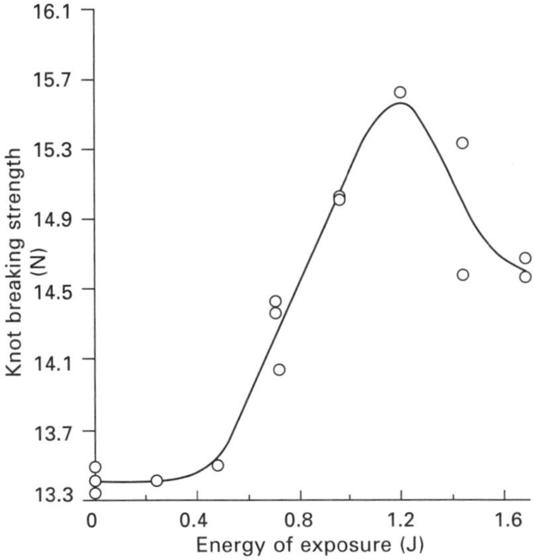

3.11 Effect of the energy of exposure to a CO_2 laser beam on the breaking strength of a 2-throw square knot in a 3-0 Mersilene® polyester suture [20].

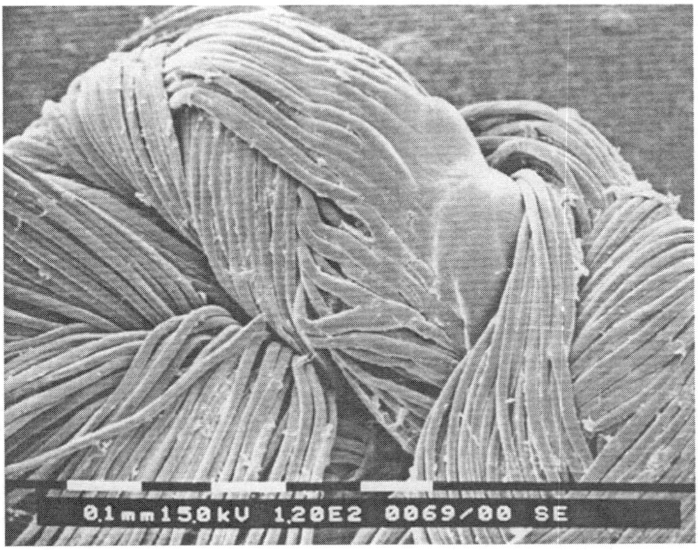

3.12 SEM of a knotted polyester 3-0 suture (Tevdek II®), exposed to CO_2 laser for 0.325 J at 0.5 W (120X) [23].

may be considered to extend from some small but finite value of normal force or stress to a value, at the upper end, that lies within the visco-elastic limit of the fiber. Most friction related studies on textiles have been conducted with forces lying in this range, and the behavior found has been characterized effectively by the model given by equation 3.3. However, there are applications, such as in ballistic products, in which a textile experiences extreme normal stresses. It is expected that in ballistic type applications of fabrics, friction at the crossover points plays a critical role in enabling translation and distribution of stresses and, therefore, in absorption of impact energy. Under such high stresses encountered, a textile material could be expected to deform largely beyond its elastic or quasi-elastic, i.e. visco-elastic, limit.

Not much work relating to measurement and study of friction at very high normal forces has been published and whatever is available is not well evaluated and understood. It should be generally clear that the model, $F = aN^n$, or $\mu = aN^{n-1}$, has limitations and its application cannot be justified except for some middle or practical range of normal stresses. In textiles, presence of normal adhesion, and/or mechanical obstructions or entanglements, lead to some measurable frictional resistance even at essentially zero normal force [24, 25]. And, although not explored, it is highly unlikely that the coefficient of friction will continue to drop, as suggested by the model, indefinitely with increase in N without a change in trend or reaching a limiting value when the normal stress has exceeded some high value.

Considering that at high normal stresses a textile will have significant plastic deformation, a plausible model for characterizing the behavior in this region may be as given by equation 3.27:

$$F = \alpha_0 N + aN^n \tag{3.27}$$

In this, α_0 is an additional empirical constant. Equation 3.27, using the classical definition of the coefficient of friction, μ, changes to equation 3.28:

$$\mu = \alpha_0 + aN^{n-1} \tag{3.28}$$

This model, initially proposed by Gralen [26] for applying to fiber friction in general, was used by Briscoe [27] for characterizing friction in ballistic textiles. Under the ballistic levels of stresses, one may expect that the material will lose its elastic recovery properties and be mostly deformed plastically. The second term of the model will then be small and friction will largely be given by the first term.

Examining friction, determining the effects of material and structural factors on it at stresses that lie in the upper region of the stress – strain curve, and modeling the behavior to apply to this domain, will close a gap that exists in the current literature. The new knowledge will be found valuable in understanding the behavior of textiles used in many load bearing high performance technical applications, including civilian and military land and air vehicles and body armor.

3.10 Summary

The deformation of a textile fiber is neither plastic, as in metals, nor elastic, as in rubbers. When a fiber is extruded from a polymer that contains chains of intermediate lengths (shorter than those of rubbers) and functional groups it has a semi-crystalline structure and a visco-elastic mechanical behavior. Accordingly, the tangential force needed to slide a fiber against another material increases non-linearly with the normal force. This has been consistently expressed by the relation $F = a\,N^n$. Using a non-linear pressure-area relation, $P = K \cdot A^{\alpha}$, it has been possible to quantify the various friction parameters, including the indices a and n, and also to identify the bulk, the morphological and the testing procedure related factors that affect a fiber's frictional response. The detailed structures given to the friction parameters provided a useful means for understanding the effects of many of the variables of the experimental studies included in this chapter. The approach adopted in the treatment gives further support to the work of Bowden and Tabor by showing that the adhesion-shearing theory applies as effectively to fibers as it does to the metals for which it was originally developed.

Much of the experimental work on friction of textiles published in the literature has been restricted to normal forces and stresses that lie in a more generally applicable range of values, i.e. the range within which the material can be considered to behave in a visco-elastic manner. With the textile materials finding increasing applications in high stress bearing technical products, and with friction playing an important role in governing their performance, a need exists for measuring and understanding the frictional properties of textiles subjected to stresses that make up the upper part of the load-deformation curve.

3.11 Glossary

a	Friction constant (see equation $F = aN^n$)
A	True area of contact
A_i	True area of contact of an asperity i
C_M	Stress distribution constant (see Table 3.1)
DFE	Differential frictional effect in hair fibers
DR	Draw ratio
f	Orientation factor ($0 \leq f \leq 1$)
F	Frictional force
K	Hardness in compression. Slope of pressure-area curve (see equation $P = KA^{\alpha}$)
KHF	Knot holding force, i.e. force required to cause a knot in a suture to slip or break
m	Number of points of contact between two bodies
M	Number of points of contact per unit area

n	Friction index (see equation $F = aN^n$); $n = (\alpha + 1)^{-1}$
N	Normal force
P	Normal pressure (see equation $P = KA^{\alpha}$)
P_o	Maximum pressure supported by asperities between two bodies
PPM	Parts per million
P_y	Yield pressure in transverse compression
r	Radius of filament or yarn for which friction is tested by the capstan or twist method
S	Specific shear strength of junctions (Nm^{-2})
SEM	Scanning electron microscope
T	Withdrawing tension required in the capstan or twist method of testing friction
T_0	Initial tension against which friction test is conducted using the capstan or twist method
W	Weight of the top body resting or sliding on another body
w_i	Load supported by an asperity i
w_x	Average load supported by asperities
x	Space co-ordinate for asperities arranged strategically in a circular area of radius x_c (see Fig. 3.4)
x_c	Radius of circle in which asperities are assumed to be arranged and over which pressure is distributed
α	Exponent determining the shape of the pressure-area curve, $P = KA^{\alpha}$ (see equation 3.4, Fig. 3.1)
α_0	Parameter, which will represent the coefficient of friction in fibrous materials when the normal stress is high enough for the material to deform plastically
β	Twist angle between angle two filament or yarns during measurement of friction by the twist method
γ	$\gamma = n = (\alpha + 1)^{-1}$
μ	Coefficient of friction, given by F/N
μ_a	Coefficient of friction in the "against-scale" sliding in hair fibers
μ_w	Coefficient of friction in the "with-scale" sliding in hair fibers
θ	Angle of wrap in the capstan method of measuring friction
τ	Number of turns of twist used in the twist method of measuring the coefficient of friction

3.12 Acknowledgements

The theoretical and experimental studies included in this chapter are based on the works of several of the writer's colleagues and graduate students. The writer thanks the following individuals: Dr. Yehia El-Mogahzy, Dr. Nancy Fair, Ms. Nancy Peneff, Dr. Ramond W. Postlethwait, Dr. Elizabeth Stone, and Dr. Kay W. Wolf.

3.13 References

1. Clark, J F and Preston J M, 'Thermoelastic properties of synthetic fibres', *J. Text. Inst.*, 1953, **44**, T 596.
2. Halsey G, White H J and Eyring H, 'Mechanical properties of Textiles, I', *Textile Res. J.*, 1945, **15**, 295.
3. Bowden F P and Young J E, 'Friction of diamond, graphite, and carbon and the influence of surface films', *Proc. R. Soc.*, 1951, **208**, 444.
4. Lindberg J and Gralen N, 'Measurement of friction between single fibers. II. Frictional properties of wool fibers measured by the fiber-twist method', *Textile Res. J.*, 1948, **18**, 287.
5. Lincoln B, 'Frictional and elastic properties of high polymeric materials', *Brit. J. Appl. Phys.*, 1952, **3**, 260.
6. Howell H G, 'The Laws of Static Friction', *Textile Res. J.*, 1953, **23**, 589.
7. Gralen N, Olofsson B and Lindberg J, 'Measurement of friction between single fibers. Part VII. Physicochemical views of interfiber friction', *Textile Res. J.*, 1953, **23**, 623.
8. Howell H G and Mazur J, 'Amontons' Law and fibre friction', *J. Text. Inst.,* 1953, **44**, T59.
9. Lodge A S and Howell H G, 'Friction of an elastic solid', *Proc. Phys. Soc. London*, 1954, **B 67**, 89.
10. Viswanathan A, 'Frictional forces in cotton and regenerated cellulosic fibres', *J. Text. Inst.,* 1966, **57**, T30.
11. El-Mogahzy Y E, Ph.D. Dissertation, *North Carolina State University, Raleigh, North Carolina, USA*, 1987.
12. Gupta B S and El-Mogahzy Y E, 'Frictional forces in fibrous materials. Part-I: Structural model', *Textile Res. J.*, 1991, **61**, 547.
13. El-Mogahzy Y E and Gupta B S, 'Friction in fibrous materials. Part-II: Experimental study of the effects of structural and morphological factors', *Textile Res. J.*, 1993, **63**, 219.
14. Gupta B S, El-Mogahzy Y E and Selivansky D, 'The effect of hot-wet draw ratio on the coefficient of friction of wet-spun Acrylic yarns', *J. App. Poly. Sci.*, 1989, **38**, 899.
15. Fair N and Gupta B S, 'Effect of Chlorine on friction and morphology of human hair', *J. Soc. Cosmet. Chem.*, 1982, **33**, 229.
16. Fair N B and Gupta B S, 'The chlorine-hair interaction. II. Effect of chlorination at varied pH levels on hair properties', *J. Soc. Cosmet. Chem.*, 1987, **38**, 371.
17. Fair N B and Gupta B S, 'The chlorine-hair interaction. III. Effect of combining chlorination with cosmetic treatments on hair properties', *J. Soc. Cosmet. Chem.*, 1988, **39**, 93.
18. Gupta B S, Wolf K W and Postlethwait R W, 'Effect of suture material and construction on frictional properties of sutures', *Surg. Gynecol. Obstet.*, (now *J. Am. Coll. Surg.* 1985, **161**, 12.
19. Gupta B S and Postlethwait R W, 'An analysis of surgical knot security in sutures', in Winter G D, Gibbons D F and Plenk H Jr., Eds., *Biomaterials 1980*, p. 661, Wiley-Interscience, New York, 1982.
20. Gupta B S, Milam B L and Patty R R, 'Use of Carbon Dioxide laser in improving knot security in polyester sutures', *J. of Applied Biomaterials*, 1990, **1**, 121.
21. Gupta B S, Peneff N L and Stone E A, 'Application of CO_2 laser in improving

mechanical performance of surgical knots', *Proceedings of the 15th Southern Biomedical Engineering Conf., IEEE*, 1996, 221.
22. Peneff N L, *M.S. Thesis*, North Carolina State University, Raleigh, North Carolina, USA, 1994.
23. Gupta B S, 'Performance of polymers, fibers and textiles in medicine', *Proceedings, Medical Textiles 2004: Advances in Biomedical Textiles and Healthcare Products,* Industrial Fabrics Association International (IFAI), Roseville, MN, 76–88, 2004.
24. Pascoe M W, *Ph.D. Dissertation*, 1955, Cambridge.
25. Briscoe B J and Kremnitzer S L, 'A study of the friction and adhesion of polyethylene terephthalate monofilaments,' *J. of Physics*, 1979, **12**(4): 505–516.
26. Gralen N, 'Friction between single fibers,' *Proc. Roy. Soc. A*, 1952, **212**, 491.
27. Briscoe B J and Motamedi F, 'The ballistic impact characteristics of aramid fabrics: the influence of interface friction,' *Wear*, 1992, **158**: 229–247.

4

Stick-slip phenomena in textiles

R KOVAR, Technical University of Liberec, Czech Republic,
B S GUPTA, North Carolina State University, USA and
Z KUS, Technical University of Liberec, Czech Republic

4.1 Introduction

Stick-slip is a commonly observed phenomenon in friction tests. It occurs when the speed of the bodies in contact oscillates between zero and some finite value. For stick-slip phenomena (SSP) to occur, two basic conditions should be fulfilled:

1. Coefficient of friction is variable, i.e. $\mu_s > \mu_k$ ($_s$ – static, $_k$ – kinetic).
2. The system is flexible enough to enable a change in the speed of the sliding body.

In the literature, the term stick-slip is used in a broader sense than implied above; for example, the term also implies a change of frictional resistance without a decrease of speed to zero. Stick-slip is a well-experienced phenomenon in life. Often it is recognized in terms of the generation of sound (chalk on board, music from some instrument, noise from doors lacking lubrication [27]). The fact that a violin produces music is a result of SSP as an oiled bow will not play.

In this chapter an attempt is made to provide basic information on SSP in textiles. The emphasis is placed on understanding the phenomenon and the behavior of textiles affected by it. Since the frictional resistance of textile fibers, yarns and fabrics is influenced by many variables, the prediction of friction and SSP is usually very difficult. Accordingly, the models used to describe SSP are simple and only in principle represent the actual behavior.

Since many different types of forces are considered in this chapter, F is used as the primary parameter to represent it. A subscript is used to differentiate one type of force from another. A major departure from other chapters in notations in this chapter is the use of F_n, instead of N, to represent normal force. Section 4.9 at the end gives the glossary used in this chapter. Also the abbreviated term SSP is used in this chapter not only to imply stick-slip behavior but also in general the friction phenomenon.

4.2 Physical basis of SSP

There are two main forms of friction of solid bodies:

1. **Sliding** friction, that is often connected with SSP. In this case we can have two set ups:
   (i) The planar, in which the normal force, F_n, is not dependent on the tangential force F_t.
   (ii) The capstan, in which the normal force arises due to the axial force of the curved material sliding over a stationary rod (Euler's friction).
   Idealized formulas for calculation of the output tensile force are in the case of planar friction $F = F_0 + \mu \cdot F_n$ and for capstan friction $F = F_0 \cdot e^{\alpha \cdot \mu}$, μ is coefficient of friction.
2. **Rolling** friction. The SSP of rolling friction is connected with a material's deformation and hysteresis, but this form of friction is not important in either processing or determining the behavior in use of textiles.

SSP is only examined in sliding of solid bodies; the behavior of liquids is not considered.

In general, the contribution to SSP can arise in contacting materials from different levels of organization within the structure:

- Nano-level – due to bonds and forces between particles (atoms, molecules, etc.).
- Micro-level – due to surface morphology of fibers.
- Macro-level – geometries of assemblies (yarn and fabric).
- Environmental level – due to influence of air, moisture, finish at the surface.

4.2.1 Principles involved in the development of static frictional resistance

Nano-level

This level is perhaps the most important. It deals with the forces between elementary particles of matter (atoms, molecules), which should translate to all levels. A general description (empirical results) of influence of distance on forces between two particles is shown in Fig. 4.1 [9]. The stability of the diatomic system is based on the fact, that at some distance d the interactive force is zero. The compression of the matter is connected with the shortening of this distance and with a relatively great repulsive force F_{rep}; the elongation, on the other hand, will depend on the attractive force, F_{att}, playing a role. The force beyond point A follows the relation (4.1), in which k is a constant.

$$F = \frac{k}{r^7} \qquad 4.1$$

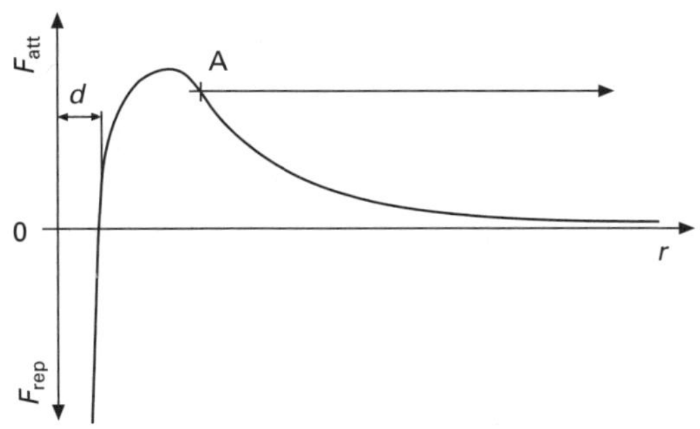

4.1 Influence of distance *r* on particle interaction *F*.

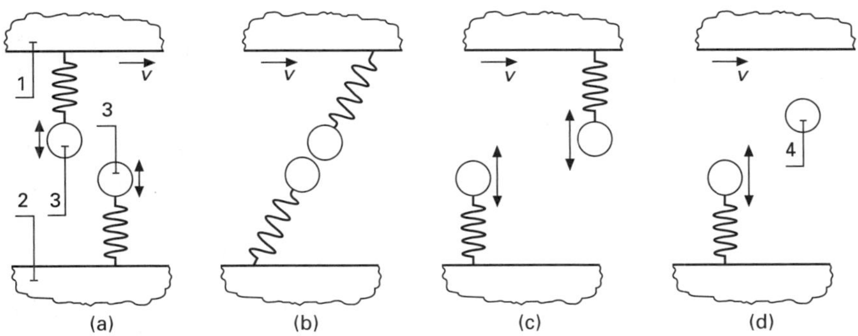

4.2 Change of macro-motion into micro-motion.

The slip of one body on another is connected with the change of motion from macro of bodies into micro of particles (molecules), in other words, into heat. The separation between particles needs to be small, i.e. with distances of the order of 0.1 nanometer (Å) so that the mutual attractive forces are sufficiently high. This condition will be met effectively in smooth surfaces that will have many such contacts. In uneven surfaces the number of such contacts will be lower. The principle of energy translation is shown in Fig. 4.2.

Let us consider that body 1 is sliding on body 2 with speed v (a). Some of the surface particles 3 may be near enough that strong attractive force exists. Continuing movement leads to dislocation of particles in relation to their "home" body (b). After overcoming the attractive force, particles 3 are separated and the amplitude of their vibration (heat) will become higher (c). It may even happen, that the original bond of a particle with its own body will be interrupted and the particle, crystal, or a portion of crystal, will get dislodged (d). Such particles may even be transferred from one body to another.

These interactions are connected with SSP because when the two bodies in contact are at rest the surfaces deform and more and stronger bonds are formed than when the two are sliding against each other. Let us describe several situations and their influence on SSP [21]:

- Two flat bodies in a vacuum. If the surfaces are ideally flat, the atoms and molecules will get near each other to within short enough distances to form bonds [30]. When the surfaces are not so ideally flat, the number of bonds formed will be relatively fewer. Frictional resistance will be proportional to the number of bonds created.
- The same two flat bodies but not in a vacuum. The surface of almost every chemical material will be changed by the deposition of an oxide film that will inhibit the development of as strong a connection on the surface as possible between atoms within the bulk. The contamination of the surface by dust is also possible. The result will be that friction will be lower but static will still be higher than kinetic. The exact results will be difficult to predict as many variables play a role.
- A layer of liquid on the solid surface. The behavior will be changed as the friction of a liquid follows different laws than that of a solid. If the film is thin and can be penetrated, then both will govern friction based on relative areas of contact between films and solids. These will vary with pressure, speed, viscosity and surface morphology.

Both static and kinetic friction depend on the real area of contact [7, 16]. This area is of course a function of normal stress and surface landscape, more specifically, of the deformational behavior of the asperities in contact. Figure 4.3 describes a situation, when the real area of contact is small relative to the apparent geometric one. Contact occurs in few regions which deform and undergo an increase in area. If the bodies in contact are at rest, then the area of contact may increase with time due to material rheology. So the value of static friction depends as well on the time of the application of normal force before sliding occurs.

Micro-level

Microstructure of fiber surfaces may cause variations in frictional resistance due to the presence of irregularities on the surface. It can result in a variation of sliding body speed and therefore in SSP, irrespective of the frictional behavior of the surface.

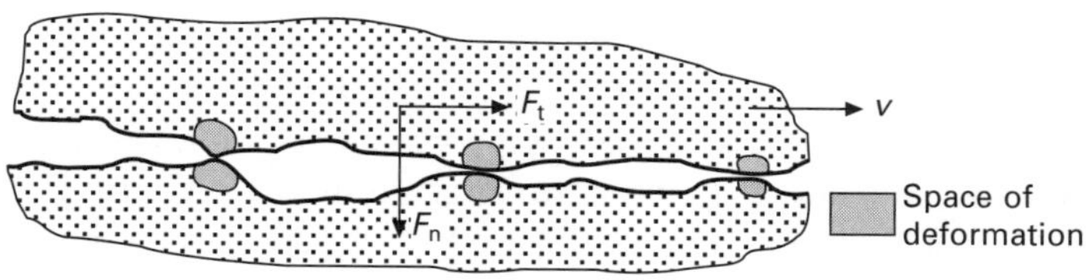

4.3 Real area of contact [22].

The effect of surface roughness is illustrated in Fig. 4.4 with the use of a simple concept of roughness [15]. The oscillation of tangential or frictional force F_t will be regular in the case (a) ($F_t = \pm F_n \cdot \tan\beta$, where β is the angle of surface to the direction of F_t) but irregular in the case (b), as expected, for example, when measuring friction in raw wool.

The effect of surface roughness may appear at a number of levels, i.e. from the crystalline or macromolecular through the fiber, yarn and fabric structural levels.

SSP at fiber-to-fiber level is important for the role internal friction plays in governing the behavior of assemblies as the deformation of a yarn or fabric is connected with the movement at the fiber level. Higher values of static friction can lead to higher initial modulus or bending rigidity and higher hysteresis; in extreme cases it may even lead to the generation of audible sound.

Macro-level

This level deals with the impact the surface unevenness of a yarn or a fabric, including the unevenness caused by the presence of hair and pile, has on friction. As compared to the degree of surface roughness of a fiber, that of a yarn or a fabric is higher and so the impact of it on friction between two such bodies is stronger. Let us consider, for example, yarn-to-yarn friction. Real contact develops between fibers with friction at nano- and micro-levels. These should translate to stick-slip at the yarn level. The impact of bending rigidity of the fibers is shown in Fig. 4.5 (a). The fibers on the surface need

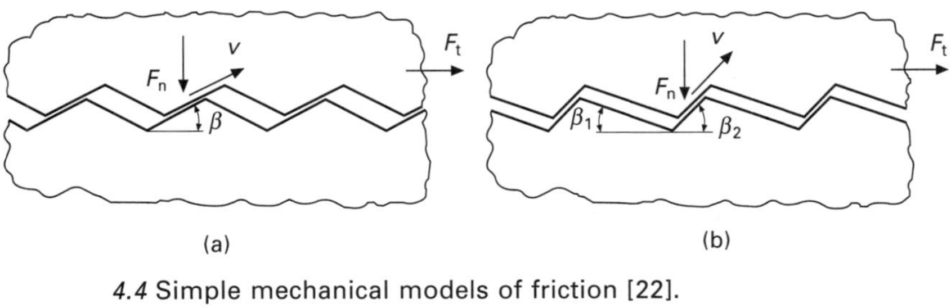

4.4 Simple mechanical models of friction [22].

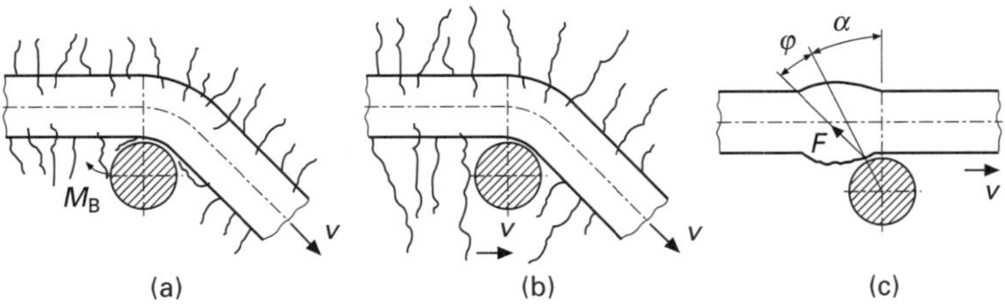

4.5 Influence of yarn unevenness on frictional resistance [22].

to be bent and so a moment M_B arises. It not only leads to the development of force in the direction opposite to the movement of yarn, but also to an increase in the real area of contact, both requiring greater force to maintain speed.

Long hair in conjunction with high speed can also cause acceleration/deceleration in surface elements that could lead to SSP (Fig. 4.5(b)). For example, part of long hair moves with speed v in the direction of yarn movement, but after encountering the friction body its speed changes direction, and, therefore, the frictional resistance changes. The geometrical unevenness of a yarn can also play a role in influencing SSP.

As shown in Fig. 4.5(c), at the irregularity, the direction of the normal force is not perpendicular to yarn axis; accordingly there will exist a force component opposite to the direction of movement of the yarn. This will appear as additional force registered as friction.

Adhesion

It is known [5, 6, 9], that some liquids are able to spread and adhere to the bodies, for example water on a hydrophilic material. Waxes and lubrications and finishing agents are applied on fibrous surfaces, which can be expected to affect friction and SSP.

4.2.2 Influence of speed on frictional resistance

Some possible models of the effect of speed on frictional force are shown in Fig. 4.6 [4]. The simplest model of Coulomb, Fig. 4.6(a), may be described by:

$$F_\mu = F_t \cdot \text{sgn}(v) = \mu \cdot F_n \cdot \text{sgn}(v), \tag{4.2}$$

where F_μ is frictional force, F_t is tangential force and F_n is normal load. Sign of the speed determines sign of the force F_μ, value of the expression sgn(v) is either +1 or –1. In Fig. 4.6(b) some viscous friction, linearly dependent on v, is incorporated. Figures 4.6(c) and (d) reflect the idea that static friction is

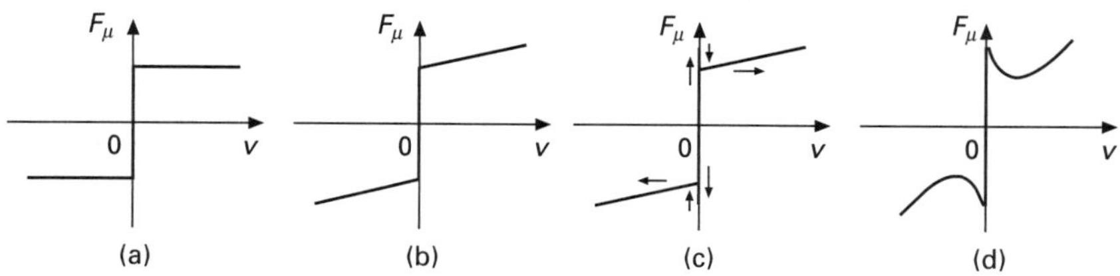

4.6 Relation between speed v and frictional force F_μ [4].

higher than kinetic, in an ideal way in Fig. 4.6(c) and with a more complex continuous function in Fig. 4.6(d).

The movement of a body in the case that enables SSP is shown in Figs 4.7 and 4.8. Spring, 4, of stiffness s, (Fig. 4.7) enables variable speed v_2 in sliding body, 5. Line 1 (Fig. 4.8) corresponds to constant speed v_1 of a drive (for example, of the cross-head of a dynamometer). If the mass m is neglected and the static friction is higher than the kinetic, the movement of sliding body will be "jerky" and described by line 2 (Fig. 4.8(a)). Owing due to the acceleration associated with mass m and the inertia imposed by it and the involvement of visco-elastic rather than the elastic effects, SSP will be suppressed. The result will be that stick-slip will not involve steep bends (Fig. 4.8(b), line 3). Sometimes the influence of inertia may be so high that stick-slip is not observed. A typical example is inclined plane (Fig. 4.17) in which once the frictional resistance is overcome, the slip will not be stopped until a barrier is encountered.

It is expected that low speed enables formation of bonds better than high speed. If the speed is zero, then the longer the dwell time, the greater and the stronger will be the links. Studies show that this is particularly true when the surface has no lubricant (Fig. 4.9) [15, 17]. Friction is known to increase with speed: this may be due to the logical effects, i.e. higher modulus and

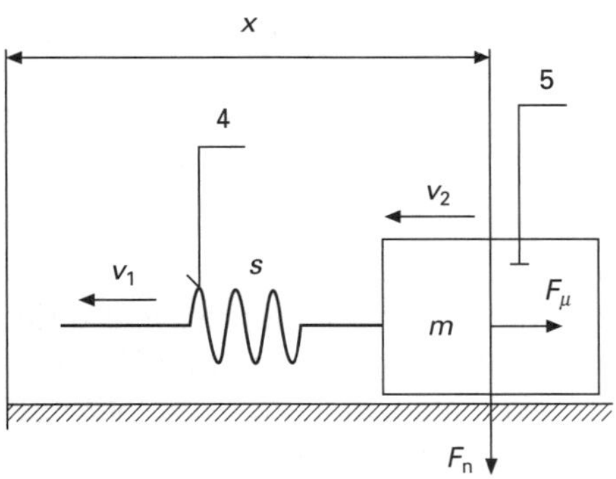

4.7 Schematic arrangement.

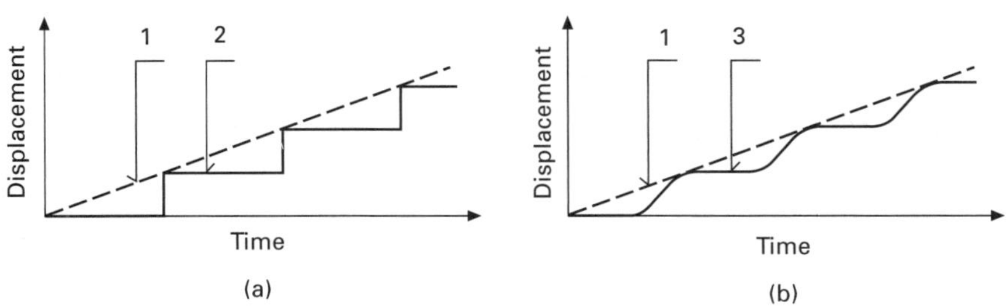

4.8 Movement of slipping body: (a) idealized, (b) real.

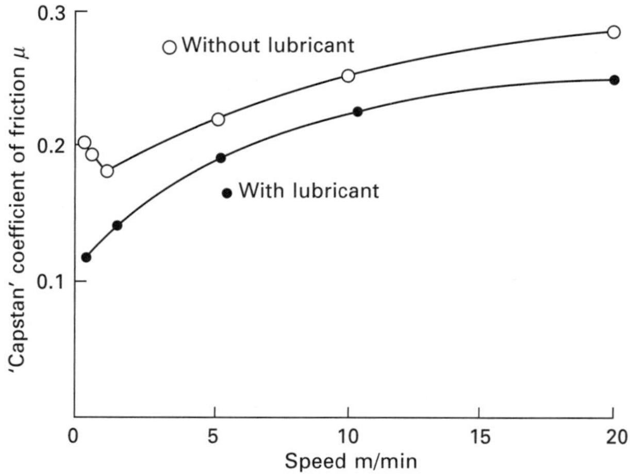

4.9 Influence of yarn speed on friction [15].

bending rigidity in unlubricated materials and increase in viscosity in lubricated systems.

4.2.3 Coulomb and Newtonian (viscous) friction

SSP is mostly observed in friction of solid bodies (Coulomb friction). The resistance to shear or flow of liquids, Newtonian friction, is dependent on speed:

$$\sigma = \eta \cdot v = \eta \cdot \frac{dx}{dt} \qquad 4.3$$

Although viscosity η in equation (4.3) is a variable and influenced by speed, at very low speed, when $v \to 0$ and $\sigma \to 0$, no SSP at nano-level could be expected. Textile fibers, however, are a composite of solid and liquid – they often have a layer of natural or applied finish on the surface – and, therefore, their behavior will lie somewhere in between. Accordingly, SSP may be observed.

4.3 Impact of SSP on textiles

4.3.1 Influence of SSP on fabric deformation

Tensile behavior

Yarn-to-yarn slip is much more prevalent in knitted than in woven fabric deformation. Elongation of knitted fabric is connected with large movement of yarns at the so-called binding (interlocking) points. This is illustrated in Fig. 4.10 [18, 22]. For proper functioning of the fabric it is important that a

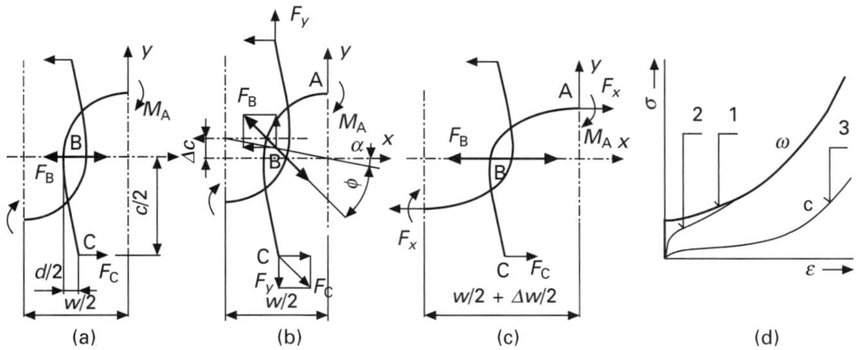

4.10 Model of knitted structural element deformation [22].

force, F_B, of a particular value exists between yarns in contact even in the relaxed fabric. As shown in Fig. 4.10, the following observations can be made:

- Load in the direction of the wales. In this case, the frictional resistance needs to be overcome at the very beginning of the fabric elongation (Fig. 4.10(b)). Model solution leads to a step increase in load σ (Fig. 4.10(d), line 1), the actual stress-strain curve of the fabric shows high initial modulus (line 2).
- Load in the direction of the courses. The same effect is not observed in this case (Fig. 4.10(d), line 3); the initial fabric elongation is enabled strictly by a change in the yarn axes geometry.

Many woven structures, plain weave in particular, have symmetrical orthogonal constructions. The result is, that fabric elongation is associated with low or zero yarn-to-yarn slip, but for any deformation to occur in yarns, slip among the fibers will be required, therefore SSP could be expected [10].

Behavior involving other forms of deformation

SSP could also occur during bending and shearing of fabrics.

Generally, the resistance to bending is supported by a combination of tensile and compressive stresses occurring on the opposite sides of the neutral plane of a product. The bending of a knitted or a woven fabric or of a yarn is a complex phenomenon. In textile assemblies, it involves several types of deformation, i.e. tensile, bending, shear, and compressional. Since all of these involve relative motion at the microscopic or the macroscopic level, friction affects the behavior. The "plate" model of a fabric subjected to a couple causing it to bend is shown in Fig. 4.11 [1, 11, 12].

Lateral stresses cause deformation of the cross-section of a textile product. This form of deformation is complex but important and is often overlooked.

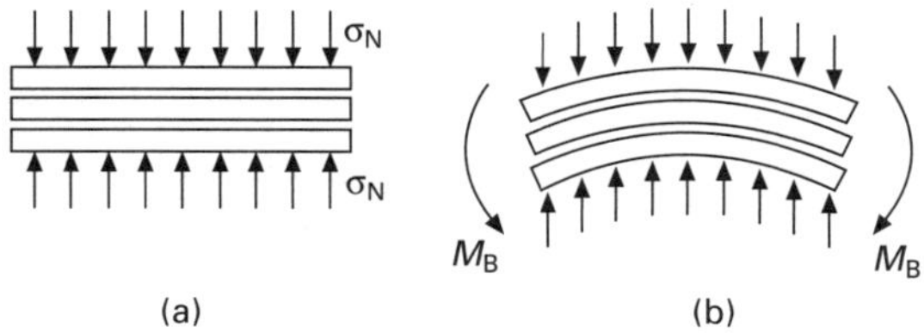

4.11 Friction model of fabric bending deformation.

It is intimately connected with fiber-to-fiber slippage and, therefore, with friction.

Initial shear deformation of a woven fabric is enabled by yarn-to-yarn rotation at the binding elements. Accordingly, friction plays a role. The prediction of such frictional resistance, however, is difficult as the shape of the contact area and distribution of tension in it are usually unknown. The shear deformation of a knitted fabric is accompanied by the yarn-to-yarn slippage in the loops, which will also depend on friction. These issues could represent interesting subjects for future research activities, particularly with the availability of today's computing power and modeling programs.

4.3.2 Influence of internal friction

Deformation of a yarn or a fabric is associated with fiber-to-fiber slip in the product. This is a form of internal friction that plays an important role in determining textiles mechanical properties and in causing hysteresis and converting mechanical into thermal energy.

To examine the nature of SSP phenomena found during tensile tests in a textile structure, a single-faced, weft-knitted fabric made from model yarn (polyamide monofilament Teflon®-coated, 0.15 mm diameter) was tested.[1] To prevent problem associated with curling, a tubular structure was used. The width of the sample was 10 wales (5 face and 5 back), length in the jaws was 50 mm (greater width and length would have suppressed SSP, as stick and slip between yarns at different binding points do not occur simultaneously). Stitch length l_s was 7.4 mm. The fabric was mounted on a dynamometer under zero load (i.e. without pretension). One cycle of deformation in the wales direction is shown in Fig. 4.12(a). As noted earlier (Fig. 4.10(b), (d)), the beginning of the curve is steep. SSP, represented by the waves on the curve, could be observed from the beginning of elongation. Detailed behavior of small portions of the curves are illustrated in Figs 4.12(b) and (c). Hysteresis is very high. For comparison, the behavior of a commercial fabric of similar

[1] Most of the results included in this chapter are based on extensive experiments conducted at the University of Liberec, Czech Republic.

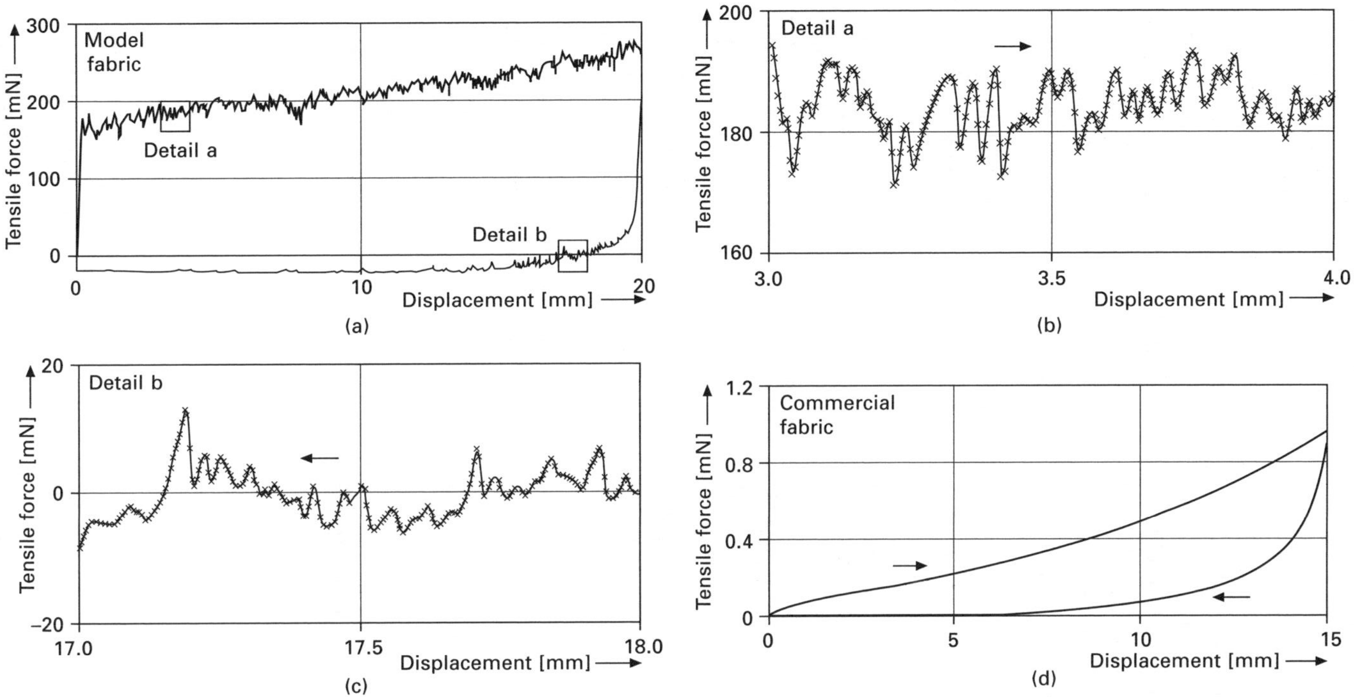

4.12 Deformation of model and real commercial fabrics (wales direction).

construction, but made with spun yarn (polyacrylic double twisted yarn 35 tex × 2), is shown in Fig. 4.12 (d). SSP is suppressed and increase of initial modulus is lower; the reasons are given later (see Section 4.3.3).

The next example shows the behavior of the same fabric but subjected to four cycles of loading and unloading with elongation varying from 0–2.5–0–5–0–10–0–20–0 mm. Test length was 50 mm and, therefore, extensions were 5, 10, 20 and 40%. The same yarn (polyamide monofilament) and tubular structure were used with the stitch length l_s of 6.8 mm. The shortest cycle (elongation 2.5 mm, Fig. 4.13(a)) shows steep curves with high initial modulus and some negative load during the return cycle. The latter is due to the use of a small pretension at the beginning of the test and the plastic deformation of the fabric.

It is due to this slack, that the beginning of the second cycle (5 mm) is not that steep. However, after reaching elongation of 2.5 mm, the load almost

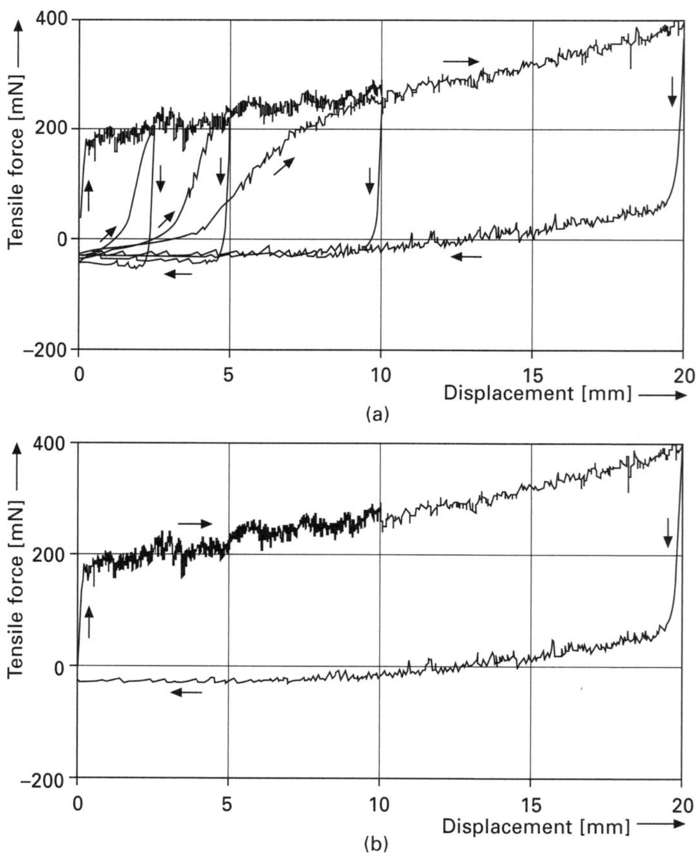

4.13 (a) Four cycles of deformation, model fabric, wales direction; (b) Reconstructed 4th cycle.

reached the magnitude found at the end of the first cycle. A similar situation is noted in the next two cycles. Figure 4.13 (b) shows reconstructed 4th cycle that includes the corresponding loading portions of the first three cycles. The different appearance of the original portion of the fourth cycle (10–20 mm) is caused by lower frequency of data acquisition: the first three cycles were set at 100 data points per mm while the fourth was set at 40 mm^{-1} to meet the limit of data points per experiment.

Figure 4.14 shows the behavior of the same fabric tested in the course direction. As expected, based on earlier discussion (Fig. 4.10 (c), (d)), SSP is not seen at the beginning of deformation but appears after about 7 mm elongation (fluctuations on the curve). The hysteresis is lower due primarily to the lesser role of yarn-to-yarn friction, but increases in later cycles when yarn-to-yarn slip starts to occur more extensively.

Four cycles of deformation of the commercial fabric mentioned earlier (see Fig. 4.12(d)) is shown in Fig. 4.15. SSP is suppressed as generally expected due to averaging effects. More on this is discussed later.

Table 4.1 shows the values of relative hysteresis in various cycles of knitted fabric deformation. In agreement with the model, introduced in Fig. 4.10,

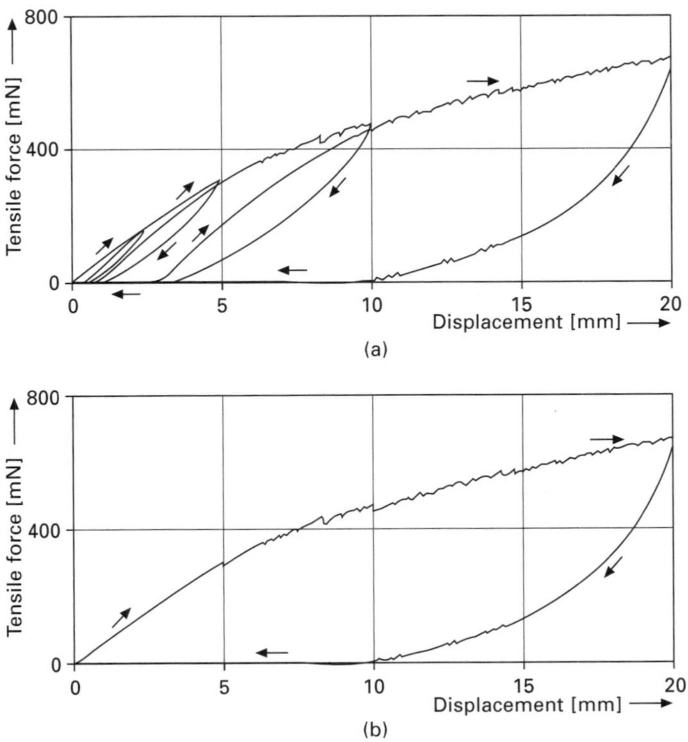

4.14 (a) Four cycles of deformation, model fabric, courses direction; (b) Reconstructed 4th cycle.

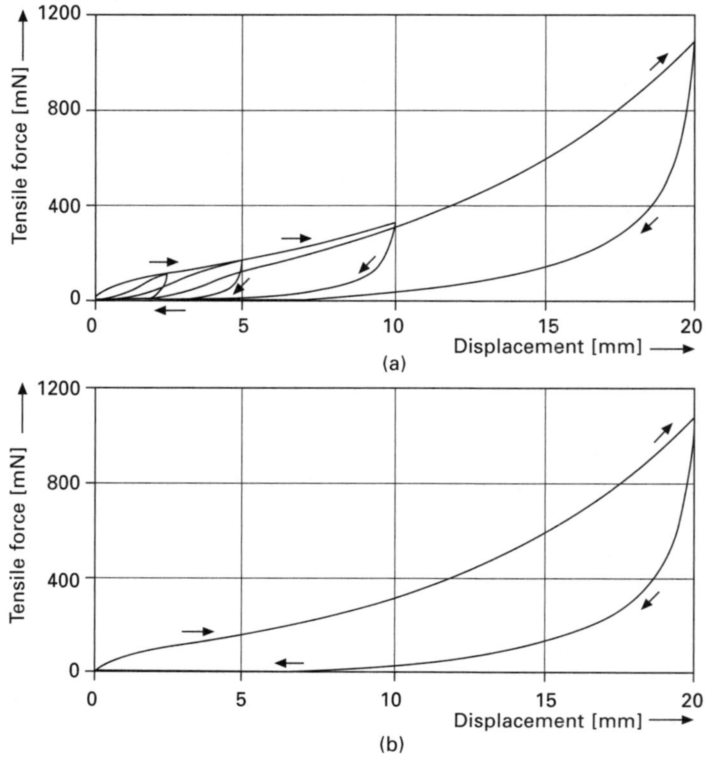

4.15 (a) Four cycles of deformation, real fabric, wales direction; (b) Reconstructed 4th cycle.

Table 4.1 Hysteresis of knitted fabrics cyclic elongation

Figure	Load direction	Yarn	Hysteresis in cycle [%]			
			0–2.5 mm	0–5 mm	0–10 mm	0–20 mm
4.13	Wales	Polyamide	128.2	118.1	111.5	99.9
4.14	Courses	Polyamide	37.8	40.1	58.8	78.2
4.15	Wales	Acrylic 35 tex × 2	98.9	93.3	83.9	73.4

the hysteresis involving deformation in wales direction is high and some values exceed 100% even when the material used is elastic. Hysteresis of the same fabric in the courses direction is lower and shows substantial increase with elongation of the cycle; this, as mentioned earlier, is because of the slip between yarns occurring later at higher elongation.

4.3.3 SSP and the hierarchy of textile structures

The behavior of textile yarns and fabrics when they contact each other is different than the behavior of solid bodies [22]. When a one-dimensional or a flat textile product is placed over a frictional surface (steel, ceramic, or another textile product) contact is made by fibers over several small regions. When a tangential force is applied, the fibers may shift or deform even before frictional resistance is overcome. Following this, after some change in position, the slip will start to occur at contact points; however, this will not happen abruptly in one step but will be gradual and spread over a series of steps. The result will be that the higher the textile structure is in hierarchy – from chain molecule (lowest), to fiber, to yarn, to fabric (highest) – the lower will be the presence of SSP in friction profile. Often SSP at nano- and micro-level becomes masked by SSP caused by structural unevenness at macro-level.

4.3.4 Influence of yarn and fabric pre-tension [22]

SSP may occur only if there is some normal force F_n that insures contact of two bodies. Examined here is the origin of such forces in textile structures.

In a spun yarn containing twist, tensile stress will give rise to lateral stresses. There are numerous points of contact in such yarns; the existence of normal forces at these points is important for yarn strength. A necessary condition for existence of normal force in a relaxed yarn is residual elastic deformation of fibers.

Additional such force in a woven or knitted fabric will arise due to change in yarn axes geometry and deformation in yarn cross-section. These will cause development of normal forces at yarn binding points. This indicates that fabric deformation of almost any form will be influenced by both the fiber-to-fiber and the yarn-to-yarn frictional resistances.

These effects in a fabric lead to an apparent paradox in tensile behavior: the higher the elasticity of fibers, the higher the pre-existing forces in structure. Due to this, the frictional resistance is high and, therefore, the loss of mechanical energy during deformation is also high. Accordingly, higher elasticity of fibers could be a reason for lower elasticity or elastic recovery of a textile product given large deformation. On the other hand, the only source of yarn and fabric elasticity is the elasticity of fibers. One can conclude that the optimum contribution by fiber elasticity to yarn and fabric elasticity is made at low extension, i.e. the extension at which slip does not occur at binding points.

4.3.5 Anisotropy of friction

Friction and SSP could be expected to have an anisotropic character. The roots are embedded in the anisotropy of the structure. At the fiber level, preferred orientation of molecules on the surface will contribute to anisotropy.

At the yarn level anisotropy is present in the structure; i.e. contributed by preferential alignment of fibers parallel to yarn axis. Therefore, the angle between two yarns in contact will influence frictional resistance. At the fabric level, yarns are preferentially looped or interlaced. In a woven fabric, yarns are arranged orthogonally; this means it is anisotropic. In a knitted fabric, although the orientation of yarn axis is somewhat more random, preferential orientation exists in a unit cell and, therefore, in the fabric.

4.4 Review of experimental principles for measuring SSP

4.4.1 General description

From the standpoint of planning experiments for the purpose of exploring the nature of SSP in textiles, one should note that there are two sources that lead to the phenomenon:

1. Periodic changes in profile of a tested material (fiber, yarn and fabric, SSP at nano-, micro- and macro-levels). These changes are material related and of greatest interest.
2. Periodic changes resulting from the testing method. It is very difficult to avoid these superimposed but undesirable variations on material-related fluctuations. Contributing in this regard are: dynamics of the measuring system (dynamic forces due to periodically moving mass connected with an elastic elements of the system) that is described later by the general equation (4.6), and imperfections of the apparatus in terms of the variation in the frequency of step motor drive, frequency of measuring head, etc.

Some examples of experimental set up are shown in Fig. 4.16. If speed v is maintained exactly constant (Fig. 4.16(a)), SSP may not appear. The mass of box 1 can also suppress SSP due to inertia. Another situation will be if only average speed v is maintained but the speed at the friction points can change due to the elastic nature of the spring of stiffness s_1 (Fig. 4.16(b)). Additional support for change in speed will occur if the mass m_2 is low and if the normal force F_n is imposed by an elastic element (stiffness s_2).

The impact of the acceleration of mass m may be reduced, if we change the primary sliding body from the mass to a belt. In Figs 4.16(c) and (d), the lower frictional surface belt 2 moves and the box 1 or 4 is relatively stable (some movement back and forth will still occur due to SSP).

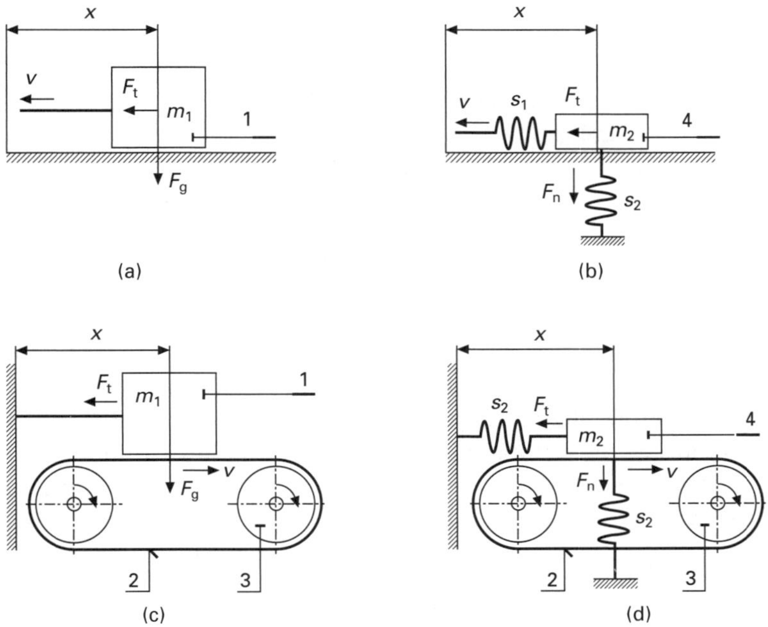

4.16 Principles involved in the measurement of sliding friction.

Any experimental set up used should simulate the performance of textiles

- during technological processes, or
- during end use.

However, as alluded to above, it is also necessary to reflect on the limits imposed by the experimental techniques involved in characterizing physical properties. For example, experiments on INSTRON 4411 (TU of Liberec, CZ) are influenced by the constant associated with the acceleration of the measuring head, by the traverse of the jaws (these influence the measured force), and by the existing limit of the frequency of the data collection points. For example, data acquisition capacity of 20 per second may not be sufficient when the speed is high.

Often, one of the techniques used for measuring friction is inclined plane (2 in Fig. 4.17). This apparatus is simple, but limited in usefulness, because only static values of friction can be measured. Angle φ is gradually increased and its value when frictional resistance is overcome is registered. The mass m_1 of sliding body 1 creates not only gravity force F_g but also dynamic force if $F_{ts} > F_{tk}$. (static and kinetic tangential frictional forces, respectively). However, inertia of the mass will usually not allow quick changes of the speed and the recording of SSP.

A similar principle is sometimes used in measuring friction in a textile yarn (4 in Fig. 4.17(b)). The angle φ of frictional surface 3 (for example,

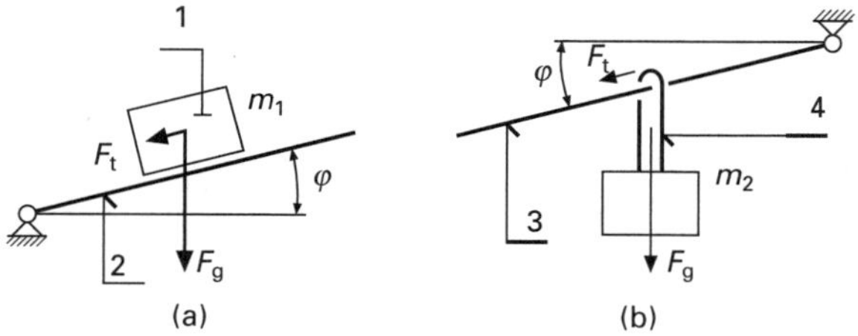

4.17 Measuring of friction using inclined plane principle.

stalk of a knitting needle) is increased till friction is overcome. In addition to friction, this result is also influenced by yarn bending properties: the greater the bending rigidity, one can expect, the lower the angle of contact and, therefore, the lower the angle φ for the same coefficient of friction.

From the discussion thus far it is clear that apart from material factors, there are many experiment-related factors that may be considered when designing experiments to study the nature of SSP. These are listed below:

- system used to impose/measure input tension
- system used to impose/measure normal force
- system used to impose/measure output tension
- nature of dynamic forces
- test configuration.

The effects of these factors are examined by considering hypothetical systems below. This is followed by the discussion of results of actual experiments conducted on textile materials involving many of these factors and concepts.

4.4.2 Textile-to-non-textile body friction measurement

Longitudinal textiles

Friction of fiber or yarn to metal, ceramic or another surface is frequently encountered in textile technology. Friction set up involving contact between a yarn (1) and a knitting needle hook (2) is illustrated in Fig. 4.18. Input yarn tensile force F_0 is generated by the weight of mass m_0. Total angle of contact is π. The output tension is measured by sensor 3 (for example, by INSTRON dynamometer).

Three speeds are important:

1 Speed of the drive (dynamometer crosshead) v_d. This is independently set and is often an experimental constant.
2 Slipping speed v_s, the relative speed of bodies at frictional points.
3 Speed of the weight v_w creating pretension.

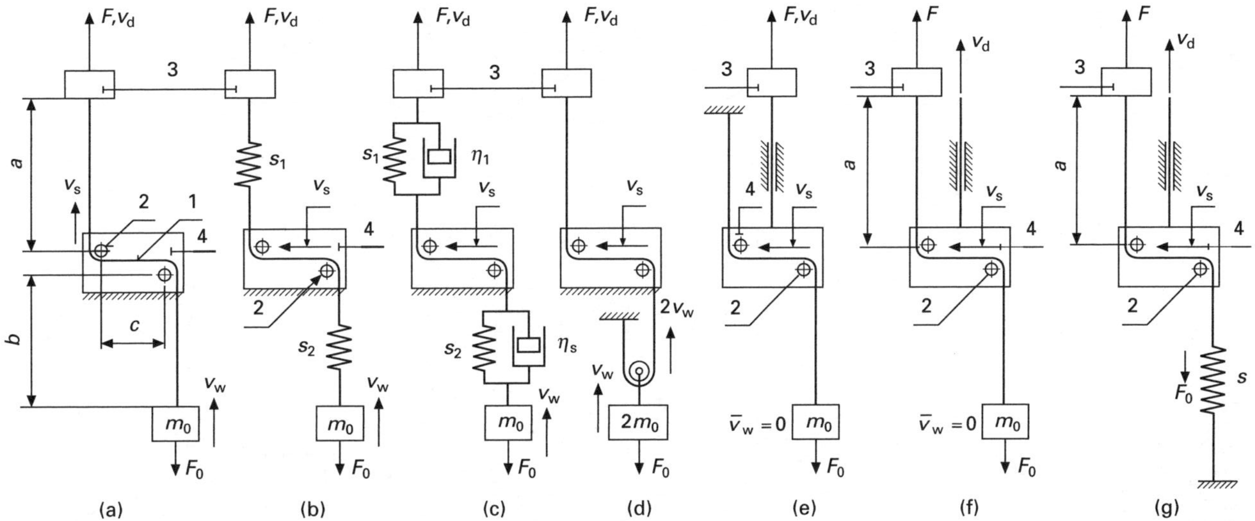

4.18 The influence of yarn deformation and inertia on SSP.

SSP could be observed only if speed of the drive is different from the sliding speed: $v_s \neq v_d$. If these are equal, the slippage will be continuous and the static friction coefficient will not be recorded, nevertheless the change of frictional resistance caused by the yarn unevenness could be observed. To enable change of sliding speed, some deformability between jaws and places of friction is necessary. It may be provided by the length of yarn *a* (influencing elongation, Fig. 4.18(a)), or by addition of spring of stiffness s_1 (Fig. 4.18(b)) or a rheological body (Fig. 4.18(c)) of parameters s_1, η_1. Arrangement (Fig. 4.18(d)) may be used to eliminate the possibility of a change in yarn twist, which may be important when measuring friction in single spun yarn.

Stick-slip effect may be influenced as well by the inertia of mass m_0. When SSP is present, slipping speed v_s varies and oscillates around an average value of the drive speed v_d. If speed v_w of mass is not constant, acceleration of m_0 will change pretension F_0 that will also not then be a constant. This is the reason why yarn lengths *b* and *c* (Fig. 4.18(a)) or elastic or visco-elastic behavior of the body represented by the parameters s_2 and η_2 (Figs 4.18(b) and (c)) influence the results of experiments.

To evaluate the effect of the yarn input axial force F_0 better, experiments with stable weight and portable plate 4 with frictional bodies 2 were designed, Figs 4.18(e), (f) and (g). In the first case (Fig. 4.18(e)) the frictional resistance is measured on plate 4; in the next examples (Figs 4.18(f) and (g)) it is measured on yarn (as in the cases Figs 4.18(a–d). Nevertheless, some impact of mass acceleration remains, caused by weight oscillation up and down due to SSP. It could be partly overcome by the last arrangement (Fig. 4.18(g)), where mass is replaced by a spring of stiffness *s*, or by a rheological body (suitable is a piece of rubber yarn of low inertia).

Different principles, which use yarn 1 tensile forces equilibrium on movable friction point 2, are shown in Fig. 4.19. The yarn is led over miniature rollers

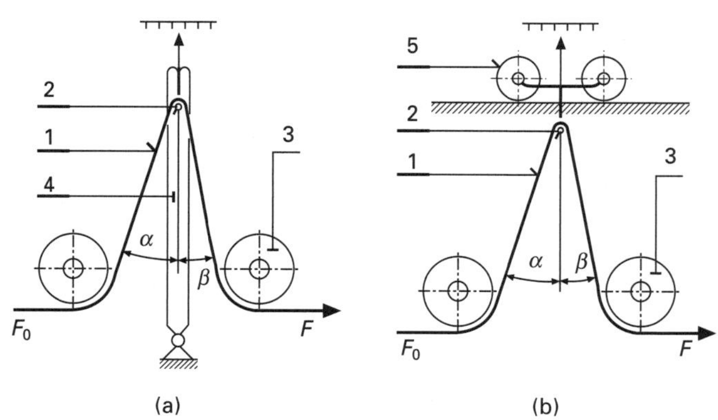

4.19 Measurement of friction using force equilibrium [22].

3 of negligible frictional resistance and low inertia. From the force equilibrium at friction body 2, one gets the following:

$$F_0 \cdot \sin\alpha = F \cdot \sin\beta\text{; and the ratio, } \frac{F}{F_0} = \frac{\sin\alpha}{\sin\beta} \qquad 4.4$$

Angles α and β are variable: the higher the frictional resistance, the greater the difference between these angles. As the inertia of the body 4 or 5 is high, these settings are usually not suited for measuring rapid changes. SSP could be measured only at extremely low speeds.

Flat textiles (fabric to another surface)

The widely practiced experimental method for fabric to another surface friction measurement is shown in Figure 4.20(a). In this case, the normal force F_g is the gravitational force due to mass m_1, and the sensor 4 (jaws of dynamometer) records tensile force of the rope 5. Frictional resistance of the roller 3 should be negligible. The use of the second mass m_2 is not necessary. However, if used, it helps to stabilize body 2, but the force due to gravity on m_2 must then be subtracted from the measured force F_t. Figure 4.20(b) uses the inclined plane method. Coefficient of friction is defined as $\mu = \tan\varphi$. The value of only static friction is measured. SSP is either not registered or greatly dampened by the inertia of mass m.

There are other possibilities as well. For example, the experiment described later in Fig. 4.26 for yarn-to-yarn friction measurement could be easily adapted to measure fabric friction. The results of all these experiments are strongly dependent on the nature of friction surface (material, unevenness, contamination by dust, etc.). Effectively comparing results obtained from different apparatus is usually difficult on account of the variables associated.

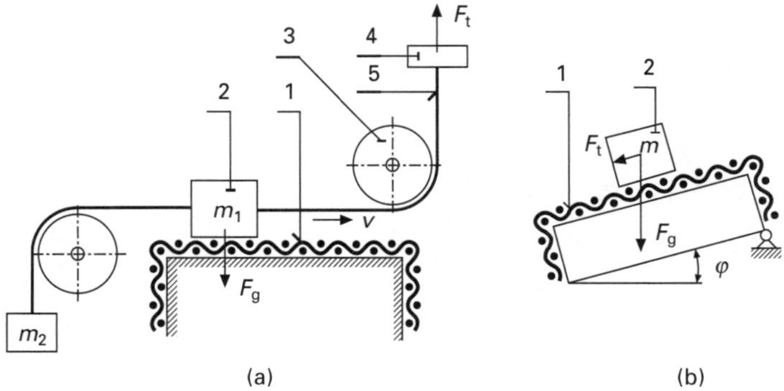

4.20 Non-textile body to fabric friction. Sled set up (a) and inclined plain (b).

4.4.3 Textile-to-textile friction measurement

Longitudinal textiles (fiber-to-fiber or yarn-to-yarn friction)

Figure 4.21 shows photographs of yarn-to-metal (a) and yarn-to-yarn (b) friction measuring set-ups. The yarn 1 is led over knitting needles hooks 2 or over rollers 6 to create friction contacts. In these figures, 3 is miniature clamp, 4 is sensor (capacity ±1 N) and 5 is weight.

The possibilities of using simulated knitting binding point 2 on yarn 1 are illustrated in Fig. 4.22 [21, 22]. Miniature rollers 3 of low friction, mounted on plate 5, are used to guide the yarn. Input tension F_0 is insured by mass m_0. In this case, the deformability of yarn between the friction point 2 and the sensor 4 has little influence on SSP, as the contact 2 can move down and up. Acceleration of mass m_0 could influence the results in case Fig. 4.22(a), not so much in case Fig. 4.22(b), as now the mass is in a relatively stable position. Used in Fig. 4.22(c) is full twist between yarns and so two weights are necessary. Now acceleration associated with each will influence results.

The system in Fig. 4.23 [15] measures friction between two stretched fibers or yarns that are kept under tension and in restraints, and at right angles to each other. Fig. 4.24 illustrates the impact of yarn twist on friction and SSP. Combination of S and S or Z and Z twist directions (Fig. 4.24(a)) will result in approximately parallel surface fibers and, therefore, in higher

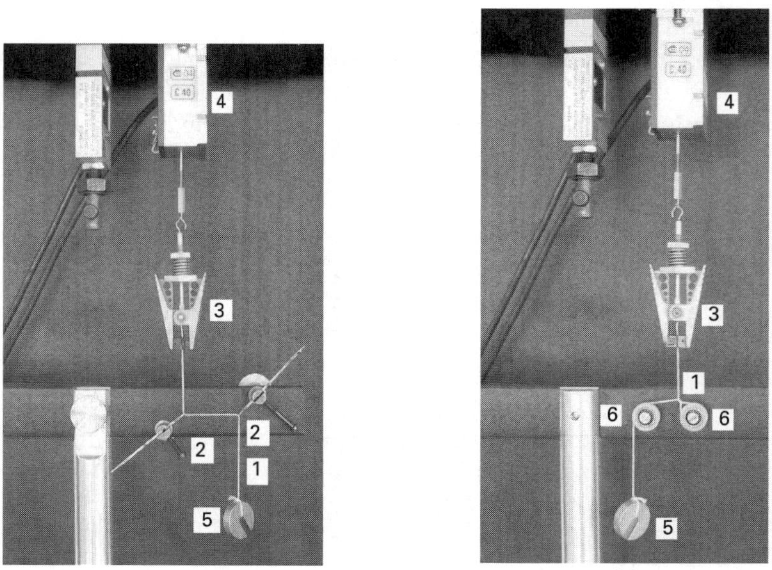

(a) (b)

4.21 Photographs of yarn-to-metal (a) and yarn-to-yarn (b) friction measurement set-ups.

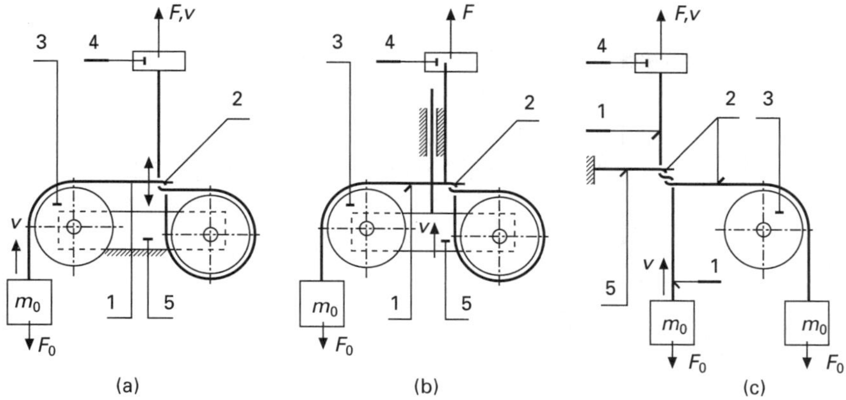

4.22 The yarn-to-yarn friction measurement.

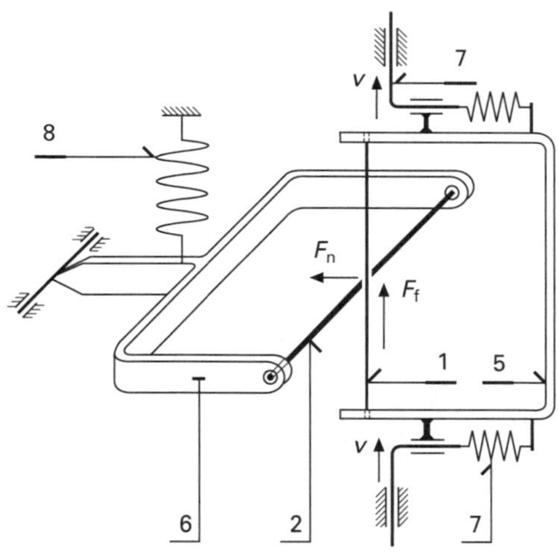

4.23 The experiments on stretched yarns or fibers.

SSP. Combination of S and Z twist (Fig. 4.24(b)) will lead to approximately perpendicular contacts between fibers and, therefore, to lower frictional resistance.

Other possibilities [8], in which twist is used to create frictional contacts between yarns or fibers, are shown in Fig. 4.25. Multiple twists between two yarns are used. Figure 4.25(a) shows asymmetrical set up in which yarn 1 moves with speed v. Figure 4.25(b) introduces similar but a symmetrical set up where both yarns move.

Another example of measuring yarn-to-yarn friction is shown in Fig. 4.26 [31, 32]. The sliding yarn 1 is led over fixed yarns 2. Pretension F_0 is imposed by mass m_0 and the output tension F is measured by sensor 4. The fixed yarns 2 are clamped on rollers 3. There are several possibilities. They could be wound parallel to the rollers' axes or at some angle. The rollers 3

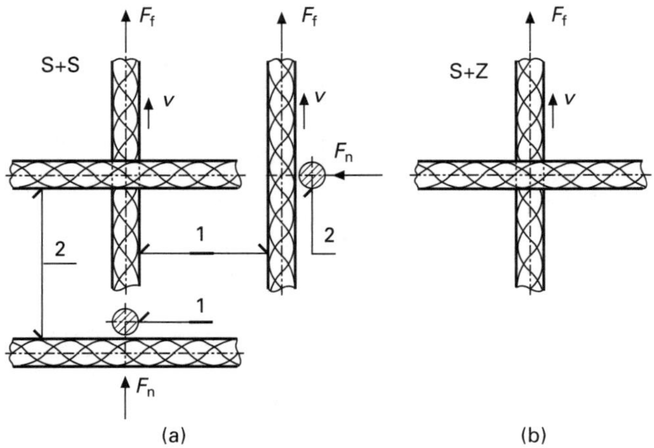

4.24 The impact of the yarn twist direction on SSP.

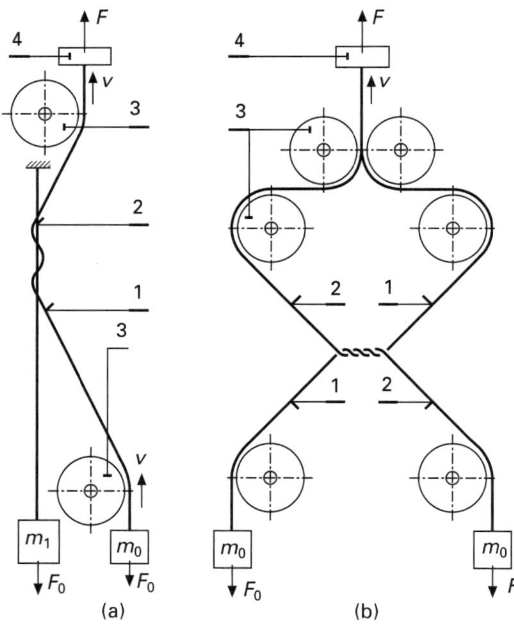

4.25 The experiments involving twist between yarns or fibers.

could be fixed (suitable for measuring SSP), free revolving, or positively driven with yarn 1 held fixed ($\overline{v} = 0$).

Flat textiles [2, 22, 24]

Principles of measuring fabric-to-fabric friction are similar to those of measuring fabric-to-metal friction. Measuring frictional force of fabric 1 on

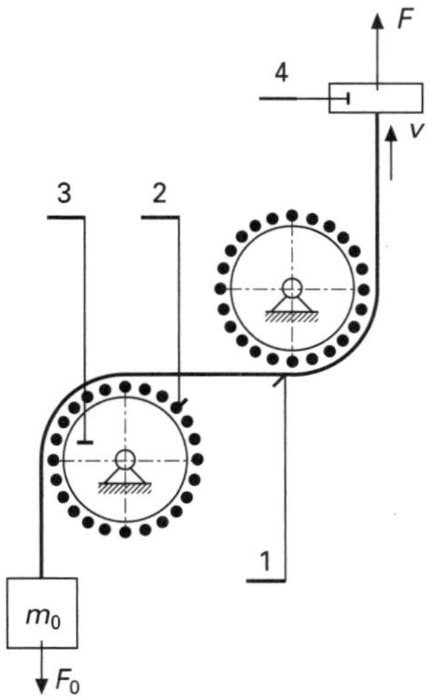

4.26 Measurement of friction with a yarn passing over rollers.

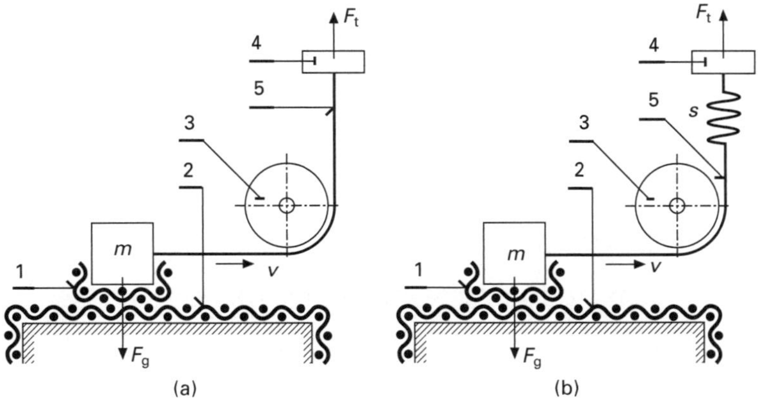

4.27 The fabric-to-fabric friction by the sled method.

2 by dynamometer sensor 4 using sled method is shown in Fig. 4.27, and measuring friction by the inclined plane method is shown in Fig. 4.28.

4.4.4 Calculation of the friction parameters from experimental results

In the cases in which the normal force is independently imposed (examples in Figs 4.16, 4.20 and 4.27) the coefficient of friction is described as $\mu = F_t/F_g$ or $\mu = F_t/F_n$. In the other cases, the capstan (Euler's) equation, $F = F_0 \cdot e^{\alpha \cdot \mu}$, is used. The value obtained is generally not accurate, as the influences of yarn or fabric deformation and of internal friction are not considered. One

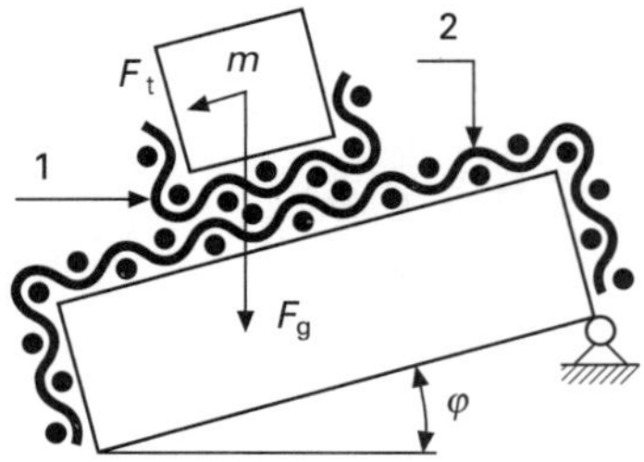

4.28 The fabric-to-fabric friction by the inclined plane method (μ = tan φ).

of the assumptions that lead to capstan equation is that the material is ideally flexible. Textiles fulfill this assumption only approximately and at relatively high stress imposed in the friction direction. The capstan formula also neglects frictional resistance that is not dependent on lateral stress (adhesion) and sometimes gives incorrect results. For these reasons, the value of static and of kinetic coefficients of friction may be considered only as approximate values. The lower the input tension, the greater are the errors. For example, even if $F_0 = 0$, output tension F would have some value due to influence of mass acceleration and adhesion. Using the capstan equation, one will get the unacceptable value of ∞ for μ.

For accurately characterizing SSP, it is necessary that one take into consideration acceleration of all masses that influence the speed of motion.

4.5 Experimental results

4.5.1 Textile-to-non-textile body friction

The influence of the elasticity of a model yarn, namely polyamide monofilament of 0.15 mm diameter, between friction point and measuring head is shown in Figs 4.29–4.31. Experiments were conducted in accordance with Fig. 4.18(a) and (b) on an Instron dynamometer 4411. Capacity of measuring head was 5 N, crosshead speed was 0.5 mm s^{-1}, and the frequency of data input was 20 s^{-1} or 40 mm^{-1}. The procedure used was as follows:

1 A length of yarn 5 mm (displacement from 0 to 5 mm) was pulled; then allowed 10 seconds rest.
2 The same yarn was pulled 5 mm further (displacement from 5 to 10 mm); allowed 10 seconds rest.
3 The yarn was returned 5 mm (displacement from 10 to 5 mm); allowed 10 seconds rest.
4 The yarn was returned 5 more mm (displacement from 5 to 0 mm).

Right charts give details of results. In the second half of experiment (return, left arrow), the roles of F and F_0 were reversed (gravity force of m_0

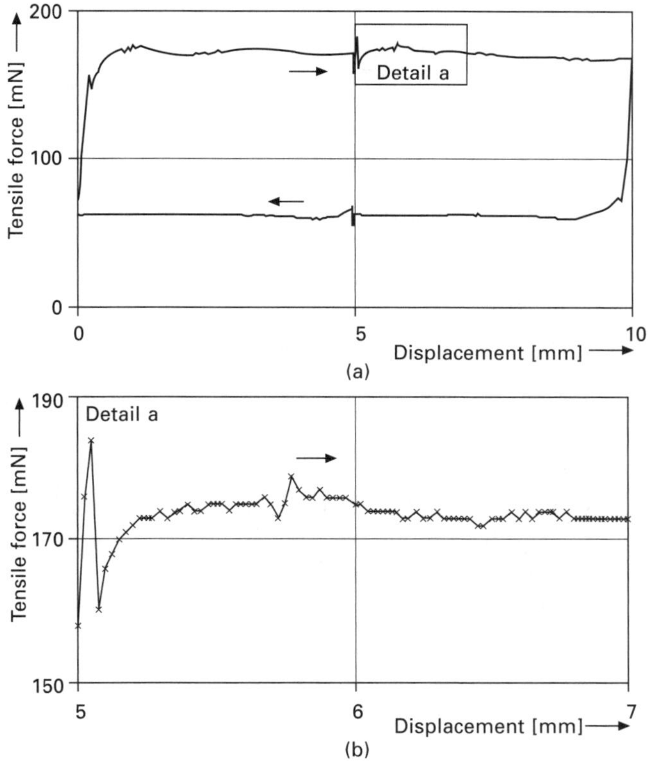

4.29 Polyamide 0.15 mm to knitting needle slip, yarn length "*a*" = 40 mm.

became the output force, while the input force *F* was measured). In the first case (Fig. 4.29) the yarn length between the jaws and the friction point *a* was 40 mm; however, during the course of the experiment, it increased to 50 mm. Elongation of such a short length of yarn was low and so the speed of yarn slip was nearly the same as that of the jaws and, therefore, SSP could not be observed except at the beginning of the traverse, i.e. at displacements 0, 5, 10 and 5 mm. When the machine was stopped at the displacement of 5 mm for 10 seconds and then restarted, the frictional force increased to what appears to be the static value. Acceleration of the mass of the jaws and weight could play some role as well. Detail "*a*" reflects insufficient sensitivity of measuring head, which was able to sense fluctuations of the order only of 1 mN.

The second example (Fig. 4.30) shows the influence of yarn deformation when the length a = 500 mm. Measured force change was quite regular with an average frequency of 5.6 and 9.0 mm^{-1} (first and second half of experiment, respectively) and amplitude of 20 and 10 mN (relatively 11 or 25%, respectively). The fall in yarn tensile force *F* was steep, due to high material

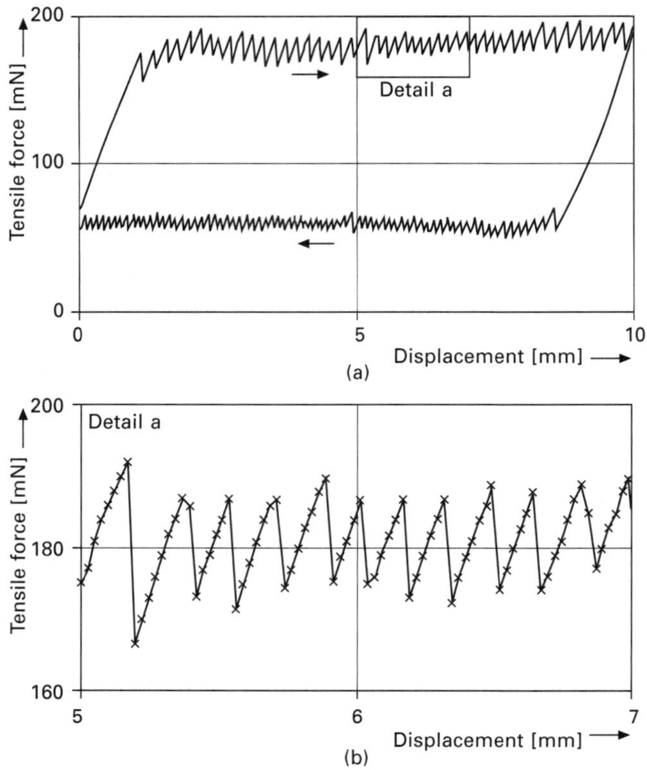

4.30 Polyamide 0.15 mm to knitting needle slip, *a* = 500 mm.

elasticity (see rheological model in Fig. 4.55(a)). Unfortunately the speed of the yarn slippage could not be measured.

In the second half of the test, the steep fall of the force F is replaced by almost equivalent steep increase. It is caused by the exchange of input and output forces. After the static friction is overcome, the mass m_0 (Fig. 4.18) falls quickly involving acceleration due to free fall till new equilibrium is reached with kinetic friction.

Figure 4.31 describes a similar experiment with the incorporation of an additional spring of stiffness $s_1 = 180\ \text{N m}^{-1}$ (Fig. 4.18(b)). This led to lower frequency and higher amplitude values of oscillation.

Table 4.2 shows influence of the length a on average values of SSP amplitudes (average differences between maximum and minimum values) and wavelengths.

As the sensitivity of the 5 N load cell and the data input frequency of 20 s^{-1} proved insufficient for SSP description, dynamometer LabTest 2.050 (Labortech, CZ) with a capacity of the sensor of ±1 N and the data input frequency of 50 s^{-1} was used in the next set of experiments [23].

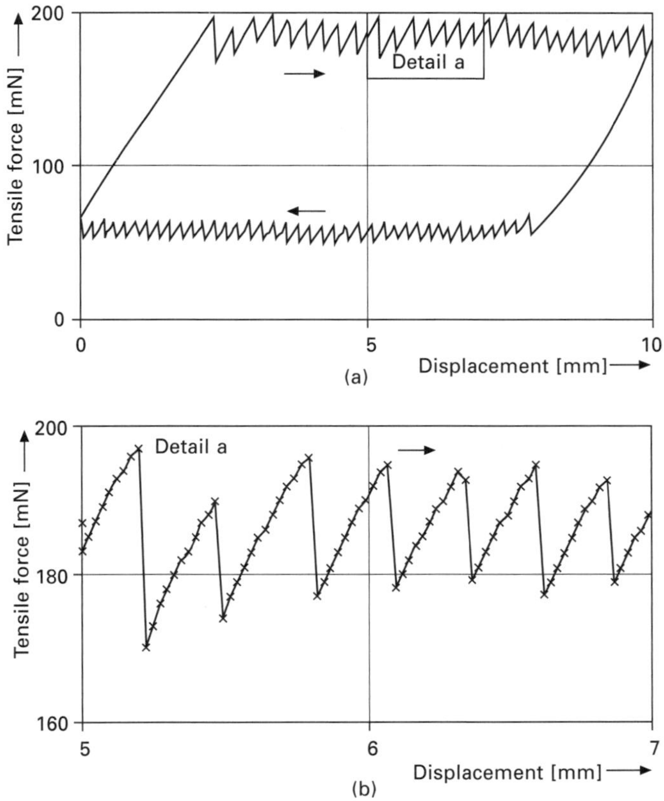

4.31 Polyamide 0.15 mm to needle slip, a = 500 mm + spring of s_1 = 180 N m^{-1}

Table 4.2 Average values of SSP amplitude and wavelength

Length a [mm]	1st half of the test		2nd half of the test	
	Amplitude [mN]	Wavelength [mN]	Amplitude [mN]	Wavelength [mN]
40	approx. 1	irregular	approx. 0.5	irregular
500	17	0.179	9	0.116
500 + spring	18	0.263	11	0.200

Examples of the results obtained from experiments on spun yarns, given in Figs 4.32 and 4.33, show the effect of material on SSP. Other conditions of the experiments were the same as described above (Fig. 4.31). The basic frequency value of tensile force oscillation of plied yarn (Fig. 4.32) corresponds with yarn twist value (twist was 380 m^{-1}, average length of one half-twist was 1.32 mm); SSP was supported mainly when a rubber spring of length l_R

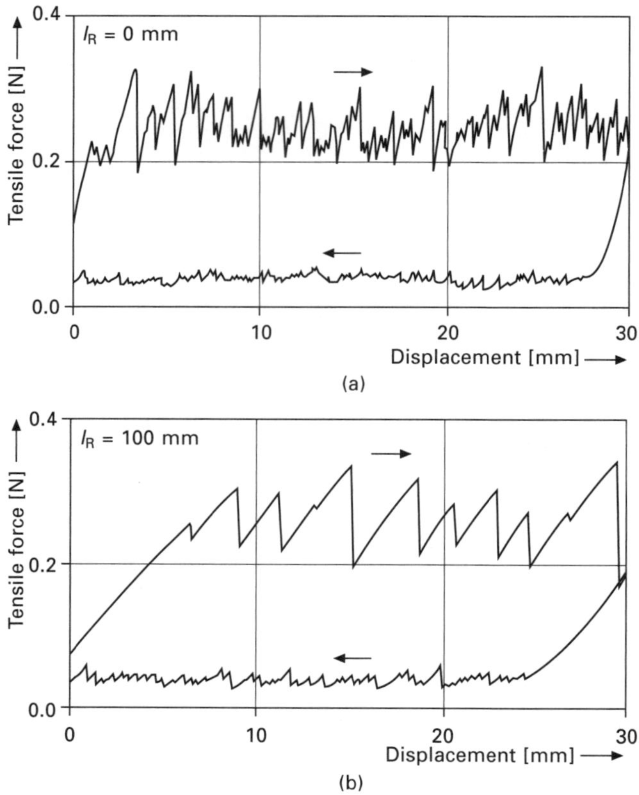

4.32 Spun yarn (acrylic, 35 tex × 2).

= 100 mm (b) was incorporated into the yarn length a (Fig. 4.18(a), (b)). The total length of part a was 200 mm.

The instability of frictional force, caused by the change in yarn length a in the course of the experiment, was investigated by conducting a "long duration" experiment, Fig. 4.34. The set up of it corresponds with Fig. 4.18(f). The details of the results obtained are shown in Fig. 4.35. Material used was polyamide monofilament of 0.15 mm diameter "Damyl Ultra Strong", input yarn tensile force F_0 was 100 mN, speed was 0.2 mm s^{-1}, and a was increased from 20 to 200 mm. Data input frequency of 50 s^{-1} means, that 250 points per each millimeter of yarn were registered.

As alluded to earlier, the length of yarn 'a' is important for registering SSP. With increase in a, the amplitude of frictional force oscillation becomes higher and the frequency of it lower. At the end of the experiment, the change of measured tensile force is composed of two parts – slow increase and almost steep fall (Fig. 4.35(f), (h)). This behavior corresponds with that of model considered later in Fig. 4.55(a), in which the material used behaves

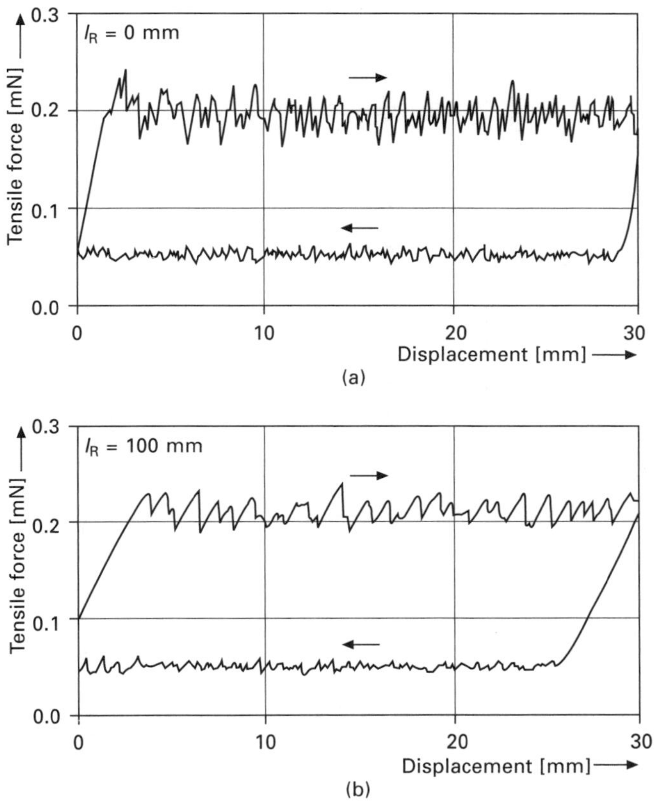

4.33 Spun yarn (cotton, 35 tex).

as an elastic body. The steep fall not being ideal is caused by yarn acceleration and by the viscous component of yarn deformation.

The influence of input yarn tension on SSP in a similar experiment (Figure 4.18(f)) is shown in Fig. 4.36. The scheme of measurement was 0–30–0 mm (displacement from 0 to 30 mm and return) with speed of 1 mm s^{-1}. In this, *a* acted as a spring (200 mm of rubber yarn). The result shows that the lower the input load, the higher the frequency of SSP. Secondary fluctuations, seen in (d) and (f), could not be explained by the properties of the material. Most probably they are connected with the internal vibration of the measuring system.

The impact of the elastic deformation of part *a* (Fig. 4.18) is illustrated in Fig. 4.37. The experimental set up was the same as in the previous case except that the length of the rubber spring l_R was varied. The input load in all cases was 100 mN and speed was 1 mm s^{-1}. In agreement with the model, briefly described later in Section 4.6.2, the higher the elastic extension of part *a*, the lower is the frequency of SSP.

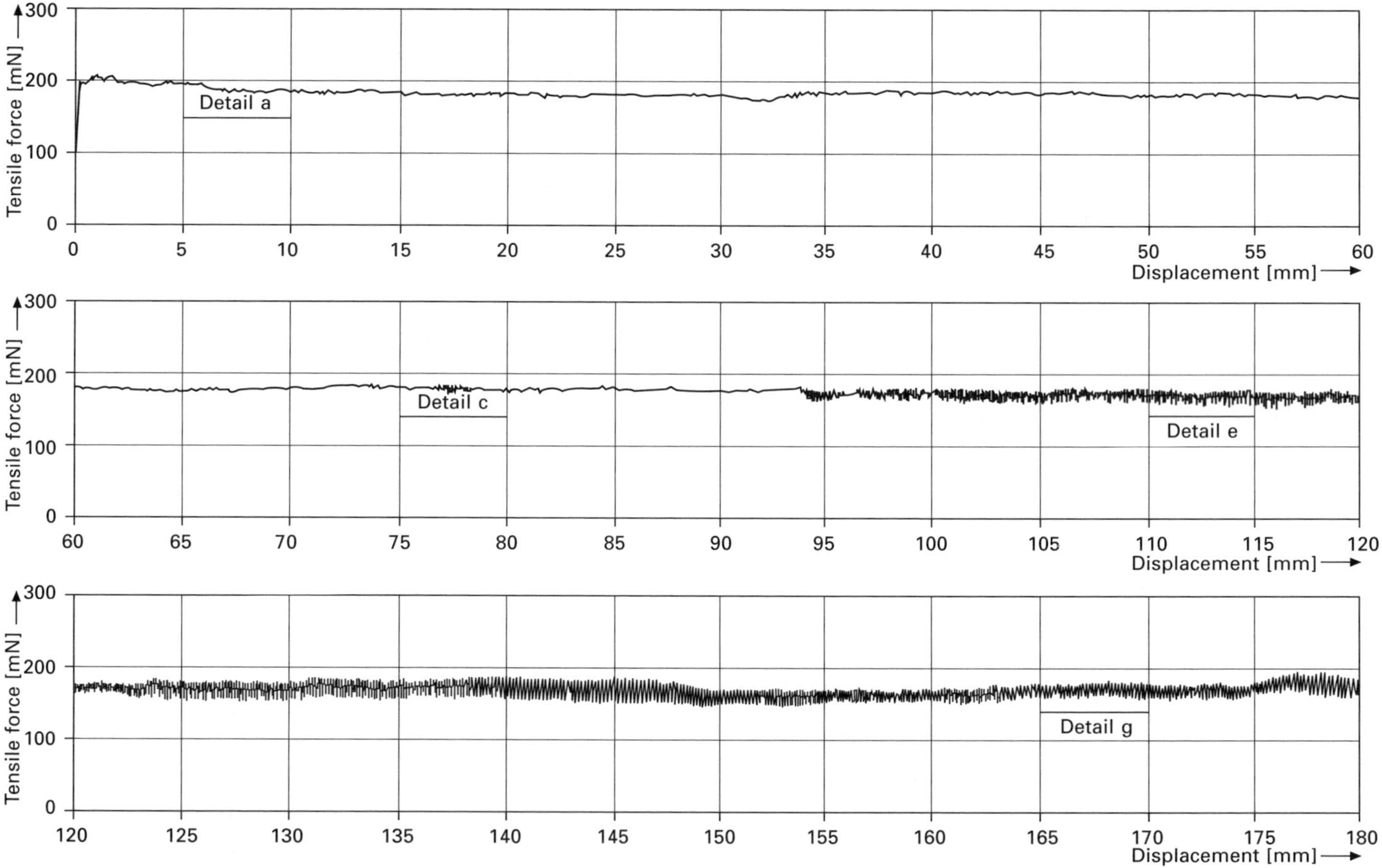

4.34 The example of results for a 0.15 mm polyamide monofilament rubbing against knitting needles.

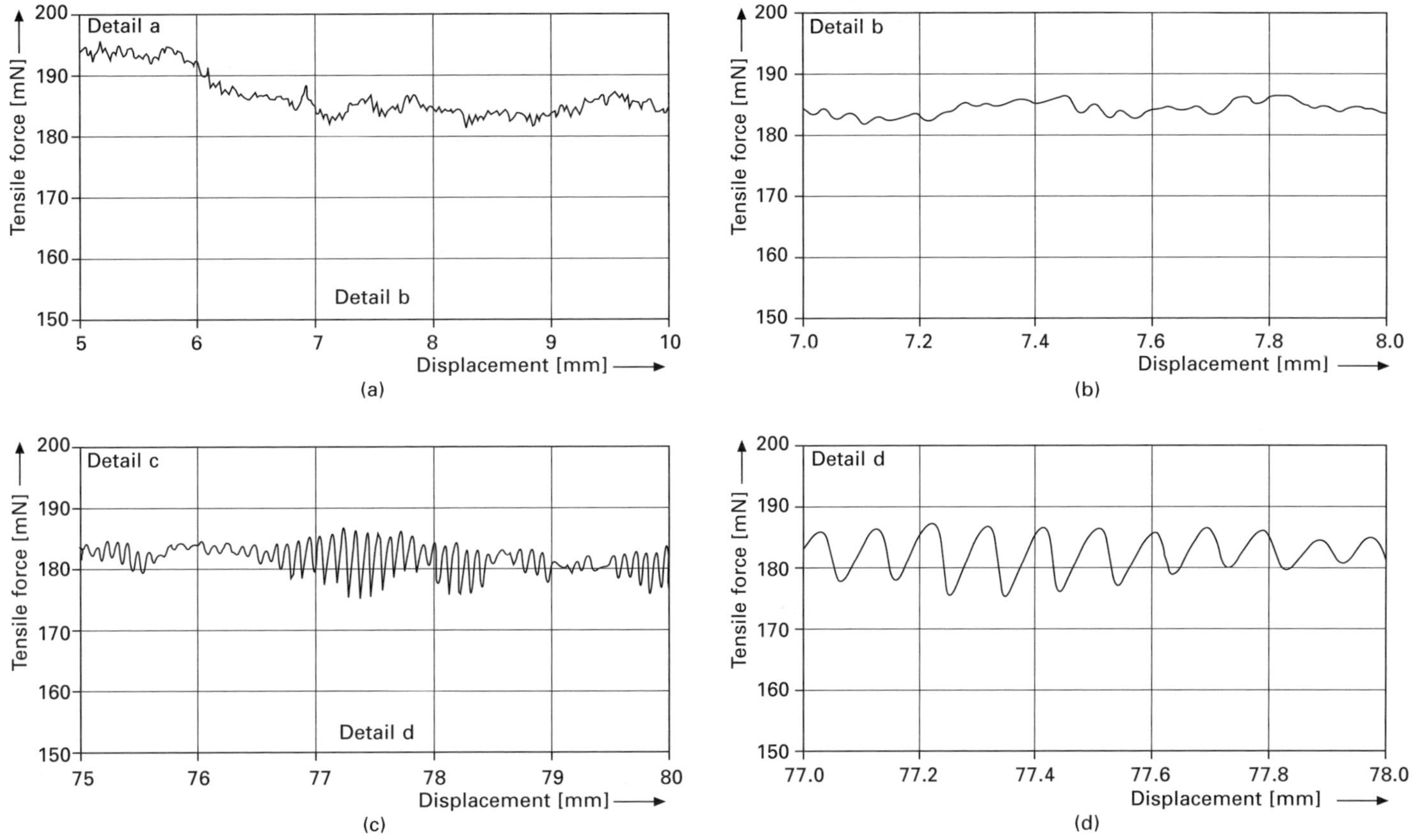

4.35 Detailed results obtained on 0.15 mm polyamide monofilament rubbing against knitting needles.

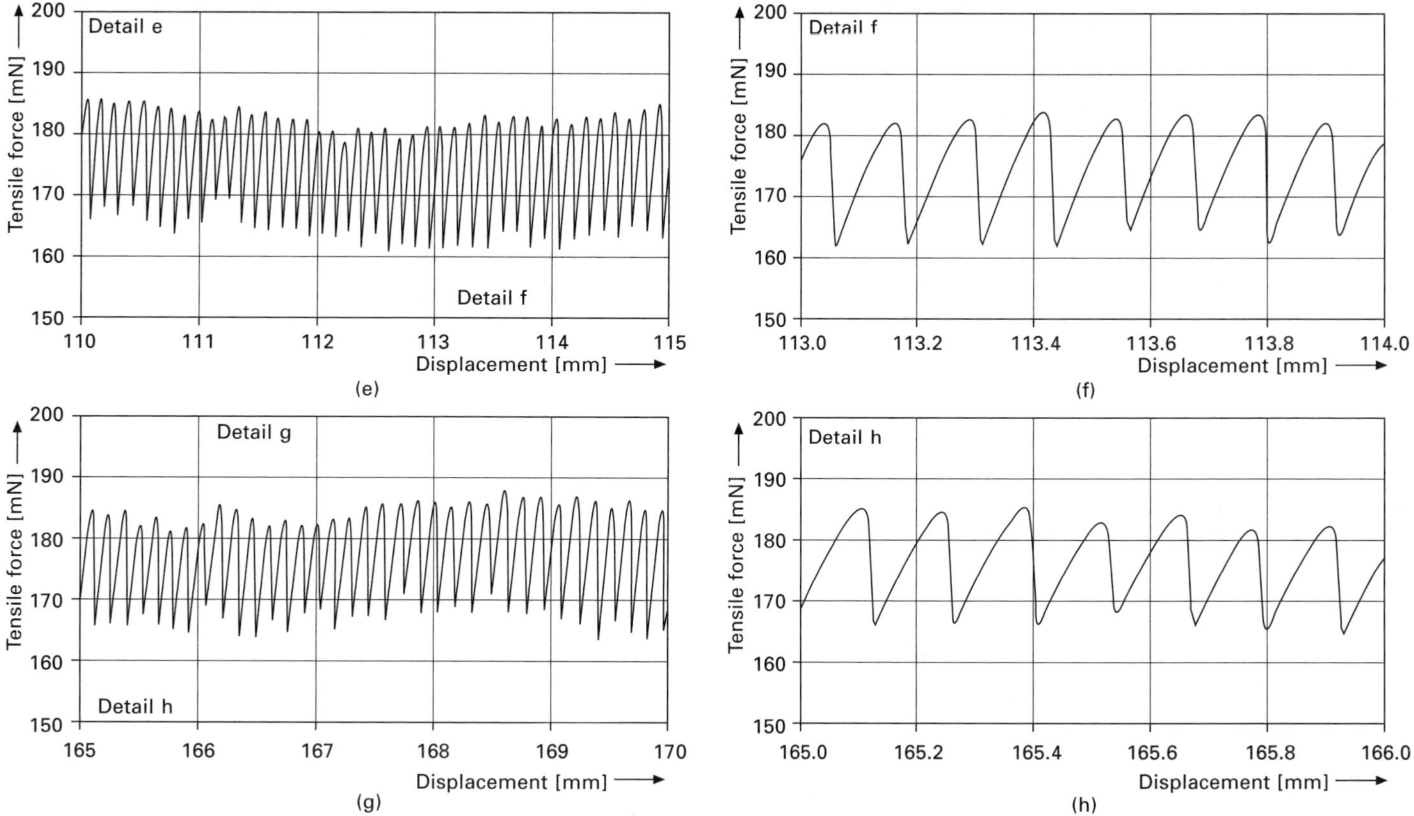

4.35 Continued

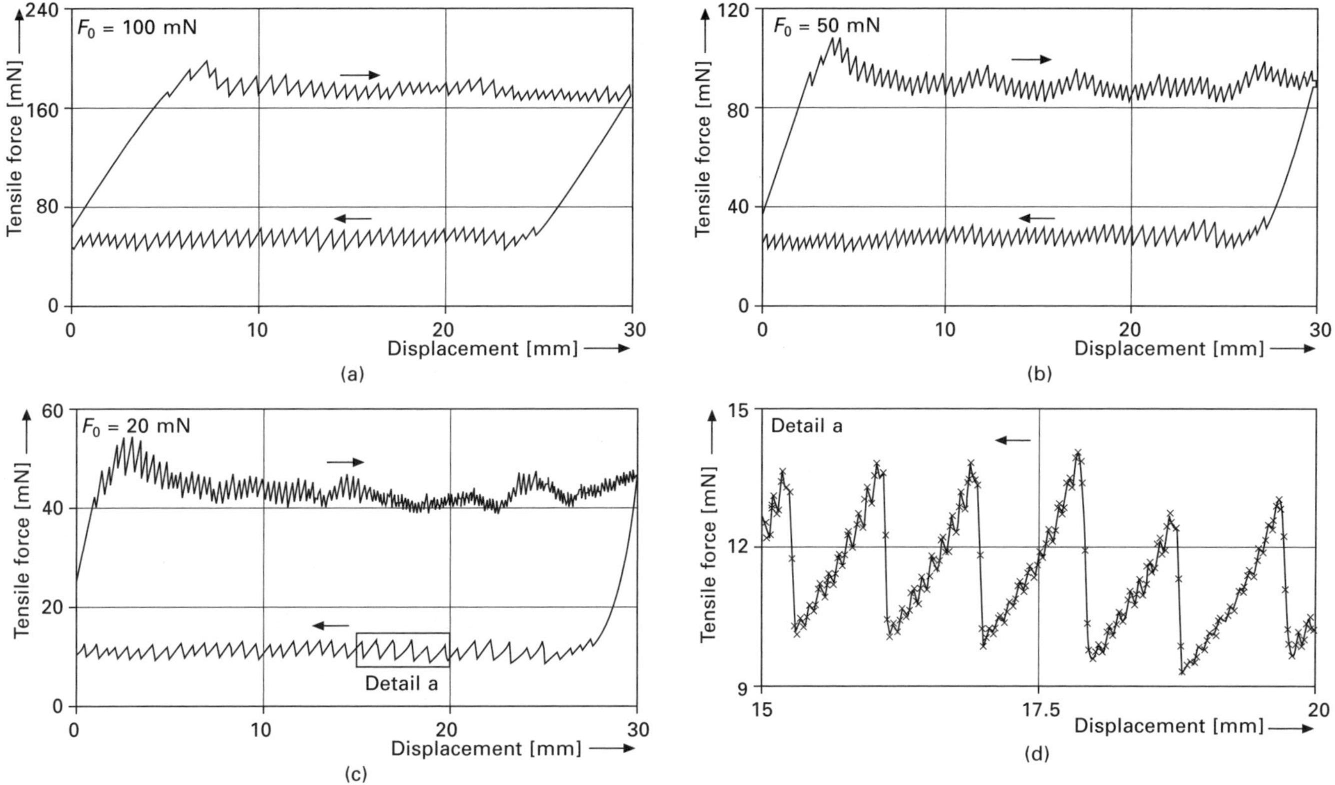

4.36 Friction of polyamide monofilament (0.15 mm) on knitting needles.

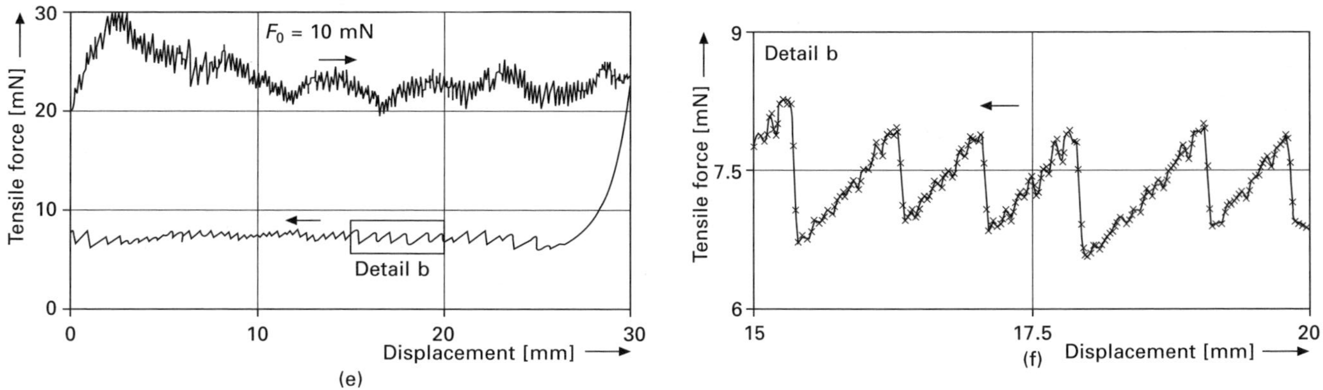

4.36 Continued

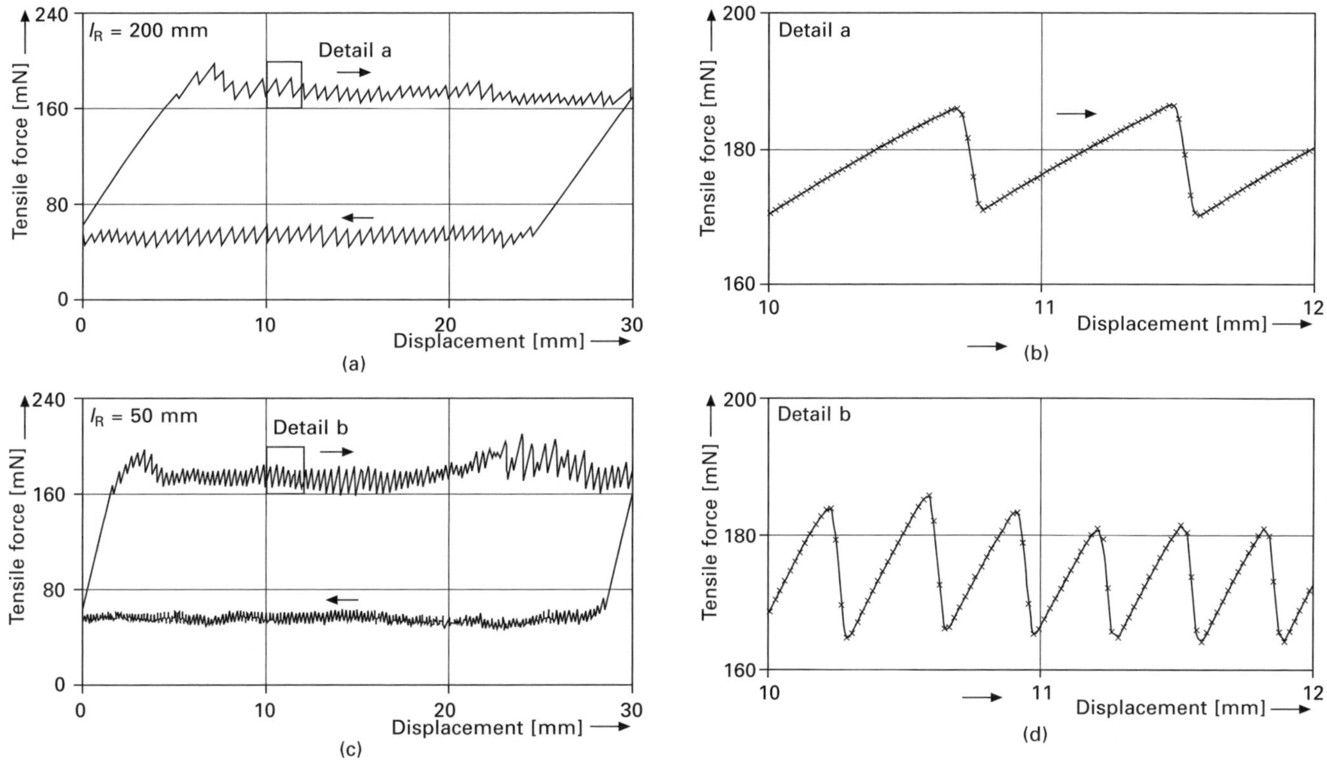

4.37 Friction of polyamide monofilament (0.15 mm) on knitting needles.

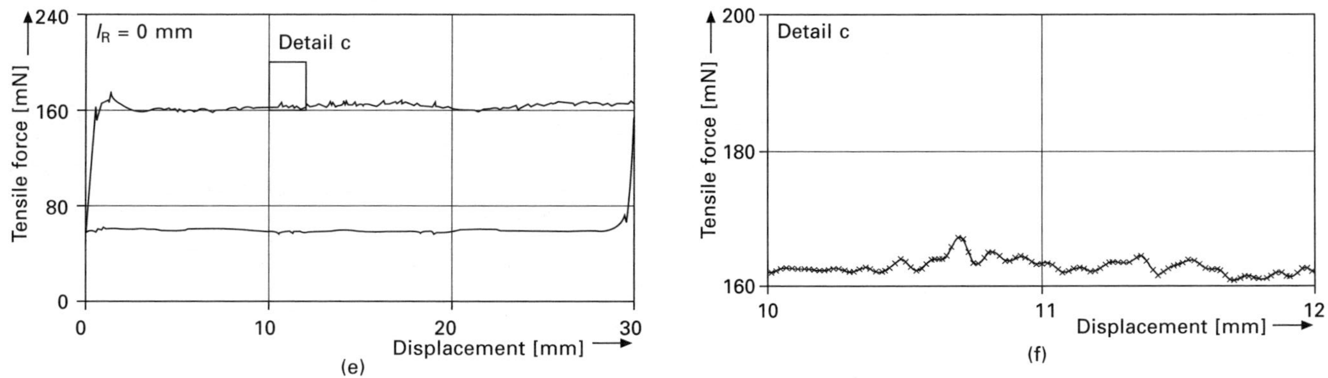

4.37 Continued

Steep or instant decrease and approximately linear increase in force in the first parts of experiments (see details in Figs 4.37(b, d)) correspond with high elastic extension of material. Second parts of experiments (Figs 4.36 and 4.37) show the behavior the other way around, i.e. steep increase and gradual decrease, with exchange between input and output forces. In other words, in the slip region, the measured force increases rapidly. Higher than basic SSP frequency, observed in Fig. 4.36 (d) and (f), is analyzed later (see Fig. 4.45).

4.5.2 Textile-to-textile friction

Studies by Ajayi [2, 3] showed the existence of linear relations between frictional (tangential) force and normal load in knitted and plain and twill weave fabrics on logarithmic scales. It was also shown, that the amplitude of F_t variation, caused by SSP, diminished as the sliding velocity increased. His experiments clearly reflected the strong influence of fabric geometry on SSP.

Examples of experiments, conducted according to the procedure illustrated in Fig. 4.22(b) (yarn-to-yarn friction), are shown in Fig. 4.38. The results are similar to those found when measuring yarn-to-steel friction. Polyamide monofilament shows relatively low and regular SSP (Fig. 4.38(a), (b)). The frictional force of double twisted acrylic spun yarn is high but the profile not as regular, although some periodical changes could be seen, caused by the yarn ply twist of 380 m^{-1} (Fig. 4.38(c), (d)). Frictional resistance of single cotton rotor spun yarn (Fig. 4.38(e), (f)) is similar to that of acrylic yarn; there are some periodical changes that could be connected with the yarn twist (650 m^{-1} or 1.54 mm/turn).

Parts *b*, describing return of the dynamometer crosshead, when the final tensile force is set by the mass of 50 or 100 mN and the input force is measured, show very low values of these forces. For these materials, low output axial forces were almost insufficient to overcome static friction.

Simple statistical evaluation of the results, noted in Fig. 4.38, was conducted and the outcome is given in Table 4.3. The data in the interval 10 to 20 mm of yarn displacement in both directions are used. High values of CV (coefficient of variation, which represents unevenness in yarns friction) on the return parts of experiments Fig. 4.38 (c, d, e and f) are connected with low values of measured yarn tensile force.

Fabric-to-fabric slip [24, 25] is influenced by the geometry of surfaces in contact. Both woven and knitted structures are composed of basic structural elements (cells), which lead to periodical changes in surface geometry. The topographical features of the fabric may fit into those of another in contact and become an important source of SSP.

The frequency of SSP should then correspond to that of the threads in the woven (sett) and the knitted (count) fabrics. Examples of experimental results

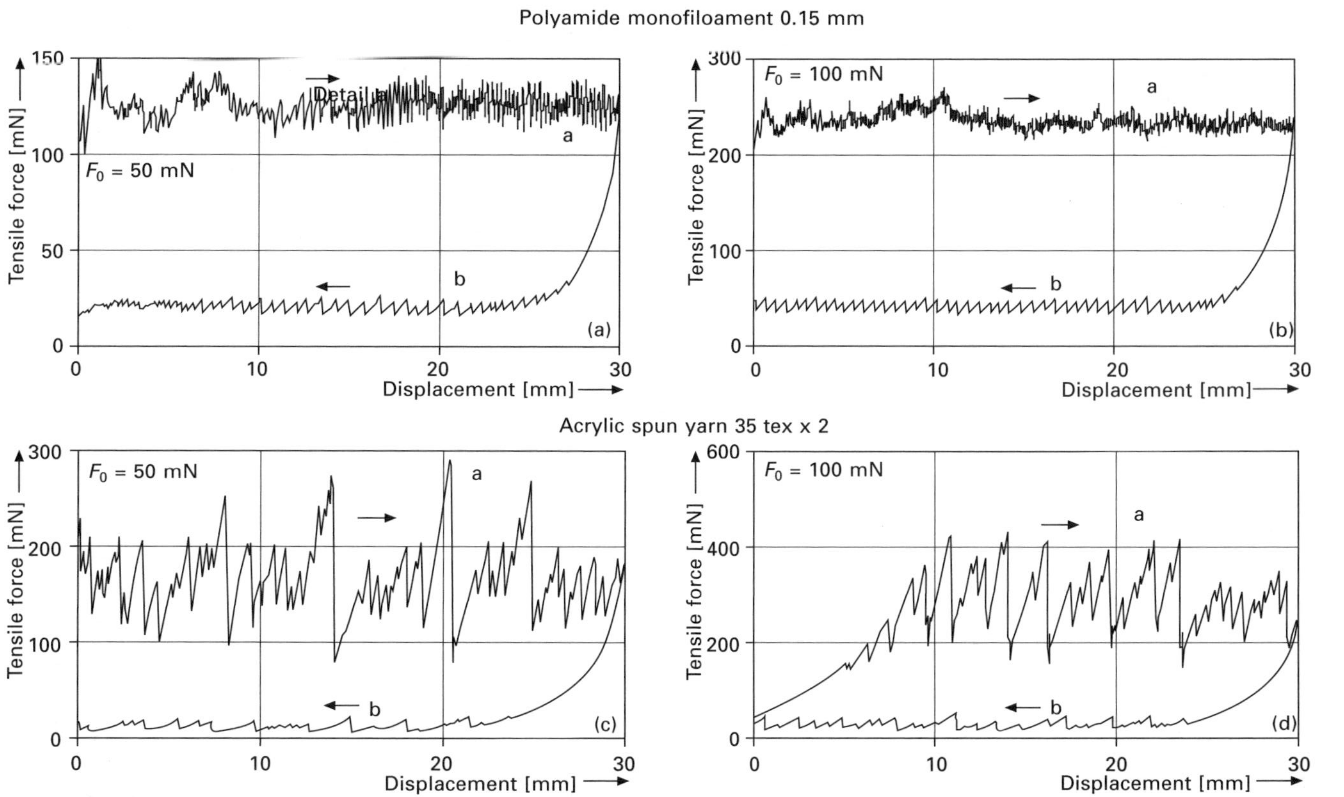

4.38 Examples of yarn-to-yarn friction. The results shown in (a, b, c and e) were recorded after a short movement of the dynamometer crosshead, and in (d and f) they were recorded from the beginning.

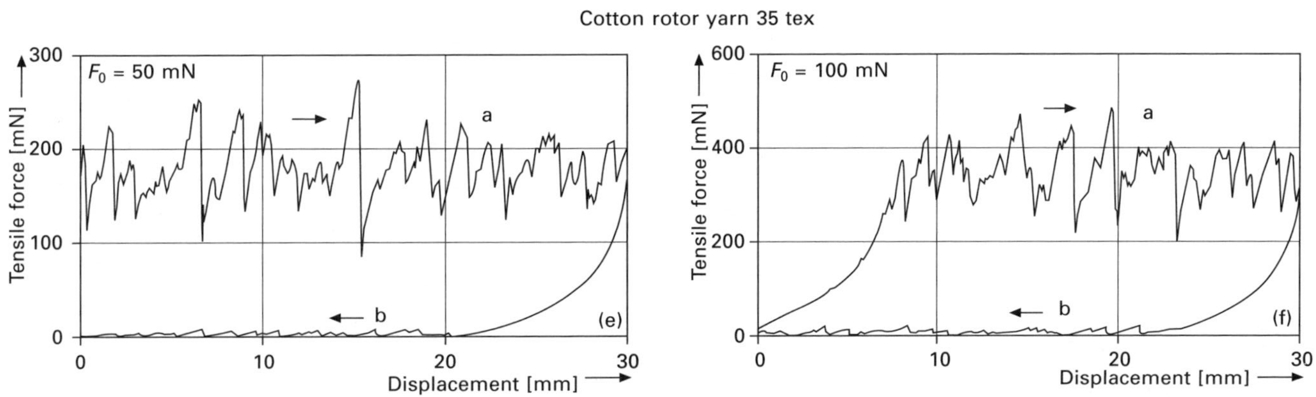

4.38 Continued

Table 4.3 Evaluation of results shown in Fig. 4.38

Figure	Yarn	F_0 [mN]	Average F [mN]	CV [%]
4.38 A–a	Polyamide 0.15 mm	50	125.3	4.5
4.38 B-a	Polyamide 0.15 mm	100	238.6	3.9
4.38 C-a	Acrylic 35 tex × 2	50	167.3	21.9
4.38 D-a	Acrylic 35 tex × 2	100	306.2	19.4
4.38 E-a	Cotton 35 tex	50	180.7	16.5
4.38 F-a	Cotton 35 tex	100	357.1	15.2
4.38 A-b	Polyamide 0.15 mm	50	20.3	11.3
4.38 B-b	Polyamide 0.15 mm	100	43.4	8.5
4.38 C-b	Acrylic 35 tex × 2	50	11.2	26.7
4.38 D-b	Acrylic 35 tex × 2	100	29.6	24.8
4.38 E-b	Cotton 35 tex	50	4.1	38.1
4.38 F-b	Cotton 35 tex	100	8.5	46.7

found on plain-woven flax yarn fabrics are shown in Fig. 4.39. Other specification were: warp flax yarn linear density 220 tex, weft flax yarn linear density 170 tex, warp sett 1120 m^{-1}, weft sett 840 m^{-1}. The tests were conducted with warp sliding parallel to warp with an average speed of 0.5 mm s^{-1}. Results show, that the higher the normal pressure F_n, the more regular are the changes in frictional resistance, caused by the fabric geometry.

The results of similar experiments on the same fabric but at lower speed (0.1 mm s^{-1}) are shown in Fig. 4.40. The results indicate higher frequency of SSP, caused by yarn unevenness and hairiness that play an increased role at low speed. And because these surface characteristics have little inherent regularity, SSP at this level is quite irregular (Figs 4.40 (c and d)).

The influence of the angle of orientation of warp yarns in warp-warp sliding on SSP is shown in Fig. 4.41. As expected, the highest amplitude of frictional force oscillation is found at 0° orientation. In Fig. 4.41(a), the warp yarns of the two surfaces are parallel to each other and therefore the nodes or asperities fit in ridges of each other and ride up and down during traverse. Accordingly, SSP has the highest amplitude. At other angles, particularly 15–90°, the nodes ride more on the surface than in the ridges and, therefore, the amplitude is lower and the profile is more irregular.

Interesting results were obtained in experiments conducted on model woven fabrics, made from polyamide monofilament with structure resembling a net used for protecting against attack by insects. The fabric was treated thermally to bind filaments at crossover points. The fabric had a warp sett of 690 m^{-1} and a weft sett of 590 m^{-1}. The results given in Fig. 4.42 for four values of normal tension σ_n show that the amplitude of SSP was very high due to regular nature of the structure and the smooth surface of the yarn (monofilament) used.

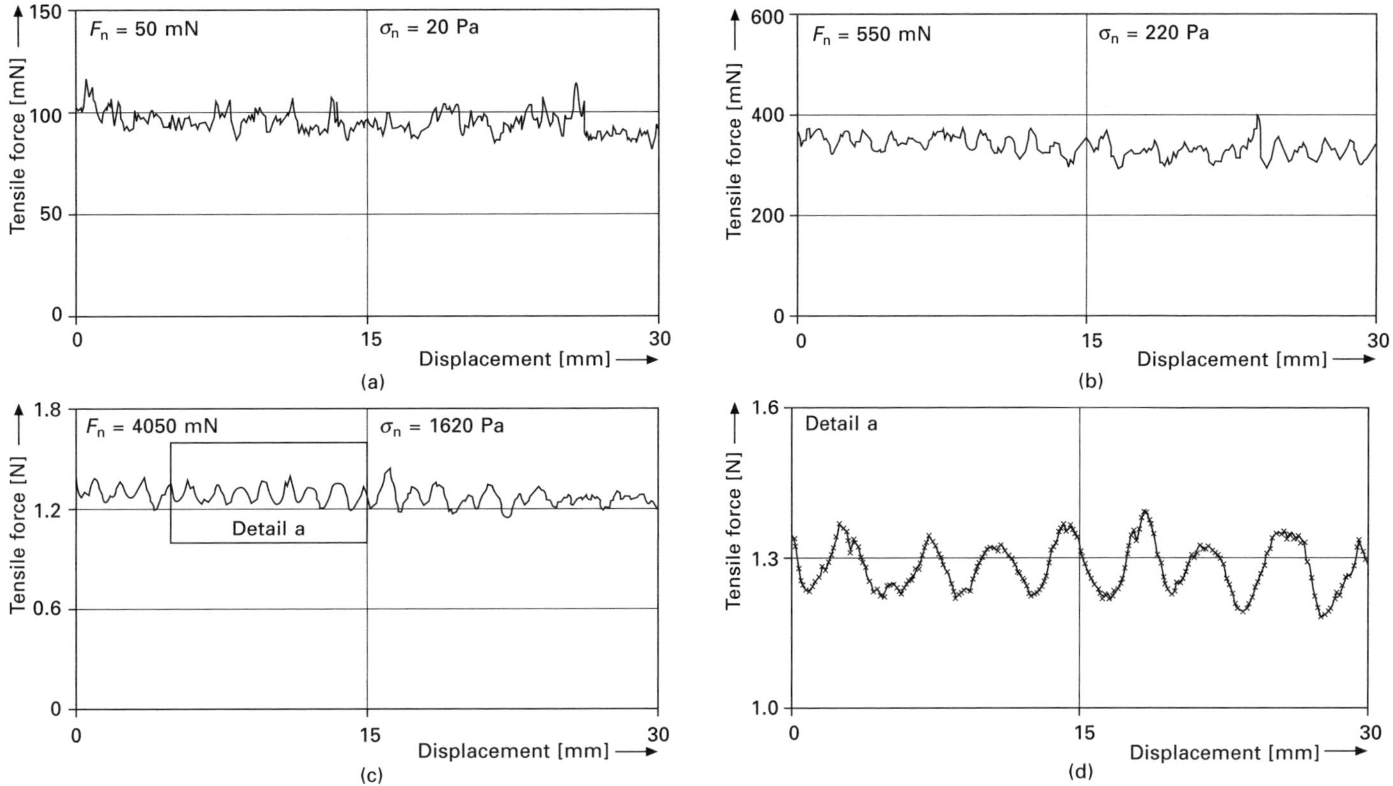

4.39 The woven flax yarn fabric-to-fabric friction (speed 0.5 mm s^{-1}, area of the sled 5 × 5 cm).

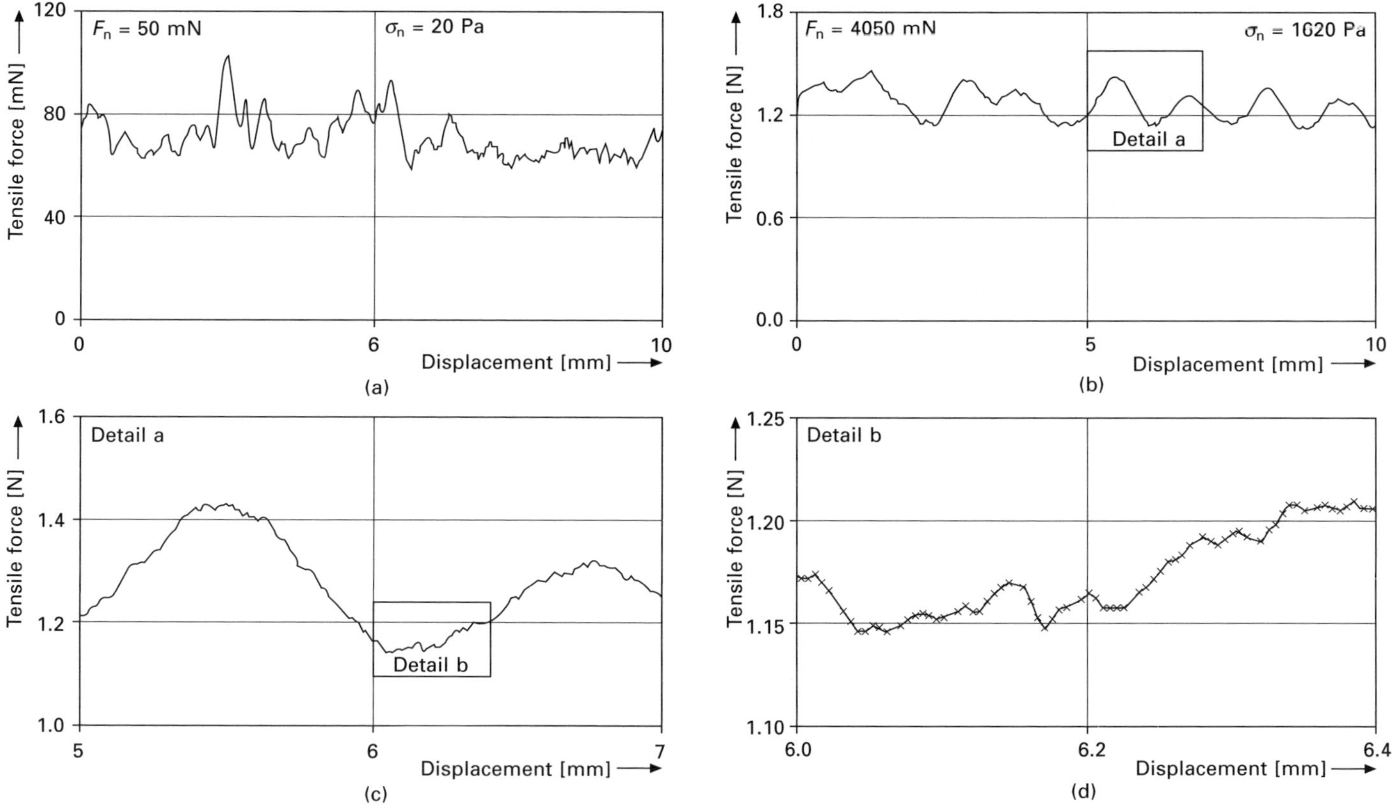

4.40 The woven flax yarn fabric-to-fabric friction (speed 0.1 mm s^{-1}).

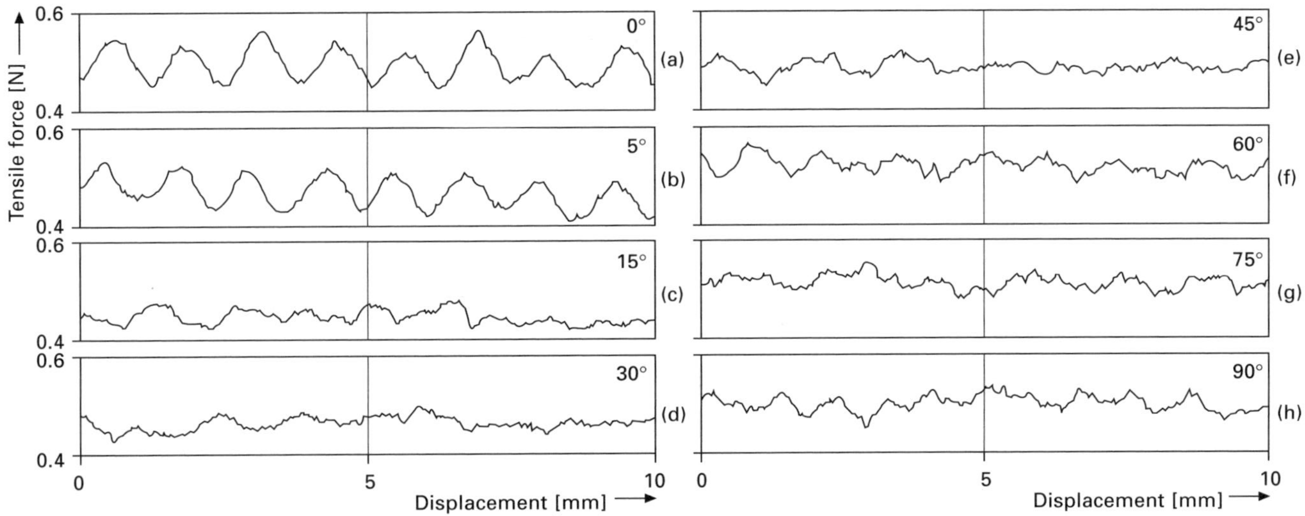

4.41 The influence of angle of traverse on SSP.

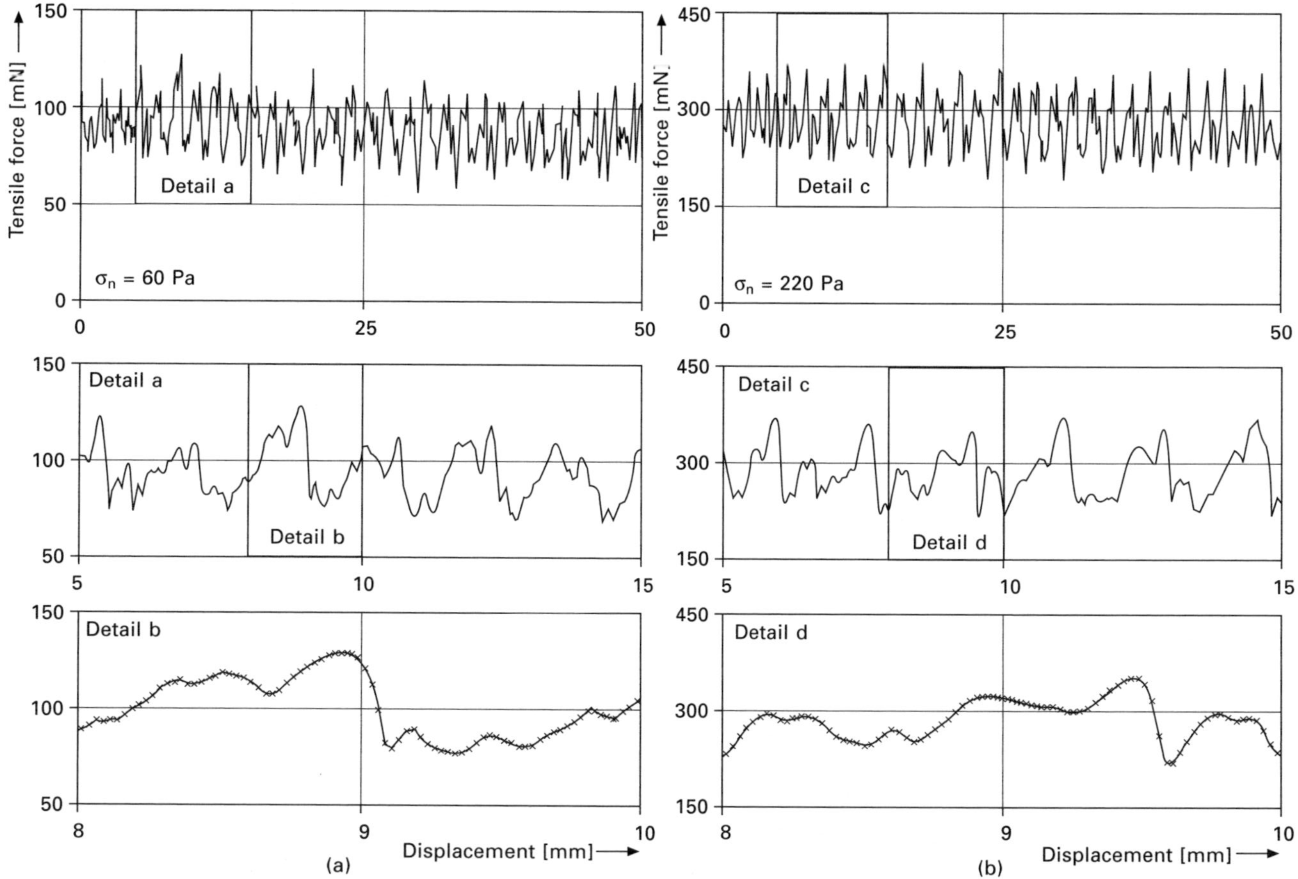

4.42 The friction behavior of model woven fabric.

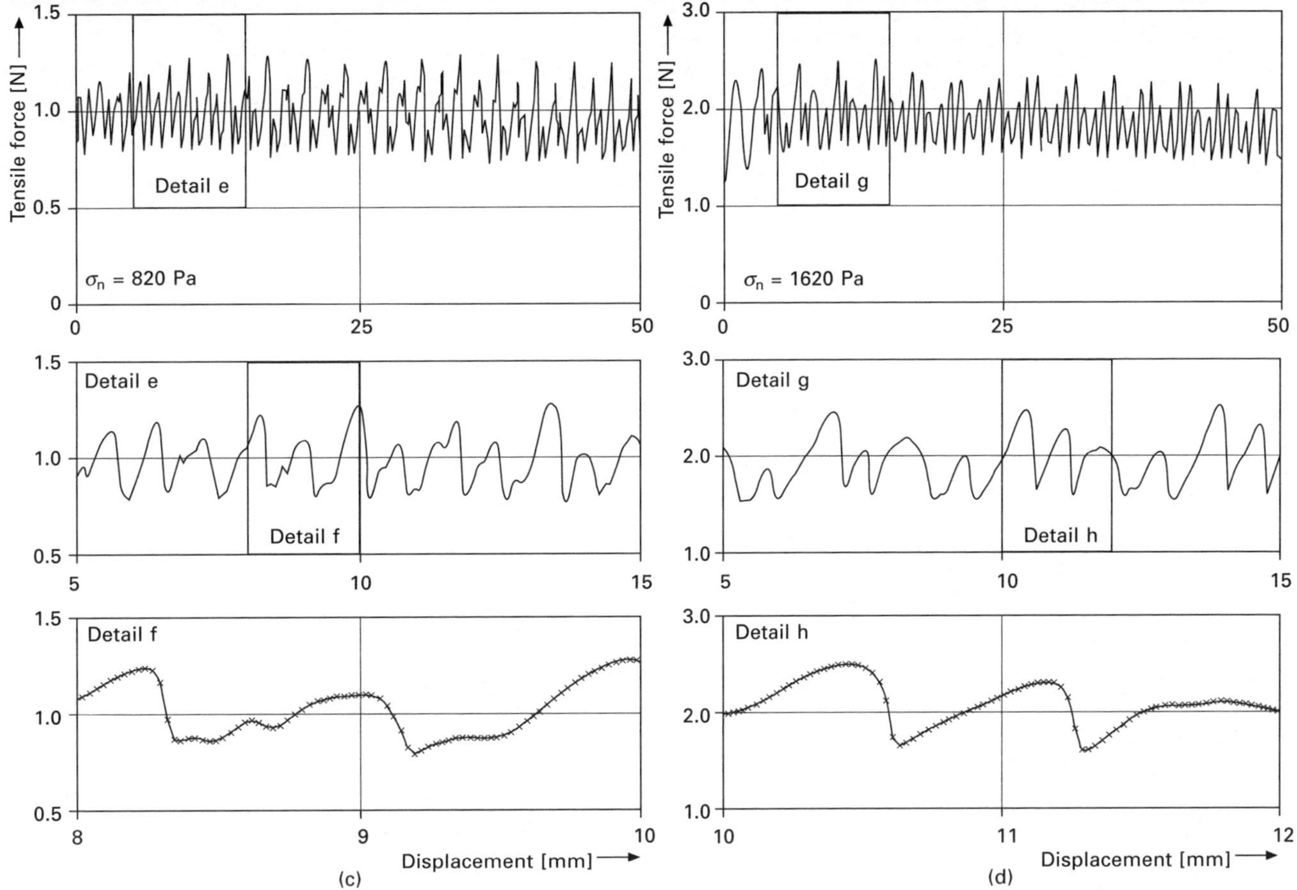

4.42 Continued

Table 4.4 shows a numerical evaluation of the data collected. While the impact of normal pressure σ_n on the coefficient of variation is negligible, the ratio F_t/F_n (friction coefficient) decreases with increase in σ_n or F_n.

Figure 4.43 shows the effect of the relative angle of orientation of warp yarns in the warp-warp test on the model fabrics conducted at a constant normal pressure of 420 Pa (normal force $F_n = 1050$ mN). As seen before, the highest amplitude is obtained for 0° test and a more irregular behavior is noted at other angles. At 90° traverse (warp-weft), the profile appears irregular due to differences in the warp and weft setts.

Considered next is the effect of the elasticity of the cord used to pull friction body on friction. For short or inextensible cord, the change in speed v (Fig. 4.27(a)) is minimal (stiffness $s = 190$ N m^{-1}) and, therefore, one could only capture change in frictional resistance (Fig. 4.44(a)). After putting a spring of stiffness $s = 25$ N m^{-1} between rope 5 and sensor 4 (experiment Fig. 4.27(b), results Fig. 4.44(b)) real SSP is enabled, i.e. speed of slip changes from zero to maximum. Wavelength in case (a) corresponds to yarn pitch (average value is approximately 1000/840 = 1.2 mm), in case (b) each second weft yarn is skipped. Energy, accumulated in the spring, causes such upper fabric acceleration, that it jumps over the next weft yarn and then, when the spring is relaxed, the fabric stops.

Sometimes unexpected results are found as noted in Fig. 4.45 for friction measured on cotton fabric (rotor spun yarn of 33 tex and weft sett of 1560 m^{-1}). Not only was the frequency corresponding to the weft sett (primary cycle) noted but also superimposed on it was a secondary frequency approximately 14 times greater. The latter is expected to be the result of some dynamic oscillation present in the testing system and not of some morphological feature of the fabric. Similar results were seen in Figs 4.36(d) and (f).

Usually an increase in lateral stress leads to a decrease in relative frictional resistance, or coefficient of friction, in textiles [13]. Impact of pressure on SSP, however, may be governed by different criteria. At low pressure, fabric hairiness, adhesion between fibers, etc., could be expected to cause high variation and high frequency of changes in frictional resistance, whereas at high pressure, SSP could be expected to be largely caused by fabric geometry. Decrease of coefficient of friction with lateral tension between two woven

Table 4.4 Evaluation of results shown in Fig. 4.42

Figure	Normal pressure σ_n [Pa]	Normal force F_n [mN]	Average F_t [mN]	CV [%]	μ (F_t/F_n)
4.42 A	60	150	90	13.0	0.597
4.42 B	220	550	282	13.5	0.513
4.42 C	820	2050	975	13.5	0.475
4.42 D	1620	4050	1902	12.9	0.470

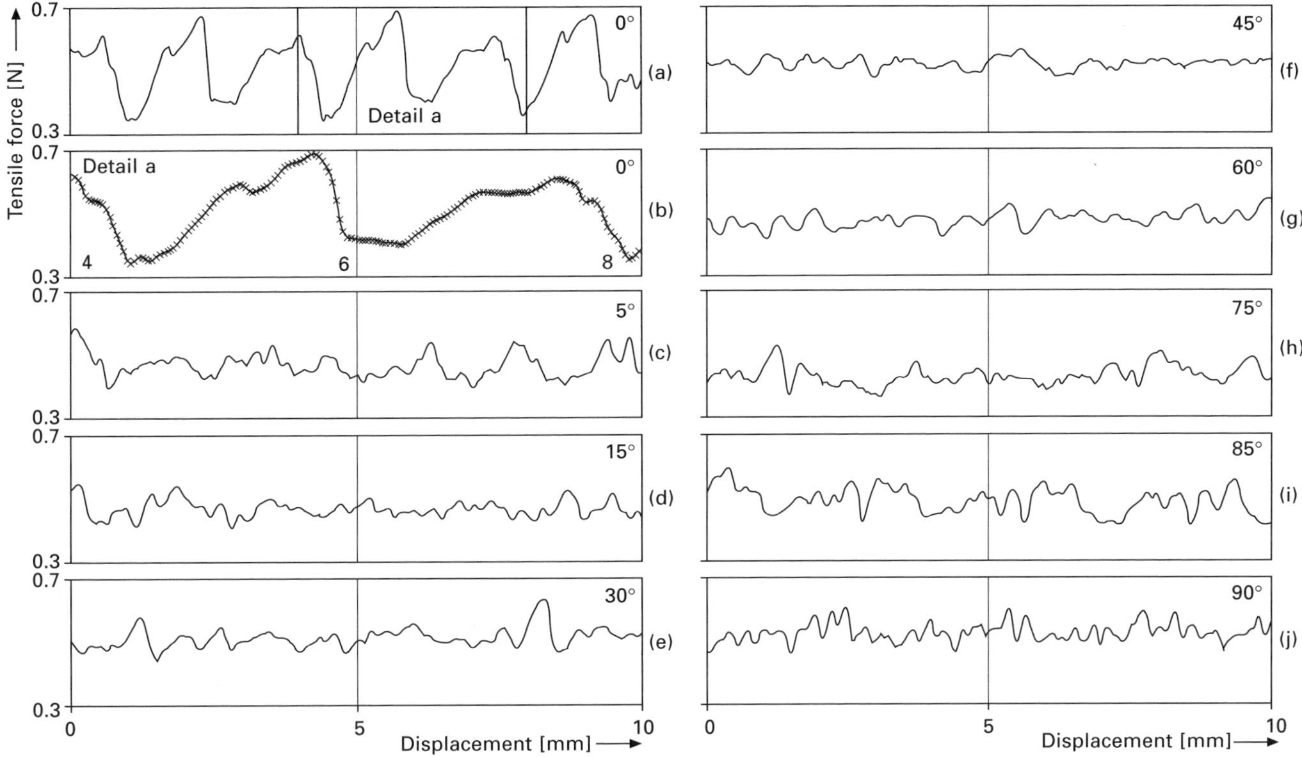

4.43 The influence of the direction of test on SSP ("model" fabric).

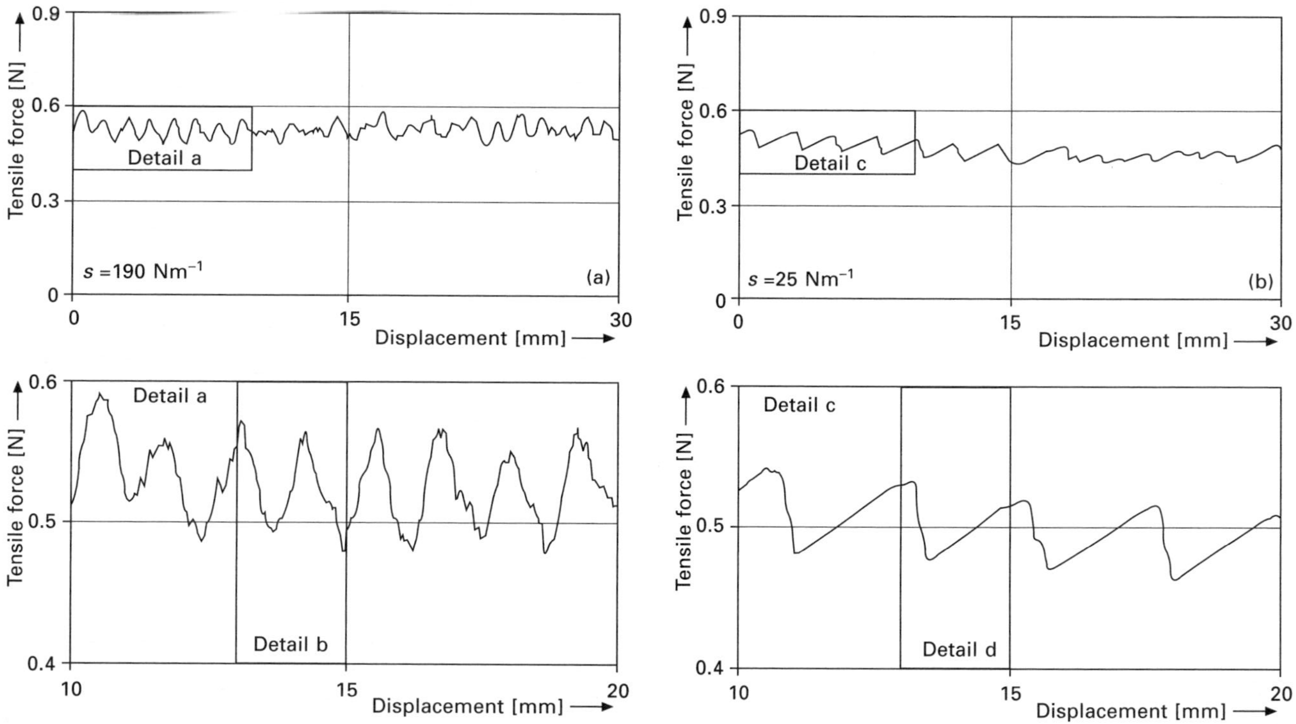

4.44 The impact of rope stiffness on SSP frequency.

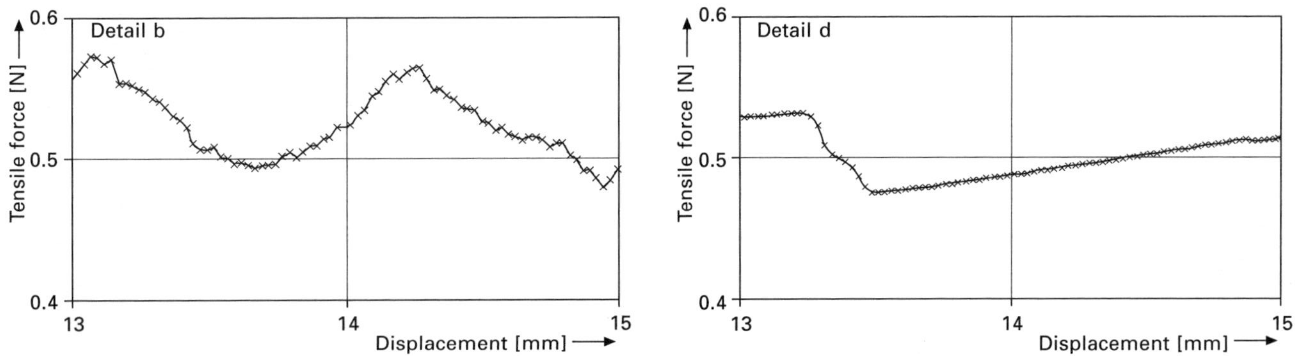

4.44 Continued

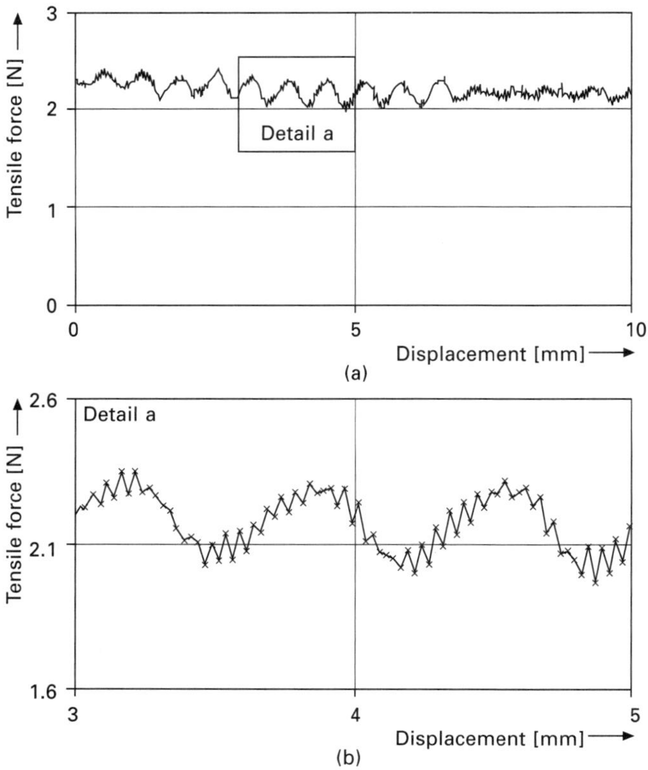

4.45 The friction of cotton fabric.

fabrics at two speeds of the sled motion is shown in Fig. 4.46. Material used was plain-woven flax fabric of which the force profiles were included in Fig. 4.39.

Changes in frictional resistance could be observed even when a fabric is slid over a glass or a leatherette (Fig. 4.47). Shown in Figs 4.47(a) and (c) are the profiles obtained for the friction of a cotton woven fabric (plain weave square fabric, yarn of linear density 53.5 tex, sett of 2550 ends per m), fastened on bottom side of a sled (Fig. 4.27), to glass (a) or leatherette (c), fixed on the platform. Similarly, in Figs. b and d are shown the results of frictional resistance of model polyamide woven fabric (the same as used in Section 4.4.2, Figs 4.42 and 4.43) against the same two friction materials. Average stiffness of the system, i.e. from the rope 5 and the pulley 3 (Fig. 4.27), was 310 Nmm^{-1}. The area of the sled was 5 × 5 cm, and the lateral pressures used were 80, 160 and 320 Pa (i.e. the normal forces used were 200, 400 and 800 mN, respectively.)

Because the lower frictional surfaces are smooth, the impact of the fabric geometry should be negligible and so the result observed may be due to pure

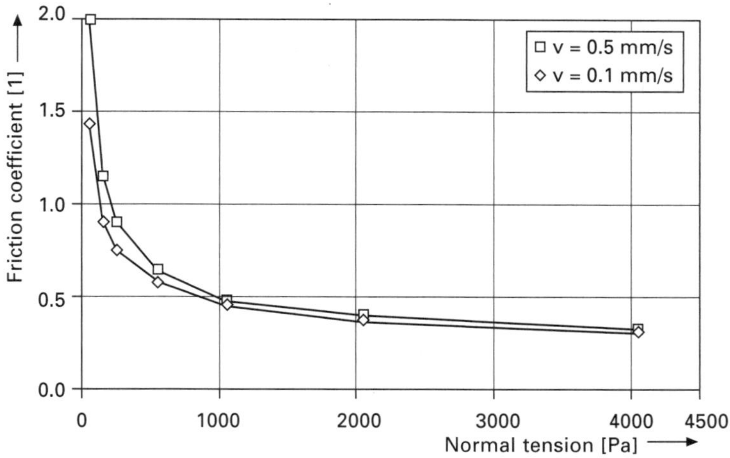

4.46 The impact of normal tension on coefficient of friction.

friction between the two different materials. One indication of this is that the fluctuations, reflecting SSP, in both fabric materials, are highly regular. The geometry of the fabric could, in principle, influence only the area of real contact. Higher amplitude in polyamide to glass or to leatherette test is probably caused by the higher value of the static friction connected with the adhesion of the two materials.

4.5.3 Spectral analysis of SSP [14, 19]

Evaluation of experimental results involving SSP is not simple. Common statistical methods, used to characterize variation in signal, depend, for example, on data collection frequency and sensor sensitivity. It is possible to use Fourier's analysis that can allow evaluation of amplitudes at different frequencies (or wavelengths). Fractal analysis does not appear suitable, as friction at different levels (nano, micro and macro) does not lead to similar results.

Standard methods are not available and changes in test dimensions (e.g., yarn lengths a and b in Fig. 4.18) can lead to changes in amplitude and frequency of SSP. Results can be instructional, however, and therefore, spectral analyses based on several experiments suggested in Figs 4.18 and 4.27, are presented.

Yarn-to-metal slip

Figure 4.48 shows the results of friction experiment involving polyamide monofilament yarn passing over needle hook (Fig. 4.18(a)) with the speed 0.2 mm s^{-1}, the input yarn tensile force F_0 100 mN and the length a 200 mm.

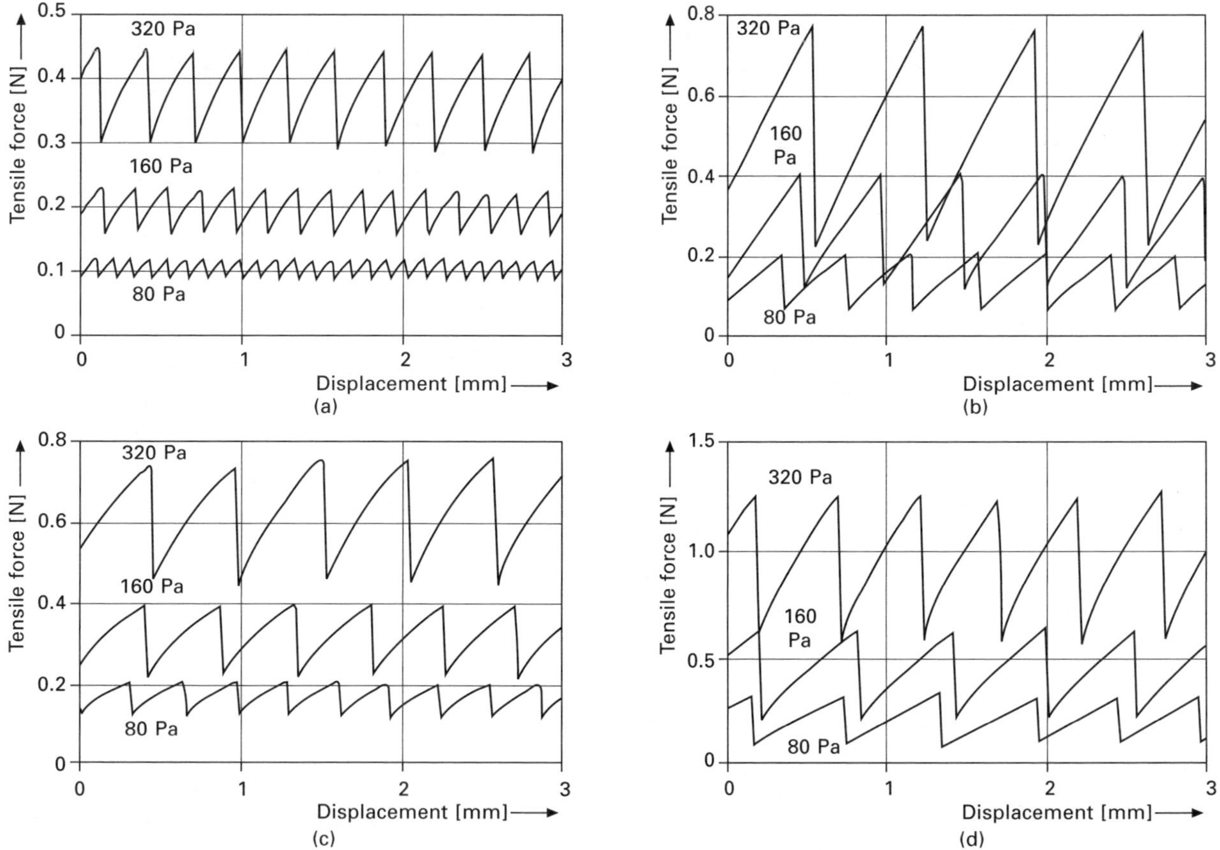

4.47 The friction of Cotton (a, c) and polyamide (b, d) woven fabrics rubbing against glass (a, b) and leatherette (c, d).

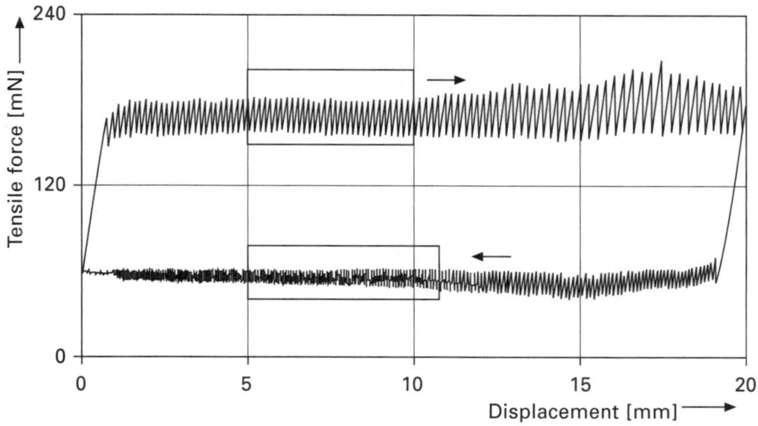

4.48 The friction results of polyamide monofilament to metal.

Fast Fourier's transformations (FFT) of chosen parts of these results are presented in Fig. 4.49. Left figures show the original results of SSP measured in the analyzed interval, and the right figures give the results of spectral analysis, conducted using MATLAB software ("Fast Fourier's Transformation").

Figure 4.49(a) shows one dominant peak at $1/\lambda = 6$ (wavelength $\lambda \sim 0.167$ mm). This frequency could vary as it depends on experimental conditions. One cycle of experiment (Fig. 4.48) gives two different results based on forward and backward traverses with exchange between yarn axial stresses. Figure 4.49(b) shows the results of the return cycle. Variation in the frequency in this case is higher and so the peak of FFT near $1/\lambda$ of 12 is wider.

An example of the results from a similar experiment with acrylic double twisted spun yarn is given in Fig. 4.50. The spectrum is broader; the only important maximum, at $1/\lambda = 0.75$ mm^{-1} (wavelength λ is approx. 1.33 mm), corresponds with the yarn plied twist of 380 m^{-1} as the average length of one half-twist is 1.32 mm.

Fabric-to-fabric slip

It is expected that maximum amplitude of Fourier's analysis will be determined in the traverse of two fabrics oriented warp-warp or weft-weft by the weft or warp sett, respectively. This is confirmed by the example of results given in Fig. 4.51 (flax plain weave fabric, warp 220 tex, weft 170 tex, warp sett 1120 m^{-1}, weft sett 840 m^{-1}, sliding in the direction of warp in both samples, sliding speed 0.5 mm s^{-1}). The results are based on the experiment suggested in Fig. 4.27(a). The weft spacing of 1000/840, or 1.20 mm, corresponds exactly with the main peak of the wavelength.

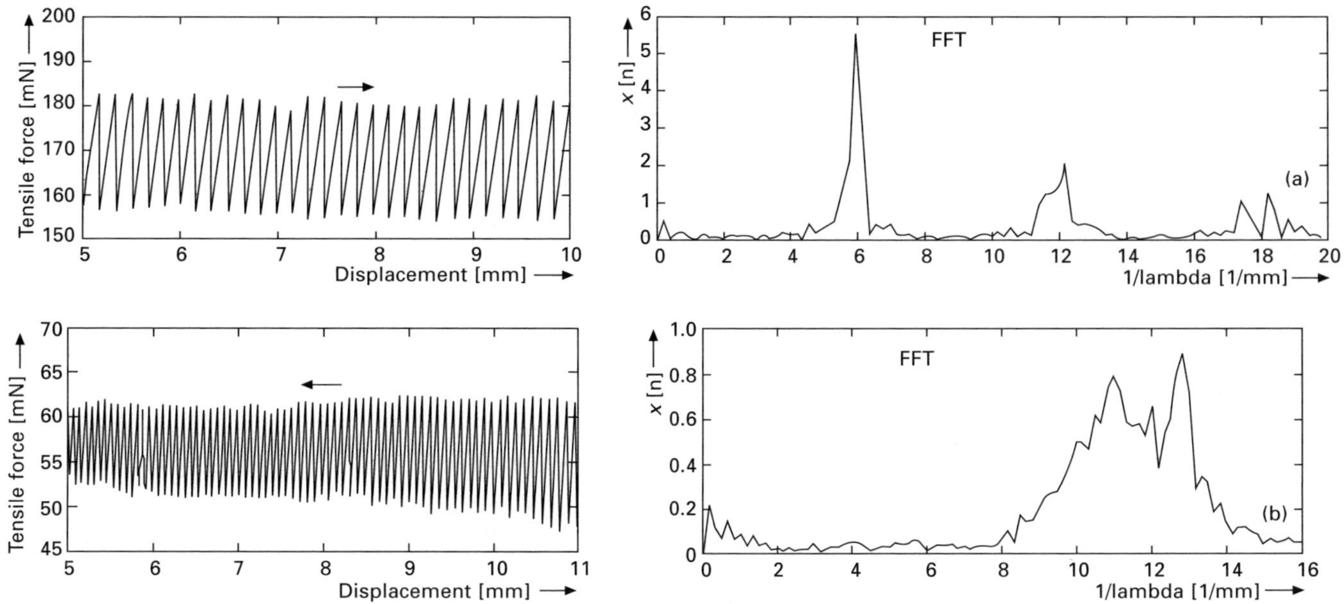

4.49 FFT of chosen parts from Figure 4.48.

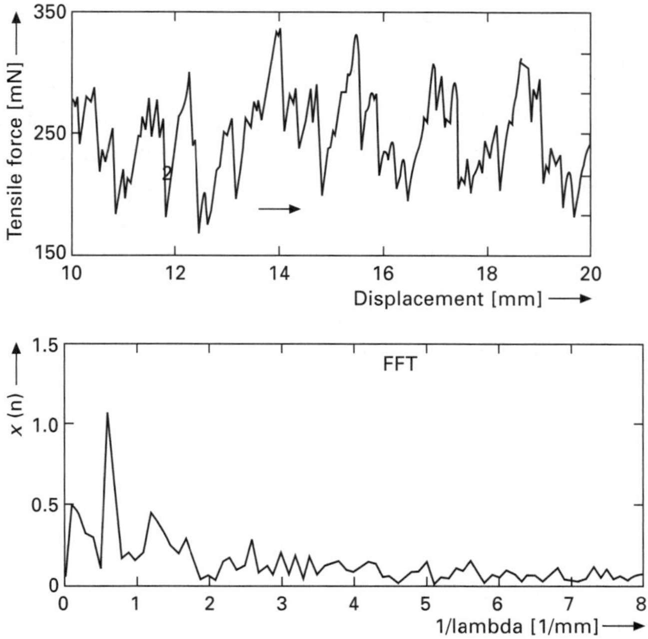

4.50 Friction results of acrylic 35 × 2 tex spun yarn to needle hook. Speed 1 mm s^{-1}, F_0 100 mN.

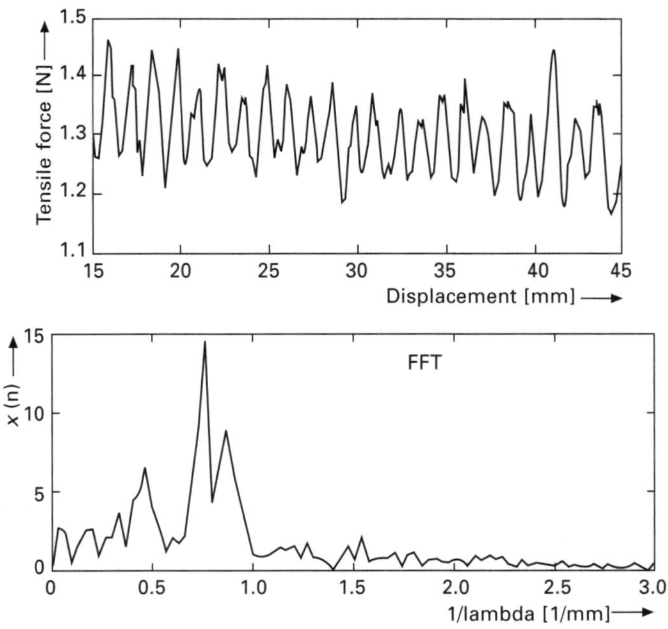

4.51 Example of fabric-to-fabric friction spectral analysis, normal tension is 1620 Pa.

Interesting results are obtained again when the "model" fabric is used (Fig. 4.52). Specifications are: plain weave, polyamide monofilament, upper fabric area 5 × 5 cm, speed 0.2 mm s^{-1}, normal stress 420 Pa, warp sett 690 m^{-1}, weft sett 590 m^{-1}. The weft sett determined the primary frequency; peaks at higher frequencies are caused by other factors, including deformation of yarn and fabric. Angle α between warp axes of the lower and upper fabrics was changed in the experiments; only three basic levels are shown (A – 0°, B – 45°, C – 90°). Example (c) shows additional periodicity, derived from difference between warp and weft setts.

Similar experiments conducted on the flax plain weave fabric are shown in Fig. 4.53. The highest amplitude of SSP is, of course, achieved again when the position of both fabrics is the same ($\alpha = 0°$, i.e. warp-warp transverse); the wavelength of the FFT corresponds with the value of the weft yarn sett.

4.5.4 Influence of fabric topography and mechanical properties

The effect of a systematic change in fabric geometry [28, 29], with several levels of organization, can be considered to be an important area of study for understanding SSP in fabric and for confirming the role the geometry of fabric can have on SSP. Another area that is of practical importance and that can also affect SSP is the tendency of a fabric to deform and wrinkle during rubbing. Factors playing a role may be high μ, high fabric extensibility, and low bending rigidity. Effects are often compounded during traverse. For example, deformation leads to wrinkle (Fig. 4.54(b)); this leads to an increase in frictional resistance, and which causes more wrinkles. Accordingly, mechanical properties of a fabric, that vary with construction (woven, knitted, braided) and material, will play a role in governing μ and SSP.

4.6 Rheological description of SSP

4.6.1 Evaluation of some possibilities

Probably the simplest possible rheological model of SSP, neglecting mass of the system, is as illustrated in Fig. 4.55(a). The model consists of only two elements – elastic stiffness s and frictional constants F_s and F_k. The latter is described by two parameters – the higher static F_s and the lower kinetic F_k frictional resistances. The relation between force F and the displacement Δ is periodical with the force growing linearly to the value F_s and then taking a step fall to F_k. The amplitude of the cycle is $F_s - F_k$, and the wavelength, λ, is given by $\lambda = \dfrac{F_s - F_k}{s}$. This equation can easily be derived from the definition of stiffness s ($F = s \cdot \Delta$), because for one period of SSP, $F_s - F_k = s \cdot \lambda$.

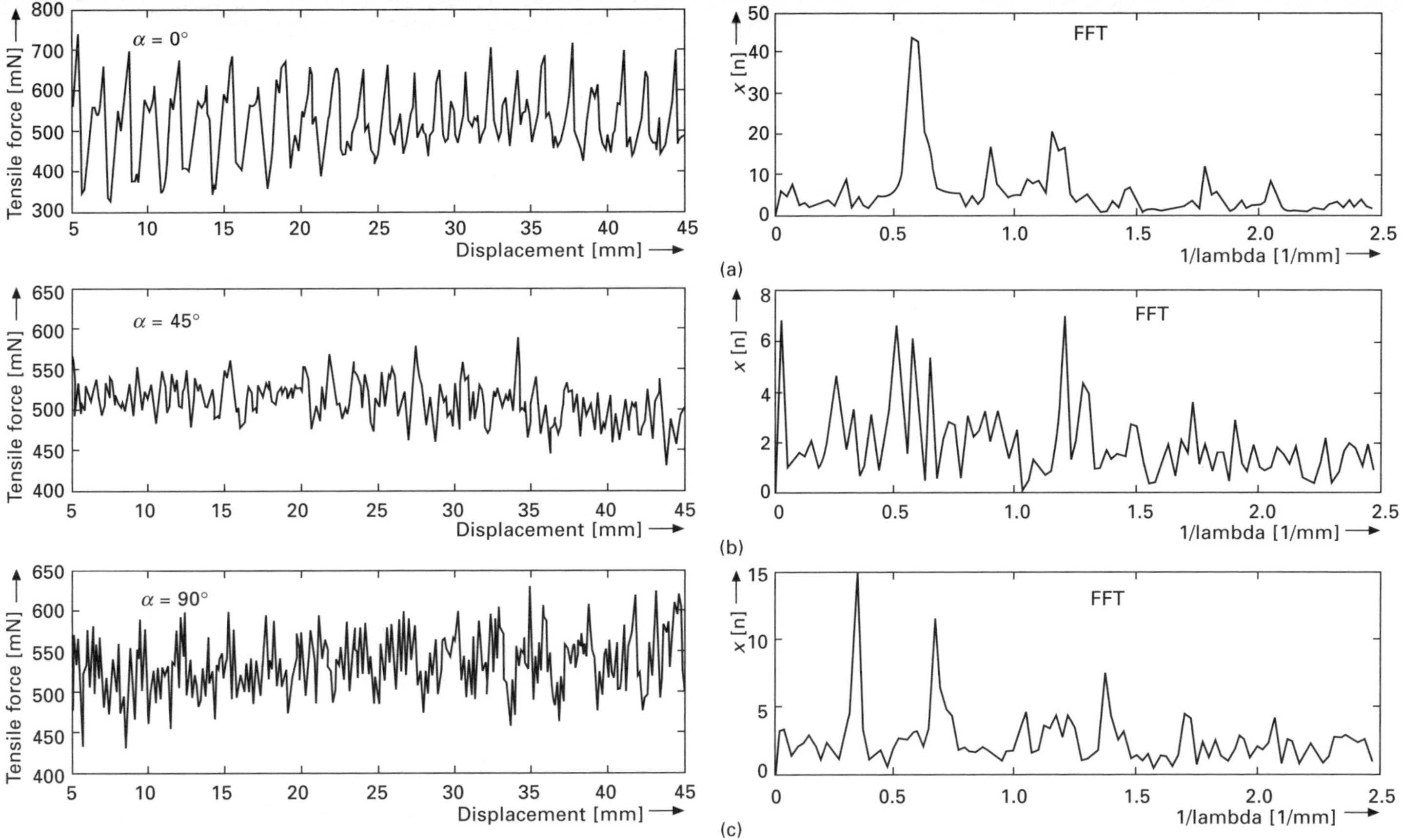

4.52 Friction and spectral analysis results of polyamide plain weave monofilament fabrics in fabric-to-fabric sliding at different angles.

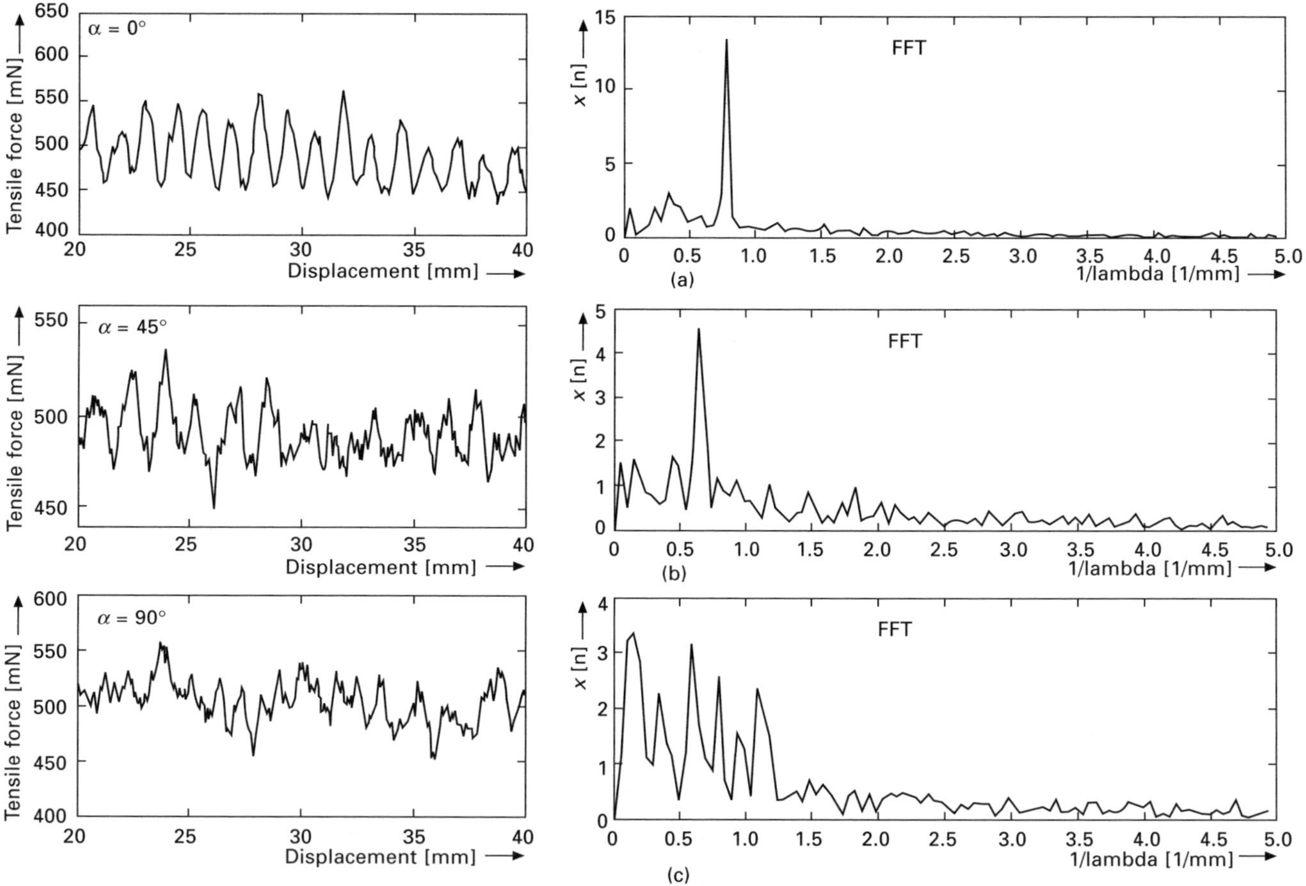

4.53 Friction and spectral analysis results of flax fabrics; normal tension 420 Pa.

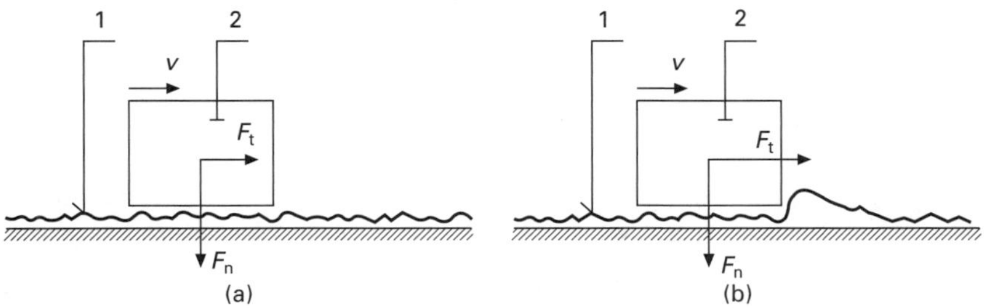

4.54 Illustration of the compounded effects of the surface topography and the mechanical properties [29].

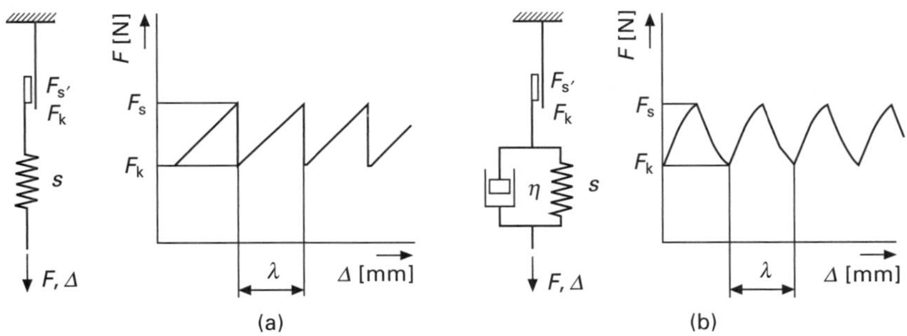

4.55 Rheological models of SSP.

A somewhat more complicated model is shown in Fig. 4.55(b) in which the elastic element is replaced by Kelvin's visco-elastic body. In this case, there is no steep change in force F; the relations involved become non-linear. The slope of the curve as well as the wavelength will depend on the values of the parameters s, η, F_s, F_k.

A dynamic model of SSP that includes frictional resistance, visco-elastic material behavior and acceleration associated with one or more masses is discussed in references [4, 26, 29]. Arrangement of elements was similar to those illustrated in Figs 4.56 and Figure 4.18(c). From equilibrium of body 4 in x direction we get:

$$F_f = F_r + m \cdot a \qquad 4.5$$

In this, F_f is the force of frictional resistance (in the slip part of movement $F_f = F_k$), F_r is the force from Kelvin's rheological body 5 (this force was experimentally measured) given as $F_r = \eta \cdot v + s \cdot x = \eta \cdot \frac{dx}{dt} + s \cdot x$, and $m \cdot a$ represents the dynamic force F_d, expressed by $m \cdot a = m \cdot \frac{d^2x}{dx^2}$. Body 4 of the mass m moves with the speed v and acceleration a with the average

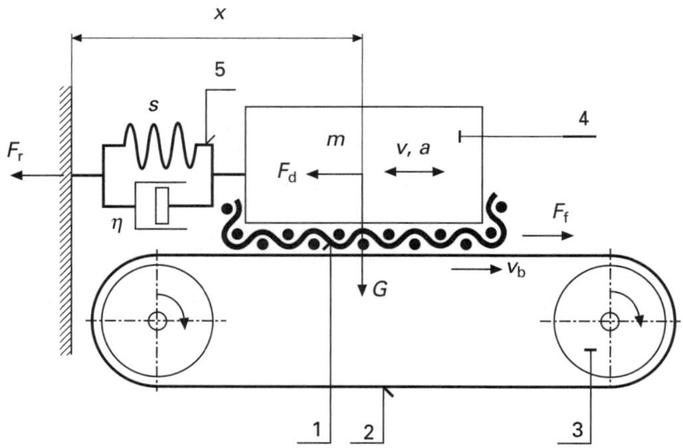

4.56 Dynamic model of SSP.

value of the speed v zero. So the movement of the sled or the mass may be described by the general equation:

$$F_f = m \cdot \frac{d^2 x}{dt^2} + \eta \cdot \frac{dx}{dt} + s \cdot x \tag{4.6}$$

In this, x is displacement, η is the parameter characterizing the property of the viscous element (damping coefficient), and s represents the stiffness of the spring. In equation (4.6), the term with the second derivative $a = \frac{d^2 x}{dt^2}$ represents dynamic force F_d connected with the mass acceleration, the term with the first derivative $v = dx/dt$ represents damping force dependent on the speed, and the linear term describes force from the spring.

4.6.2 Solution of differential equation

The main parameters of the system, described by Fig. 4.56 and by equation (4.6), are:

- x – the position (displacement) of the slipping body 4,
- m – the mass of the slipping body 4,
- v – the speed of the slipping body (the first derivative of displacement x),
- v_b – the speed of the belt 2,
- s – the stiffness of the elastic body (ideal spring),
- η – the viscosity of the damping body,
- μ_s – the static coefficient of friction (for static friction, $F_f = \mu_s \cdot G$, $G = m \cdot g$, g is the acceleration due to gravity, i.e. 9.81 m/sec^2),
- μ_k – the kinetic coefficient of friction (for kinetic friction, $F_f = \mu_k \cdot G$).

The assumption, used in the analysis, is that the values of the static and the kinetic coefficients of friction are constant and not dependent on the speed v. The result is, that the static coefficient of friction μ_s could be incorporated only in the case when the relative speed of the body 4 and the belt 2 (Fig. 4.56) is zero. At this stage of friction in each cycle, the speed of the slipping body will be the same as that of the belt and the movement will be described by the linear function, $x = v_b \cdot t$.

To make the solution of equation (4.6) easier, both sides are divided by the mass m. By making the substitutions: $p = \eta/m$, $q = s/m$ and $F = F_f/m = F_k/m$, one gets,

$$\frac{d^2x}{dt^2} + p \cdot \frac{dx}{dt} + q \cdot x = F \qquad 4.7$$

The solution [20] leads to a function whose derivative contains the original function, i.e. $x = e^{\chi \cdot t}$, $x' = \chi e^{\chi \cdot t}$, $x'' = \chi^2 e^{\chi \cdot t}$. After incorporating this exponential function into (4.7) and setting the right side equal to zero ($F = 0$), we get:

$$\chi^2 \cdot e^{\chi \cdot t} + \chi \cdot p \cdot e^{\chi \cdot t} + q \cdot e^{\chi \cdot t} = 0 \quad \text{or} \quad \chi^2 + \chi \cdot p + q = 0 \qquad 4.8$$

The solution of this quadratic formula is

$$\chi_{1,2} = \frac{-p \pm \sqrt{p^2 - 4q}}{2} = -\frac{p}{2} \pm u, \quad \text{where} \quad u = \frac{\sqrt{p^2 - 4q}}{2} \qquad 4.9$$

For our purpose, there are two main possibilities (the discriminant, $p^2 - 4q$, can be either a positive or a negative number):

Positive discriminant, $p^2 - 4q > 0$

In this case, there are two real roots of equation (4.8), i.e. χ_1 and χ_2, and the solution, ignoring details, will be:

$$x = x_g + x_p = A \cdot e^{\chi_1 t} + B \cdot e^{\chi_2 t} + \frac{F}{q}$$

$$= e^{-\frac{p}{2} \cdot t} \cdot (A \cdot e^{u \cdot t} + B \cdot e^{-u \cdot t}) + \frac{F}{q} \qquad 4.10$$

in which the "total" solution x of differential equation is the sum of the "general" solution $x_g = A \cdot e^{\chi_1 t} + B \cdot e^{\chi_1 t}$ of equation (4.7), without the right side (i.e. with $F = 0$), and of the "particular" solution x_p, found as any solution with the right side included. The latter in equation (4.10) was easily found from (4.7) when the displacement was set as constant (x = const.). In this case, all derivatives will be zero ($x' = x'' = 0$) and particular solution will be $x_p = \dfrac{F}{q} = \dfrac{F_k}{m \cdot q}$.

The first derivative of (4.10), speed v, is:

$$x' = -\frac{p}{2} \cdot e^{-\frac{p}{2} \cdot t} \cdot (A \cdot e^{u \cdot t} + B \cdot e^{-u \cdot t}) + e^{-\frac{p}{2} \cdot t} \cdot u \cdot (A \cdot e^{u \cdot t} - B \cdot e^{-u \cdot t}) \tag{4.11}$$

The unknown parameters A and B could be determined by specifying the initial conditions; for example, by:

$x(0) = 0$, i.e. the beginning of the curve is set at point $t = 0$ and $x = 0$, and the derivative $x'(0) = v$; it insures, that the slope at the initial part of the curve will correspond with the slope of the linear movement (v_b is the speed of the belt, Fig. 4.56). Higher value of the static frictional force $F_s > F_r$ will be operative only when no mutual slip of the body 4 and the belt 2 occurs.

The function (4.10) is not periodic and so this case does not support SSP. The speed of the slipping body v would reach v_b, the speed of the belt, only once, i.e. at the beginning of the process, after it, the speed v will be decreasing. The behavior of the function (4.10) is shown in Fig. 4.57. The main parameters for this were set as: mass $m = 1$, speed of the belt $v = 1$, static friction coefficient $\mu_s = 0.5$, kinetic friction coefficient $\mu_k = 0.3$, and stiffness of the spring $s = 1$. Viscosity η was considered a variable and set at 2.01 (to meet the condition, $p^2 - 4q > 0$, or $\eta^2 - 4s > 0$), 4, 8, 16 and 32.

The aim in this treatment was only to display the behavior given by equation (4.7) qualitatively. Accordingly, no dimensions for the parameters were used.

Negative discriminant, $p^2 - 4q < 0$

This case leads to irrational roots of (4.8) and to solution (4.12):

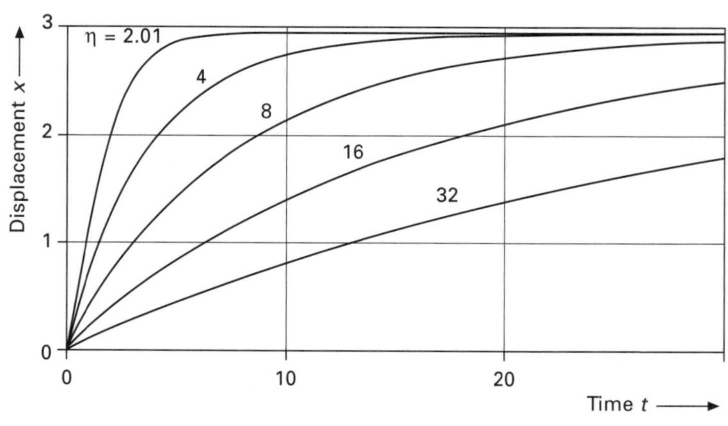

4.57 Influence of viscosity η on function $x = f(t)$ when $p^2 - 4q > 0$ or $\eta^2 - 4s > 0$.

$$\chi_{1,2} = \frac{-p \pm \sqrt{p^2 - 4q}}{2} = -\frac{p}{2} \pm r,$$

where $$r = \sqrt{-1} \cdot \frac{\sqrt{4q - p^2}}{2} = i \cdot \frac{\sqrt{4q - p^2}}{2} \qquad 4.12$$

The solution of equation (4.7) in this case will be as given below ($F/q = x_p$ is again the "particular" solution, the same as in previous case):

$$x = e^{-\frac{p}{2} \cdot t} \cdot (A \cdot \cos(r \cdot t) + B \cdot \sin(r \cdot t)) + \frac{F_k}{q} \qquad 4.13$$

with the first derivative being:

$$x' = e^{-\frac{p}{2} \cdot t} \times \left(-\frac{p}{2}(A \cdot \cos(r \cdot t) + B \cdot \sin(r \cdot t)) + r(-A \cdot \sin(r \cdot t) + B \cdot \cos(r \cdot t)) \right) \qquad 4.14$$

The function (4.13) is periodical and, under suitable conditions, should enable SSP. The necessary condition is that in the course of oscillation the speed of the body (the first derivative of the displacement $v = x'$) must reach the speed v_b of the belt. Only then could the slipping body be captured by the belt and move with it at constant speed v_b till the force F_r from the rheological body (Fig. 4.56) reaches the value of the static frictional force F_s and overcomes it. So the SSP oscillation is composed of two characteristic parts:

1. Linear motion with the friction body moving with the speed v_b, with a linear increase of the force from the rheological body F_r, and
2. Harmonic motion with dampening (continuously decreasing) amplitude. The kinetic frictional force F_k is constant and corresponds with the value of the kinetic friction coefficient μ_k. The force F_r is variable due to variable dynamic force $m \cdot a$.

4.6.3 Construction of the stick-slip curve

General

In Fig. 4.58, the two mentioned characteristic parts of the displacement of the frictional body, reflecting SSP, are shown. In the initial linear part MN, starting at displacement $x = 0$, the static friction coefficient μ_s is involved. The second harmonic part NO involves the kinetic friction coefficient μ_k. The linear part OP follows, which begins at the value of x, corresponding with the conditions described later. Thinner dashed line shows a plot of the continuous function (4.13), which reflects damped vibrations. Its use is made

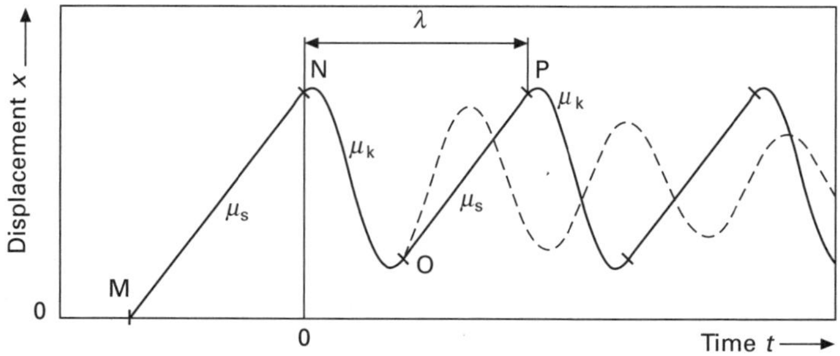

4.58 Characteristic parts of SSP.

only for establishing the coordinates of the point O, needed for constructing SSP. For establishing the position of point O, it is necessary that at the end of the first harmonic cycle, the slipping body reaches the speed (equation 4.14) equaling that of the belt v_b. At this point the body sticks to the belt and the next stick-slip cycle commences.

Co-ordinates of points for SSP and values of the parameters A and B

A crucial problem was the calculation of the values of the parameters A and B in (4.13) using initial conditions of our virtual experiment. It is clear, that the first derivative of x with respect to time (the speed v), involved at the points where the linear and the harmonic parts of the movement meet, i.e. at the points N, O, P, etc., is identical (Fig. 4.58). Any difference between the speeds of the slipping body and the belt would be connected with a steep change in force. So the part of the curve NO, described by equation (4.13), must begin and end with $x' = v_b$.

The second necessary initial condition was rather problematic, as one needed to determine where to set the origin of the coordinates. It was found necessary to set $t = 0$ corresponding the point N (Fig. 4.58).

Co-ordinates of the point N

The x-co-ordinate of this point could now easily be calculated from the linear part MN: the force F_r from the rheological body is $F_r = s \cdot x + \eta \cdot v_b$ and at the point N it must just overcome static frictional resistance, or F_s, which is $m \cdot g \cdot \mu_s$, where g is the gravitational constant ($g = 9.81$). For $F_r = F_s$, displacement of the point N will be

$$x(\mathrm{N}) = \frac{m \cdot g \cdot \mu_s - \eta \cdot v_b}{s} \qquad 4.15$$

Parameters A and B

From these conditions one can now solve equation (4.13) for the value of x at $t = 0$ and obtain the values of the parameters A and B.

- Calculation of the parameter A from (4.13, 4.15) at time $t = 0$:

$$e^0 = 1, \cos(0) = 1, \sin(0) = 0:$$

$$x(0) = 1 \cdot (A \cdot 1 + B \cdot 0) + \frac{F_k}{m \cdot q} = A + \frac{F_k}{m \cdot q} = \frac{(m \cdot g \cdot \mu_s - \eta \cdot v_b)}{s} \qquad 4.16$$

or, $$A = \frac{(m \cdot g \cdot \mu_s - \eta \cdot v_b)}{s} - \frac{F_k}{m \cdot q}$$

- Calculation of the parameter B from (4.14) at time $t = 0$ and speed $x' = v_b$:

$$x'(0) = 1 \cdot \left(-\frac{p}{2} \cdot (A \cdot 1 + B \cdot 0) + r \cdot (-A \cdot 0 + B \cdot 1) \right) = v_b;$$

hence $$-A \cdot \frac{p}{2} + B \cdot r = v_b \qquad 4.17$$

or, $$B = \frac{v_b + A \cdot \frac{p}{2}}{r} = \frac{v_b + \left(\frac{(m \cdot g \cdot \mu_s - \eta \cdot v_b)}{s} - \frac{F_k}{m \cdot q} \right) \cdot \frac{p}{2}}{r}$$

4.6.4 The dependence of SSP oscillations on system parameters

The figures that follow illustrate the influence of the basic variables on the movement of the frictional body in time. The parameters whose effects were examined are listed in Table 4.5.

For each of the six parameters, two figures are presented. Shown in the first are the variations in x and x' as function of t for all four values of a particular parameter. In time interval ⟨–10, 0⟩ the movement is linear with

Table 4.5 Parameters whose effects on $x(t)$ and $x'(t)$ are examined

Parameter	Standard value	Variable values			
(a) Mass of frictional body m	1	0.25	0.5	1	2
(b) Stiffness of the spring s	1	0.5	1	2	4
(c) Viscosity (damping coefficient) η:	0.1	0.02	0.05	0.1	0.5
(d) Speed of the belt v_b:	1	0.5	1	2	4
(e) Static friction coefficient μ_s:	0.5	0.3	0.4	0.6	0.8
(f) Kinetic friction coefficient μ_k:	0.3	0.0	0.1	0.3	0.5

the constant speed, i.e., v_b (the speeds of body 4, Fig. 4.56 and the belt are same). In time interval ⟨0, 30⟩ the curves of damped vibrations are generated in accordance with (4.13), for the displacement x, and (4.14), for the first derivative x'.

In the second figure, the curves showing periodical movement are drawn separately for each different value of the selected parameter. The values are generated for the time increasing in steps of 0.1 second (300 points for each line). SSP is also constructed if the conditions favored it. For it, it was necessary that at some point during the first harmonic phase, the speed of the slipping body coincided with that of the belt. This would lead to sticking and to the repetition of the stick-slip cycle.

Illustrated in the second figure are separately introduced lines, describing periodical movement if the conditions allowed SSP (see description of Fig. 4.58). The step of computation was 0.1 second (300 points at each line from $t = 0$).

The effect of the mass of the frictional body, m (Figs 4.59 and 4.60)

Increase of the mass is connected with higher amplitude, because F_s, given by $m \cdot g \cdot \mu_s$ (4.15), increases and so also with lower frequency or higher

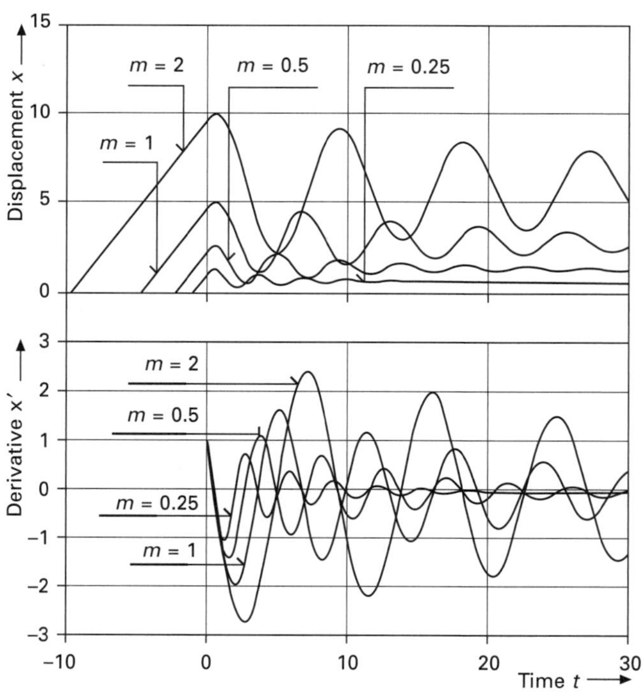

4.59 The effect of the mass m on x and x'.

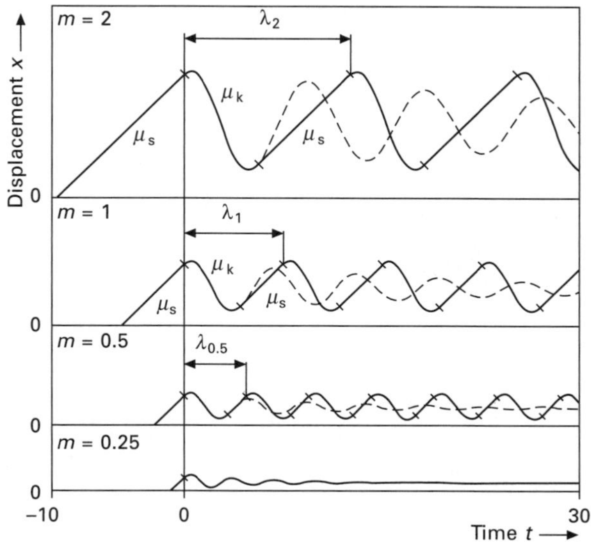

4.60 The effect of the mass *m* on *x* reflecting SSP.

wavelength λ of vibrations. The mass of 0.25 is below the critical level required. In other words, the speed of the oscillating body does not reach the selected belt speed (the speed of the belt is 1 but the maximum value of x' at $m = 0.25$ is only 0.765). Accordingly, Fig. 4.60 ($m = 0.25$) only shows the curve of damped vibration.

The effect of the stiffness of the spring, s (Figs 4.61 and 4.62)

As expected, the higher the stiffness of the spring the lower the amplitude and the wave length λ (higher frequency) of oscillation.

The effect of the viscosity η of the damping element (Figs 4.63 and 4.64)

High viscosity causes a reduction in the amplitude of oscillation and can exclude SSP, as the required speed may not be reached due to strongly damped vibrations. In our case, if $\eta = 0.5$, the first derivative x' of the displacement does not reach the required speed of 1 even in the first cycle.

The effect of the speed of the belt, v (Figs 4.65 and 4.66)

This parameter has a strong effect on the linear part of the traverse, a low influence on the amplitude of the harmonic part, and only a negligible effect on the frequency of oscillation. The highest speed v of the belt used, i.e. $v = 4$, does not allow SSP as x' does not reach the value 4.

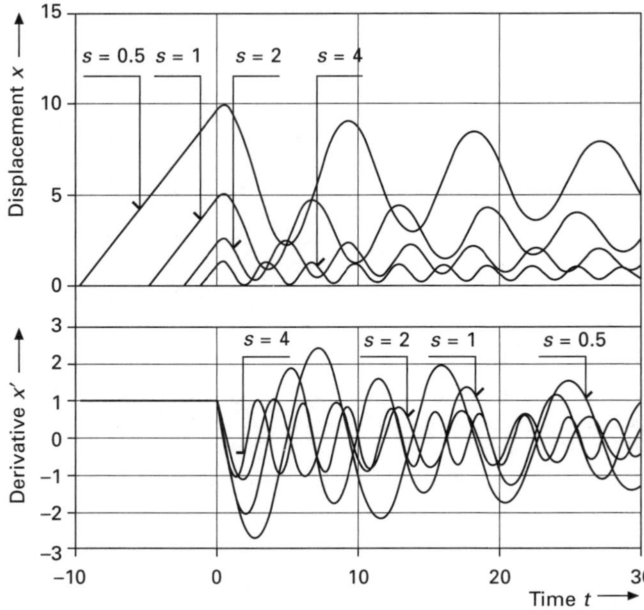

4.61 The effect of the magnitude of the spring stiffness s on x and x'.

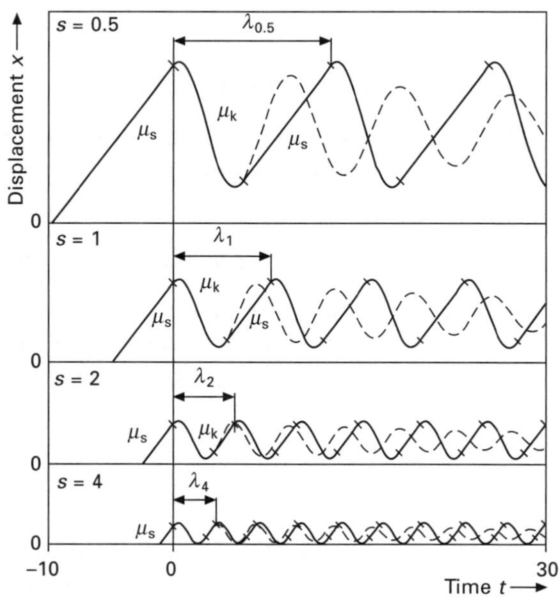

4.62 The effect of the spring stiffness s on x reflecting SSP.

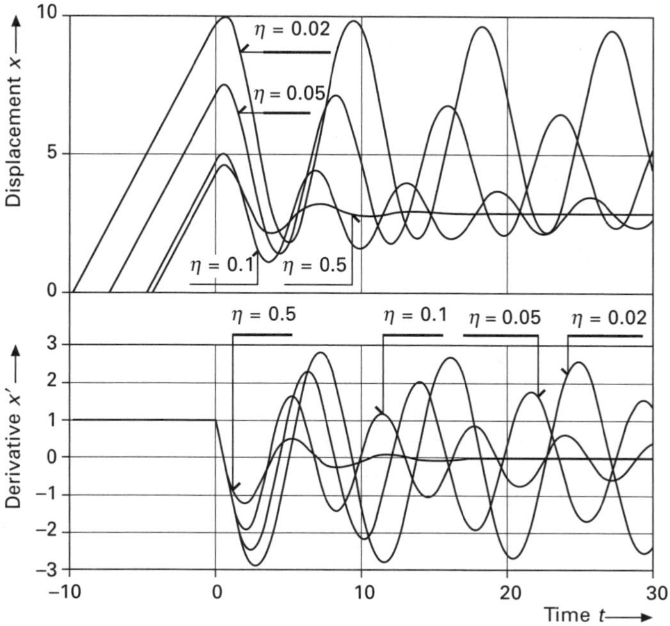

4.63 The effect of the damping coefficient η on x and x'.

The effect of static friction coefficient, μ_s (Figs 4.67 and 4.68)

The static friction coefficient affects the displacement x(N) that corresponds the force needed to overcome the frictional resistance F_s and, therefore, significantly influences the amplitude of oscillation. The lowest value, $\mu_s = 0.3$, does not allow SSP.

The effect of kinetic friction coefficient, μ_k (Figs 4.69 and 4.70)

An increase in the value of the kinetic friction, with the static maintained constant, causes a decrease in the amplitude and moves the second stick point (O, Fig. 4.58) closer to the stick point (N) in subsequent cycles. If the difference, $\mu_s - \mu_k = 0$, i.e. $\mu_k = 0.5$, then the slipping body oscillates around the position x(N) at which the frictional resistance is initially overcome. If, instead, $\mu_k = 0$, the axis of oscillation coincides with the position, $x = 0$.

4.6.5 The SSP model and the results in Section 4.5

The examples of computations given in the previous Section 4.6.4, support the experimental results presented in Section 4.5 to varying degrees. The primary reason for any departure noted would obviously be the idealistic nature of the assumptions made in modeling the behavior. However, in spite

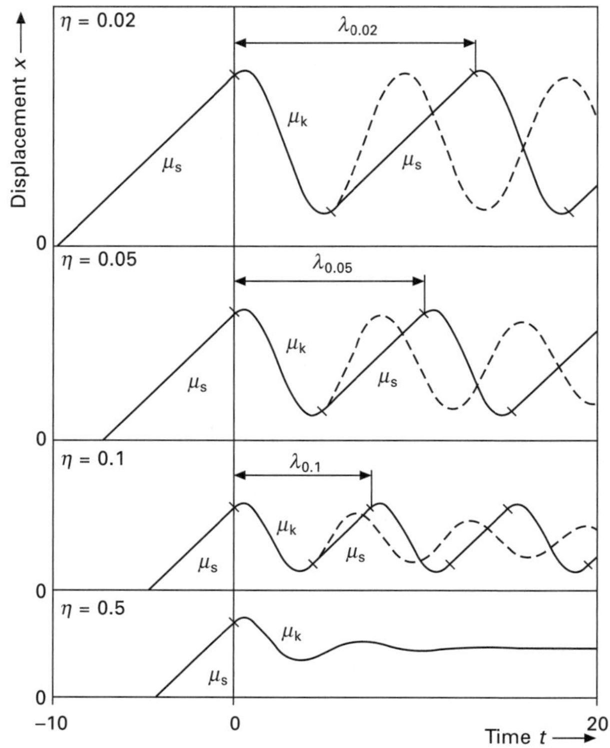

4.64 The effect of the damping coefficient η on x reflecting SSP.

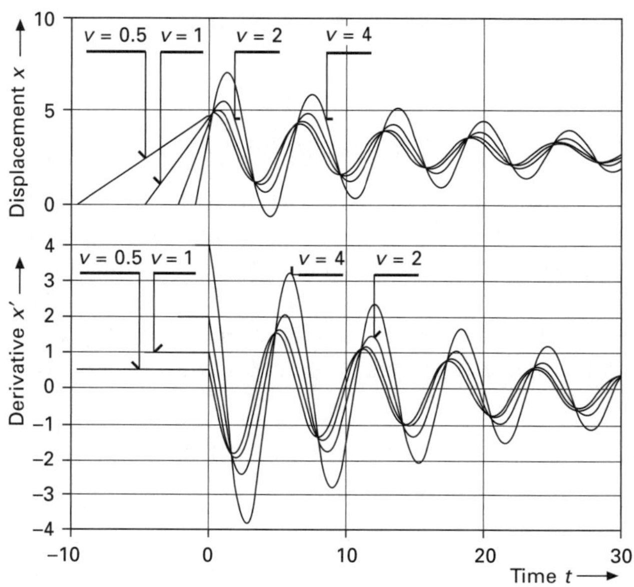

4.65 The effect of speed v on x and x'.

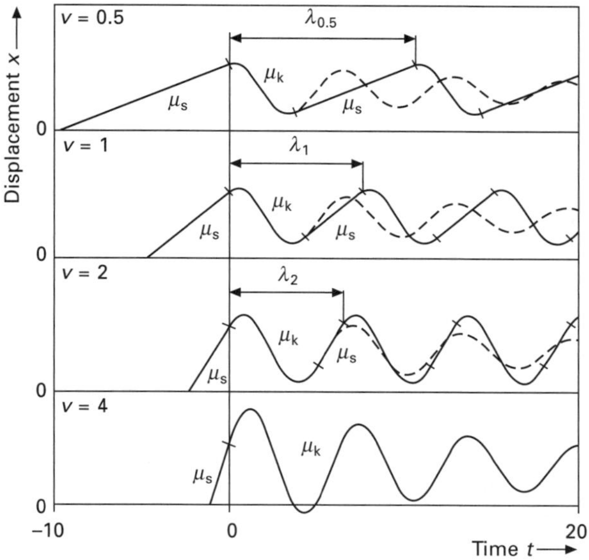

4.66 The effect of speed *v* on *x* reflecting SSP.

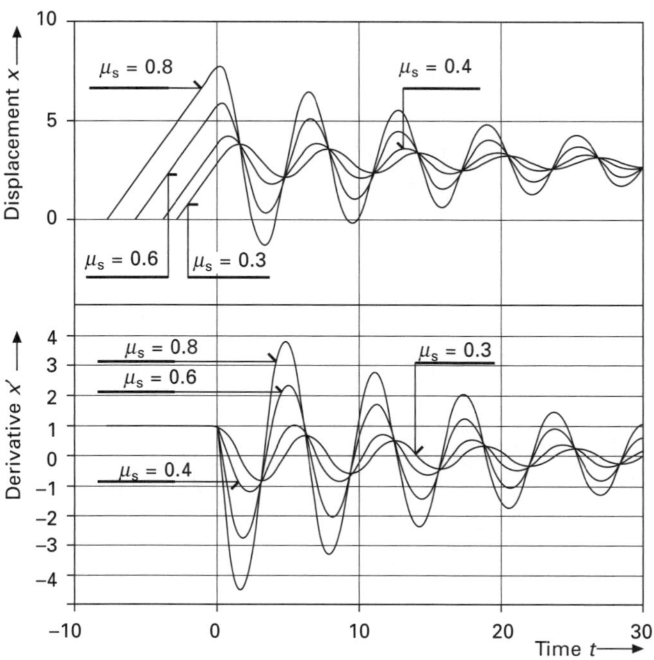

4.67 The effect of the static friction coefficient, μ_s, on *x* and *x′*.

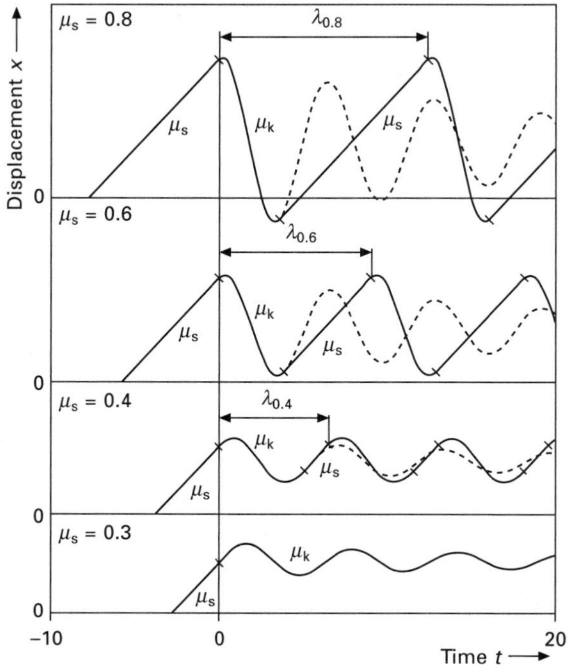

4.68 The effect of the static friction coefficient, μ_s, on x reflecting SSP.

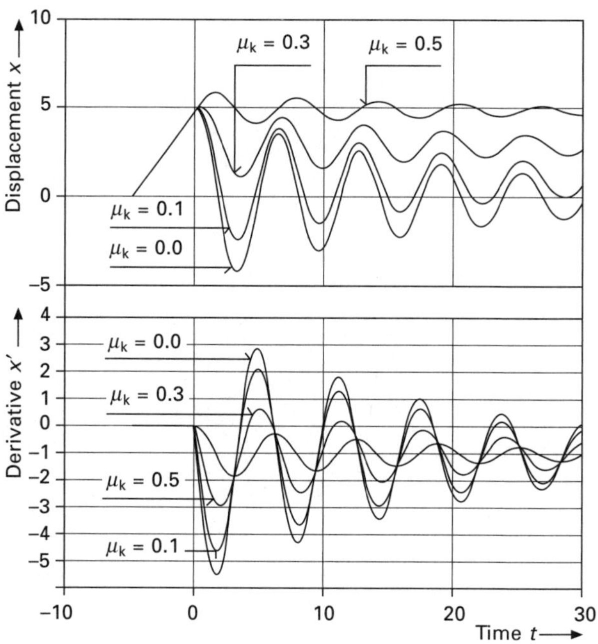

4.69 The effect of the kinetic friction coefficient, μ_k, on x and x'.

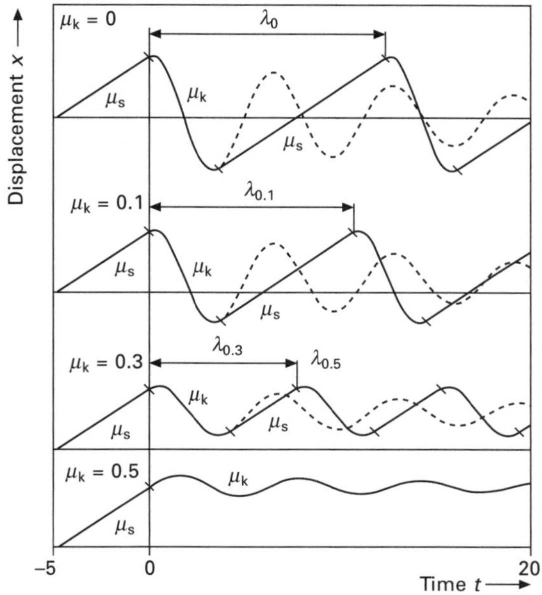

4.70 The effect of the kinetic friction coefficient, μ_k, on x reflecting SSP.

of the simplistic approach taken, the model considered gives insight into the nature of the effects and in many instances yields trends that support those found experimentally. This is illustrated for one set of conditions (m = 2, s = 1, $\eta = 0.2$, $v_b = 0.3$, $\mu_s = 0.5$ and $\mu_k = 0.3$) in Fig. 4.71. In part (a), the force F_r from rheological body 5 (Fig. 4.56) is illustrated, calculated in accordance with $F_r = x \cdot s + x' \cdot \eta$. This force can be measured on a dynamometer. In part (b) the overall frictional force F_f is shown; step fall at $t = 0$ is caused by the transition of static to kinetic friction. The actual measurement of this force is difficult. Shown in part (c), is the resultant force F_{SSP}, illustrating SSP.

The parameters were selected to provide trends noted experimentally (e.g. Figs 4.37 and 4.44). The speed v_b is responsible for the gradient of the first part of the F_{SSP} curve. This is lower than the gradient of the second part (slip). Features such as these, found in the curve computed from the model (Fig. 4.71), correspond with those noted in experiments.

4.7 Scope for future research on SSP in textiles

A wealth of information about textile surfaces, bulk properties and geometries, can be gained by measuring frictional properties and carefully analyzing the resulting stick-slip profile. The latter is influenced by a material's bulk and surface properties, topology, surface finish or contamination, and the geometry and characteristics of the measuring device. Precise prediction of a friction

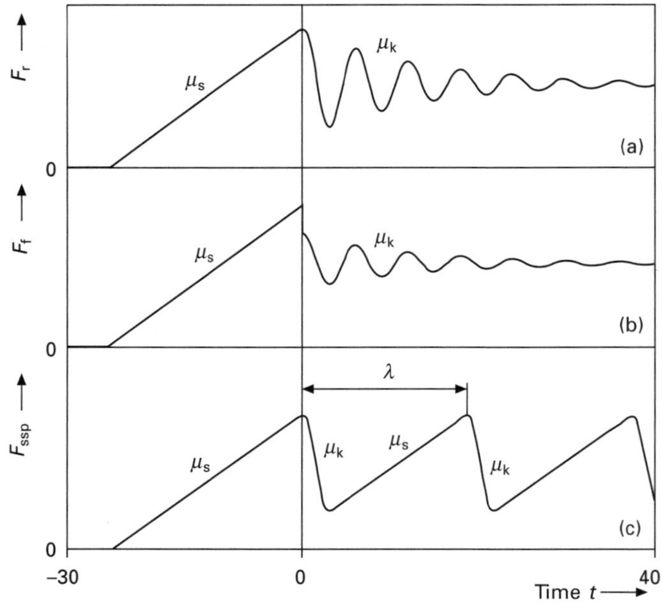

4.71 An example illustrating the construction of SSP.

profile, a goal in any research, is, however, almost impossible. This is because of the following reasons:

- Surface changes with time due to oxidation, abrasion, UV degradation.
- Bulk properties change with environment (moisture and humidity), handling and prior history of loading.
- The geometries, the dynamics of testing, and the electronics of the measuring devices play significant roles in influencing the recorded behavior.

It is clearly understood that friction profile or SSP has its root at the nano level; the other levels, i.e. the micro and the macro, modify the results by superimposing their own effects. New avenues for research on SSP may become available with the advancement of the work in nanotechnology.

Since friction, an important phenomenon in textiles, is dependent on such a large number of variables, increased attention should be placed on developing new methods that evaluate forces under conditions that accurately simulate an intended process or application.

Friction is said to play both a positive and a negative role in processing and in the use of products. An example of negative role may be the sound emission, which is significantly connected with SSP. An understanding and control of SSP will provide a basis for controlling undesirable acoustics during processing and use of textiles.

4.8 Sources of further information

There are numerous papers published on friction but only a few have focused on an understanding and evaluation of the stick-slip phenomenon. The reader is directed to the books by Bowden and Tabor, the articles on friction in the Journal of the Textile Institute, the Textile Research Journal, the Journal of Tribology, and to the list of the references that follow.

4.9 Glossary

Symbol	Definition, unit	Dimension	Description
c, w		m	Courses and wales spacing
E	Pa	m^{-1} kg s^{-2}	Modulus of elasticity
F, F_g	N	m kg s^{-2}	Force, gravity force
F_a, F_n	N	m kg s^{-2}	Axial force, normal force
F_s, F_k	N	m kg s^{-2}	Static, kinetic frictional force
F_t, F_μ	N	m kg s^{-2}	Tangential force, frictional force
l		m	Length
m		kg	Mass
p		m	Yarn spacing, pitch
s	N · m^{-1}	kg s^{-2}	Stiffness
SSP	–	–	Stick–slip phenomenon
t		s	Time
v		m s^{-1}	Speed
x, y		m	Co-ordinates
α, β	rad	1	Angle of frictional contact, angle
ε	$\Delta l/l_0$	1	Strain, relative elongation
η		m^{-1} kg s^{-1}	Viscosity
λ		m	Wavelength
μ	F_T/F_N	1	Coefficient of the slip friction
σ	Pa	m^{-1} kg s^{-2}	Stress, tension
φ	arctg μ, rad	1	Angle of friction, angle

4.10 Acknowledgements

A vast majority of the results included in this chapter were based on elaborate experiments conducted at the Technical University of Liberec, Czech Republic. This work was supported by the research project "Research Centre Textile" No. 1M4674788501 of Czech Ministry of Education. The authors thank the Czech Ministry of Education for its support.

4.11 References

1. Abbott G M, Grosberg P and Leaf G A, 'The mechanical properties of woven fabrics. Part VII: The hysteresis during bending of woven fabrics', *Textile Res. J.*, 1971, **41**(4/12), 345.

2. Ajayi J O, 'Fabric smoothness, friction, and handle', *Textile Res. J.*, 1992, **62**, 52
3. Ajayi J O, 'Effects of fabric structure on frictional properties', *Textile Res. J.*, 1992, **62**(2/12), 87.
4. Astrom K J, 'Control of systems with friction', *Lund Institute of Technology, Sweden*, WWW: http://www.control.lth.se.
5. Behary N, Caze C, Perwuelz A and El Achari A, 'Tribology of sized glass fibers. Part II: Using an electric microbalance technique to study stick-lip behavior', *Textile Res. J.*, 2001, **71** (3/12), 187.
6. Benenson W, Harris J W, Stocker H and Lutz H, *Handbook of Physics*, Springer, 2002.
7. El Mogahzy Y E and Broughton M R, 'A new approach for evaluating the frictional behavior of cotton fibers. Part I: Fundamental aspects and measuring techniques', *Textile Res. J.*, 1993, **63**(8/12), 465.
8. El Mogahzy Y E and Gupta B S, 'Friction in fibrous materials. Part II: Experimental study of the effects of structural and morphological factors', *Textile Res. J.*, 1993, **63**(4/12), 219.
9. Feynman R P, Leighton R B and Sands M, *The Feynman Lectures on Physics*, 4th Edition, London, Addison-Wesley, Paolo Alto, 1966.
10. Grosberg P and Swani N M, 'The mechanical properties of woven fabrics. Part IV: The determination of the bending rigidity and frictional restraint in woven fabrics', *Textile Res. J.*, 1966, **36**(4/12), 338.
11. Grosberg P, 'The mechanical properties of woven fabrics. Part II: The bending of woven fabrics', *Textile Res. J.*, 1966, **36**(3/12), 205.
12. Grosberg P and Kedia S, 'The mechanical properties of woven fabrics. Part I: The initial load extension modulus of woven fabrics', *Textile Res. J.*, 1966, **36**(1/12), 71.
13. Gupta B S. and El Mogahzy Y E, 'Friction in fibrous materials. Part I: Structural model', *Text. Res. J.*, 1991, **61**(9/12), 547.
14. Hosseini Ravandi S A, Toriumi K and Matsumoto Y, 'Spectral analysis of the stick – slip motion of dynamic friction in the fabric surface', *Textile Res. J.*, 1994, **64**(4/12), 224.
15. Howell H G, Mieszkis K W and Tabor D, *Friction in Textiles*, Butterworths Scientific Publications, London, 1959.
16. Jong H G, 'Yarn-to-yarn friction in relation to some properties of fiber materials', *Text. Res. J.*, 1993, **63**(1/12), 14.
17. Kalyanaraman A R, 'Yarn-friction studies with the SITRA friction-measuring device', *J. Text. Inst.*, 1988, **79**(1/12), 147.
18. Kovar R, 'Stick-slip effect in knitted fabric deformation', *International Textile Clothing and Design Conference*, University of Zagreb, Dubrovnik, Croatia, 2002.
19. Kovar R and Kus Z, 'Periodicity in textile friction', *Proceedings of Conference Strutex*, Technical University of Liberec, 2003.
20. Kovar R, Kovarova A, 'Rheological analysis of stick-slip phenomenon', *Proceedings of Conference Strutex*, Technical University of Liberec, 2005.
21. Kovar R and Rajkumar R, 'Low-speed friction of textiles', *Materials Engineering*, 2004, **11**, No. 1, 109, ISSN 1335–0803.
22. Kovar R, *Structure and properties of flat textiles* (in Czech), Lecture notes, Technical University of Liberec, 2003.
23. Kovar R, Kus Z and Rajkumar R, Unpublished results of experiments pursued at the Technical University of Liberec, 2003.

24. Kovar R and Konecny F, 'Stick-slip phenomenon on woven fabrics friction', *Proceedings of Conference Strutex*, Technical University of Liberec, 2003.
25. Koza W M, 'A new instrument to measure textile fiber friction by a yarn-to-yarn technique', *Textile Res. J.*, 1975, **45**(9/12), 639.
26. McMillan A J, 'A non-linear friction model for self-excited vibrations', *Journal of Sound and Vibration*, 1997, **205**(3), 323.
27. Sodomka L and Kus Z, 'Acoustic emission as a new tool for the textile structure', IFFTI International Conference, Hong-Kong, November 2002, CD-ROM.
28. Virto L and Naik A, 'Frictional behavior of textile fabrics. Part I: Sliding phenomena of fabrics on metallic and polymeric solid surfaces', *Textile Res. J.*, 1997, **67**(11/12), 793.
29. Virto L and Naik A, 'Frictional behavior of textile fabrics. Part II: Dynamic response for sliding friction', *Textile Res. J.*, 2000, **70**(3/12), 256.
30. Yam P, 'There's the Rub', *Scientific American*, 1991 (June) 14.
31. Zurek W, Jankowiak D and Frydrych I, 'Surface frictional resistance of fabrics woven from filament yarns', *Textile Res. J.*, 1985, **55**(2/12), 113.
32. Zurek W and Frydrych I, 'Comparative analysis of frictional resistance of wool yarns', *Textile Res. J.*, 1993, **63**(6/12), 322.

5

Experimental methods for analyzing friction in textiles

B S GUPTA, North Carolina State University, USA,
J O AJAYI, Abubakar University, Nigeria and
M KUTSENKO, Goulston Technologies, USA

5.1 Introduction

Textile to textile and textile to other surface friction plays an important role in governing the behavior of textiles in processing, wear and use. Whether the material is a fabric with complex structure or a single fiber, the contact in friction process can be considered to occur between two fibers or between a fiber and another surface. Accordingly, although frictional studies have been carried out at different levels in textile hierarchy (fiber, yarn, fabric and product), it has been of fundamental interest to be able to measure and understand the frictional behavior of single fibers.

A large number of test methods for measuring friction have been described in published literature. They differ from each other primarily in terms of:

- the nature of the contact maintained between sliding bodies – point, line or area
- the environment in which the test is conducted – air, water or lubricant
- the type of textile used – fiber, yarn or fabric, the latter including braided, nonwoven, knitted and woven structures
- the method by which the relative motion is actuated, the normal force is imposed and the friction force is measured.

The variables of the study, thus, have been the mode of contact, the type and morphology of the contacting materials, the magnitude of the normal force, the speed of sliding and the environment around the contact region. Howell, Mieszkis and Tabor have given a review of the early methods in their book published in 1959 [1]. A more recent monograph, by Hong and Jayaraman [2], also gives a timely review of the various methods used for testing friction in textiles. It should be noted that much effort has been devoted during the past several decades in automating the procedures, increasing the versatility of the device, and improving the precision by which the variables are maintained and the responses are recorded. Since friction during processing of fibers, especially of man-made origin, is controlled extensively with the use of topically applied lubricants, some of the methods incorporate a lubricant as a primary variable, with the speed of sliding and the nature of the second material as the associated variables.

In this chapter, the discussion presented is in three parts. In the first part (Sections 5.2–5.5), the general methods of testing textiles, along with the associated equations, are presented. In the next part (Sections 5.6–5.9), a detailed discussion is given on the testing and characterization of the frictional properties of the fabrics. Also included in this is a discussion on the characterization of the topographical features of fabrics and their correlations with friction and with the subjective ranking of hand. In the last part (Sections 5.10–5.12), commercially available devices and standard test methods for testing of friction in yarns or filaments and on fabrics, including nonwoven, are described.

5.2 Relevance of test methods to processing and use of textiles

The three major modes of contact used during measurements are the point, the line and the area. These correlate with the nature of contacts found in textiles during processing and use. Point contacts exist between metallic wires and fibers during carding, between needles and fibers during needle punching in the preparation of nonwoven, between needle and fibers during sewing in apparel manufacturing, and between fibers at the crossover points during stretching and bending of woven and knitted fabrics. It should be noted, however, that since textiles are soft and deformable, the seemingly point contact actually involves a small area of some shape. The line type of contact can be expected to exist between fibers during drafting in yarn formation, and during stretching and bending of yarns and yarn-based products in processing and use. Just as a point contact involves contact over some area, so does the line contact. Examples of area contact, i.e. contacts over significant geometrical areas, include those between two items of clothing worn by an individual, clothing and skin, clothing and upholstery, clothing and bed linen, stocking and shoe liner, and between footwear and flooring, including carpet.

5.3 Friction models and parameters

5.3.1 Amonton's Law and the coefficient of friction

All methods utilize the classical Amonton's Law [3] in calculating the value of the coefficient of friction:

$$F = \mu N \qquad 5.1$$

The friction force, F, is measured between two surfaces against a known value of the normal force, N. This allows the estimation of the value of the coefficient of friction, μ. If the value of the friction force varies during sliding, so does the computed value of μ. In most systems, at least two values are obtained: the static at the initiation of relative motion or at the start of slipping, and the kinetic that exists during sliding. Usually, the kinetic value is considered to lie approximately halfway between the slipping and the sticking values. The plot of force against time obtained is commonly stick-slip in nature, which is a characteristic of most materials. Generally, the softer and more viscose-elastic the material, the more prominent is the stick-slip pattern. The textile fibers fit this description. The profiles of yarns and fabrics turn out to be more complex than obtained on single fibers; this is because, in these assembly structures, the macroscopic roughness and bulk deformations in individual components play additional roles and influence the character of the trace.

5.3.2 The general model and the friction indices (a, a', n)

As discussed in detail in Chapter 3, in textile materials (fibers, yarns and fabrics) the friction force or stress does not vary linearly with normal force or stress, as is assumed in the Amonton's model, but non-linearly according to:

$$F = a N^n \text{ (with } F \text{ and } N \text{ in the units of force, N)} \quad 5.2$$

$$F = a' N^n \text{ (with } F \text{ and } N \text{ in the units of stress, Pa)} \quad 5.3$$

However, in friction tests on most materials, including textiles, the normal force is usually maintained constant. In these tests, therefore, the primary parameter assessed is still the coefficient of friction, μ, which may have two values, μ_s and μ_k. On the other hand, when the interest is in characterizing the behavior of a material in terms of the values of the constants a or a' and n, then it will be necessary that the tests for the friction force F are conducted at several values of N and the data of the logarithm of F is fitted against the logarithm of N by the least square method. If the tests are on a filament or a yarn, the values of F and N used are in the units of force (N), and the model used is the one given by Equation 5.2. The units of the constant a are $(\mathrm{N})^{1-n}$. If, on the other hand, the tests involve fabrics, then the values of F and N are more appropriately in the units of stress (Pa). The units of the constant a' are then $(\mathrm{Pa})^{1-n}$. In either case, the procedure for assessing the values of the associated constants a or a' and n is the same.

The above discussion is based on the models applying to the basic planar methods of measuring friction (Figs 5.3 and 5.13). Different equations and procedures apply for determining the values of the constants if the tests are conducted by the capstan or by the twist methods (Fig. 5.14 and 5.15). These are discussed in Section 5.5.4 for the capstan and 5.5.5 for the twist methods.

5.3.3 Friction parameters based on stick-slip profile

It has been noted repeatedly that a plot of tension against time during sliding reflects fluctuations. The first peak usually registers the highest force and corresponds to the static friction force, since at this point the adhesion junctions are broken for the first time before sliding commences. Most materials being visco-elastic, which tend to flow or creep, the area of contact and, therefore, the force required to break the junctions will depend on the time of contact prior to sliding. Once the relative motion has begun, the trace obtained generally oscillates but the cyclic variations may be irregular. As clear from the discussions in previous chapters, in particular Chapter 4, the profile found can vary from material to material, method to method, and morphology to morphology. The environment (fluid or air, relative humidity and temperature) and the testing conditions (normal force, size and nature of contact, and speed of sliding) influence the trace. Accordingly, it is useful to characterize the profile in terms of a number of parameters whose values can be used to understand and rationalize the behavior of a material in greater details. A hypothetical friction curve is shown in Fig. 5.1. The following parameters may be used to characterize its nature.

Frictional resistance (Fs_0, Fs, F_k)

Fs_0 may be called the static friction force, or the force at the first instance of sliding. This corresponds to the first peak in the profile and usually represents the highest value of the force. Following this, the force oscillates between peaks (registered at the instant of slips) and troughs (registered at the instant

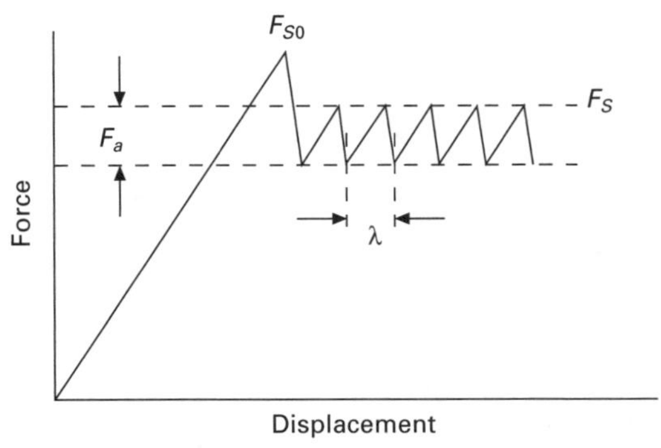

5.1 Hypothetical friction trace for a textile material.

of sticks), these at lower values than Fs_0. The mean of peaks, excluding the first, is the static value, or Fs. The average value of the force, i.e. the average value of the peaks and the troughs is taken to mean the kinetic value (see Section 2.4). This force will equal $Fs - 0.5\ Fa$, where Fa is the amplitude of vibrations (excluding first peak).

Amplitude of frictional resistance (F_a)

F_a is taken as the average height of the stick-slip pulses, excluding the first peak. This value is also sometimes referred to as the "scroop" since it relates to the subjective feeling of "scroopiness" of a surface.

Frequency of peaks (F_f)

The frequency F_f represents the average number of peaks per unit length of traverse. This will equal λ^{-1}, where λ is the average wavelength of the fluctuations.

5.4 Classification of test methods

Figure 5.2 gives schematics of the principles involved in the measurement of friction by various techniques. Its examination will indicate that the methods can be classified and discussed in three different groups:

1 The fundamental study methods based on Amonton's Law in which the motion of one body with respect to the other is rectilinear. This group accounts for the largest number of apparatuses developed and used for much of the fundamental studies of the property.
2 The capstan method in which one material, usually a textile, passes over a length of the curved surface of a cylindrical body, usually a non-textile material. These methods are suited for tests at high speed and often involve a lubricant at the interface.

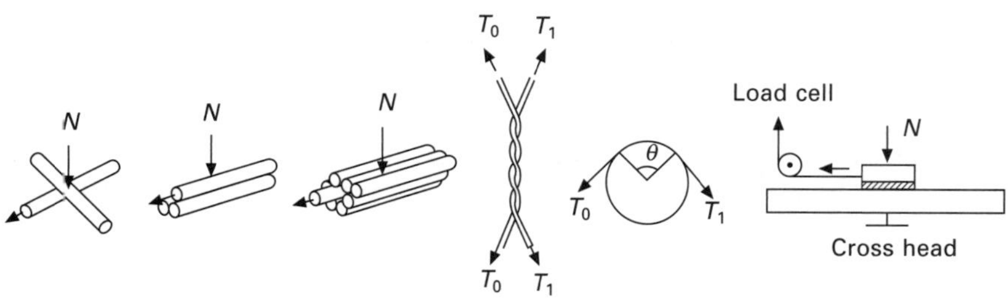

5.2 Illustration of the principles involved in various types of friction tests.

3 The twist method, in which two materials, yarns or filaments, are twisted together. Although in this the path of each body is helical, the actual contact is a line parallel to the axis of each material (see Figure 3.7 in Chapter 3).

A general discussion of the essential features of the methods in each group is given below.

5.5 Fundamental test methods based on rectilinear motion

The device of Bowden and Leben [4] (Fig. 5.3) represents the most fundamental general method for the study of friction between two surfaces. A rider under a load *N* presses onto a flat surface that traverses at a constant speed. It is connected to a tension measuring system, and rides with the moving body till the tension, *T*, developed just exceeds the static friction force. At this point, the rider starts to slip, and it continues to do so by accelerating back till the force has decreased from the static to the kinetic value. The rider continues to slip further till it has decelerated and come to rest. At this point it sticks and moves with the platform till it reaches the static or the slip value again. A plot of tension against time gives a typical stick-slip trace. If the damping is little or none, one can show that the kinetic friction force is equal to the mean force recorded during the slip. Consequently, values of both the static and the kinetic friction forces can be determined. This set up represents the basic principle of the classical method that makes use of the Amonton's Law given by Equation 5.1. With the value of *N* known and that of *F* measured, one can calculate the value of the coefficient of friction, μ. It will have two values; the static, μ_s, and the kinetic μ_k.

This versatile method has been adopted for tests involving point, line and area contacts and on fibers, yarns, and fabrics.

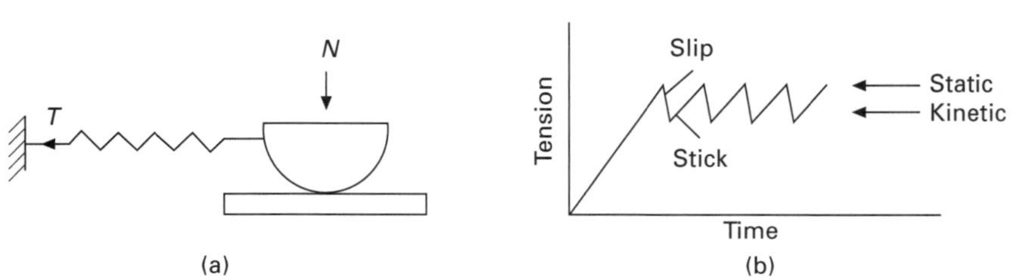

5.3 Bowden and Leben's apparatus which represents the principle of friction measurement by most devices [4]; (a) the schematic of the apparatus, (b) the friction trace.

5.5.1 Point contact

For work on single fibers of short length, the fibers are mounted on frames under light tension and then pressed against each other under known force. Several authors have used this technique [5–7], with one of the first being that due to Mercer and Makinson [5] illustrated in Fig. 5.4. In this technique, one fiber with the frame is traversed along a certain direction while a torsion wire, leaf spring, or other suitable tension-sensing device, restrains the movement of the other. An ingenious procedure, utilizing the principle of cantilever and suitable for use with very small loads (~ 10^{-6} N), was described by Pascoe and Tabor in 1956 [8]. Briscoe, Adams, and co-workers have used this method for conducting fundamental studies of friction between monofilaments [9–11]. The essential features of the apparatus are shown in Fig. 5.5. In this, a fiber is mounted at one end and the other end rests on a second fiber mounted in a frame that can be traversed in a horizontal plane. The upper fiber acts as a cantilever whose displacement in the vertical plane gives an estimate of the normal force and that in the horizontal an estimate of the friction force. The displacements of the free end are determined microscopically. The normal force N is calculated from vertical deflection Δh_1 (Fig. 5.6(a)) as:

$$\Delta h_1 = N\ell^3/3EI \qquad 5.4$$

In this, I is the second moment of inertia of the cross-section of fiber, which, with r as the radius, is given by $\pi r^4/4$, and E is the longitudinal elastic modulus of fiber. The free length from the point of contact is ℓ and is of the order of 1mm. Lateral movement causes the top fiber to bend because of the frictional force, F, which can be calculated from Δf (Fig. 5.6(c)) as:

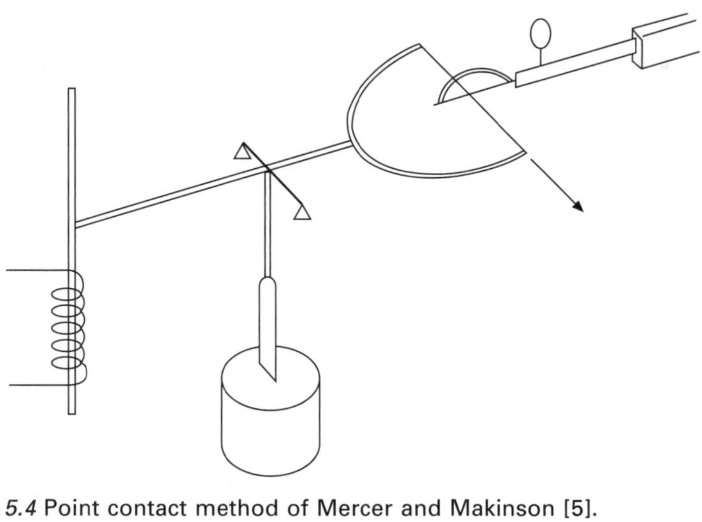

5.4 Point contact method of Mercer and Makinson [5].

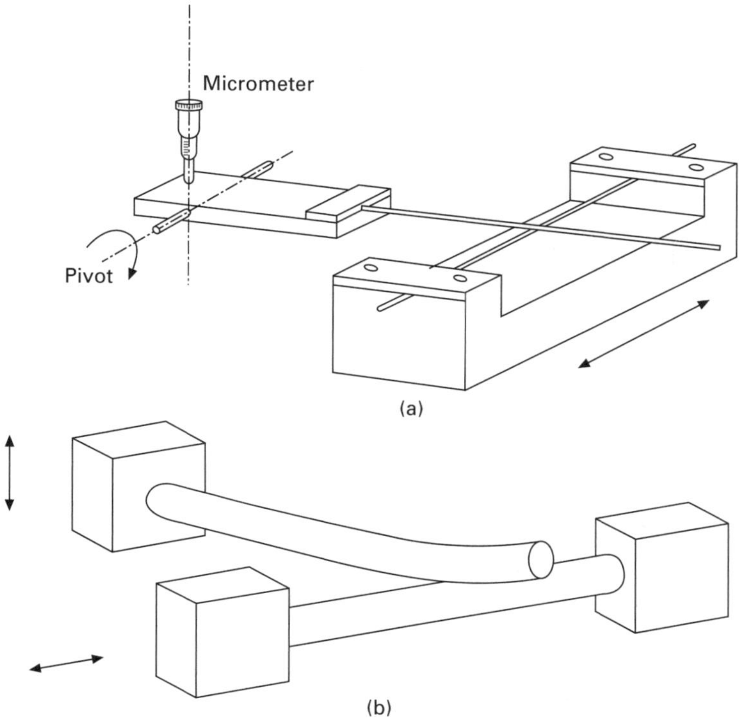

5.5 Pascoe and Tabor's apparatus involving cantilever system [8]

$$\Delta f = F\ell^3/3EI \quad 5.5$$

Briscoe *et al.* have used this method on a number of fibers and, in addition to studying the effects of many other factors, evaluated the role of surface energetics in influencing friction in dry and fluid environments. They determined the contribution of adhesive force to the total normal force. They assumed that the adhesive force due to attraction between molecules of two surfaces augmented normal force and determined the magnitude of the former by the procedure shown in Fig. 5.6(b). The top fiber, just touching the mounted fiber initially, is pulled up gradually till the two separate. The separation distance Δh_2, when substituted in Equation 5.4, allows calculation of the adhesive force. In their study, the authors were able to use normal forces in the range 10^{-6} to 10^{-8} N, which covered the magnitude of the surface interaction forces.

On long fibers or filaments, the procedure in which a fiber tensioned by deadweight on one end is suspended over a fiber mounted in a bow lends to convenient measurements of friction. The method, due to Howell [12], is illustrated in Fig. 5.7. The fiber in bow is advanced into the suspended fiber to achieve the desired normal force and then traversed as shown. Due to

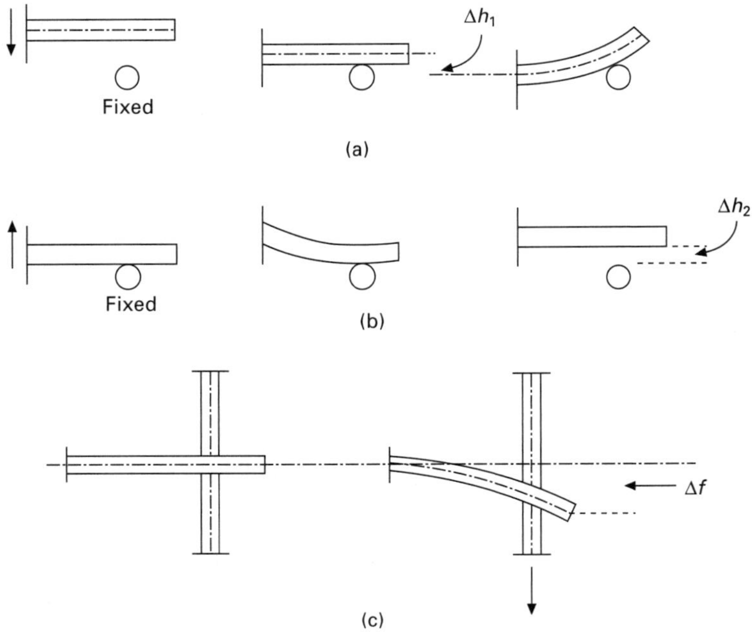

5.6 Adaptation of the cantilever device for measurement of frictional and adhesive forces between single filaments at light loads by Briscoe *et al.* [9]. (a) the normal force on the specimen ($\propto \Delta h_1$), (b) the adhesive force between the fibers ($\propto \Delta h_2$), and (c) the frictional force between the fibers ($\propto \Delta f$).

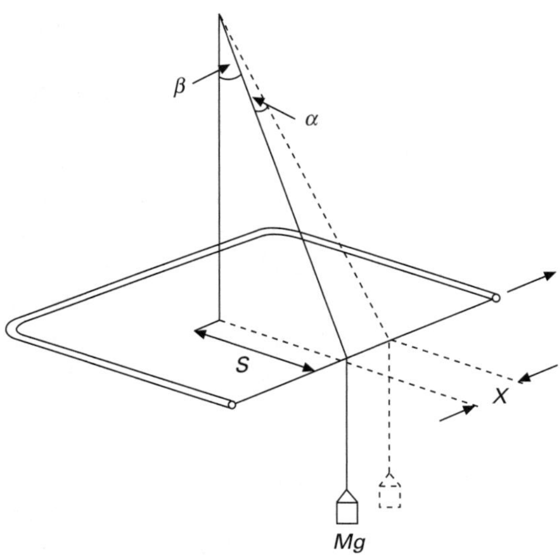

5.7 Point contact method of Howell involving hanging fiber [12].

friction, it carries the suspended fiber with it till the restoring force of the tension of the hanging fiber balances the force of friction. At this moment, slip occurs and the length X of the displacement is determined. From the forces acting at the contact, it can be shown that for small angles and displacements:

$$\mu = (Mg \sin \alpha)/(Mg \sin \beta) = X/S \tag{5.6}$$

An even simpler and more convenient method, illustrated in Fig. 5.8, is used by El Mogahzy and Gupta [13]. The suspended fiber is led over a frictionless pulley and attached directly to the load cell of a tensile tester. A frame holding the bow (with the second fiber mounted in it), the advancing mechanism, and the pulley, is fixed on the crosshead. When the crosshead moves down, the suspended fiber traverses over the mounted and the resistance offered is recorded on a chart. Two methods can be used to estimate μ: the force analysis and the capstan analysis [14]. The details of the former are illustrated in Fig. 5.8(c). The fiber diameter being small, the contact is assumed to be a point. T_0 and T_1 are the tensions existing on the free end and the withdrawing end, respectively, and N is the normal force acting at the midpoint of the contact and at right angle to the tangent (Fig. 5.8(b)). Resolving forces in the direction of the normal force, one gets:

$$N = (T_0 + \Delta T)\sin\theta/2 + T_0 \sin\theta/2 = 2T_0 \sin\frac{\theta}{2} + \Delta T \sin\frac{\theta}{2} \tag{5.7}$$

The second term, as compared to first being small, is neglected. This leads to a value of μ as follows:

$$\mu = (T_1 - T_0)/(2\, T_0 \sin\theta/2) \tag{5.8}$$

With the contact being nearly a point, the angle θ can be conveniently given by $\tan^{-1}(S/L)$.

For applying the capstan method, discussed in detail in a later section, one assumes that the fiber is cylindrical (Fig. 5.8(b)). One fiber passes over another at right angle with the angle of contact at the center given by the geometry as θ. With the value of T_0 known, of T_1 measured and that of θ very nearly given by $\tan^{-1}(S/L)$, one can calculate μ by the capstan equation:

$$T_1 = T_0\, e^{\mu\theta} \tag{5.9}$$

Calculations show that for reasonably low angles (< 20°), the differences in the values of μ obtained from Equations 5.8 and 5.9 are small.

5.5.2 Line contact

A number of methods have been used to characterize friction when fibers or yarns make line type of contacts. One approach used by several authors,

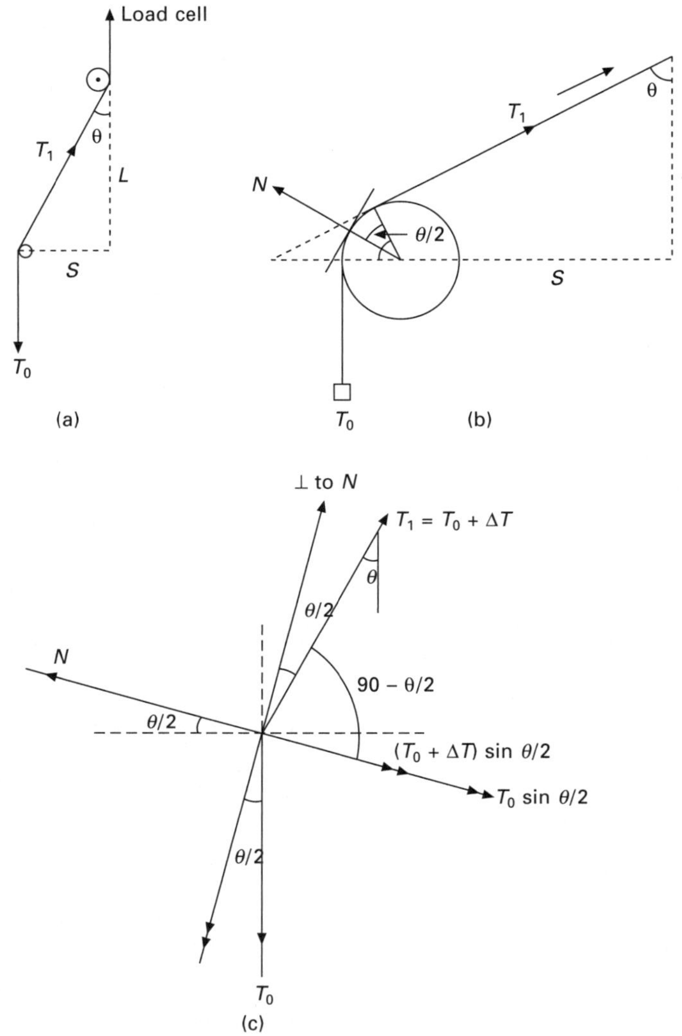

5.8 Adaptation of hanging fiber concept by El Mogahzy and Gupta to measure friction using a tensile tester [13]: (a) essential features of the method, (b) magnified view of the geometry at the contact region, (c) analysis of forces at the contact point.

including Adderley [15], Wood [16], Postle *et al.* [17], and Taylor [18], has been to measure the force needed to pull a fiber from a bundle in which it is held under pressure (Fig. 5.9). The challenge in a method like this lies in estimating the magnitude of normal force exerted on the fiber being withdrawn. In another method, used by Lord [19] and Hearle and Husain [20], the force needed to pull apart two interlocking fringes of fibers is measured. The most recent application of this technique is due to El Mogahzy and Broughton

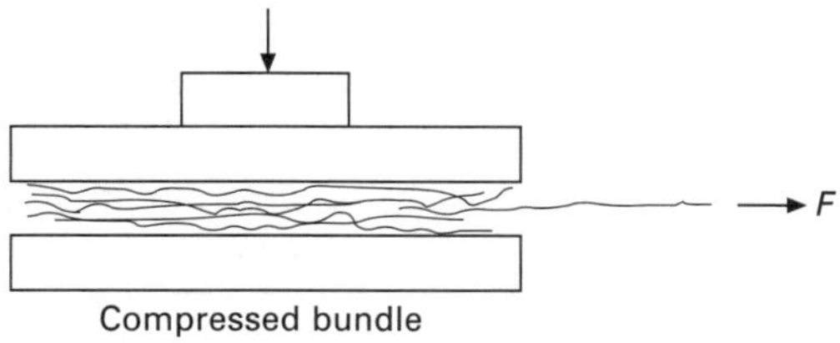

5.9 The force of withdrawing a fiber from a compressed mass.

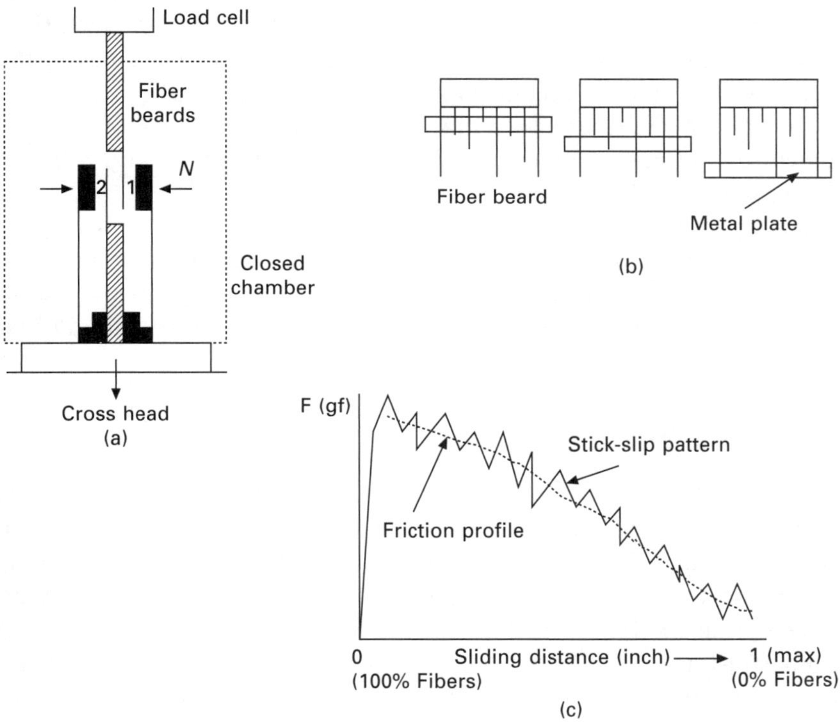

5.10 The fiber beard method of El Mogahzy and Broughton [21]: (a) schematic of the device, (b) illustration of change in number of contacts at different positions in the beard, (c) typical result on cotton.

[21] who proposed the so-called "Auburn Beard" method illustrated in Fig. 5.10. It requires preparation of two fiber beards using the High Volume Instrument (HVI®) developed by the Uster Technologies. The beards are clamped in tensile tester, one to crosshead and the other to load cell, and pressed together by suitable metallic surfaces of given area with the desired force. The lower beard and the pressing metal plates move down over the top beard. As this happens, the number of contacts varies from the highest at the top to the lowest at the bottom of the traverse. A typical plot of the friction

force against traverse is shown in Fig. 5.10(c). By conducting tests with only one beard (top) one assesses fiber to metal friction and by conducting tests with both beards one can get a measure of fiber-to-fiber friction.

Another method that has been used for measuring frictional properties of filaments oriented parallel and perpendicular to each other, as well as at other angles, is illustrated in Fig. 5.11 [22]. The material is wrapped carefully on two blocks such that filaments lie close to each other. The bottom block is fixed on a moving platform pulled by a linear motor. The top resting on it is attached to a tension-sensing system. The relative angle of traverse can be adjusted to a value between 0° and 90° by rotating the bottom block (Fig. 5.11(a)). The method also lends to measurements when the fibers making the line contact are given the relative axial or the relative radial traverse (Fig. 5.11(b)). Gupta and Chang used this technique to examine the effect fiber friction had on fiber migration, which occurs when a yarn is formed by the twisting process [22].

During yarn formation, the fibers, which nearly lie parallel to each other, migrate from one radial position to another to equalize their tensions. The frictional properties were changed by the use of lubricants. A typical result illustrating the effect of angle of contact on friction force is given in Fig. 5.11(c).

5.5.3 Area contact

In this procedure, friction is usually measured between two fabrics or a fabric and another material. The simplest technique is the inclined plane method, shown in Fig. 5.12. Friction is measured between the materials covering the bottom surface of the block and the top surface of the platform. One end of the platform, which is initially horizontal, is slowly raised until the block resting on it starts to slide. At this point angle θ, causing slippage, is established. From the forces shown in the figure, the value of the coefficient of friction, μ, can be assessed as follows:

$$\mathrm{F} = W \sin \theta$$

$$N = W \cos \theta$$

$$\mu = F/N = \tan \theta \qquad 5.10$$

Although the method is simple to use on planar materials, Speakman and Stott have successfully utilized it on keratin fibers [23]. The limitation of the technique is that only one value of μ, the static, can be obtained.

Other equally simple but more commonly used and versatile are the sled-type procedures, used by several workers [24–28] and illustrated in Fig. 5.13. For fabric-to-fabric friction, specimens of fabric cover the bottom surface of the sled and the top surface of the platform. Friction can be

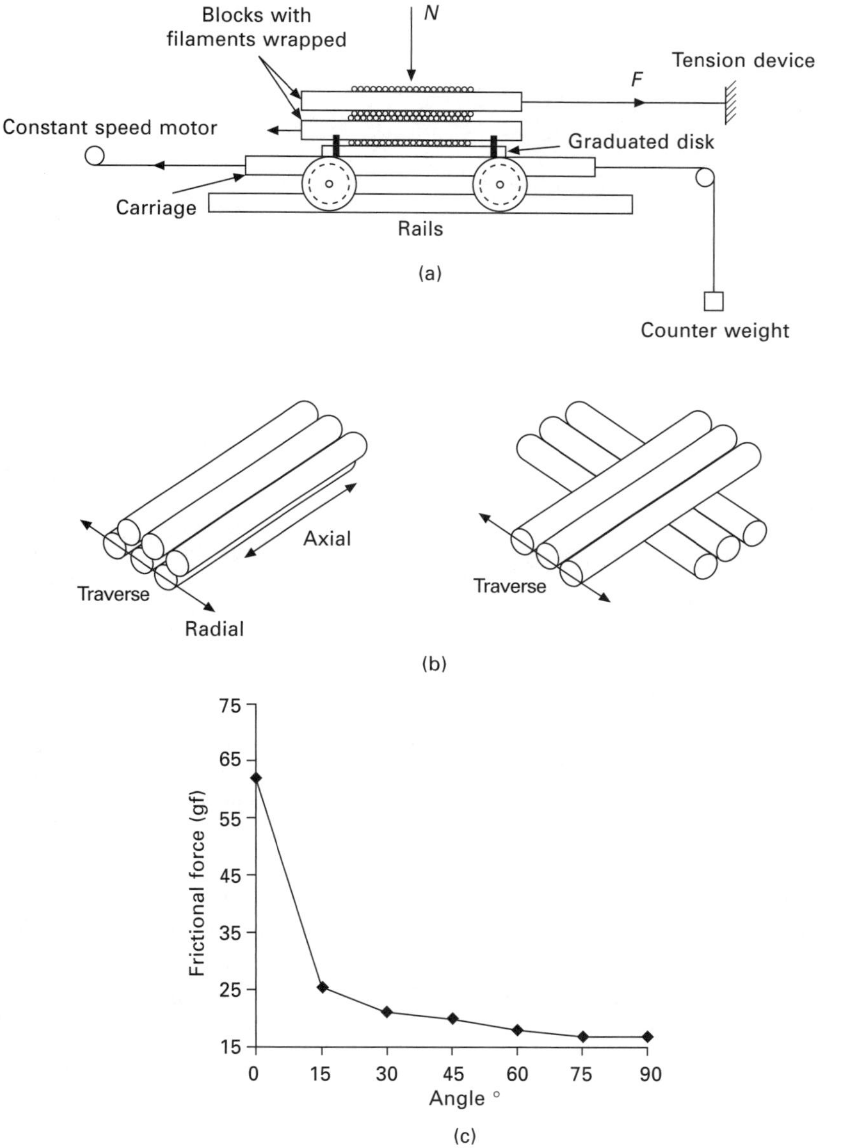

5.11 Method by Gupta and Chang for friction between filaments [22]: (a) the device, (b) illustration of different types of rubbing among fibers, (c) a typical result showing effect of relative angle of sliding on friction force.

measured with warp sliding parallel to warp, if woven, or course sliding parallel to course, if knitted, or at any angle to these between 0° and 90°. When the interest is to measure friction between a textile fabric and another material, then only one surface has the fabric.

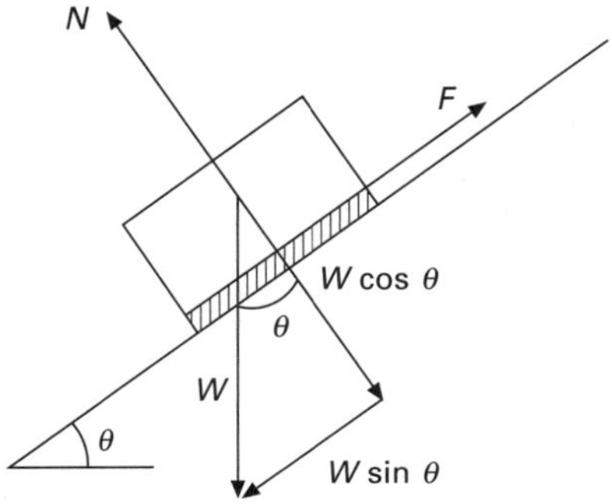

5.12 The inclined plane method of measuring friction.

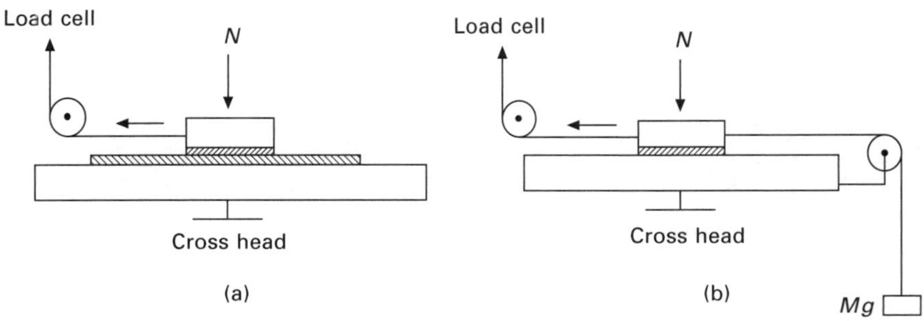

5.13 Schematics of devices used for measuring fabric to fabric and fabric to other surface friction: (a) the basic method, (b) the method involving reciprocating motion suitable for measuring friction over cycles of traverse [28].

In addition to measurements on woven and knitted materials, the method has been used effectively also on staple fibers and nonwoven webs [26, 27]. A detailed discussion of the procedure is given in Section 5.11.

An interesting variation of the device is used by Ramkumar *et al.* [28] (Fig. 5.13(b)). The sledge is attached to a weight (Mg) on the free end. By satisfying the condition, $Mg > F$, the movement of the sledge can be reversed and frictional behavior over a number of cycles of reciprocating traverse can be examined. These authors also developed a probe (made of polyvinylsiloxane) whose surface features are claimed to match nearly those of a human index finger. This was used as the sled. The geometrical area of contact between the finger-shaped probe and the fabric, needed for analysis, was determined with an ink impression of the former on a plane surface.

5.5.4 The capstan method

When a textile material is pulled over a cylindrical body (Fig. 5.14), a frictional force develops between material and cylinder surface due to normal

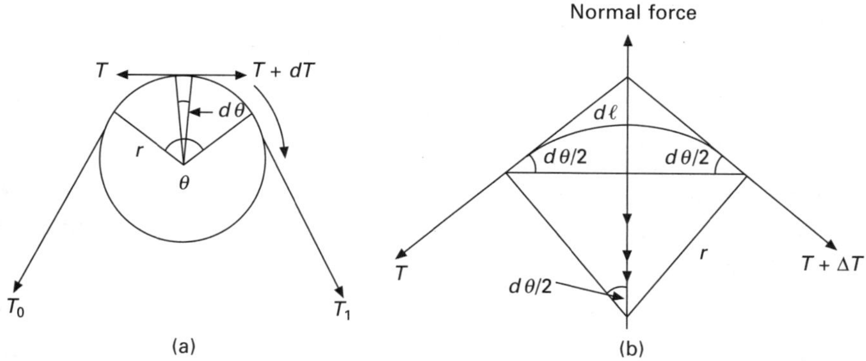

5.14 Characterization of friction by the capstan method: (a) geometry of the capstan method, (b) evaluation of forces in a small element.

force generated by tensions T_0 and T_1 exerted on the ends. The contact angle θ sweeps out that portion of circumference over which the contact occurs. An element involving contact over a small angle $d\theta$ is analyzed in Fig. 5.14(b). The tensions T and $T + dT$ on the ends of element lead to the following value of the normal force:

$$dN = T \sin (d\theta /2) + (T + dT) \sin (d\theta /2) \quad 5.11$$

With $d\theta$ being small, $\sin d\theta/2 \cong d\theta/2$, Equation 5.11, after neglecting the small term $dT \sin d\theta /2$, becomes:

$$dN \cong T\, d\theta \quad 5.12$$

Since the increase in tension dT is due to frictional drag, one can write the classical relation as:

$$dT = \mu\, dN, \text{ or}$$

$$dT = \mu T\, d\theta \quad 5.13$$

Integration using appropriate limits gives the following well known expression:

$$T_1 = T_0\, e^{\mu\theta} \quad 5.14$$

The capstan Equation 5.14 has been the subject of many theoretical and experimental studies in fiber friction. Howell [29] carried out an analysis to modify this equation by basing it on the general expression, $F = aN^n$, which accounts for the visco-elastic effects. Let the length of contact involving small element of angle $d\theta$ be $d\ell$ (Fig. 5.14(b)). If N' is the reaction to normal force per unit length and F' is the frictional force per unit length, then the usual equilibrium equations can be written:

$$N'\, d\ell = T\, d\theta \quad 5.15$$

$$F'\, d\ell = dT \quad 5.16$$

Substituting for F' from Equation 5.17 into 5.16, determining the value of length $d\ell$, and substituting this in 5.15, we get Equation 5.18.

$$F' = a\, N'^{n} \qquad 5.17$$

$$dT/T = a\, N'^{n-1}\, d\theta \qquad 5.18$$

From Equation 5.15 and $d\ell = r\, d\theta$, where r is the cylinder radius, we have $N' = T/r$. Substituting this in Equation 5.18 and integrating over appropriate limits, we get:

$$(T_1/T_0)^{1-n} = 1 + (1-n)\, a\, (r/T_0)^{1-n}\, \theta \qquad 5.19$$

(except in the special case, $n = 1$). Equation 5.20 gives solution of Equation 5.19 in the region $n \to 1$.

$$T_1 = T_0 \bullet \exp\{a\, \theta\, (r/T_0)^{1-n}\} \qquad 5.20$$

In this, a and n are constants. The value of n has been shown in earlier chapters to generally lie between 0.8 and 0.96 for textile fibers. Comparison of Equations 5.14 and 5.20 gives the following relation between μ and the constants, a and n:

$$\mu = a\, (r/T_0)^{1-n} \qquad 5.21$$

It is clear that, in the special case $n = 1$, μ becomes equal to a and independent of tension T_0, which is the classical case. By conducting tests at different values of T_0, calculating the values of μ by Equation 5.14 and fitting the values of μ and T_0 to the following model, one can estimate the values of the constants a and n:

$$Ln\, \mu = Ln\, a + (1 - n)\, Ln\, (r/T_0) \qquad 5.22$$

5.5.5 The fiber twist method

A convenient method for measuring inter-filament or inter-yarn friction is the twist method, proposed by Lindberg and Gralen [30]. Two fibers are twisted together by a certain number of turns (Fig. 5.15) and known weights, T_0, are attached to one end of each. If the tension on the other end is increased, the inter-fiber friction will support the difference in tension until the critical value, T_1, is reached and the fibers slip. The twisted fibers form a doubly wound helix, with the pitch angle of the helix as $\pi/2 - \beta/2$. If R is the radius of curvature of the helix, the relation between R and filament or yarn radius r can be written as follows:

$$R = r/(\sin^2\beta/2) \qquad 5.23$$

From Fig. 5.15(b), the following relation can be derived:

$$\text{Sin}\, \beta/2 = r\, d\alpha/d\ell \qquad 5.24$$

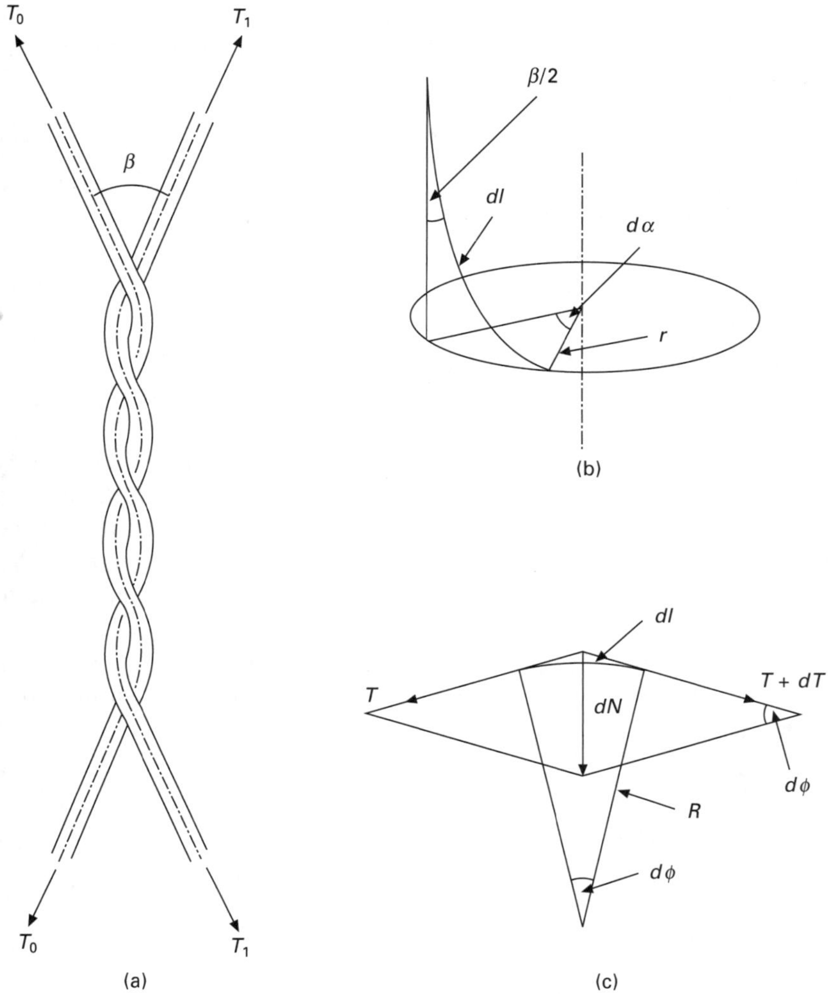

5.15 The twist method of Lindberg and Gralen [30]: (a) the geometry of the method, (b) a small element dl of the helix, (c) the forces acting on the element.

Let $d\phi$ represent the angle at the center of curvature of the element $d\ell$ (Fig. 5.15(c)). This gives, $d\ell = R\,d\phi$. Substitution in this for $d\ell$ from Equation 5.24 and R from Equation 5.23 yields:

$$d\phi = d\alpha \sin \beta/2 \qquad 5.25$$

Considering Fig. 5.15(c) where T is fiber tension (localized along the fiber axis) and N is normal force, one has the relation:

$$dN = T\,d\phi \quad \text{or} \quad dN = T\,d\alpha \sin \beta/2 \qquad 5.26$$

The normal force over a small length $d\ell$ of twist is dN, and assuming that the frictional force over this length is dT, one has the relation based on the classical equation:

$$dT = \mu\, dN \quad \text{or} \quad dT = \mu T\, d\alpha \sin \beta/2 \qquad 5.27$$

Integration over the limits T_0 to T_1, for tension T, and 0 to $2\pi\tau$, for azimuthal angle α, where τ is the number of turns of twist, gives:

$$\int_{T_0}^{T_1} \frac{dT}{T} = \mu \sin \beta/2 \int_0^{2\pi\tau} d\alpha \quad \text{or} \quad \ln (T_1/T_0) = 2\pi\tau\, \mu \sin \beta/2 \qquad 5.28$$

The above equation, for small values of the angle β, simplifies to:

$$T_1 = T_0\, e^{\mu\pi\tau\beta} \qquad 5.29$$

Experimental studies by Fair and Gupta [31] on human hair clearly demonstrated that although each fiber followed a helical path, the contact between the fibers was a straight line, parallel to the axis. El-Mogahzy and Gupta [13, 14] have modified Equation 5.29 by using the relation, $F = a\, N^n$, as the base to come up with a model that is more generally applicable to fibrous materials.

Considering Fig. 5.15(c), if it is assumed as before that N' is the reaction per unit length and F' is the friction force per unit length, then the usual equilibrium equations can be written as:

$$dN = N'\, d\ell = T\, d\phi \qquad 5.30$$

$$dF = F'\, d\ell = dT \qquad 5.31$$

From 5.24, 5.25, 5.30 and 5.31, one can write the following equations:

$$N' = T \sin^2 (\beta/2)/r \qquad 5.32$$

$$F' = dT \sin (\beta/2)/r d\alpha \qquad 5.33$$

Equation 5.17 is the general equation relating F' and N'. Substitutions for F' and N' in 5.17 gives:

$$\frac{\sin \beta/2}{r\, d\alpha}\, dT = a\left(\frac{T \sin^2 \beta/2}{r}\right)^n \quad \text{or} \quad \frac{dT}{T^n} = \frac{a(\sin \beta/2)^{2n-1}}{r^{n-1}}\, d\alpha \qquad 5.34$$

Integrating over the limits T_0 and T_1 for T, and 0 and $2\pi\tau$ for α, and replacing $\sin \beta/2$ with $\beta/2$, for small values of β, one gets:

$$\int_{T_0}^{T_1} \frac{dT}{T^n} = \frac{a}{r^{n-1}} (\sin \beta/2)^{2n-1} \int_0^{2\pi\tau} d\alpha$$

$$\text{or} \quad \left(\frac{T_1}{T_0}\right)^{1-n} = 1 + \pi\tau a\beta(1-n)\left\{\frac{T_0\beta^2}{4r}\right\}^{n-1} \tag{5.35}$$

(except for the special case when $n = 1$).

As n tends to unity, the above equation, as in the case of the capstan model, yields the result:

$$\frac{T_1}{T_0} = \exp\left\{\pi\tau a\beta\left(\frac{T_0\beta^2}{4r}\right)^{n-1}\right\} \tag{5.36}$$

Comparing Equations 5.36 and 5.29 one gets the following relation between μ and a and n:

$$\mu = a\,(T_0\beta^2/4r)^{n-1} \tag{5.37}$$

As noted with the general capstan equation, when $n = 1$, μ becomes equal to a and independent of T_0.

Figure 5.16 illustrates a device by Gupta *et al.* [31, 32] that conveniently adapted the twist method for use on a standard tensile tester. It consists of six frictionless pulleys and is clamped onto the cross head. Equal weights, T_0, are tied to the ends of two fibers that are attached to the load cell at the free ends. Twist is inserted by crossing the fibers over each other the desired number of times. The fibers are then arranged around pulleys and fitted into

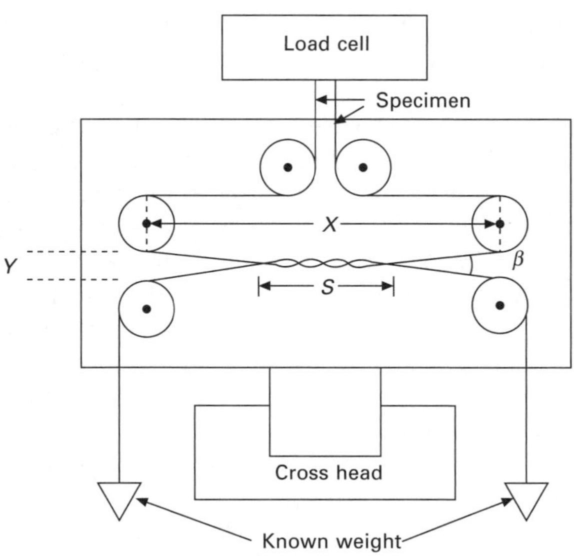

5.16 Schematic of twist method device suitable for use with a standard tensile tester [31].

grooves to prevent slippage over edges. As the cross head moves down, carrying the device with it, the tension builds in the fibers until it overcomes the force of friction and the slippage starts to occur. The tension at this point corresponds to the first peak on the chart. The fibers hold again, then slip, creating a stick slip pattern from which the values of T_1 for the static and the kinetic forces can be obtained. These along with the values of the number of turns of twist, τ, and the twist angle β, given by "2 arctan$\{Y/(X\text{-}S)\}$" (Fig. 5.16), are substituted in Equation 5.29 to estimate the values of μ_s and μ_k. By conducting tests at different values of T_0, calculating the value of μ by Equation 5.29, and fitting the data to the model shown below, one can estimate the values of the constants a and n:

$$Ln\ \mu = Ln\ a + (n - 1)\ Ln\ (T_0\ \beta^2/4r) \qquad 5.38$$

5.5.6 Tests in various media

Little published literature exists in which friction has been measured in a medium other than air, primarily because of the difficulties involved in conducting such tests. In 1976 Landwehr *et al.* [33] described the working principles of an instrument that allowed measurements in both air and fluid media. Wet measurements were achieved by putting the desired liquid in a container that could be placed on a block situated under the contact area. In his method, two fibers rubbed together while held as parallel to each other as possible (~15°). Since Landwehr was primarily interested in the felting properties of wool, the near parallel rubbing of the fibers facilitated detection of scale interaction. In order to maintain continuous fiber-fiber contact during a rub at low normal forces (~1 mg), very low traverse speeds (2 μm/sec) were used.

El-Mogahzy and Gupta adapted the line and point contact devices for tests in fluid media [13]. Figures 5.17(a) and (b) illustrate the techniques of which the principles have been discussed earlier. For wet measurements, the lower two pulleys of the twist device are replaced by pins (shaded) coated with Teflon® to give minimum friction. Plate P, the main component, is attached to the cross head of the tensile tester through brackets. For measurements in liquid, the plate is raised, the container holding fluid is positioned under it, and then the plate is lowered and clamped in place. This arrangement facilitated easy threading and tensioning of fibers before immersing the contact region in fluid and testing.

The schematic of the point contact device suited for wet tests is shown in Fig. 5.17(b). A fiber is mounted horizontally in bow with desired tension. It is then moved laterally with the use of micrometer M for contact with vertical fiber, which is suspended from load cell and tensioned with weight, T_0. The frame containing the assembly is attached to cross head. A vessel containing

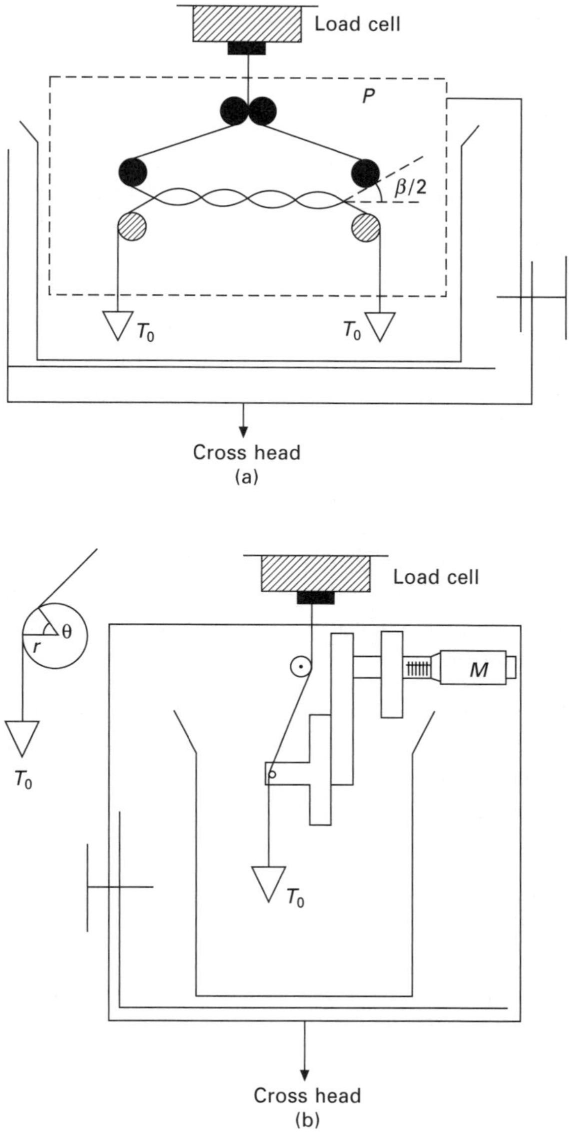

5.17 Schematics of the line and point contact devices set up for tests in a fluid media [13]: (a) test involving twist method, (b) test involving hanging fiber method.

fluid and resting on the base is raised so that the fluid level is above contact region. The procedures worked well on filaments and yarns and for a range of tensions T_0. The main limitation of the method was the speed of testing that had to be necessarily low (~ 1 cm/min).

5.6 Measurement and characterization of friction in fabrics

5.6.1 Introduction

Textiles are used largely as fabrics and being able to measure, characterize, and understand their frictional behavior is important. A fabric may rub against a fabric or against another surface. Usually a horizontal platform method is used for characterizing fabric-to-fabric and fabric-to-other surface frictions. An interest also lies in characterizing the hand or feel of a fabric, but for this a correlation must first be established on a set of fabrics between the properties measured by instruments and the rankings from touching and feeling by judges. The objective measure in the case of friction usually involves a probe that can characterize a surface's contour (landscape) and another probe that is designed to simulate a human finger and assess friction as a fabric is pulled against it. The objective measure characterizing surface contour or undulations can often be related to fabric geometry, for example fabric sett, spacing between cords, or ribs, etc. The objective measure of friction, on the other hand, is influenced by not only the geometry of the fabric but also the material of which it is composed. Described in the next four Sections (5.6 to 5.9) are the methods used for:

- measuring fabric to fabric friction using a horizontal platform
- characterizing surface morphology using a number of techniques (stylus, roller, microscope)
- assessing hand and correlating the subjective ranking with the objective measurements.

Although not used in the studies being described here, perhaps because of the lack of access to the equipment at the sites, the Kawabata KES system introduced in the 1980s was designed to characterize friction and other properties of a fabric that could correlate with its hand. Section 5.12, gives a description of the KES method.

5.6.2 Horizontal platform

The most popular method of fabric friction measurement is based on the principle of rectilinear motion of a sled over a horizontal platform. The assembly is made up of a platform, and a sled connected to a constant rate of elongation tester as illustrated in Fig. 5.13(a). In the work by Ajayi [34] that typifies the studies on fabrics, and the care one may take to enhance sensitivity and reproducibility of results, the platform measured 52 cm × 5 cm and the sled 8 cm × 5 cm. The weight of the latter was only 25 gf (~ 25 cN), or about 61 Pa, which minimized surface compaction, compression and polishing that could result from the use of a heavier sled. Before starting the test, the sled was allowed to stand on the platform for a specified time i.e. the time of loading (3 min). The sled was pulled at a constant speed, (5 cm/min) by a constant rate of traverse tester, using an inextensible towing yarn passing over a frictionless pulley. Both the chart speed (5 cm/min) and full-scale deflection (50–200 gf) were adjusted to give the maximum sensitivity and reproducibility to results.

For fabric-on-fabric friction measurement, rectangular specimen (30 cm × 10 cm) was mounted on the horizontal platform. Over this, a sled covered with an identical fabric was traversed [34]. The fabric on the platform was smoothened by hand and held under a slight tension by bulldog clips, which ensured that the test fabric did not wrinkle or buckle during tests. The fabric specimen mounted on the sled was also slightly tensioned and fastened on the upper surface of the sled with the use of a double-sided tape.

Two directions of orientation during sliding, i.e. warp-on-warp, and weft-on-weft, are generally recommended for tests. It is also crucial to employ fresh samples for each different test. Both the face and the back of a fabric may be tested in order to obtain a more complete friction behavior of a given structure. Generally, the mean and standard deviation of five tests are recommended for friction characterization.

5.6.3 Expression of results

The friction parameters needed to describe the surface properties of fabrics are determined from the friction trace and presented in Section 5.3.3. These parameters conform to the procedures of several investigators, namely: Ajayi [34], Carr *et al.* [35], Grosberg *et al.* [36], Hearle and Husain [20], Ohsawa *et al.* [37] and Ajayi *et al.* [38]. The proposed parameters are generally more acceptable in view of the controversy surrounding the use of coefficient of friction and its possible correlation with the tactual assessment of roughness or smoothness.

5.7 Fabric surface roughness characterization

5.7.1 Introduction

Apart from friction measurements, there are methods developed for surface characterization of fabrics. These are usually based on the hypothesis that the morphological features of the fabric will influence the vertical displacement of a probe (roller or stylus), resting on the fabric surface. Generally, there are two types of fluctuations: systematic variations, arising from construction pattern (for example, cords or ribs), and random variations, caused by uneven features of the structure (for example, variations in thread spacing).

Theoretically, a smooth plane surface would be expected to give a smooth trace. If the fabric has a rippled surface of distinct periodicity, a regular and repetitive signal would be expected. The number, amplitude and pitch of such a signal would be dependent on the number, height and width of surface undulations, respectively. Any random irregularities in or the displacement of yarns, or the presence of surface hairs, or uneven deposition of finishing agents, might be expected to produce irregularity in signals. Changes in amplitude would be expected at the transition between thin and thick places. In addition, the amplitude of irregularity may be associated with the variation in the fabric bulk density.

The mode of contact, the probe mass, and the surface characteristics of fabrics can influence the level of discrimination. For example, a stylus-type probe may be expected to be more sensitive and responsive to small variations or protuberances on fabric surfaces, which may not be detected by a roller probe because of usually higher pressure and, therefore, flattening of surface asperities.

In the assessment of surface topography of fabrics, it should be possible to establish some correlation between different objective measurements and between an objective measurement and subjective hand or tactile evaluation made by judges. The objective measurements of surface characteristics include friction profile, morphological features from roller and/or stylus profiles, and topological assessment from image analysis using a microscope. In view of the usefulness of these methods, they are considered here in more detail.

5.7.2 Roller (tensile tester) method

The accessory used is shown in Fig. 5.18 [39]. Basically, it consists of a tapered brass rod fitted with a freely rotating roller. The roller surface is

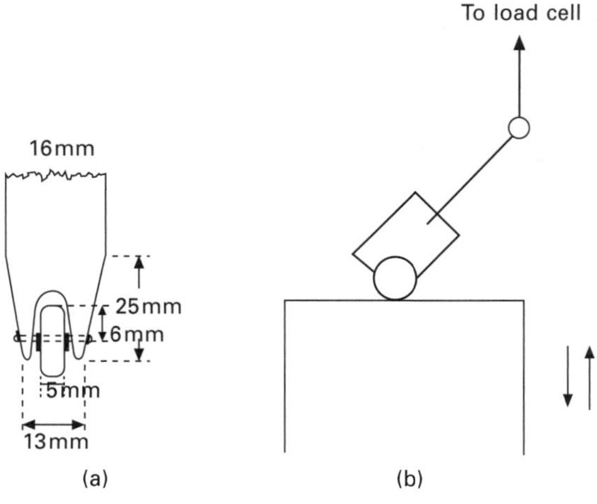

5.18 (a) Sketch of roller probe, (b) roller probe attachment to a tensile tester cross head.

made of polished stainless steel in order to minimize friction between the roller and the test surface.

The brass arm is connected to a constant rate of elongation tester in such a manner that it is inclined to the horizontal, i.e. the test surface, at an angle (45°), as shown in Fig. 5.18(b). As the tensile cross head traverses toward load cell, the roller moves forward. A reverse traverse of the cross head corresponds to the backward movement.

The length of the stroke in each direction is predetermined (5 cm). Both the choice of the angle of inclination (45°) and the length of stroke are made partly for convenience and partly to minimize the geometric effect of the apparatus. The angle of inclination of the arm to the fabric surface changes as the cross head traverses up and down. A high-speed pen sensitivity response enables the deflection of the roller by surface irregularities to be accurately recorded.

5.7.3 Stylus method

In order to assess the short-term irregularities on a fabric surface, measurements of thickness and changes in it along the line of probe and under a constant pressure, a stylus is used. In a study by Ajayi [39], the main component of this apparatus is a stainless steel pin (e), shown in Fig. 5.19. The latter is fastened onto a rubber strap, which in turn is glued onto a transducer finger (c). The finger is mounted on the tension sensing head of a Rothschild tension meter, in such a way that the stylus is perpendicular to the fabric surface, mounted on the Perspex plate (f). The latter is driven by the constant rate of elongation tester via a towing cord (i). In order to exercise some control over the vertical movement of the probe, and hence on the pressure exerted on the fabric underneath, the cylinder of the tension sensor (c) is mounted on a traveling retort (d). This can be raised or lowered manually as

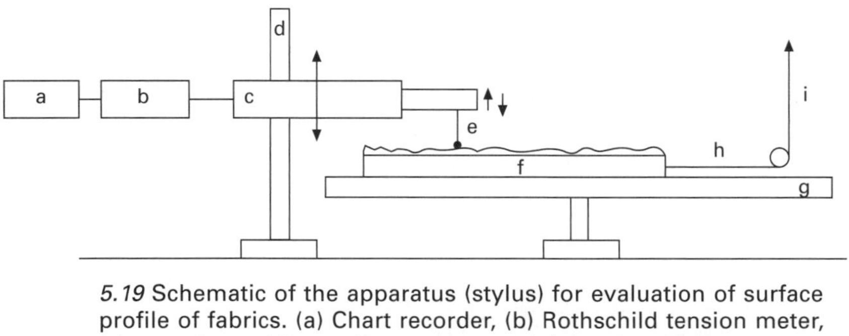

5.19 Schematic of the apparatus (stylus) for evaluation of surface profile of fabrics. (a) Chart recorder, (b) Rothschild tension meter, (c) transducer, (d) traveling retort stand, (e) stylus, (f) Perspex sample holder and fabric, (g) horizontal platform, (h) towing cord, and (i) cord to constant rate of elongation tester or any suitable drive.

required. A built-in damping facility in the Rothschild device serves to filter out any extraneous signal arising from a possible vibration of the stylus. The vertical displacement of the stylus is fed into the tension meter and recorder (b) and (a), respectively. A calibration curve of the stylus obtain by lowering the stylus in steps onto a weighing balance, is shown in Fig. 5.20.

5.7.4 Analysis of roller and stylus signals

The signals from two different woven fabrics, obtained by the stylus and roller probes, are shown in Figs 5.21 and 5.22, respectively. The vertical displacement represents the surface contours of fabrics, i.e. surface protrusions such as yarn crowns, fiber piles, etc. Two repeats of the stylus trace, corresponding to 5 cm displacement of fabrics, in the same direction, are shown. As speculated, the stylus trace captures short-term features and leads to relatively higher frequency fluctuations for smoother of the two surfaces.

In the case of the roller test, a repeat consisted of one forward and one backward movement of the roller over the test surface. Thus a repeat corresponded to an upward and a downward movement of the cross head. The start of a cycle (s) also corresponded with the finish (f) of the previous cycle. The roller test profile is seen to by pass short-term irregularity (compare the profiles for fabric C_6 in Figs 5.21 and 5.22) but captures longer-term

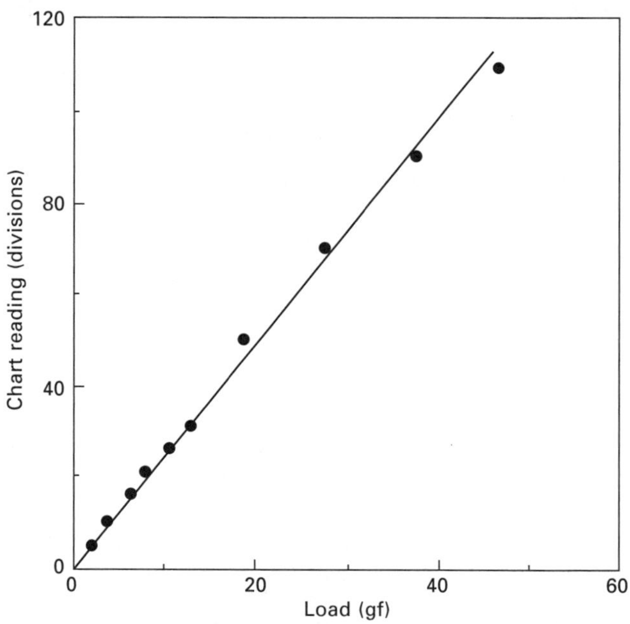

5.20 Calibration curve of the stylus probe (calibration equation; $Y = 1.24\ X$, where Y is chart reading and X is load (gf) [40]).

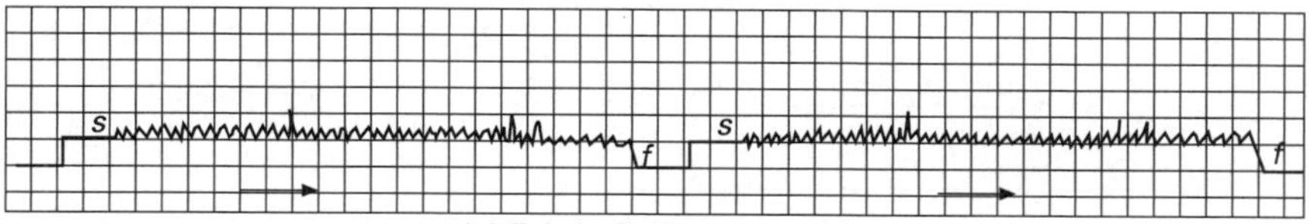

(a) Fabric C_6 (less smooth)

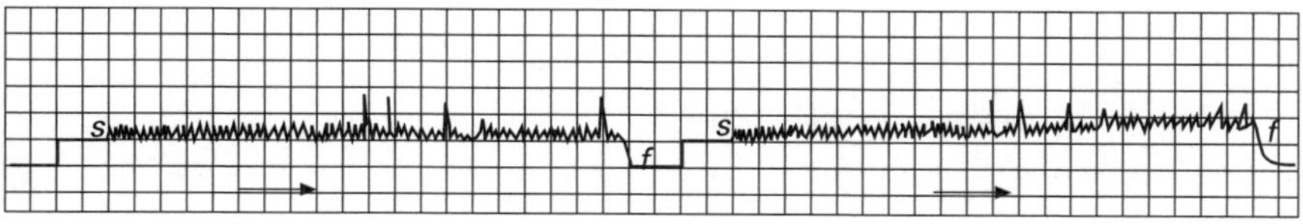

(b) Fabric C_{10} (smooth)

5.21 Typical stylus traces for woven fabrics: (a) fabric C_6, low sett, and (b) fabric C_{10}, high sett. (See Table 10.1 for details of C_6 and C_{10} fabrics.)

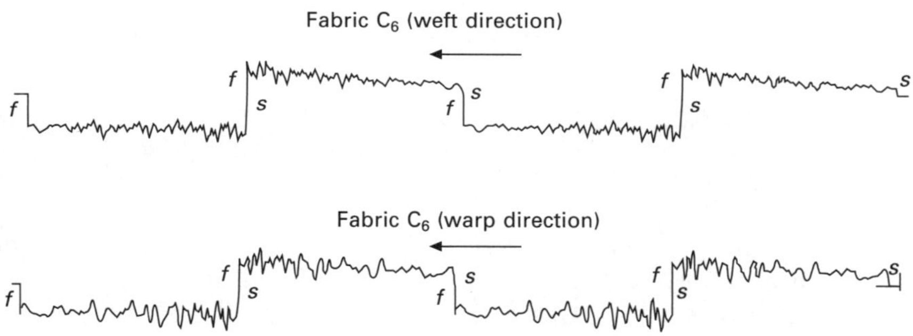

5.22 Typical roller traces of woven fabric C_6: top, weft direction; bottom, warp direction; s start of a cycle; f finish of cycle [34, 39]. (See Table 10.1 for details of fabric C_6.)

features. In the weft direction test, closer fluctuations reflect higher warp thread density than the wider fluctuations seen in the warp direction test, which are caused by lower weft thread density.

5.7.5 Microscopic evaluation of surface

The relative protrusion of yarn crowns, fiber pile, and other features of structure, form the plane of fabric surface and can be determined by cross-sectional examination of the fabric on a projection microscope. The technique entails sandwiching a fabric specimen between two faces of Perspex blocks in such a manner that the fabric section is uniformly displayed as shown in Fig. 5.23.

This method is useful for surface examination and can be employed for an estimate of the surface protrusions. Some results obtained by Ajayi [34] and

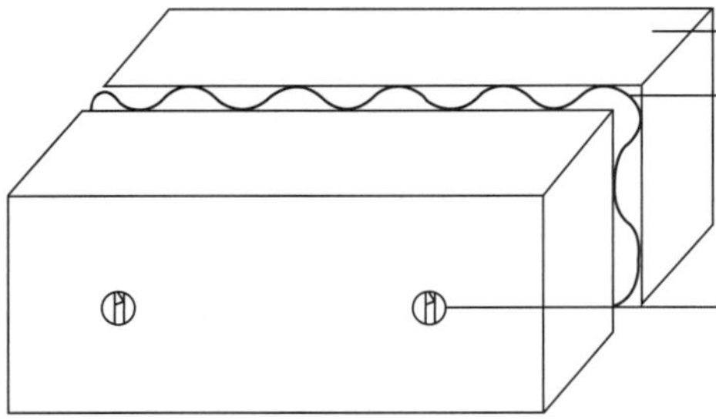

5.23 A specimen sandwiched between two Perspex blocks.

(a) Smooth (limbric) fabric C_2

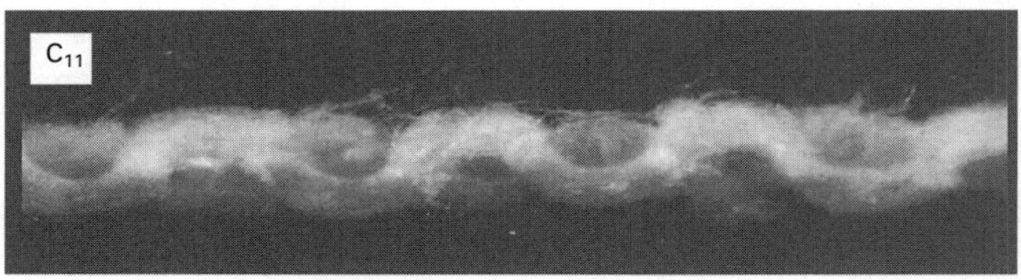

(b) Rough (canvas) fabric C_{11}

5.24 Examples of smooth limbric cotton fabric surface (a), and rough canvas surface (b). Details of fabrics are as follows. C_2: yarn sett, warp 28, weft 39; count (tex), warp 10, weft 15; crimp (%), warp 0.8, weft 8; mass 83 g/m^2; thickness 0.22 mm. C_{11}: yarn sett, warp 18, weft 12; count (tex), warp 160, weft 180; crimp (%), warp 23, weft 4; mass 500 g/m^2, thickness 1.01 mm.

Ajayi and Elder [40, 41] are shown in Figs 5.24 to 5.26 for various fabric surfaces.

Apart from the use of this technique in the estimation of structural parameters, such as crown heights in plain weave fabrics and cord heights and widths in weft pile fabrics, some qualitative information on surface characteristics of fabrics (flatness, ridginess and hairiness) could also be obtained. As shown in Fig. 5.24, a cotton limbric fabric C_2 is thinner, planer, lighter and perhaps less hairy than a canvas fabric (C_{11}). Tactually, the canvas was rougher and coarser than the cotton limbric fabric. Another example is given in Fig. 5.25, wherein a velveteen (C_{12}) and a corduroy (C_{14}) are compared. It can be seen that although the fiber pile effectively concealed the base

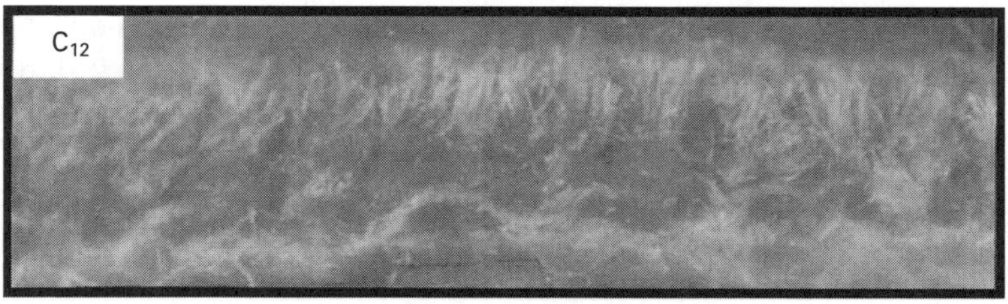

(a) Smooth (velveteen)

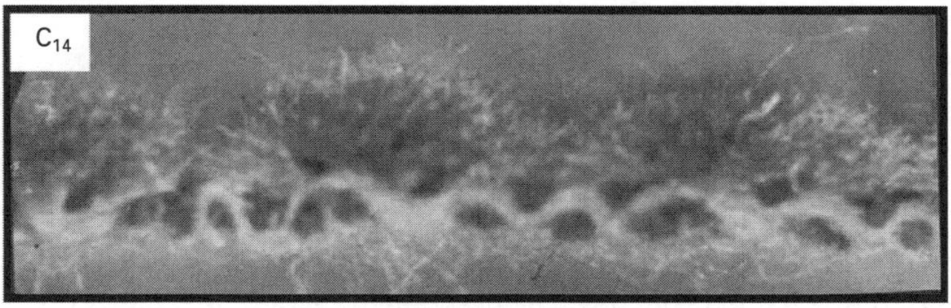

(b) Less smooth (corduroy)

5.25 Examples of smooth velvet fabric surface (a), and less smooth corduroy surface (b). (Brief details of fabrics C_{12} and C_{14} are given in Table 10.2, Chapter 10.)

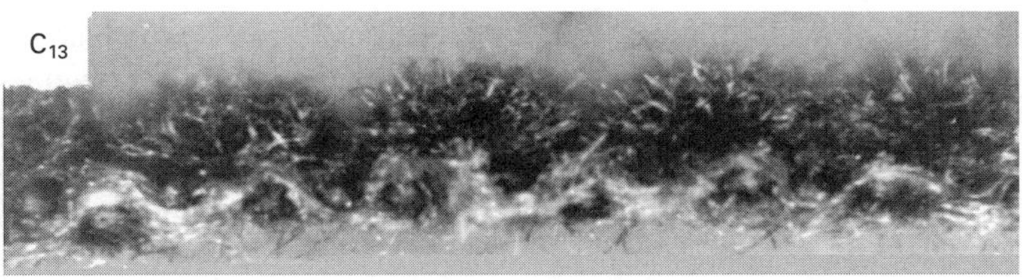

Shallower ridge (smooth)

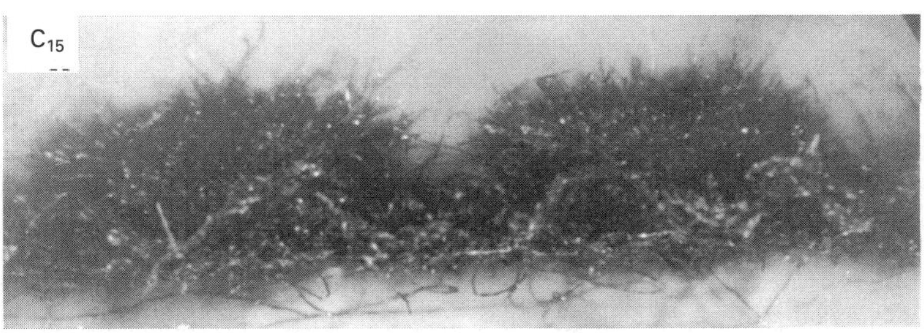

Deeper ridge (less smooth)

5.26 Examples of smooth (a), and less smooth (b) corduroy surfaces. (Brief details of fabrics C_{13} and C_{15} are given in Table 10.2, Chapter 10.)

weaves of both fabrics, the corduroy appears more ridgy and hence rougher than the velveteen. Similarly, a comparison between two corduroy fabrics, Fig. 5.26, also reveals significant differences in the size, number and depth of cords. The size and depth are greater in fabric C_{15}, which leads to a texture that is rougher and coarser. This kind of roughness appeared to be associated with the ridginess of the corduroys.

5.7.6 Correlation between different surface measurements

The surface contour of a fabric has been defined as the divergence of a fabric surface from plane-ness [42]. Quantitatively, surface irregularities may be expressed as the number of peaks, amplitude of pulses, and standard deviation of the pulses [41]. These parameters were determined from the roller and stylus traces in the same manner as those from the friction traces described in Section 5.3.3. From the consideration of plain weave fabrics, it was shown that the parameters derived from the friction traces, such as the frequency, amplitude and average friction forces, are readily identifiable with structural factors such as the sett, cords and ribs.

Measurements of surface irregularities of fabrics by various methods in addition to friction tests have been reported [34, 39, 40, 43]. A plot of number of cords on the surface of a weft pile fabric and the number of pulses obtained from various methods, i.e. friction, stylus, roller and visual observation from the projection microscope, is given in Fig. 5.27, which shows a linear relationship. It is demonstrated that the number of peaks obtained from the roller, stylus, and the friction traces may be the best indicators of fabric smoothness. The results suggest that the fabrics that yield, a larger number of peaks for a given length, are usually perceived smoother than those that yield a smaller number of peaks. This also agrees broadly with the results of lateral airflow in which a higher flow rate is associated with a rougher surface.

5.8 Subjective assessment of friction

5.8.1 Paired comparison of fabric surfaces

The consumers touch, stroke and feel a fabric and then make a decision based on tactile feeling. In this section, a method used for ranking fabrics by a subjective method is discussed along with a comparison of subjective ranking of a set of fabrics with objective measurements of friction and other assessments discussed in Section 5.7.

For subjective assessment of surface contours, or the degree of smoothness or roughness, certain attributes are considered; these include the number, the

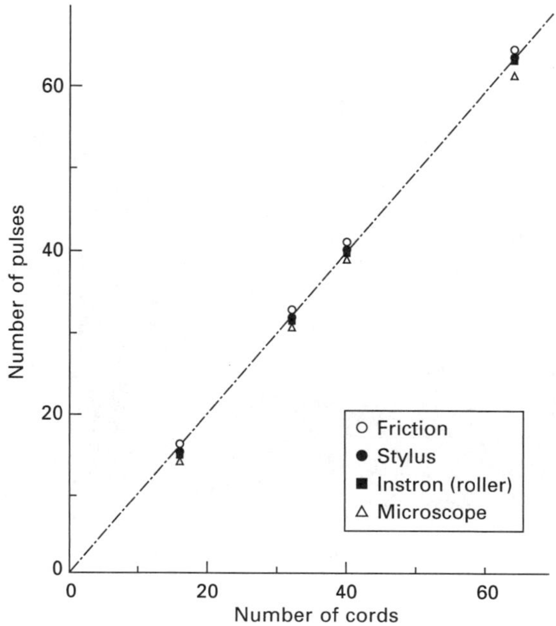

5.27 Comparison between the friction and the surface contour measurements [34].

height and the width of asperities. In the paired comparison test, judges evaluate the stimulus perceived. In a test, a panel of 10 female judges, all university students within the age group of 21–30 years, was asked to stroke the fabric surface [43]. The tests were carried out in a quiet room, maintained at a temperature of about 22°C and relative humidity of 65%. Fabric samples, measuring 10 cm by 30 cm, and mounted on a stiff card of an inclined plane at an angle of about 45° to the horizontal, were stroked in this position as depicted in Fig. 5.28.

Two separate tests are usually required; in one, the judges are allowed to see the fabrics as happens commercially; in the other, a pillory box is used to prevent the judges from seeing the fabrics and thus the decisions are based on touch alone. A period of two weeks is normally allowed between the tests to eliminate the effect of any possible memory of the earlier experience.

In paired comparison assessment, the judges are asked to compare two fabrics and answer pertinent questions relating to the surface characteristics, such as the roughness, smoothness, and ridginess, of fabrics. For example, judges are asked to compare two fabrics and decide which fabric is rougher. If fabrics C_6 and C_{10} are being compared, and sample C_6 is chosen as rougher, then 1 is recorded in the box in column C_{10} row C_6. If C_{10} is chosen 0 is recorded in this box. If no choice is made $^1/_2$ is recorded in this box. This is exemplified in Table 5.1.

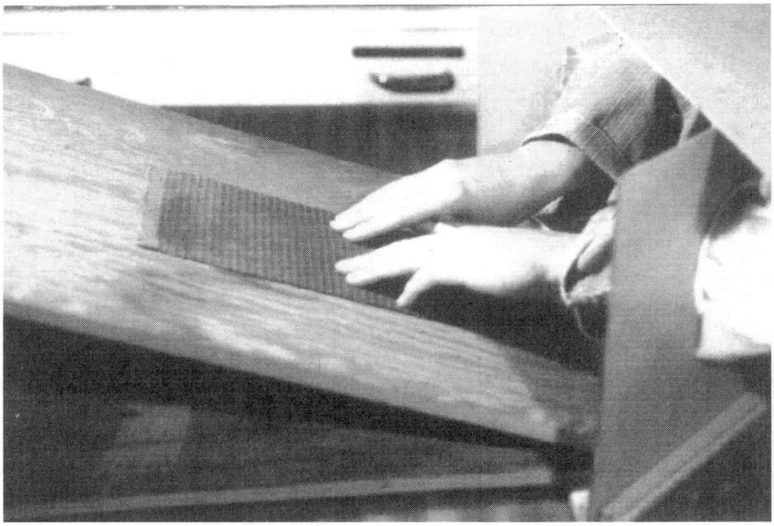

5.28 Subjective assessment of fabric surface using a pillory box [43].

Table 5.1 Paired comparison of fabrics by a judge

Fabrics	C_6	C_7	C_8	C_9	C_{10}
C_6	–	0	1	1	1
C_7	0	–	1	1	1
C_8	0	1	–	1/2	1
C_9	0	0	1/2	–	1
C_{10}	0	0	0	0	–
Total	0	1	$2^1/_2$	$2^1/_2$	4
Rank order	5	4	2	2	1

The test is repeated for all other pairs of specimen selected randomly. In the example taken, recording a zero (0) in column C_6 row C_{10} completes the lower triangle. The entries in each column are summed to derive the overall ranking of the degree of roughness. The overall ranking of all judges are collated and then summed. The magnitude of the subjective assessment is again ranked to give the composite rank order. The specimen with the highest score then ranks the smoothest and the one with the lowest score the roughest (Table 5.2 for the current fabrics). The results of subjective assessment can be compared with objective measurement of kinetic frictional resistance shown in Table 5.3.

5.8.2 Statistical analysis

From the final ranking of the subjective and objective scores, the Spearman rank correlation coefficient r_s can be calculated from the following relation [44]:

Table 5.2 Tactile rankings by judges

Fabric code judges	C_6	C_7	C_8	C_9	C_{10}
1	0	1	2	3	4
2	0	1	3	2	4
3	0	1	2	3	4
4	0	1	2	3	4
5	0	1	2	3	4
6	1	0	3	2	4
7	0	1	2	2	4
8	0	1	2	4	3
9	1	0	2	3	4
10	0	1	2	3	4
11	1	0	2	3	4
Total	3	8	24	32	43
Composite ranking	5	4	3	2	1

Table 5.3 Table of objective measurements of frictional resistance of fabrics

Fabric code parameters	C_6	C_7	C_8	C_9	C_{10}
Kinetic frictional resistance (cN)	49	53	58	56	60
Objective ranking	5	4	2	3	1

See Table 10.1 (Chapter 10) for fabric details.

$$r_s = 1 - \frac{6\,\Sigma\, d^2}{N(N^2 - 1)}$$

Where N is the number of fabrics ranked and d is the difference between objective measurement and subjective assessment.

For the fabrics, then, the correlation between the subjective and objective rankings can be calculated using the equation given above. To test for the significance of the results obtained, it is customary to test for the null hypothesis. This means that the two sets of results are not associated with each other unless the value of r_s is equal to or greater than the value given in the table for the null hypothesis. If the value is equal or greater, the hypothesis is rejected, and the calculated value is significant at the level indicated, either 0.05 or 0.01.

5.8.3 Magnitude of estimation of a surface attribute

For characterizing an attribute of the surface subjectively, the pillory box was used. The controller demonstrated the method to be used. Three fingers of both hands were to be simultaneously drawn over the surface of the specimen mounted on an inclined plane as shown in Fig. 5.28. A light pressure of about 20 cN/cm^2, chosen based on the work of Elder *et al.* [45] for the assessment of the magnitude of the fabrics' softness, was to be applied on the fabric samples. The judges were to concentrate on a particular attribute being evaluated, for example the number of ridges or their height in corded surfaces. A standard sample, which the controller considered to be at the middle of each set, was assigned a magnitude. This standard was mounted alongside with the specimen (not shown in Fig. 5.28). The task of each judge was to compare a specimen with the standard and assign a number proportionately higher, lower or the same to the test fabric as felt appropriate. Geometric mean rather than arithmetic mean was considered because of the relative nature of the values.

In the five fabrics considered as examples, a good linear relation was found between the values of the objective measurements and the subjective assessments of fabric roughness on a logarithmic scale [34, 43].

In the tactile evaluation of surface irregularities, care must be taken to ensure that the parameter being investigated was clearly spelled out and that the controller demonstrated the method to be used. This method provided a very useful tool for establishing correlation between subjective assessment and instrumental measurement of a characteristic.

5.9 Summary: testing friction in fabrics

Friction results of fabrics are sensitive to variations in experimental conditions (normal pressure, speed of sliding, nature of surface, time of loading and number of traverses). Processing variables such as finishing treatment, yarn and fabric construction and fiber content also affect results of fabric friction. Therefore, these factors must be noted when specifying results. A single value function, such as the coefficient of friction, is not adequate in describing the frictional behavior of fabrics. It is more valuable to characterize the friction profile in fuller details using the parameters mentioned (F_{so}, F_s, F_k, F_a, F_f). It is also useful to assess other surface topographical features using an appropriate probe and image analysis in order to obtain a more complete representation of a fabric surface profile.

5.10 Standard test methods, and commercially available friction meters for yarns

5.10.1 Introduction

Described in this section are standard test methods designed for use on yarns under speeds, pressures, and nature of contacts encountered during processing. Yarns rub against surfaces of guides and rollers during continuous filament yarn processing and fabric formations, as well as the yarns rub against each other during weaving, knitting and braiding. Yarns also rub and slip against each other when fabrics are stretched, bent or sheared during handling and use. The processes mentioned involve a broad range of speeds, pressures and contact surfaces. In spite of the fact that a very large number of measuring devices have been designed and used in investigating friction, few standard methods have been established. No such method exists for characterizing friction in single fibers. This is because single fibers are small denier and, therefore, delicate materials that require extreme care in preparing specimens and ultra sensitive devices for imposing and measuring low forces involved. In contrast, yarns are macroscopic in size and much stronger and easier to handle. Accordingly, standard test methods are available for measuring yarn-to-yarn and yarn-to-another surface forces. A standard test method is also available for use on fiber webs and nonwovens.

5.10.2 Standard test method

A standard test configuration for a yarn-to-solid surface friction is available as ASTM D 3108 [46]. The graphical representation of this test is shown in Fig. 5.29. The method utilizes capstan type yarn to counter body configuration and the data generated in this test is presented in terms of the coefficient of friction calculated by the capstan Equation 5.9 given earlier:

$$\mu = (1/\theta)\text{Ln}(T_1/T_0)$$

The work of Rubinstein in 1958 [47] had shown that coefficient of friction measured by such tests was influenced by a number of factors, such as the

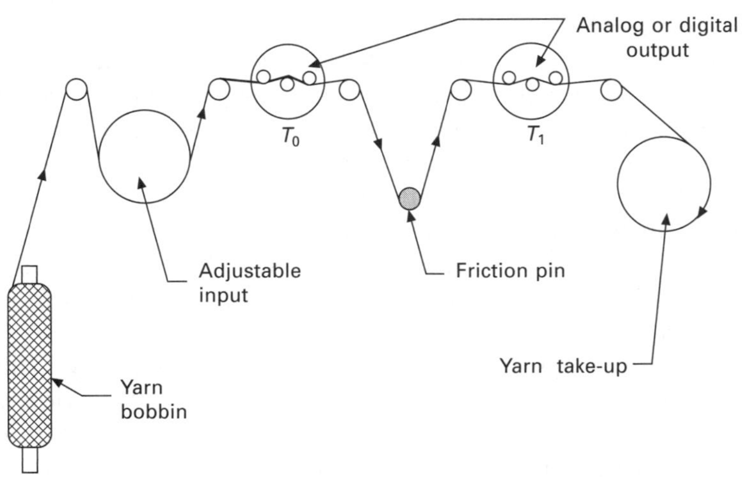

5.29 Schematic of the standard yarn-to-solid surface test, according to ASTM D 3108.

diameter of friction pin, the wrap angle, the input tension, and the running speed. This indicates that the value of the coefficient of friction could be used only for comparative purposes, assuming that all other test variables were kept constant. This also means that the calculation of friction coefficient was not always necessary, and the friction behavior of different yarns could be compared using simply the values of the output tension T_1, if input tension T_0 was maintained constant. There is also the commonly used practice of reporting friction test results in terms of the difference between the input and the output tensions as $(T_1 - T_0)$.

The apparatus has a number of components that are worth mentioning. One of these is the input tension meter. Because μ is a function of T_0, the latter must be set as required and controlled during friction tests. In other words, the equipment should have an appropriate adjustable input tension device. A variety of input tension devices have been used and are available to select from. The simple stationary friction body, such as an array of ceramic pins, is not acceptable due to frictional differences at various speeds and differences caused by yarn size (decitex). An alternative is a spring-loaded friction brake, which is a spring-loaded rotating wheel. The tension generated at this wheel can be adjusted by changing the spring tension. The shortcomings of this are the inability to provide smooth adjustments as well as the usually poor maintenance of set value and generally poor durability. The valuable alternative is the so-called magnetic brake where the resistance of rotating wheel is controlled by the distance between two magnets. This system is much more accurate and durable as compared to the one involving spring. Problems remain, however, which are again the inability of obtaining a smooth control on tension and the availability of somewhat limited range of input tensions due to relatively small size of the magnets and the corresponding forces. An advance in this regard is a friction brake that utilizes electrical magnets to provide necessary resistance to wheel rotation. These have much wider range of tensions and are capable of providing smoother adjustments and greater control on T_0. Currently, the best available input tension controllers combine electrical magnet friction brakes with the load cell and the feed back controllers. Although relatively expensive, they allow accurate setting and maintenance of the input tensions for various yarns in a wide range of speeds. The digital I/O readout display makes tension settings user friendly. These components can be connected to a computer for data collection, analysis and printing.

The second component to be concerned with is the electronic load cell, capable of measuring tension of a moving yarn. These devices must be equipped with very nearly frictionless, low torque, pulleys in order to minimize their own friction interfering with frictional measurements. These cells must be capable of operating at reasonably high data collection rates to avoid unintentional damping of the signal and, therefore, undesirable loss of the data. An adequate load cell will have digital/analog output that will be easily accessible for routine calibration. A wide variety of such cells are available.

The next critical part of friction apparatus is the counter or friction body itself, i.e. the pin against which friction is measured. ASTM standard recommends a chrome-plated steel rod with diameter of 12.7 mm (0.5 in) with an average surface roughness (RMS) of 4–6 microns. The total wrap angle can be set between $\pi/6$ (30°) to π (180°). This standard also does not exclude a possibility of using multiple friction bodies to provide a combined final wrap angle. The schematic of a device with multiple pins is presented in Fig. 5.30. The advantages of using multiple pins are:

- smaller friction on each pin results in less heat generation and better heat dissipation
- lesser abrasion effect leading to improved durability of the friction counter body.

The analysis given below illustrates that in the capstan equation, the angle $\theta = \sum\theta_i$.

$$T_1 = T_1^{**} e^{\mu\theta 3} = T_1^{*} e^{\mu(\theta 2+\theta 3)} = T_0 e^{\mu(\theta 1+\theta 2+\theta 3)} = T_0 e^{\mu\Sigma\theta i} \qquad 5.39$$

The other important part of the friction measuring system is the take up drive. It should be able to provide a wide range of speeds without the loss of torque. The older style direct current variable drive utilized gearbox coupled with timing belt to change the speed. These drives are reasonably easy to control, although their torque tends to drop rapidly with an increase of speed. Contemporary stepper motor or compumotor drives allow a wide range of speeds (normally from 10 to 10,000 rpm) without loss of torque. The advantage of the latter drives is one's ability to integrate them into complete computer controlled packages, which allows greater flexibility and unlimited possibilities for tests' design. Torque is important for both measuring friction and controlling test speed. Friction tests can be run at high speeds of 1000 m/min and above. At these speeds, the instrument is actually measuring the hydrodynamic

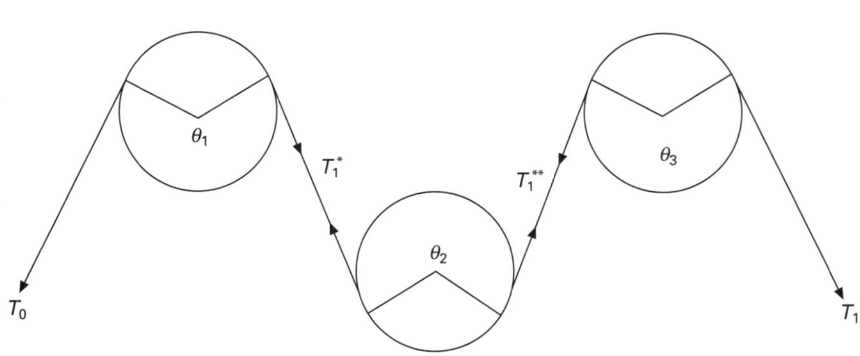

5.30 Friction build-up on multiple pins.

drag, meaning the resistance of the lubricating film to shear. Normally this resistance increases with speed and if the tester's drive is not capable of maintaining the set speed (the higher the speed the higher the resistance and, therefore, the lower the torque) then the frictional response is measured at a speed lower than intended. This leads to errors, the most common of which is the leveling of friction force with speed.

The last major component of a friction meter to be considered is the so-called frictionless pulley. For example, in the classic friction meter, presented in Fig. 5.29, there are five such pulleys to provide desired yarn path. Unfortunately, these pulleys will generate some friction and affect measurements. The friction imposed by these in line before the input load cell can be easily taken into account, unless their friction exceeds the desired value of the input tension. On the other hand, the presence of two parasite pulleys (Fig. 5.29) before and after the friction body can affect friction data. Such pulleys can, however be omitted as illustrated in the arrangement shown in Fig. 5.31.

The presence of multiple pins with the advantages discussed earlier and the two pulleys makes this configuration a preferred one. The use of multiple pins is also not in contradiction with the requirements of ASTM D3108 standard, which allows the usage of such friction bodies.

5.10.3 Commercially available yarn-to-solid surface friction meters

Despite well-established procedures for the yarn-to-solid surface friction tests, only few friction meters are available commercially. One of the most

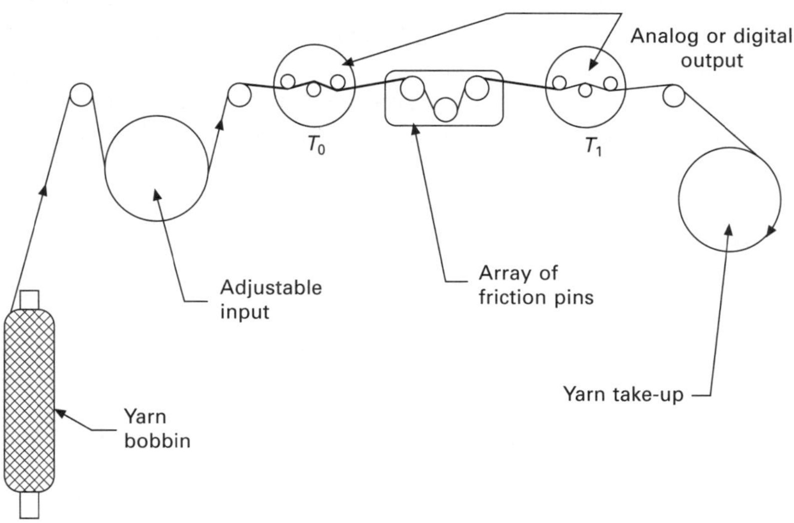

5.31 Yarn-to-solid surface test device without parasite pulleys.

widely used and earliest is the friction meter produced by Rothschild. This apparatus has computerized output, although it still has a belt driven variable drive system with analog controllers. The device utilizes a spring-type manually controllable input tension system, electro capacitance load cells and single pin design (Fig. 5.29). A wide range of testing speeds can be achieved by changing the gear ratio in the timing belt gearbox. Although the instrument has analog controllers and minimal computerization, it is nevertheless the most widely used apparatus that over the years has found its place in many commercial textile labs and institutions.

Another commercially available friction meter is Attrifil II, marketed in the USA through Mesdan® Lab. The apparatus in design is very similar to the classic Rothschild friction meter and uses the yarn path recommended in ASTM standard (see Fig. 5.29). It uses a permanent magnet friction brake for manual input tension control, piezoelectric load cell with integrated pulleys, and nip roll computer controlled variable drive as a take up. The design avoids pulleys between the friction pin and the load cells by using, instead, integrated load cell pulleys. Test speed is set using computer and all data acquisition, calibration, and calculations are performed by the proprietary software package. Tested yarn is aspirated into a wastebasket.

Atlas/SDL markets a device similar to Attrifill II, but it also has an optical sensing cell for characterizing hairiness.

Another friction meter, by Honigmann (Germany), combines a wide range of speeds (1–800 m/min) with solid electronic controllers, including electronically controlled input tension. It offers interchangeable friction bodies, including heated friction body option. This device does incorporate extra pulleys that must be considered. The instrument offers both manual and computer controlled modes of operations.

To complete the description of commercial yarn-to-solid body friction meters, one should also mention a few rather simple and relatively inexpensive instruments that are available. A hand-held friction meter that is capable of on-line friction measurements is available from Lawson-Hemphill Inc. (Fig. 5.32). This instrument can be operated with Windows® XP-compatible software and used with a laptop via a USB connector. It is designed for use by manufacturers of any type of yarn, as well as by knitters. The mechanical version of a very similar instrument is marketed by SDL Atlas.

Another simple friction meter is the one offered by WIRA Inc. This device (Fig. 5.33) has a simple design along with solid electronic controllers, operated through microprocessors. It has a variable speed direct drive, and simple mechanical input tension controller. Frictional data is reported on digital readout in terms of tension build up, i.e. $(T_1 - T_0)$.

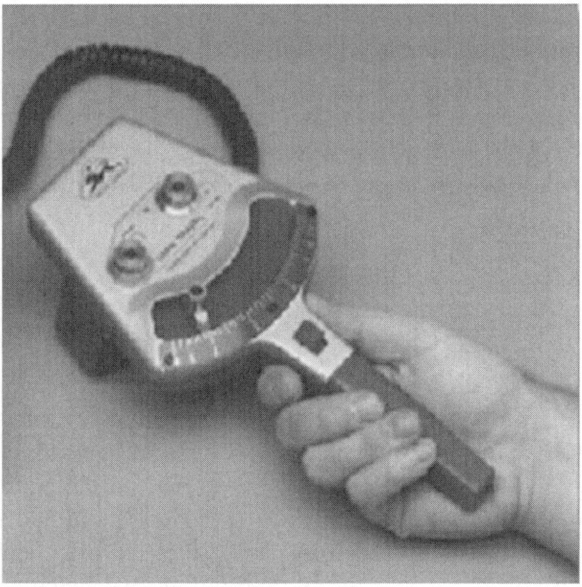

5.32 Hand-held friction meter.

5.33 WIRA friction meter.

5.10.4 Standard test methods for yarn-to-yarn (fiber-to-fiber) friction

A separate standard test method is available for yarn-to-yarn friction: ASTM D 3412 "Test method for coefficient of friction, yarn to yarn" [48]. It involves the twisting of yarn with itself and the measuring of the input and the output tensions along with the value of the twist angle β. The value of the coefficient of friction is calculated using Equation 5.41, given in the manual. The method is illustrated in Fig. 5.34, which has a layout very similar to the one for the standard test method presented in Fig. 5.29. Although the standard method recommends testing speeds in the range between 20 and 100 mm/min, much lower speeds, such as below 1 mm/min, can be employed. At such low speeds, there is little need to use the adjustable input tension device; it can be successfully replaced by a dead weight that significantly improves the accuracy of the test. The apex angle β between the input and the output yarns axes is calculated using Equation 5.40:

$$\beta = 2 \arctan (H/(V - W) \qquad 5.40$$

where: H, V and W are geometrical dimensions, illustrated in Fig. 5.34. The coefficient of friction μ is calculated according to Equation 5.41 [48]:

$$\mu = \{\mathrm{Ln}((T_1 - \Delta T/2)/(T_0 + \Delta T/2))\}/2\pi m\beta \qquad 5.41$$

where, T_0 is the mean input tension, T_1 is the mean output tension, ΔT is the zero twist tension, m is the number of wraps, and β is the apex angle calculated according to Equation 5.40.

The correction factor ΔT is incorporated in the equation to account for friction added by the presence of additional pulleys, necessary in this case. This factor is measured by letting a yarn pass through the apparatus without the twist.

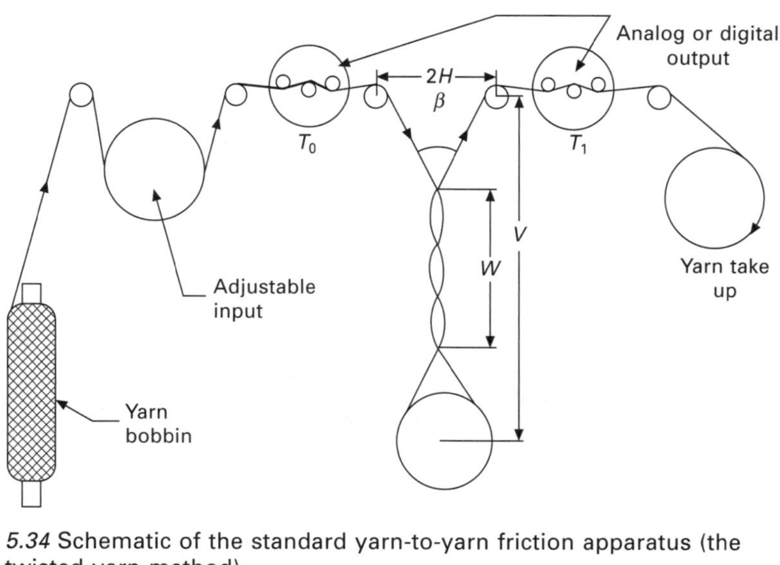

5.34 Schematic of the standard yarn-to-yarn friction apparatus (the twisted yarn method).

5.11 Standard friction tests for staple fibers and nonwovens

Although the cotton and staple fiber-based products constitute a major part of the primary textile industry and the efficiency of processing and the quality of product depends on a control of friction (see Chapter 9), no standard test exists for measuring friction in such fibers. The only reasonable approach to measure friction in staple fibers is the so-called staple pad friction test. In this, a certain amount of staple fiber is made into a pad, this pad placed on the platform, and a sled of a certain weight is placed on the top of fiber pad (Fig. 5.35). The platform is installed on the frame of the tensile tester and the sled is dragged using the movement of the tensile tester cross head. In certain cases, the bottom surface of the sled and the top surface of the platform are covered with sandpaper to avoid slippage at the fiber/solid surface interface and facilitate the shearing of the pad along the intermingled fibers. If there is no sandpaper on the platform surface, then the friction between the staple fibers and platform material (usually steel) can be measured. A staple pad can be prepared by a variety of ways using available carding equipment. Testing speeds are usually kept reasonably low to improve test resolution. The weight of the sled can be varied depending on the specific requirements but it is usually kept between 10 and 20 N.

Based on a similar method of mounting and conducting tests, a standard test method, i.e. INDA Standard IST 140.1 [49], is available for use on

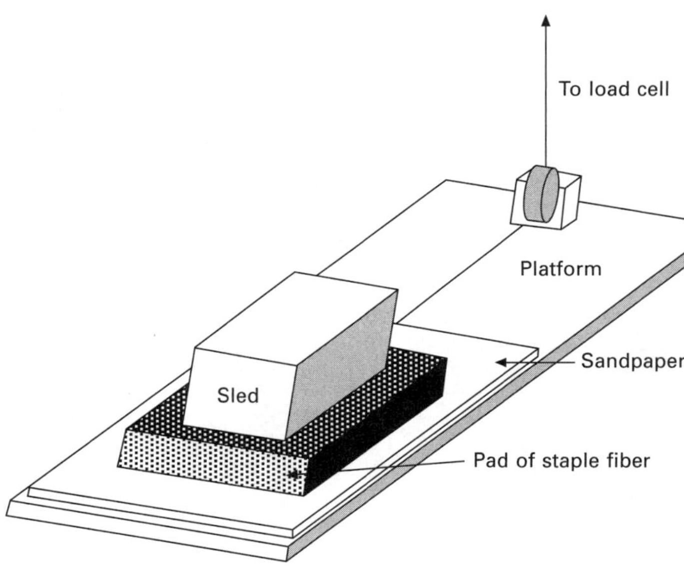

5.35 Staple pad method friction test.

nonwoven fabrics. The method gives the detailed procedure for the test, including dimensions of the sample and the sled, and the testing speed.

5.12 Standard friction test and commercial apparatus for fabrics: the KES system

The KES-F system was developed and commercialized by the Hand Evaluation and Standardization Committee of the Textile Machinery Society, Japan, under the chairmanship of Dr. S. Kawabata in 1975 [42]. The system was designed to characterize hand as perceived by a consumer. Among the many properties assessed are the surface roughness and the coefficient of friction. In the characterization of the former, a U-shaped 0.5 mm diameter steel piano wire is pressed at a force of 10 g_f over a tautly clamped fabric (Fig. 5.36a). The latter is traversed at a speed of 1 mm/sec. The landscape of the fabric is recorded by the vertical movement of the probe as a function of the distance along the fabric (Fig. 5.36b). A parameter, mean deviation of the thickness (SMD), calculated according to Equation 5.42, gives a measure of the fabric roughness:

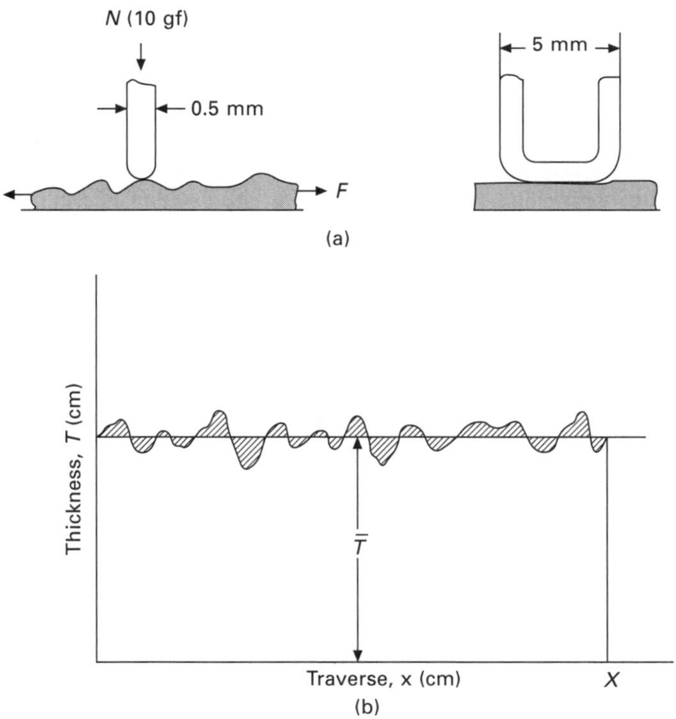

5.36 KES test for surface roughness [42]: (a) smooth steel piano wire roughness probe, (b) thickness profile. (SMD = hatched area/X).

$$\text{SMD} = \frac{1}{X}\int_0^x |T - \overline{T}| \bullet dx \qquad 5.42$$

The presence of cords or ribs in a fabric can be expected to affect the value of SMD as well as the frequency of peaks, as discussed in detail in Section 5.7.

The coefficient of friction is measured using the same procedure as given for SMD except the probe is made of 10 elements of the same wire, joined side by side, resulting in a width of 5 mm (Fig. 5.37a). The probe is pressed onto the fabric with a constant force of 50 g_f. The coefficient of friction, μ, given by the ratio of the force registered in the transducer, attached to the friction probe, to the normal force, is plotted as a function of the traverse distance (Fig. 5.37b). The mean value of μ (MIU in the KES system) is given by the average height of the curve:

$$\text{MIU} = \frac{1}{X}\int \mu \bullet dx \qquad 5.43$$

In the KES tests, mean deviation of μ from the average, MMD, is also assessed to obtain an index of variation in the value:

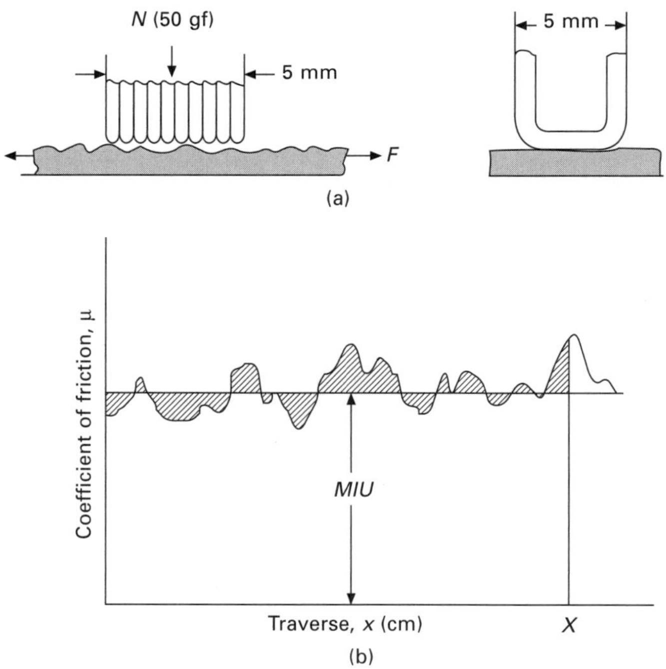

5.37 KES test for friction (metal to fabric) [42]: (a) smooth steel piano wire friction probe, (b) coefficient of friction profile. (MMD = hatched area/X).

$$\text{MMD} = \frac{1}{X}\int |\mu - \bar{\mu}| \bullet dx \qquad 5.44$$

From the description of the test method, it becomes clear that the KES system, involving fixed low forces and low speeds, is a suitable method for characterizing a component of the hand perceived by consumers of apparel when the fabric is picked and stroked between fingers. Since the magnitudes of the forces and speeds involved are of about the same order of magnitudes as assumed to be encountered during actual touching and stroking of fabrics, the ranking obtained from the KES friction and roughness parameters has frequently been found to correlate with the subjective rankings of properties such as smoothness, softness, roughness, etc., by a panel of judges [50–52].

5.13 References

1. Howell H G, Mieszkis and Tabor D, *Friction in Textiles*, The Text. Inst. and Butterworth, Manchester and London, 1959.
2. Hong J and Jayaraman S, 'Friction in Textiles,' *Textile Progress*, **34**(1/2), The Textile Institute, Manchester, UK, 2003.
3. Amontons G, *Historie de L'Academie Royale des Sciences avec Memoires, de Mathematique et de Physique*, 1699.
4. Bowden F P and Leben L, 'The nature of sliding and the analysis of friction', *Proc. Roy. Soc. A*, 1939, **A169**, 371–391.
5. Mercer E H and Makinson K R, 'The frictional properties of wool and other textile fibers', *J. Text. Inst.*, 1947, **38**, T227–240.
6. Olofsson B and Gralen N, 'Measurement of friction between single fibers', *Textile Res. J.*, 1947, **17**, 488–496.
7. Guthrie J C and Oliver P H, 'Inter-fibre friction of viscose rayon', *J. Text. Inst.*, 1952, **43**, T579–T594.
8. Pascoe M W and Tabor D, 'The friction and deformation of polymers', *Proc. Roy. Soc. A*, 1956, **A235**, 210–224.
9. Briscoe B J and Kremnitzer S L, 'A study of the friction and adhesion of Polyethylene terephthalate monofilaments', *J. Phys. D: Appl. Phys.*, 1979, **5**, 505–516.
10. Briscoe B J, Winkler A and Adams M J, 'A statistical analysis of the frictional forces generated between monofilaments during intermittent sliding', *J. Phys. D: Appl. Phys.*, 1985, **18**, 2143–2167.
11. Adams M J, Briscoe B J and Wee T K, 'The differential friction effect of keratin fibers', *J. Phys. D: Appl. Phys.*, 1990, **23**, 406–414.
12. Howell H G, 'Interfiber friction', *J. Text. Inst.*, 1951, **42**, T521–533.
13. El Mohahzy Y E and Gupta B S, 'Friction in fibrous materials. Part II: Experimental study of the effects of structural and morphological factors', *Textile Res. J.*, 1993, **63**(4), 219–230.
14. El Mogahzy Y E, *A study of the nature of friction in fibrous materials*, PhD Thesis, Fiber and Polymer Science, North Carolina State University, USA, 1987.
15. Adderley A, 'The clinging power of single cotton hairs', *J. Text. Inst.*, 1922, **13**, T249–T255.
16. Wood C, 'Dynamic friction of viscose fibers and relative humidity', *J. Text. Inst.*, 1954, **45**, T794–T802.

17. Postle J L, Ingham J and Cox D R, 'The measurement of inter-fibre friction in slivers', *J. Text. Inst.*, 1952, **43**, T77–T90.
18. Taylor D S, 'The measurement of fiber friction and its application to drafting force and fiber control calculations', *J. Text. Inst.*, 1955, **46**, 59–83.
19. Lord E, 'Frictional forces between fringes of fibers', *J. Text. Inst.*, 1955, **46**, 41–58.
20. Hearle J W S and Husain A K M M, 'Studies in needled fabrics. Part VIII: The effect of friction on the processing and properties of needle-bonded fabrics', *J. Text. Inst.*, 1971, **62**, 83–107.
21. El-Mogahzy Y E and Broughton R M, 'A new approach for evaluating the frictional behavior of cotton fibers. Part I: Fundamental aspects and measuring techniques', *Textile Res. J.*, 1993, **63**(8), 465–475.
22. Gupta B S and Pei-Tsing C, 'Structural studies in continuous filament yarns. Part III: Studies with fiber friction as a variable', *Textile. Res. J.*, 1976, **46**(2), 90–99.
23. Speakman J B and Stott E, 'A contribution to the Theory of Milling. Part I: A method for measuring the scaliness of wool fibers', *J. Text. Inst.*, 1931, **22**, T339–T348.
24. Ajayi J O, 'Fabric smoothness, friction and handle', *Textile Res. J.*, 1992, **62**, 52–59.
25. Clapp T G, Timble N B and Gupta B S, 'The frictional behavior of textile fabrics', *J. App. Poly. Sci.: Applied Polymer Symposium*, 1991, **47**, 373–382.
26. Broughton R M and El-Mohahzy Y E, 'The measurement of fiber friction and its importance in manufacturing cotton nonwovens', *Tappi Journal*, 1993, **76**(2), 178–182.
27. Ramkumar S S and Roedel C, 'Study of needle penetration speeds on frictional properties of nonwoven webs: A new approach', *App. Poly. Sci.*, 2003, **89**, 3626–3631.
28. Ramkumar S S, Wood D J, Fox K and Harlock S C, 'Developing a polymeric human finger sensor to study the frictional properties of textiles. Part I: Artificial finger development', *Textile Res. J.*, 2003, **73**(6), 469–473.
29. Howell H G, 'The general case of friction of a string round a cylinder', *J. Text. Inst.*, 1953, **44**, T359–362.
30. Lindberg J and Gralen N, 'Measurement of friction between single fibers. II: Frictional properties of wool fibers measured by the fiber-twist method', *Textile Res. J.*, 1948, **18**, 287–301.
31. Fair N and Gupta B S, 'Effect of chlorine on friction and morphology of human hair', *J. Soc. Cosm. Chem.*, 1982, **33**, 229–242.
32. Gupta B S, Wolf K W and Postlethwait R W, 'Effect of structural material and construction on frictional properties of sutures', *Surg. Gynecol. Obstet.*, 1985, **161**, 12–16.
33. Landwehr R, Menefee E and Stigler D, 'A sensitive method of measuring the friction between single fibers in liquid', *Proceedings. 5th International Wool Tex. Res. Conference VII*, 1976, 328–336.
34. Ajayi J O, *Some studies of frictional properties of fabrics*, PhD Thesis, University of Strathclyde, Glasgow, 1988.
35. Carr W W, Possey J E and Ticher W C, 'Frictional characteristics of apparel fabrics', *Textile Res. J.*, 1988, **58**, 129–138.
36. Grosberg P, Holme I and Ngai M C, *6th Quinquennial International Wool Textile Conference*, Cape Town, 1980, **IV**, Page 81.
37. Ohsawa M, Namiki S and Kodaka H, 'Relationship between fabric balance and surface friction in plain-woven fabric', *Journal of Textile Machinery Society of Japan*, 1969, **15**(3), 98–105.

38. Ajayi J O, Elder H M, Kolawole E G, Bello K A and Darma M U, 'Resolution of stick-slip friction traces of fabrics', *J. Text. Inst.*, 1995, **86**, 600–609.
39. Ajayi J O, 'An attachment to the constant rate of elongation tester for estimating surface irregularities of fabrics', *Textile Res. J.*, 1994, **64**, 475–479.
40. Ajayi J O and Elder H M, 'Comparative studies of yarn and fabric friction', *Journal of Testing and Evaluation*, 1994, **22**, 463–467.
41. Ajayi J O and Elder H M, 'Effects of finishing treatments on fabric friction', *Journal of Testing and Evaluation*, 1995, **23**, 55–58.
42. Kawabata S, 'Standards for hand evaluation', *Textile Machinery Society of Japan*, Kyoto, pp 1–25.
43. Ajayi J O and Elder H M, 'Tactile and instrumental evaluation of surface contour of fabrics', *Journal of Textile Machinery Society of Japan*, 1996, **49**, 98–107.
44. Siegel S and Castellan J N Jr., *Nonparametric statistics for the behavioral sciences*, McGraw Hill, New York, 1988.
45. Elder H M, Fisher S, Armstrong K and Hutchison G, 'Fabric softness, handle, and compression', *J. Text. Inst.*, 1984, **75**(1), 37–46.
46. *ASTM D 3108 Test Method* for Coefficient of Friction – Yarn to solid material.
47. Rubinstein C, 'The friction of a yarn lapping a cylinder', *J. Text. Inst.*, 1958, **49**, T181.
48. *ASTM D 3412 Test Method* for Coefficient of Friction – Yarn to yarn.
49. *INDA IST 140.1 Test Method* for friction in nonwovens.
50. Chen P, Barker R L, Smith G W, and Scruggs B., 'Handle of weft knit fabrics,' *Textile Res. J.*, 1992, **62**, 200–211.
51. Cheng K P S, How Y L and Yick K L, 'Surface properties in shirting materials,' *Text. Asia*, 1994, **25**, No. 12, 52–54.
52. Hu J and Chan Y., 'Effect of fabric mechanical properties on drape,' *Textile Res. J.*, 1998, **68**, 57–64.

Part II

Friction in particular textile fibers, structures and processes

6
Friction and surface characteristics of cotton fibers

Y E EL MOGAHZY, Auburn University, USA

6.1 Introduction

Cotton is a natural cellulose fiber obtained from the mature capsule of the plant. The incredible demand and attraction to cotton products that have lasted hundreds of years primarily stem from the touch and the feel of the fiber, a unique characteristic that remains uncontested by any other material. Cotton is inherently a replica of nature. In a blind test, one can easily identify the surface of the fiber by touching or handling the fibrous assembly. More interestingly, touching and feeling of cotton fibers results in an instantaneous acquaintance and comfortable bonding, much like touching soft human skin.

The cotton fiber structure consists of a network of variously sized pores or capillary spaces surrounded by numerous variously sized fibrils in the different layers of the fiber. This makes cotton fibers accessible to liquids and vapors. The capillary action of the fibrils pulls in the liquid where it is held in pores between the fibrils. This explains the unique wicking and absorption combination characteristics of the fiber that give a feeling of comfort when in contact with human skin. The tapered shape of the fiber provides a pleasant and soft touch since there are no sharply cut ends as there are with synthetic staple fibers.

When cotton is classed in the market place, surface characteristics typically dominate the subjective judgment. An expert classer, during handling and subjectively grading a certain type of cotton, will typically judge the fiber based on external appearance, brightness, color, purity, and character. All these factors are in essence surface-related. More importantly, the classer touches and feels the fibers and passes a judgment on whether it will perform smoothly or harshly during textile processing. For this reason, and despite the use of instrumental classing involving the High Volume Instrument (HVI) system, the traditional manual classing of cotton still represents an accepted purchasing and grading means around the world. In other words, the advanced instrumental approach has not fully replaced traditional cotton classing. This calls for a unique classing instrument that can truly reflect the inherent touch and feel of the fiber.

It is also very true that two cotton types of approximately the same values of standard characteristics (e.g., length, fineness, strength, etc.) may perform differently during processing. Indeed, we have witnessed situations where some cottons of normal value standard characteristics resulted in significant processing problems such as excessive drafting forces, irregular card web, and frequent failure during spinning. Diagnostic analysis of the causes of these problems revealed surface-related factors that resulted in high inter-fiber friction leading to disrupted fiber flow during processing and failure of fiber cluster breakdown [1].

The effect of cotton fiber friction is more pronounced when fibers, yarns, or fabrics undergo chemical processes such as scouring, bleaching, or mercerization. These treatments significantly alter the surface characteristics by removing the natural wax from the fiber outer layer and altering surface morphology [2, 3].

When cotton is blended with synthetic fibers, a key blending aspect is the interactive mode of blending [4]. This implies the mechanical interaction between fibers within the same fiber type and between different fiber types during the blending process. This mode of blending is largely determined by the surface compatibility of the materials in the blend, which often determines the extent of bias in the blended assembly.

In summary, the critical importance of cotton surface and cotton fiber friction stems from the following points:

- Fiber cohesion is one of the three basic qualifying characteristics of any fiber; the other two are flexibility and slenderness ratio [1].
- Fiber friction is a key factor in determining the processing performance of fibers. This is a direct result of the fact that during processing, fibers are subjected to a great deal of rubbing against other fibers and against metals.
- The cohesive integrity of intermediate fiber strands such as fiber webs, and slivers is directly influenced by surface characteristics and the extent of inter-fiber friction.
- The drafting process, which is the primary process in the fiber-to-yarn conversion system, is based on sliding of fibers against one another [1].
- The strength and the elongation of a spun yarn is a function of two main factors: fiber strength and inter-fiber friction [5].
- The weaving and knitting processes involve a great deal of inter-yarn contact and yarn-to-metal friction.
- During the dyeing and finishing process, surface characteristics play critical roles in fabric handling and transport and in the extent to which some of the chemicals are used in this process.

- In the nonwoven industry, surface morphology and fiber friction are perhaps the most critical aspects that determine processing performance. Accordingly, the cotton purification process that involves scouring and bleaching of the fiber requires an addition of an appropriate finish after the treatments to insure efficient processing.

Over the years, research in the area of cotton fiber friction has been extensive. However, a universal standard method for measuring fiber friction has not yet been established. As a result, friction values do not exist in the cotton database. In the absence of a standard testing technique, it is quite difficult to compare research findings obtained by different investigators except for determining general common trends and correlations. Some of the obstacles facing the development of a standard testing method for cotton fiber friction are as follows:

- The frictional behavior of cotton fiber is largely dependent on other fiber properties such as fiber length and fiber fineness [2, 3].
- The high variability of cotton fibers makes it difficult to obtain reproducible results.
- Friction values are highly sensitive to many external factors including the normal load, the sliding speed, moisture content, and the presence of impurities in the interface between the two surfaces in contact.
- The need to characterize fiber friction often requires that the test simulates a specific interactive mode. Since different types of contact produce different friction values [6], a standard friction test involving a single mode of friction will not be representative of all modes of friction.

The objectives in this chapter are to discuss the surface characteristics of cotton and the nature of cotton fiber friction. The focus will be to highlight some of the key aspects of friction, common research findings, and some recent methods of testing fiber friction that have standardization potential.

6.2 Cotton fiber surface and frictional characteristics at the microscopic level

Understanding of the nature of friction is greatly aided by examination of the structure of the fiber. In general, cotton fiber structure may be described by three main features: (i) the surface morphology, (ii) the taper character of the fiber and (iii) the molecular orientation.

6.2.1 Surface morphology

The structure of a mature cotton fiber may be viewed as consisting of six main parts [2, 6–9]. As shown in Fig. 6.1, the first is the cuticle, or the “skin”

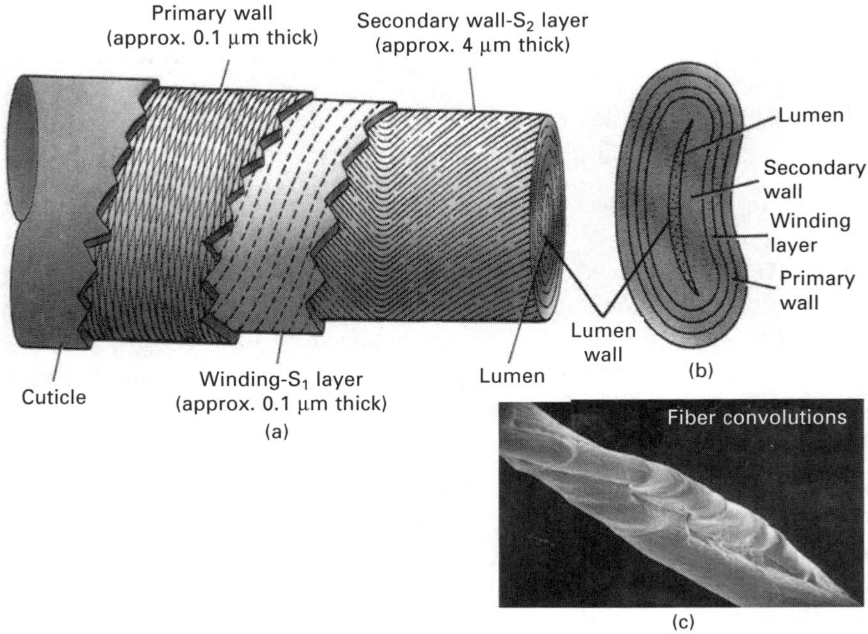

6.1 Structural features of cotton fiber [2, 6–9].

of the fiber. This waxy and smooth layer contains pectin and proteinaceous materials. The presence of this layer has a significant impact on the smoothness and the handling of cotton during processing. However, the fact that it is a very thin layer, only a few molecules thick, makes it vulnerable to environmental effects, such as heavy rain and high temperature. Upon scouring, this layer is removed, which explains the increase in fiber/fiber friction [2, 6, 7].

The second part is the primary wall. This is the original thin cell wall and is mainly cellulose made up of a network of fine fibrils. The primary wall may be visualized as a sheath of spiraling fibrils where each layer spirals 20–30° to the fiber axis. The thickness of this wall correlates with the extent of maturity of cotton fiber, the thicker the wall, the higher the maturity. The primary wall makes for a well-organized system of continuous very fine capillaries. These fine capillaries "rob" liquids from coarse capillaries; an action that contributes greatly to a cotton material's wipe-dry performance.

The third part is called the winding layer or S1 layer. This is the first layer of secondary thickening and it differs in structure from either the primary wall or the remainder of the secondary wall. It is an open "netting" pattern of fibrils that are aligned at 40–70° [7] angles to the fiber axis. The fourth part is the secondary wall, which consists of concentric layers of cellulose constituting the main portion of the cotton fiber (also called S2 layer). During the growth period, a new layer of cellulose is added to the secondary wall.

The fibrils are deposited at angles of 70–80° [7] with points along the length where the angles are reversed. The fibrils are packed close together, again forming small capillaries.

The fifth part is the lumen wall. This wall separates the secondary wall from the lumen, which represents the sixth part. It appears to be more resistant to certain reagents than the secondary wall layers. The lumen is a hollow canal that runs the length of the fiber. It is filled with living protoplasts during the growth period. After the fiber matures and the boll opens, the protoplast dries up and the lumen will naturally collapse.

Geometrically, the cotton fiber has a twisted-ribbon shape along the length of the fiber and a kidney-shaped cross section. These features are shown in Figs 6.1(c) and 6.1(b), respectively. This convoluted surface of the cotton fiber is largely attributed to the lumen collapse. Typically, there are about 20–40 twists/cm. SEM analysis [5] reveals additional distinct surface features including: (i) the ribbon-like shape, with nearly elliptical transverse surface, (ii) the existence of folds on the surface of fibers and (iii) the presence of reversals of fibrillar texture prominent on the fiber surface.

6.2.2 The taper feature of cotton fiber

The cotton fiber is tapered at one end and fibrillated at the other end where it is joined to the seed. In a study by Seshan [9], the taper effect on fiber friction was examined and a number of interesting points were made. Perhaps the most important point was that the cotton fiber exhibited a differential friction along its length. Friction at the root was found to be higher than that at the tip of the fiber. The difference was found to be significant, particularly at low levels of normal loads. The presence of differential friction was supported by two experimental observations: (i) a lower convolution angle at the tip than at the root of the fiber (the difference in this angle was as high as 175% within a fiber) and (ii) a lower surface area at the tip than at the root.

6.2.3 Fiber molecular orientation

Cotton fiber molecular orientation is commonly examined using the birefringence index, $\Delta n = n_{\Pi} - n_{\perp}$, in which n_{Π} is the refractive index with the polarized light oscillating in a plane parallel to the fiber axis, and $n_{\perp}$ is the refractive index with the light oscillating in a plane perpendicular to the fiber axis. A typical birefringence index of cotton may range from 0.04 to 0.09 [10]. Within a given fiber, this index increases from the tip (0–0.008) to the root of the fiber (above 0.04).

In relation to fiber friction, studies on synthetic fibers reveal that the higher the orientation, the more intimate the contact between fibers, and the higher the friction [11–13]. With cotton fibers, such analysis will be difficult to replicate since orientation will be directly associated with the growth rate of fibers (a difficult factor to control). In addition, cotton fibers of different levels of orientation will likely be different in many other surface-related physical characteristics. Nevertheless, the assumption that higher cotton fiber orientation will result in higher interfiber friction seems to be reasonable.

6.3 Cotton fiber surface and frictional characteristics at the macroscopic level

During processing, cotton fibers may flow as single segments or as clusters of fibers depending on the stage of processing and the extent of opening. Accordingly, the analysis of the frictional behavior of fibers should include the following factors:

- fiber arrangement
- number of fibers in contact
- extent of simulation
- fiber properties.

the roles of these factors are examined in the sections that follow. In addition, important concepts such as methods of friction testing and the classic laws of friction are discussed.

6.3.1 Fiber arrangement

The simplest fiber arrangement consists of two single fibers sliding against each other or a single fiber sliding against a solid surface (guide, roller, etc.). Figure 6.2 illustrates two modes of contact that are commonly considered in characterizing the arrangement, namely: point contact and line contact [11–15]. These are analyzed on the basis of the familiar capstan setup in which the sliding fiber is subject to an initial tension, T_0, at one end, and the tension required to initiate sliding, T_1, is measured. The relationship between the two tensions is governed by the capstan equation:

$$\frac{T_1}{T_0} = e^{\mu\theta}$$

where θ is the contact angle and μ is the coefficient of friction.

The point-contact arrangement consists of two fibers in contact at right angles or a fiber and another surface (often the horizontal one). The fiber held in the vertical plane is mounted to a load cell at one end while the other end is subjected to an initial tension imposed by a suspended weight. As the horizontal fiber moves downward, resistance to sliding at the contact point is encountered and measured by the tension required to initiate sliding. This allows a direct use of the above equation.

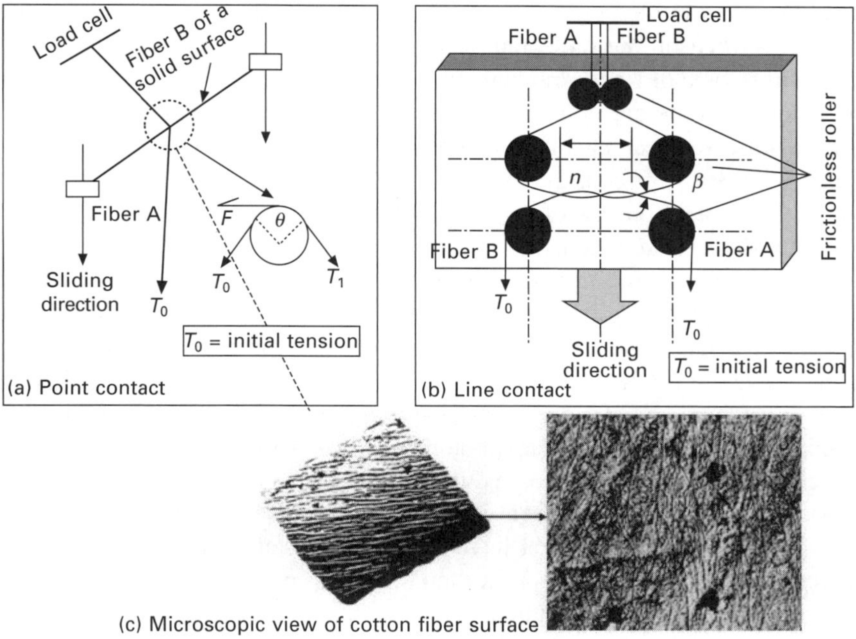

6.2 Single-fiber friction arrangement [11–15].

The line-contact arrangement is achieved by twisting two fibers together with a certain number of turns of twist, normally two, to allow a line path of contact as shown in Fig. 6.2(b). This method was originated by Gralen and Lindberg [15]. In this case, the above equation is modified to accommodate the geometrical arrangement of the two fibers in contact:

$$\frac{T_1}{T_0} = e^{\mu \pi n \beta}$$

where β is the angle of twist, and n is the number of turns of twist.

In a more recent study, El Mogahzy and Gupta [11, 13] modified the line-contact method to allow measurements in liquid media. This modification involved ways to overcome problems associated with wet testing including buoyancy and fluid drag.

It is important to point out that what may be considered as a point or line contact at the macroscopic level is indeed an area contact at the microscopic level with thousands of fibrils rubbing against each other or against another surface. This point is illustrated in Fig. 6.2(c). It is necessary, therefore, to couple the friction measurements with microscopic analysis of the surfaces in contact, perhaps before and after sliding. In recent years, advanced techniques such as scanning tunneling microscopes (STM) and atomic force microscopy

(AFM) have provided great potential for characterizing surface textures and frictional forces at the molecular scale [16, 17].

Another type of fiber arrangement is that associated with a layer of fibers forming a surface area that comes into contact with another fiber layer or with a solid surface. This type of contact is commonly called area-contact (see Fig. 6.3(a)). In contrast to the simple contact modes discussed above, the area contact consists of multiple fiber contacts with a wide variety of possible fiber arrangements.

The key factor influencing the area-contact mode is the type of fiber assembly being tested. Fig. 6.3(b) shows different cotton fiber assemblies that are typically obtained from different stages of processing in the fiber-to-yarn conversion. Clearly, at different stages, assemblies involve different fiber arrangements at the surface. A card web or a nonwoven assembly will exhibit random fiber arrangement on the surface. This will typically result in radical variation in friction force during sliding against another surface, fiber or solid. This variation is often witnessed in the stick-slip pattern (friction-time profile). Assemblies that have fibers largely aligned along the sliding direction exhibit less variation in their stick-slip patterns.

The issue of fiber arrangement discussed above is critical in developing a standard test method for cotton fiber friction. The author's experience suggests that there is a significant trade-off between having a simulative device of fiber friction that may not be highly reproducible due to the high variability involved, and a highly reproducible test which is, however, less simulative

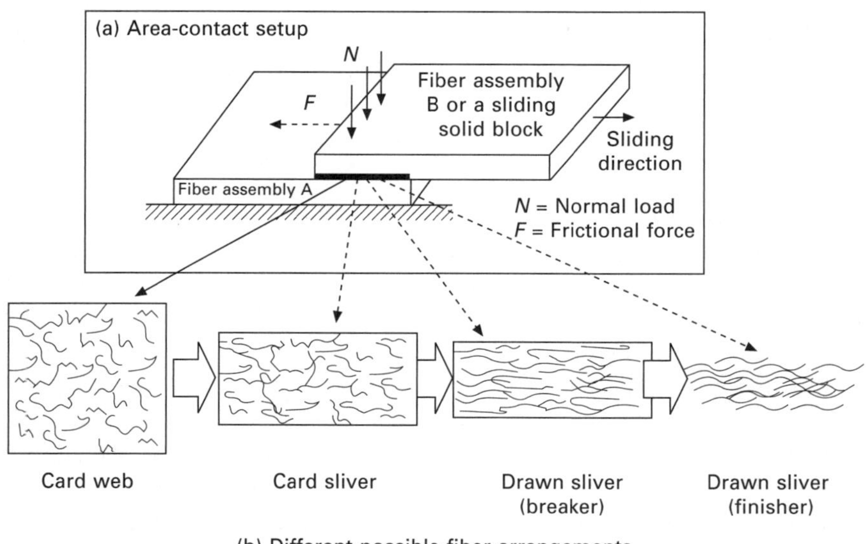

6.3 Area contact and possible fiber arrangements [2, 27–29, 31–33].

of actual fiber arrangement. This is one of the critical challenges facing the development of a universal standard test for staple fiber friction. Other challenges include calibration and sample preparation.

Area contact simulates the classic setup of sliding friction used in most physics literature. It also makes a direct use of the classical laws of friction established by distinguished scientists, among them Leonardo da Vinci (1452–1519), Amontons (1699), and Charles Coulomb (1781). Therefore, it will be useful at this point to briefly review these laws.

Friction laws are essentially statements about the relationship between the friction force and the normal force, defined by the coefficient of friction, μ:

$$F = \mu \cdot N$$

The classic laws of friction are summarized as follows [18, 19]:

- The friction coefficient of a certain contact interface is constant and it is independent of the normal force applied on the contact area.
- The friction force is independent of the apparent area of contact.
- The friction coefficient is independent of the sliding speed.

As recognized by Coulomb [19], the friction between surfaces at rest is often about 20% greater than that between the same surfaces when sliding. Static friction, or the friction required to initiate sliding, has thus been accepted as being generally larger.

Over the years, the validity of the classical laws of friction has been questioned by numerous investigations in various fields. The common finding in these investigations can be summarized in the following main points [20–25]:

- The validity of the classic laws of friction depends greatly on the external parameters influencing the contacting surfaces. These include the normal loads applied, sliding speed, the microscopic media in the interface, the shape and size of the asperities in contact, and the deformational behavior of the asperities.
- Material type is a key factor in determining the extent of validity of the classical laws of friction.
- In general, these laws are fairly valid for some materials such as the metals. For rigid polymers, friction is largely governed by the adhesion between surface atoms. The latter form bonds similar in strength to those that hold the molecules in the bulk, and friction represents the force required to break these bonds. For rubber-like materials, the mobility of polymer chains is high, and this determines the general rules for friction.
- For flexible polymers, a great departure from the classical laws of friction is witnessed and viscoelasticity becomes a dominant factor in determining the frictional behavior of this type of material.

In light of the above discussion, it follows that fibers, being flexible polymers and behaving viscoelastically under lateral pressure, are expected to depart from the basic laws of friction. In case of cotton fibers, this point was repeatedly supported by experimental results produced by many investigators [26–32]. One of the common empirical relationships between friction force and normal force that accurately represented the frictional behavior of cotton is as follows:

$$F = aN^n$$

where a, and n are called the friction indices. Note that at $n = 1$, this equation becomes identical with the classical friction equation, $F = \mu N$.

The physical meanings of the friction indexes a and n were discussed in a more recent study by Gupta and El Mogahzy [12] on the basis of a mathematical structural model of fiber friction. Using this model, the authors developed expressions that gave structural meanings to the two indices [20, 25].

On a macroscopic scale, fiber arrangement in the cotton sample also contributes to the departure of the behavior from that given by the classical laws by virtue of its direct effect on the true area of contact. Surfaces consisting of a network of randomly arranged fibers are expected to involve contacting asperities that are different in shape, size, and distribution from surfaces consisting of fibers aligned in one direction [30–32].

6.3.2 Number of fibers in contact

The effect of the number of fibers in the contact surface can be evaluated by using the two simple cases shown in Fig. 6.4. In both, the fibers are assumed to be identical, except for the number, and the weight applied on the area of contact is the same. In case A, the number of fibers in contact is n_A, this

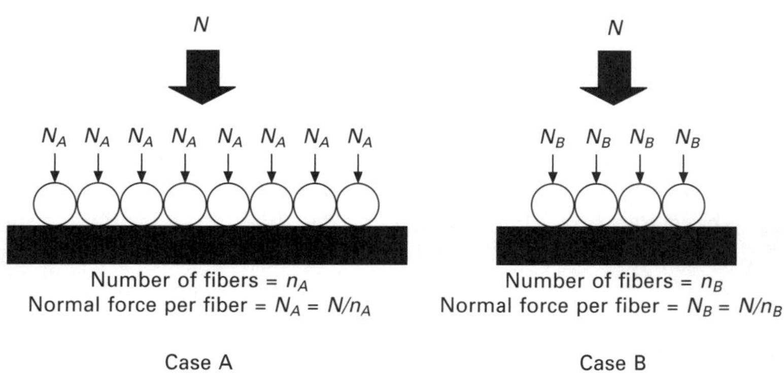

6.4 Two cases of area of contact [30].

yields a normal force per fiber of $N_A = N/n_A$. Similarly, in case B, the number of fibers in contact is n_B, this yields a normal force per fiber of $N_B = N/n_B$. If the classical Amontons' Law, $F = \mu N$, holds, the frictional force per fiber in the two cases will be as follows:

$$f_A = \mu N_A = \mu\left(\frac{N}{n_A}\right)$$

$$f_B = \mu N_B = \mu\left(\frac{N}{n_B}\right)$$

The total friction force will be:

$$F_A = \sum_{1}^{n_A} f_A = n_A\,\mu N_A = n_A\,\mu\left(\frac{N}{n_A}\right) = \mu N$$

$$F_B = \sum_{1}^{n_B} f_B = n_B\,\mu N_B = n_B\,\mu\left(\frac{N}{n_B}\right) = \mu N$$

Thus,

$$F_A = F_B$$

This equation represents the basis for the second Amontons' Law of friction, i.e. for a given normal force, friction force is independent of the area of contact.

Suppose that the empirical relationship, $F = aN^n$ holds. In this case, the total friction force will be as follows:

$$F_A = \sum_{1}^{n_A} f_A = \sum_{1}^{n_A} aN_A^n = n_A a N_A^n = n_A a\left(\frac{N}{n_A}\right)^n = a n_A^{1-n} N^n$$

$$F_B = \sum_{1}^{n_B} f_B = \sum_{1}^{n_B} aN_B^n = n_B a N_B^n = n_B a\left(\frac{N}{n_B}\right)^n = a n_B^{1-n} N^n$$

Thus, the frictional force will be generally expressed by the following equation:

$$F = a \cdot (n_f)^{1-n} \cdot N^n$$

where n_f is the number of fibers in contact, N is the normal force, and a and n are the friction indices defined earlier.

The above equation simply indicates that if the classic law of friction does not hold, which is the case for cotton fibers, the frictional force will be dependent on the number of fibers in contact, for a given normal force. Since n is usually less than one for fibrous materials, it follows that the increase in the number of fibers will result in an increase in the frictional force governed by the above equation.

The above equation was used as the basis for a friction testing device designed and constructed by El Mogahzy *et al.* in 1992 [30–32]. The main components of the device are shown in Fig. 6.5. It is called the "beard friction test" because the fiber sample used is in the form of a flat beard that can be prepared automatically using the fiber-preparation station of the High Volume Instrument (HVI®) developed by Uster Technologies. As indicated earlier, sample preparation is a key aspect of consistency and reproducibility of friction results. In this case, a random cotton sample consisting of a few thousand fibers (about 4000) is automatically extracted from a larger sample using an HVI-Motion Control® standard fiber clamp. The sample is automatically combed and brushed on an HVI length-station to form a fiber beard. The number of combing and brushing cycles can be changed to examine the effect of fiber crimp and fiber alignment on friction. This is because an increase in cycles should result in decrease of fiber crimp and increase of fiber alignment.

The device (Fig. 6.5(a)) consists of three main parts: the top fiber clamp, two bottom fiber clamps and the lateral pressure bar activated by a hydraulic piston. The top clamp is an HVI clamp attached to a load cell. The two bottom clamps are also HVI clamps mounted at an angle on a movable platen, which is driven up and down by a step-motor programmed and controlled by a motion controller through a computer terminal. The tension from the load cell is converted into voltage. The latter is digitized and displayed as friction data point. A Labview® program acquires, analyzes, and displays the data in chart form as well as saving it in a spreadsheet file.

Figure 6.5(b) shows how the contact surfaces are formed. The two bottom beard clamps are mounted on either side of the top beard, attached to the platen, and then clapped together to form a friction surface on both sides of the top beard. The lateral pressure is then applied using a pressure bar activated pneumatically. Different levels of pressure can be selected. The area of the pressure bar determines the apparent area of contact between the fiber surfaces. As the base holding the two bottom beards start to move downward, friction between both sides of the top beard and the inner sides of two bottom beards is generated.

Figure 6.5(b) also shows the contact area positions (1, 2, 3, …k) with the number 1 indicating the starting position and 2 indicating the next position, and so on. The point here is that the contact surface formed by the fibers mounted in the bottom beards will remain constant throughout the test but that of the top beard will change as the platen moves downward, from maximum at the start to minimum or zero at the end of the test. When fiber-to-metal friction is of interest, the bottom fiber clamps are removed and the contact media is restricted to the top fiber surface and the metal or other desirable surface pressure.

Figure 6.6 shows the outcome profile of the test, called "the beard friction

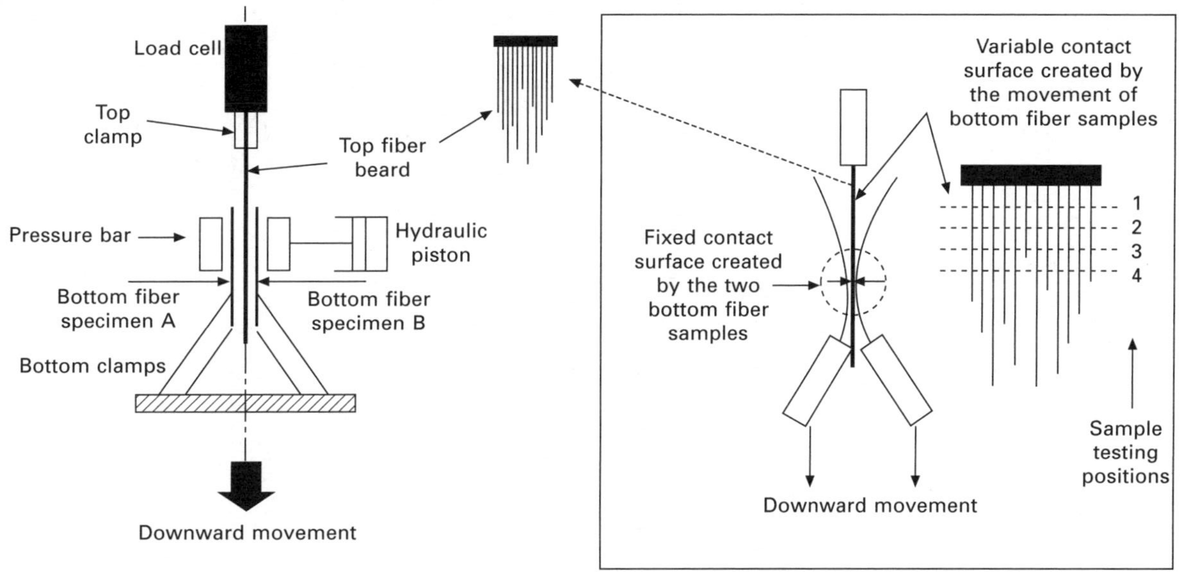

(a) Device components

(b) Contact surface setup

6.5 Basic components of beard friction device [30–32].

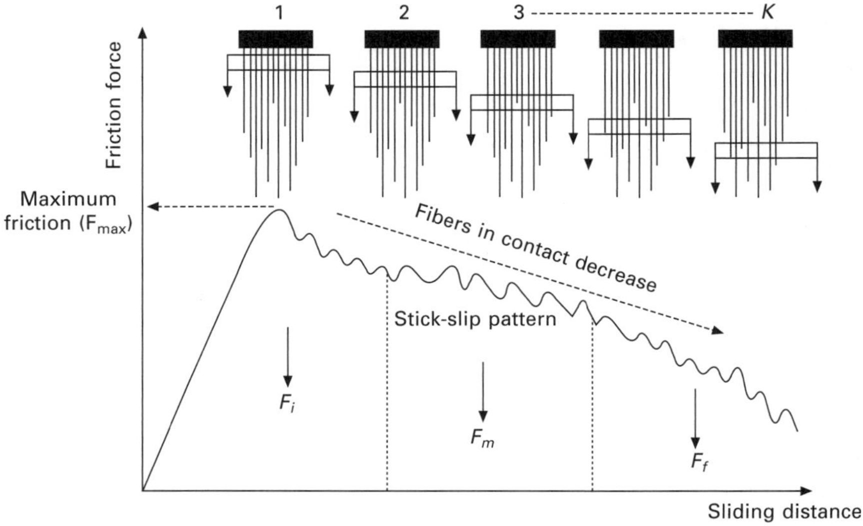

6.6 The beard friction profile [30–32].

profile". It consists of a peak expressing the maximum initial friction force F_{max} followed by a stick-slip pattern superimposed on a downward nonlinear trend corresponding to the change in the number of fibers in contact. This number at different points along the length of the beard can be estimated from the length friction profile.

The profile is divided into three equal regions (Fig. 6.6), each represented by the average value of the friction force of the region (F_i, F_m, and F_f). The initial force, F_i, has been found to be highly correlated to the surface morphology of the fibers since it involves rubbing of the densest region of the beard. The middle force F_m has been found to be correlated to both the surface and the dimensional parameters of the beard. The final force F_f has been found to be influenced largely by the dimensional parameters of the fibers in the beard (fineness, fiber length, and crimp). The friction indices a and n can be estimated either by changing the lateral pressure and developing the relationship between F and N, or from the relationship between the friction force and the number of fibers in contact at a given normal force.

6.3.3 Extent of simulation

The different contact modes discussed earlier can occur more or less exactly during processing. Cotton fibers as they flow in the process pipeline can cross each other at one or more points (point contact) or rub along the fiber axis (line contact). The macroscopic area-contact mode involves sliding of a fiber assembly against another fiber assembly or a surface of solid material.

Thus, in addition to the surface properties, the bulk properties of the assembly play a role in determining the frictional behavior during processing.

Figure 6.7 shows some of the common fiber flow modes that simulate those used in testing. In general, a fiber flow that reflects a friction phenomenon will involve: surface contacts, lateral pressure, and relative movement (sliding or rolling). This flow will encounter a resistance that can be partially or totally related to fiber surface characteristics. Figure 6.7(a) illustrates flow through a roller drafting system, which is an integral part of the fiber-to-yarn conversion process. A method simulating this is the one used for measuring sliver cohesion, such as the ASTM-D2612-99 method used for testing fiber cohesion in sliver and top in static modes. In this method fiber cohesion is given by the force required to cause initial drafting. The observed force required to separate the fibers is converted into cohesive tenacity by dividing it with the linear density of the specimen. The linear density used in this conversion is the one of the assembly at the initiation of drafting.

For determining of fiber cohesion in dynamic tests, the ASTM test method D 4120 is used, in which fiber cohesion is determined as the dynamic cohesive force or tenacity required to maintain drafting in rovings, slivers or tops as they pass between pairs of drafting rolls of different surface speeds.

Figure 6.7(b) illustrates action during opening of fibers in the carding zone. No standard test method exists that simulates this mode of friction. The closest system available for measuring force in this mode is the so-called "rotor-ring" system, developed by the Institute for Textile Technology, Reutlingen, Germany. This system has been used in a number of studies [e.g. 33–35]. A description of the system is given below.

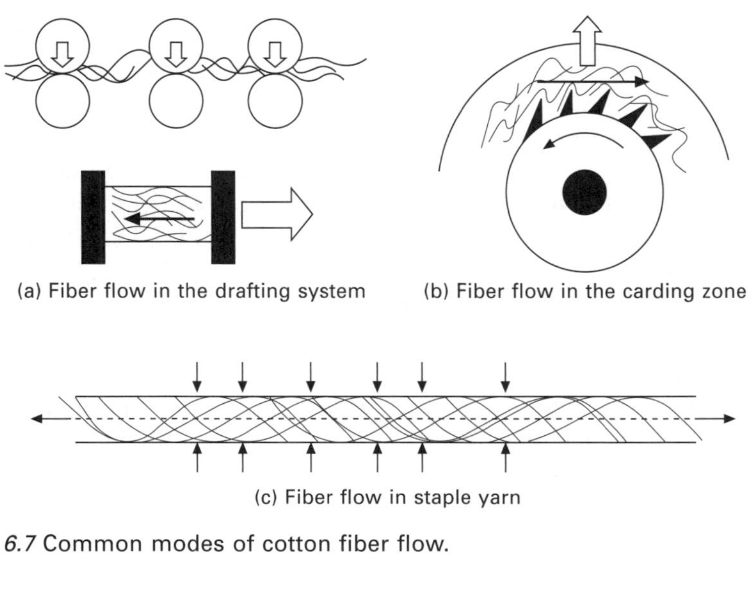

(a) Fiber flow in the drafting system (b) Fiber flow in the carding zone

(c) Fiber flow in staple yarn

6.7 Common modes of cotton fiber flow.

This system (see Fig 6.8) consists of a toothed-covered feed roll (0–24 rpm), an opening roll (0–6000 rpm), and a rotor (0–12,000 rpm). The tester can be used to formulate a fiber ring inside the rotor wall that can be treated as a short sliver for retesting on the rotor-ring or for sliver cohesion testing. Fibers fed by the feeding roll (about 2–3 g) are opened by the opening roll; and they are then accelerated in an air-stream to the inside wall of the rotor where they are condensed into a fiber ring. This fiber ring can be taken out, cut at one point, and unfolded to obtain a fiber web of a certain width.

Since the length of the rotor-ring fiber assembly is constant (equal to the rotor internal circumference), the width of the fiber web can be used as an index of the spreadability of fibers during opening, or the fiber propensity to opening. This generally means that fibers of low inter-fiber friction lead to a web of larger width. Obviously, this is an oversimplification since the width can be influenced by not only fiber friction but also other factors, including bulk resilience and fiber flexibility.

The dynamic fiber-to-fiber (or fiber-to-metal) friction is determined by measuring the energy required to open fibers. Typically, fibers are fed at a speed of 5 rpm via a feeding roll to an opening roller rotating at speed of 4000 to 5000 rpm. They are then condensed onto the inside wall of a rotor rotating at 10,000 rpm. The higher opening roll speed is normally used for fiber-to-metal friction, and the lower one for fiber-to-fiber friction. It has been also recommended that only fibers of the same average length (up to 100 mm) and the same fineness should be compared using this method. Repeated opening of fibers using the same ring is also recommended.

In a study by the present author [2, 3], the rotor-ring system was modified to impose more frictional effects on the fibers. The change involved an addition of sandpaper of pre-determined roughness in the opening roll chamber to simulate the effect of carding flats. The sandpaper acted as a fiber bulk holder that momentarily held a layer of fibers while carding or shearing was performed by the opening roll. The idea was to simulate a pure fiber-to-fiber rubbing action resulting from shearing a fibrous bundle. This modification provided a greater insight into the frictional behavior of fibers during carding.

In a more recent study aimed at using the rotor ring for investigating blending efficiency of different fibers, El Mogahzy *et al.* [4, 36] made another modification to the rotor-ring system, also shown in Fig. 6.8. In this, a torque couple was mounted on the opening roll shaft to allow measurement of the resistance of fiber flow during opening. Torque and speed signals were acquired using a data acquisition system. Signal processing and analysis resulted in a torque profile or a stick-slip pattern characterizing the behavior of fiber flow during opening and blending.

When friction testing is used to simulate certain applications, it is important to also consider other factors influencing the friction environment. As indicated above, the rotor-ring method does not fully reflect a pure friction phenomenon.

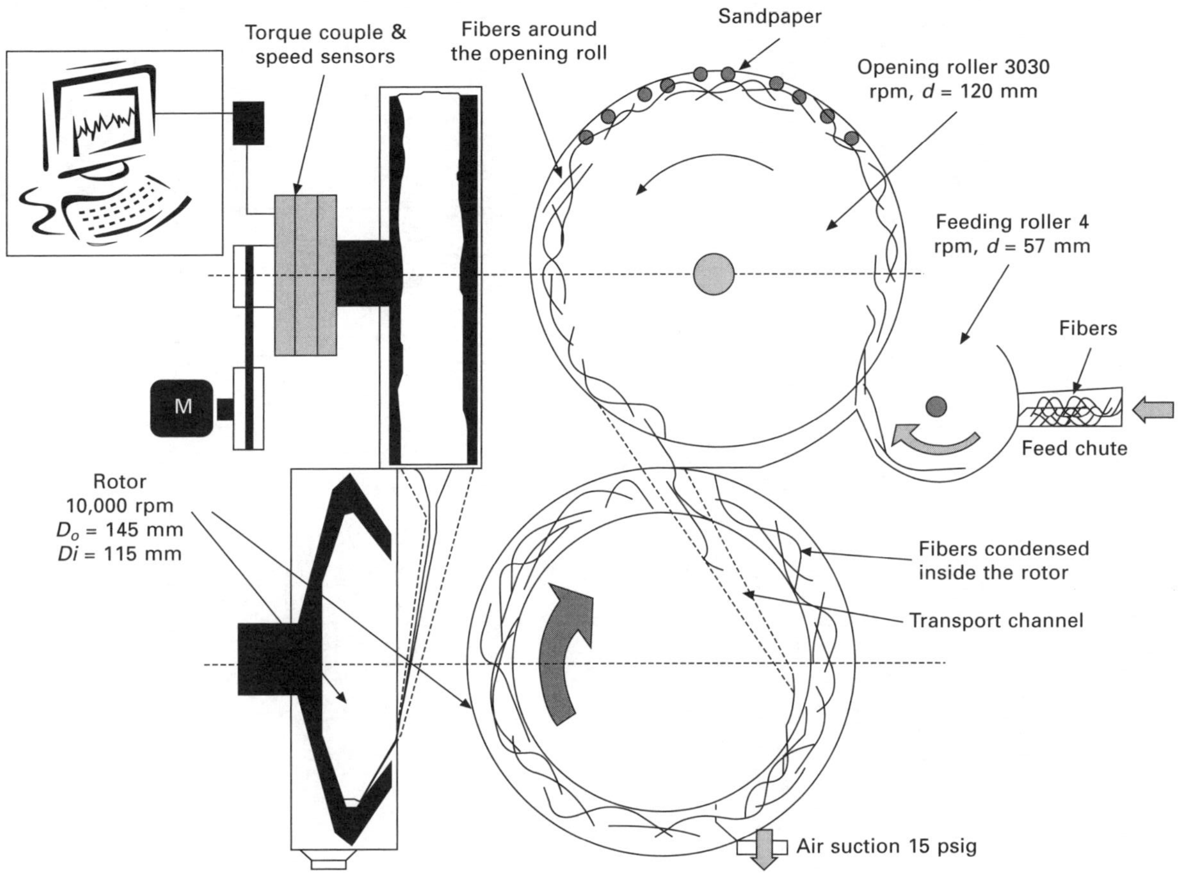

6.8 Modified rotor ring [2, 3].

Instead, it reflects the propensity to opening by the fibers, with friction being one of many aspects of the opening and carding process. This point is illustrated in Fig. 6.9, which shows some of the forces involved in opening a fiber cluster [31, 37, 38].

Fundamentally, the carding process involves the following types of fiber interaction: (i) fiber/fiber interaction, (ii) fiber/metal interaction, and (iii) fiber/fluid interaction. These are imposed by a number of forces governing the process. The first is the centrifugal force exerted against the fiber flock on the carding cylinder (Fig. 6.9(a)). This force is a function of the mass of the fiber flock, m, the radius of the cylinder, R, and the square of the cylinder speed. In high speed carding, the fiber flock will have a tendency to fly away from the cylinder clothing toward the slowly rotating flats. The second main type of force is the friction forces between the fiber flock and the cylinder cloth wire (Fig. 6.9(b)). These forces are generated by the lateral pressure between the fiber flock and the wire surface. This pressure is a result of the compacting effect of the fiber volume supported by the wire configuration. The higher the carding rate, the higher the lateral pressure and the higher the corresponding generated friction stresses.

The third type of force is the air friction force or air drag, which is a result of the narrow slot between the cylinder and the flats. Because of the difference

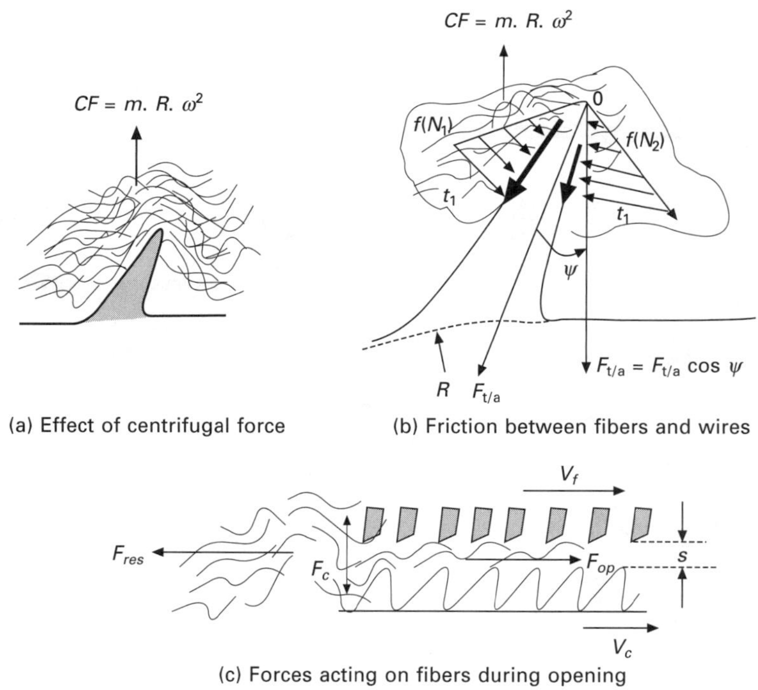

6.9 Some of the forces involved in opening a fiber cluster [31, 37, 38].

between cylinder and flats speeds, the air friction force is expected to result in a relative displacement between different fiber layers in the flock with fibers on the cylinder surface exhibiting higher speeds than those closer to the flats (i.e., friction between fiber layers). The air friction force will depend on the shape of the flock, the air viscosity, the air speed, and the flock density. Finally, the opening force (Fig. 6.9(c)) represents the actual carding force, which depends on the level of compactness of the flock at any point during carding. Accordingly, a fiber compression force F, will exist that depends on the fiber mass for a given setting. Since the flat and cylinder continue to rotate with greatly different speeds, attachment of fibers to both clothings will result in an opening force, F_{op}. This force is greatest at the first flat and it gradually decreases as the opening process continues. The opening force is, therefore, dependent on the fiber orientation and the fiber mass within the flock.

In light of the above discussion, one can see that a true simulative device of the carding process should account for many factors including the geometrical features, the carding speed, the fiber mass, and the number of carding cycles. In addition, any interpretation of results should be based not only on the frictional characteristics of fibers but also on other important factors such as the fiber bulk resiliency, fiber crimp, fiber fineness and fiber length.

Another important mode of fiber flow that also deserves simulative characterization is that of fibers inside a spun yarn (Fig. 6.7(c)). The importance of this particular flow stems from the fact that a spun yarn primarily fails due to a combination of fiber rupture and fiber slippage. This means that factors such as fiber strength, fiber elongation, and inter-fiber friction are expected to play vital roles in determining yarn strength.

Fibers are typically consolidated into a yarn either by twisting (e.g., ring-spun yarn and rotor-spun yarn) or by wrapping (e.g., air-jet spun yarn). These binding mechanisms impose different fiber alignments within a given yarn type and between different yarn types. In addition, they exert lateral pressure that typically varies from one of the co-axial layers of the yarn to another. These factors make evaluation of the relative contributions of fiber rupture and fiber slippage to failure of spun yarn a difficult task. Indeed, the correlation between fiber friction and yarn strength is often diluted by the complexity of interaction between these factors. An interesting study by Broughton and El Mogahzy [5] discussed this issue using actual spun yarns of different levels of fiber friction.

6.3.4 Effects of fiber properties

On a macroscopic scale, fiber dimensions such as length and fineness influence frictional behavior of fibers by virtue of their effects on the surface area of fiber assembly and on surface hardness [30–32]. Longer and finer fibers result in larger area of contact and higher surface friction. In some situations, the effects of fiber dimensions can overshadow the effect of surface parameters on friction [31].

One of the benefits of using the beard friction method is the possibility of measuring friction parameters that are largely surface related (e.g. F_{max} and F_i) and those that are largely dimension related (e.g., F_m and F_f). This feature is a direct result of the use of the beard shaped sample, which allows measuring friction over the fiber length distribution. Table 6.1 shows some of the results obtained by this method for fiber-to-fiber friction [31]. Two cotton types of different lengths are compared; the long-staple Pima and the medium-staple DPL. The maximum friction, F_{max}, and the initial friction, F_i, of the DPL-90 cotton were higher than those of the Pima cotton. This trend is reversed when the other two friction parameters are considered, namely F_m, and F_f. This reversal is a direct result of the effect of fiber length.

Table 6.2 shows the values of the beard friction parameters for the two cottons discussed above, but for fiber-to-metal tests. In these tests, fiber surface properties govern the value of F. Accordingly, all parameters, F_{max}, F_i, F_m, and F_f are higher for DPL-90, indicating higher coefficient of friction in this fiber.

In a study by El Mogahzy *et al.* [2], 34 different Upland cotton types produced in the 1993 and 1994 crop seasons were used to develop correlations between fiber properties measured by the High Volume Instrument (HVI) and fiber friction parameters measured by the three different methods, namely: the beard friction test, the modified rotor-ring test and the sliver cohesion method. Table 6.3 shows the range of values of fiber properties and wax contents (assessed by the Taylor NIR method [39]) and Table 6.4 shows

Table 6.1 Beard fiber-fiber friction parameters of two different cotton types at lateral pressure of 13.84 KPa and sliding speed of 0.25 cm/min [31]

Cotton type	UHML (cm)	Micronaire	Fiber strength (g_f/tex)	F_{max} (g_f)	F_i (g_f)	F_m (g_f)	F_f (g_f)
Pima 5-6	3.42	4.0	39.0	82.6	67.9	64.5	58.4
DPL-90	2.79	4.1	27.0	92.3	75.5	62.7	25.9

Table 6.2 Beard fiber-metal friction parameters of two different cotton types at lateral pressure of 13.84 KPa and sliding speed of 0.25 cm/min [31]

Cotton type	UHML (cm)	Micronaire	Fiber strength (g_f/tex)	F_{max} (g_f)	F_i (g_f)	F_m (g_f)	F_f (g_f)
Pima 5-6	3.42	4.0	39.0	33.6	29.3	24.8	22.8
DPL-90	2.79	4.1	27.0	36.3	31.2	27.6	24.3

Table 6.3 Ranges of values of cotton fiber properties [2]

Fiber property	Minimum	Maximum
HVI Properties:		
Strength (g_f/tex)	25.3	33.9
UHML (cm)	2.56	2.87
Micronaire	3.2	4.9
Color Rd	62.1	81.7
Color +b	6.5	12.8
Micromat:		
Maturity (%)	63.7	88.0
Maturity ratio	0.71	1.0
Wax content (%)	0.98	2.23

Table 6.4 Correlations between friction parameters and fiber properties [2]

	NIR wax (%)	Beard friction (F_{max}, g_f)	Rotor-ring torque (lb_f.in)	Sliver cohesion (lb_f)
NIR wax %	1.000	–0.882	–0.742	0.392
UHML (inch)	NS	NS	NS	0.320
Micronaire	–0.570	0.334	0.265	0.360
Color Rd	0.462	–0.570	–0.250	–0.320
Color +b	0.470	–0.610	–0.400	–0.254
Fiber strength (g_f/tex)	0.672	–0.470	–0.218	–0.488
Maturity ratio	–0.518	0.322	0.314	0.205

some of the correlations between the results found. The results of Table 6.4 are summarized as follows:

- Correlations of friction parameters with wax content were highly negative for the beard and the modified rotor-ring tests and positive but weak for the sliver cohesion test.
- The only correlation of some significance observed with fiber length was that with sliver cohesion.
- Fiber micronaire was negatively correlated to wax content. In a more recent study by Brushwood [40] using different types of cottons and using a different technique of wax measurement (Soxhlet extractions), a negative correlation between micronaire and wax content of 0.90 was found.
- Friction assessed by all methods exhibited positive correlations with micronaire.
- The parameters, characterizing color (reflectance or whiteness, *Rd*, and yellowness, +*b*) exhibited positive correlations with wax content. This trend of correlation was also found in Brushwood's study [40]. Friction

parameters obtained by all methods exhibited consistent negative correlations with these two parameters.
- Fiber strength was positively correlated with wax content, and this was also supported by Brushwood's study [40]. Fiber strength exhibited a consistent negative correlation with friction.
- Finally, fiber maturity followed the same trends as did fiber micronaire. It is generally accepted that micronaire is an index of both fineness and maturity.

We should point out that statistical coefficients of correlations do not necessarily imply physical correlations; they simply reveal the extent of linear association between two parameters, which may or may not be physically justifiable. However, when statistical correlations are consistent, particularly from one study to another, or when independent measurements are made, they often gain physical validity.

6.4 Cotton surface treatments

In the context of fiber friction, the most critical surface treatment applied to cotton is purification, which involves scouring or boiling in an alkali, followed by an oxidation bleaching [7]. Figure 6.10 shows the changes in the surface resulting from the scouring process. The cuticle layer with its oily and waxy surface is removed. Principally, the purification or scouring process aims at removing the noncellulosic components of the cuticle, or the primary wall. These components are generally characterized as plant waxes, various low

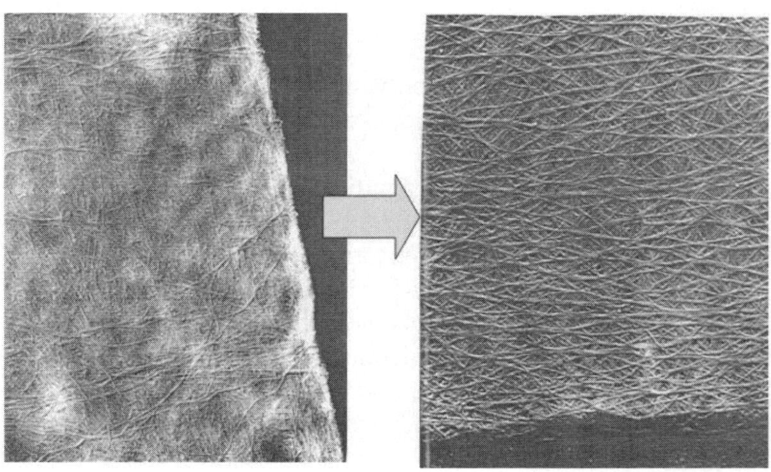

The cuticle

The primary wall after scouring (cuticle removed)

6.10 Effect of scouring on cotton fiber surface [6].

molecular weight organic acids, and metal oxides [40]. The removal of these by scouring allows more rapid wetting of the surface during subsequent chemical treatments. This typically results in more uniform application and, hence, in a more homogeneously finished product.

Another commercial chemical treatment given to cotton is mercerization. This process aims at enhancing dyeability and luster of cotton through immersing the fabric in 18 to 25% solution of sodium hydroxide. This results in swelling of the cellulosic material, which is retained after the chemical is washed out. The cellulose is so modified that the absorption and adsorption properties of the fiber are greatly improved. The swelling causes an increase in the specific surface areas available for chemical reactivity. The actual surface of mercerized fibers is not a great deal smoother than that of native fibers [7]. Mercerization removes many of the gross concavities of fiber surface, but the submicroscopic roughness remains.

Effects of chemical treatments on the frictional behavior of cotton have been the subject of many investigations. In a study by El Mogahzy *et al.* [31, 32] the effects of scouring, bleaching, and finishing on the frictional behavior of cotton were examined. The beard friction test was used on fibers and the sled test (ASTM 1894-78), normally used for plastic film sheeting, was used on nonwoven webs made from the fibers.

The sled test, shown in Fig. 6.11, measures the resistance to relative displacement between fiber layers in a web. Accordingly, it requires a preparation of a fiber web from which specimens of approximately 7.5 × 10 cm are cut

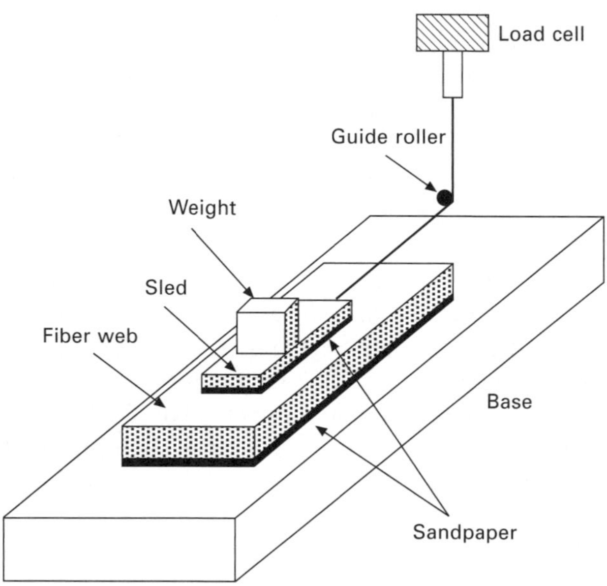

6.11 Friction sled test [31].

and mounted on the plate under a weighted block (sled). The apparatus is attached to Instron® cross head and the force required to pull the sled (consequently displacing the top fiber layer) is measured. The fiber web is laid on the top of a sandpaper base to restrict the bottom fiber layers from movement. Another sandpaper is placed between the sled and the top fiber layer. The normal force is applied by placing a dead weight (2 kg) on top of the sled.

The reliability of the results from the sled test will depend on factors such as sample preparation, condition of the web, fiber orientation, web density, and web thickness. The stick-slip patterns produced by the sled test are typically more pronounced than those produced by other methods (e.g., the beard test) under the same external conditions. This may be due to a relatively greater mobility of fibers expected in the specimen of the test: the fibers may stretch, slip and snap back in both the interface and the layers lying in the bulk. It is important, therefore, that sled stick-slip patterns be evaluated with careful attention to these factors to avoid misleading conclusions. The fundamental mechanisms underlying the stick-slip profile have been discussed in Chapter 4 and will be briefly alluded to again in Chapter 8.

The results by El Mogahzy *et al.* [31, 32] obtained by the beard method showed that the F_{max} value increased by about 182% with scouring. The corresponding average increase for the fiber-to-metal friction was about 49%. The sled test gave an average increase in friction of about 112%.

Typically, the purification of cotton by scouring and bleaching is followed by an application of finishing agents (principally, lubricants) to facilitate processing of fibers through different machinery and to recover some of the smooth hand of the cotton web. In the study by El Mogahzy *et al.* [31, 32], different types of finishes including both liquid and solid, or semi-solid (waxy) were used. The purpose of the study was to determine the best finishing agent and optimum finishing conditions. In general, the solid lubricants resulted in lower interfiber friction than liquid lubricants. Details on the effects of lubrication are discussed in Chapter 8.

One of the effective finishing agents that were tested was the so-called ATEC25 (Atebin C25, polyolefin solid lubricant). Figure 6.12 shows the beard friction profiles associated with the application of this type of finish. There are definitive differences between the friction profiles of raw, scoured, and finished fibers. The raw cotton exhibited the lowest maximum friction. Upon scouring, it increased substantially. The effect of ATEC25 agent was to reduce the friction level after scouring to a level slightly above the raw cotton value. In general, no fiber finish is available that can perfectly replicate the effects of the natural surface of cotton fibers.

In addition to the above basic treatments, chemical agents may be applied to the cotton surface to enhance or impart certain properties such as abrasion resistance, flame resistance, and soil and stain resistance. Another type of

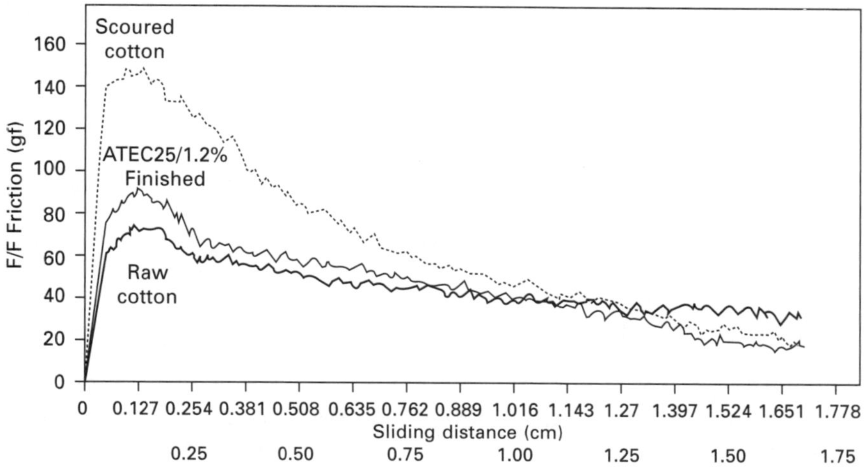

6.12 Beard friction profiles of raw, scoured and finished cotton [31].

treatment that has gained popularity in recent years is enzymatic scouring. This treatment is effective in removing non-cellulosic material from cotton; however degradation of the primary wall is one of the greatest concerns associated with this treatment [41–44].

6.5 Summary

In this chapter, we discussed some of the important aspects related to surface and frictional characteristics of cotton fibers. Obviously, there is a significant body of knowledge on the subject available in the literature. However, many unresolved issues remain that are yet to be addressed. We close this chapter with a number of points that deserve potential future research:

- In this globally competitive market, fiber source identity in end products will become a key issue. In this regard, standard fiber properties can only provide very broad identity categories that are easily disputable. Surface characteristics, on the other hand, particularly at the microscopic scale, can reflect the true cotton variety and provide an effective source for cotton fiber identity. It is important, therefore, to conduct an in-depth analysis of surface using the advanced surface characterization techniques available today.
- Current testing techniques do not fully replace the traditional cotton classing. This calls for a unique classing instrument that can truly reflect the inherent touch and feel behavior of cotton based on surface and frictional characteristics.
- Finally, there is a need for friction and surface-related values in the cotton database. This calls for a "high-volume surface characterization"

system that is both reliable and simulative of cotton processing and performance characteristics.

6.6 References

1. El Mogahzy Y E and Chewning C Jr., *Cotton fiber to yarn manufacturing technology*, Cotton Incorporated, Cary, NC, 2001.
2. El Mogahzy Y E, Broughton R, Guo H and Taylor R A, 'Evaluating staple fiber processing propensity. Part I: Processing propensity of cotton fibers', *Textile Res. J.*, 1998, **68**(11), 835–840.
3. El Mogahzy Y E, Broughton R, Guo H and Rollins C, 'Evaluating staple fiber processing propensity, Part II: Processing propensity of cotton/polyester blends', *Textile Res. J.*, 1998, **68**(12), 907–912.
4. El Mogahzy Y E, 'An integrated statistical approach to the analysis of the nature of multi-component fiber blending. Part I: Theory of multi-component blending', *Textile Res. J.*, 2004, **74**(8), 701–712.
5. Broughton R M, El Mogahzy Y E and Hall D M, 'Mechanism of yarn failure', *Textile Res. J.*, 1992, **62**(3), 131–134.
6. Allen H C Jr., 'The cotton fiber process and fiber characteristics', *Cotton Incorporated Technical Reports*, 1, 1993.
7. Duckett K E, 'Surface properties of cotton fibers', Fiber science series, in M.J. Schick, Ed., *Surface Characteristics of Fibers and Textiles*, Part I, Marcel Dekker, Inc. NY and Basel, 64–70, 1975.
8. El Gaiar M N, Cusick G E, 'A study of the morphology of cotton-fiber fracture in abrasion tests in relation to coefficient of friction between the fabric tested and the abradant', *J. Tex. Inst.*, 1976, **67**(4), 141–145.
9. Seshan K N, 'An investigation of the taper of cotton fibers. Part V: Differential friction in cotton fibers', *J. Text. Inst.*, 1978, **69**(7), 214–219.
10. Zhukov V I, Yakovlev V V and Kharitonova N F, 'The relation between the birefringence and ripeness of cotton fibers', *Tech. of Textile Industry*, U.S.S.R., 1971, **5**(3), 26–28.
11. El Mogahzy Y E, *A study of the nature of fiber friction*, Ph.D. dissertation, North Carolina State University, Raleigh, NC, 1987.
12. Gupta B S and El Mogahzy Y, 'Friction in fibrous materials. Part I: Structural model', *Textile Res. J.*, 1991, **61**(9), 547–555.
13. El Mogahzy Y E and Gupta B S, 'Friction in fibrous materials, Part II: Experimental study of the effects of structural and morphological factors', *Textile Res. J.*, 1993, **63**(4), 219–230.
14. Howell H G, 'Interfiber friction', *J. Tex. Inst.*, 1951, **42**, T521–533.
15. Gralen N and Lindberg J, 'Measurement of friction between single fibers. II: Frictional properties of wool fibers measured by the fiber-twist method', *Textile Res. J.*, 1948, **18**, 287–301.
16. Carpick R W and Salmeron M, 'Scratching the surface: Fundamental investigations of tribology with atomic force microscopy', *Chem. Rev.*, 1997, **97**, 1163–1194.
17. Mate C M, McClelland G M, Erlandsson R and Chiang S, 'Atomic-scale friction of a tungsten tip on a graphite surface', *Phys. Rev. Lett.*, October 1987, **59**(17), 1942–1945.
18. Amonton G, *Hisorie de L'Academie Royale des Sciences avec Memoires, de Mathematique et de Physique*, 206, 1699.

19. *Encyclopedia of science and technology*, 5th Edition, McGraw-Hill Company, NY, St. Louis, San Francisco, 1982, 732–735.
20. Bowden F P and Tabor D, *The friction and lubrication of solids*, Oxford University Press, London, 1950, 1964.
21. Archard J F, 'Elastic deformation and the laws of friction', *Proc. Roy. Soc. A*, 1957, **243**, 190.
22. Archard J F, 'Single contacts and multiple encounters', *J. Appl. Phys.*, 1961, **32**, 1420.
23. Howell H G, Mieszkis L and Tabor D, *Friction in textiles,* The Textile Institute and Butterworth, Manchester and London, 1959.
24. Hong J and Jayaraman S, 'Friction in textiles', Layton J M, Ed., *Textile Progress*, **34**(1/2), The Textile Institute, 2003.
25. Tabor D, 'Friction, Adhesion, and boundary lubrication of polymers', *Advances in polymer friction and wear*, Vol. 5A, Plenum Press, NY, and London, 1974, 5–28.
26. Howell H G and Mazur J, 'Amonton's law and fiber friction', *J. Text. Inst.*, 1953, **44**, T59–69.
27. Lord E, 'Frictional forces between fringes of fibers', *J. Text. Inst.*, 1955, **46**, 41–58.
28. Viswanathan A, 'Some experiments on the friction of cotton fibers', *J. Text. Inst.*, 1973, **64**(10), 553–557.
29. Viswanathan A, 'Frictional forces in cotton and regenerated cellulosic fibers', *J. Text. Inst.*, 1966, **57**(1), T30–40.
30. El Mogahzy Y E and Broughton R, 'A new approach for evaluating the frictional behavior of cotton fibers, Part I: Fundamental aspects and measuring techniques', *Textile Res. J.*, 1993, **63**(8), 465–475.
31. El Mogahzy Y E, Broughton R and Wang Qin, 'The friction profile of cotton fibers and its importance in determining fiber performance in the nonwoven process. Part I: Fundamental aspects of fiber friction and lubrication', *International Nonwovens Journal (INJ)*, 1995, **6**(4), 35–42.
32. El Mogahzy Y E, Broughton R and Wang Qin, 'The friction profile of cotton fibers and its importance in determining fiber performance in the nonwoven process. Part II: Experimental observations', *International Nonwovens Journal (INJ)*, 1995, **7**(1), 26–33.
33. Gosh S, Rodgers J E and Ortega A E, 'Rotor ring measurement of fiber cohesion and bulk properties of staple fibers', *Textile Res. J.*, 1992, **62**(10), 608–613.
34. Moore G W, *The relationship between spin lap rotor cohesion measurements and sliver and open-end rotor-spun yarn qualities using 100% acrylic fibers*, M. S. thesis, ITT 1989.
35. Watson D L, *The relationship between spin lap rotor cohesion measurements and sliver and ring spun yarn qualities using 100% polyester fibers*, M. S. thesis, ITT 1987.
36. El Mogahzy Y E, Farag R, Abdelhady F and Mohamed A, 'An integrated approach to the analysis of the nature of multi-component fiber blending. Part II: Evaluation of structural and attributive blending of cotton/polyester blends', *Textile Res. J.*, 2004, **74**(9), 767–775.
37. Artzt P and Schreiber O, 'Determination and parameters of the carding forces of high speed cards at production rates above 30 kg/hr', *Melliand Textilberichte International*, 4th edn., 1974, **3**(4), 224–230.
38. Belogolovtsev S D and Surkova T A, 'The length of the doffing arc on the card', *Tech. of Textile Industry,* U.S.S.R, 1971, **5**, 33–38.

39. Taylor R A, 'Measuring natural waxes on cotton using NIR absorbance', *Proc. Beltwide Cotton Conference*, National Cotton Council, 1997.
40. Brushwood D E, 'Noncellulosic constituents on raw cotton and their relationship to fiber physical properties', *Textile Res. J.*, 2003, **73**(10), 912–916.
41. Hartzell M M and Hsieh Y L, 'Removal of non-cellulosic wall components from cotton fibers for improved wettabilily', *213th ACS National Meeting*, San Francisco, CA, April 13–17, 1997.
42. Li Y and Hardin I, 'Enzymatic scouring of cotton', *AATCC International Conference and Exhibition* 1996, Nashville, TN, September 15–18, 1996.
43. Rossner U, 'Enzymatic degradation of impurities in cotton', *Melliand Textilberichte*, 1993, **74**, 144–148.
44. Buschle-Diller G, El Mogahzy Y E, Inglesby M K and Zeronian S H, 'Effects of scouring with enzymes, organic solvents, and caustic soda on the properties of hydrogen peroxide bleached cotton yarn', *Textile Res. J.*, 1998, **68**(12), 920–929.

7
Friction, felting and shrink-proofing of wool

J A RIPPON, Deakin University, Australia

7.1 Introduction

Of the textile fibers obtained from animals, wool from sheep is commercially the most important. Natural fibers are biodegradable and although few examples of ancient textiles have survived to the present time, wool was probably the earliest fiber to be used for the production of cloth [1,2]. Early breeds of sheep were covered in a brownish coat that was shed annually. Selective breeding resulted in the off-white fiber that we know today.

The most important breed of sheep for producing the premium fine wool used for apparel is the merino. This breed originated in Spain during the Middle Ages, from where it was exported to other countries in the eighteenth century. The most important of these is Australia, where the breed has been developed to produce wool with highly valued properties, of fineness, length, color, luster and crimp.

A disadvantage of wool, compared with other textile fibers, is its tendency to felt during aqueous washing. This property, which can lead to severe shrinkage of wool products, is the result of the surface frictional properties of the fibers.

In order to understand the frictional behavior of wool and the type of treatments required to produce shrink resistance, it is necessary that certain aspects of the physical and chemical structure of the fiber are understood. This forms the first part of the chapter. Mechanisms by which frictional properties cause felting shrinkage are then discussed, followed by treatments used for imparting shrink resistance. The final section outlines problems associated with the current commercial shrink-resist methods and discusses some of the directions that are being investigated for future industrial application.

7.2 The chemical and physical structure of wool

Wool belongs to a group of proteins known as keratins. Keratins are very diverse with respect to their structure and occurrence [3] and have been classified as hard or soft, according to their tactile properties [4]. A characteristic feature of hard keratins, such as wool, hair, horns, hooves, beaks, claws and feathers is a higher concentration of sulphur than is found in soft keratins, such as those in skin [5]. The sulphur in keratins is mainly present in the form of disulphide crosslinks, which are derived from the amino acid, cystine. The disulphide crosslinks exist either between different protein chains, or between different parts of the same chain. These crosslinks, which have been compared with the rungs of a ladder, are largely responsible for the lower solubility and greater stability of keratins compared with other proteins [6]. Keratins have also been classified according to their X-ray diffraction patterns as α-keratins or β-keratins [4, 5]. Normal wool gives a pattern characteristic of α-keratin, whereas stretched wool gives a different pattern similar to that produced by β-keratin, such as feathers.

Keratin fibers are not chemically homogeneous and it has been estimated that about 170 different proteins, composed of 18 amino acids, are present in wool [7]. Only approximately 82% of the proteins contain the high concentration of cystine that characterizes keratins. The remaining proteins have been termed 'nonkeratinous' [8,9]. They constitute around 17% of the fiber and as a result of their lower cystine concentration have a much lower concentration of disulphide crosslinks than the true keratinous proteins. The lower crosslink density of nonkeratinous proteins makes them labile and susceptible to chemical attack, for example when wool is dyed and also during treatments such as chemical shrinkproofing [9, 10].

Wool fibers also contain approximately 1% by mass of non-protein matter. This is mainly a complex mixture of waxy lipids plus a very small amount of polysaccharide material [10, 11]. Wool lipids, which are both structural and free, are not distributed evenly throughout the fiber. They are concentrated in specific regions where, despite their very low concentration, they play an essential role in determining important fiber properties. In particular, the lipids associated with the fiber surface control wettability [12–14], tactile properties [14] and felting behavior [14].

In common with other keratin fibers, wool is a biological composite material. Wool fibers consist of regions that vary from each other in physical and chemical composition. The complex morphological structure of fine wool fibers is shown in Fig. 7.1. The various regions behave differently, both in their physical properties and chemical behavior [8, 10, 15–17].

Fine wool fibers, such as those from merino sheep, are composed of two types of cell: the internal cells of the fiber cortex and external cuticle cells that form a sheath around the cortex. The chemical and physical structure of the cortex has been reviewed elsewhere [10]. Cuticle cells control the surface properties of wool, such as friction, handle, wettability, felting and shrinkproofing. This morphological component of wool has been examined by a number of chemical and surface analytical techniques [18].

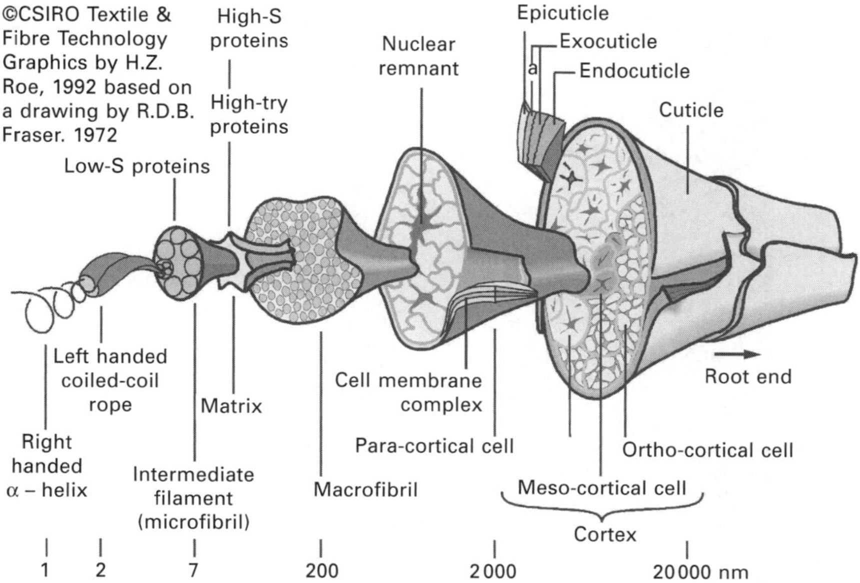

7.1 Schematic diagram of a fine wool fiber. (Diagram courtesy of CSIRO).

7.2.1 Structure and properties of the cuticle

Cuticle cells, or scales, comprise approximately 10% of the mass of fine wool fibers and can be seen clearly in the light or scanning electron microscope (Fig. 7.2). For fibers of 20 microns diameter, cuticle cells range in thickness from 0.3 to 0.5 microns and are about 30 microns in length and 20 microns wide [19]. They overlap both along and around the circumference of each fiber, like tiles on a roof, with their exposed edges pointing from the fiber root towards the tip. Except where two cells overlap, the cuticle of fine wool is normally one cell thick. The cuticle layer of other keratin fibers, however, is often composed of multiple layers of cells. For example, human hair may contain 10 layers and pig bristle up to 35 layers [20]. For wool, the amount of scale overlap varies with fiber diameter; for fine fibers the overlap in length is approximately 15% [21]. In addition to the true scale edges that correspond to the ends of the cuticle cells, shoulders or 'false' scale edges also occur on around 25% of merino cuticle cells [19]. Both true and false scale edges are believed to be produced by the inner root sheath of the hair follicle imprinting its serrations onto the fiber prior to the keratinization process that produces hardening [19]. An important function of cuticle cell edges is to anchor fibers in the skin of the animal [22]. Wool cuticle cells are separated from the underlying fiber cortex by the cell membrane complex. This component forms a continuous region that extends throughout the whole

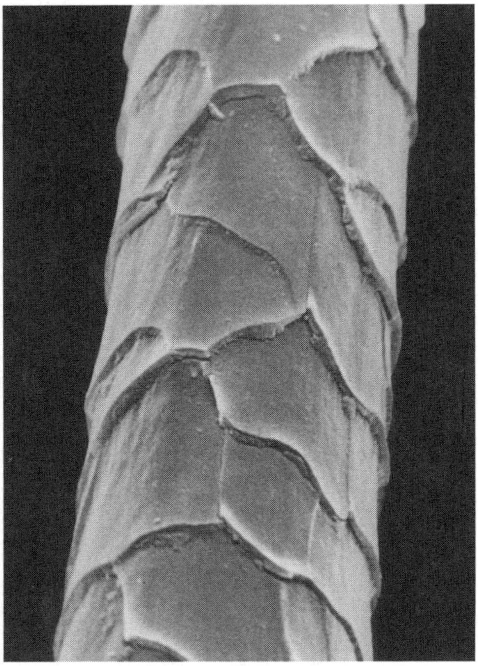

7.2 Scanning electron micrograph of merino wool fiber.

cortex. The structure, properties and importance of the cell membrane complex has been discussed in detail elsewhere [10, 16].

The substructure of the cuticle has been studied extensively by many workers, using a wide range of chemical and physical techniques [15, 19]. Chemical analysis of separated cuticle cells has shown some differences compared with whole wool (Table 7.1). In particular, it has higher cystine content. Compared with the cortex, it is poorer in the amino acids that favor an α-helical structure and richer in those that are generally non-helix forming. It has, therefore, been concluded that the cuticle has a more amorphous structure than the rest of the fiber [23, 24].

Electron microscopy, coupled with heavy-metal staining techniques, has shown that the cuticle has two distinct layers, called the exocuticle and the endocuticle [15]. These are shown schematically in Fig. 7.3.

In merino wool, the exocuticle is approximately 0.3 microns thick and represents around 60% of the total cuticle cell [26]. The amino acid analyses in Table 7.1 show that most of the cystine in the cuticle is contained in the exocuticle. The endocuticle, which contains relatively little cystine, is classified as one of the nonkeratinous regions of wool [10]. It has been shown to be important as a preferred pathway for the diffusion of dyes into the fiber [28].

Staining techniques have shown that the exocuticle consists of two poorly defined subcomponents; the A-layer and the B-layer [29]. As shown in

Table 7.1 Amino acid composition of whole wool and wool cuticle (mol%)

Amino acid	Whole wool [24, 25]	Cuticle			
		Whole- [26]	Exo- [26]	Endo- [26]	Epi- [24, 27]
Alanine	5.4	5.8	6.4	6.7	4.6
Arginine	6.9	4.3	4.8	5.0	4.3
Aspartic acid[(a)]	6.5	3.5	2.1	7.4	5.8
Citrulline [(b)]	–	–	–	–	0.9
1/2 –Cystine[(c)]	10.3	15.6	19.9	3.1	11.9
Glutamic Acid[(d)]	11.9	8.7	8.5	10.3	10.7
Glycine	8.4	8.2	8.7	8.2	15.4
Histidine	0.9	0.8	0.5	1.1	1.0
Isoleucine	3.1	2.7	2.9	3.9	2.5
Leucine	7.7	6.1	4.6	9.3	5.5
Lysine	2.9	2.7	2.1	4.2	4.8
Methionine	0.5	0.3	0.2	0.8	–
Phenylalanine	2.9	1.7	1.2	3.9	1.9
Proline	6.6	10.5	12.3	8.9	5.8
Serine	10.4	14.3	11.9	10.7	13.7
Threonine	6.4	4.4	3.9	5.5	3.6
Tyrosine	3.8	2.8	2.0	3.6	2.1
Valine	5.6	7.5	8.2	7.5	5.7

(a) Includes asparagine residues
(b) Includes hydrolysis by-products
(c) Includes the oxidation by-product, cysteic acid
(d) Includes glutamine residues

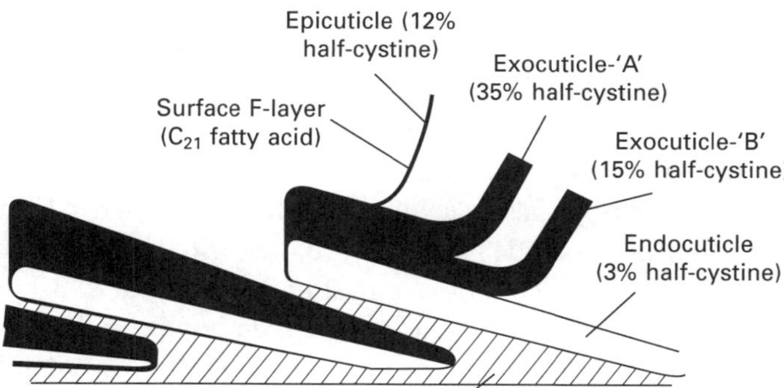

7.3 Schematic diagram of the cuticle of a merino wool fiber [17].

Fig. 7.3, the A-layer, which accounts for 30–50% of the total thickness of the exocuticle, is believed to be much richer in cystine than the B-layer [4, 15, 26]. The high cystine content of the exocuticle, particularly the A-layer, makes it the most highly crosslinked region in wool. This property is probably responsible for its lower extensibility compared with the cortex, which results in cracking of cuticle cells when the fiber is stretched [30].

Epicuticle

Treatment of wool fibers with chlorine water raises a series of bubbles, which appear as sacs apparently surrounded by a membrane [10, 15, 16, 31, 32]. This phenomenon is called the Allwörden reaction after its discoverer [33]. The mechanism of this reaction has been extensively studied because of the importance of chlorine in the shrink-resist treatment of wool.

The membrane raised by chlorine water has been called the epicuticle [34]. It has been estimated to be 3–6 nm thick and to account for approximately 0.1% of the mass of the fiber [15, 16, 31, 32]. The epicuticle, which was originally believed to be a continuous membrane surrounding each fiber [31], was shown by Leeder and Bradbury to surround each individual cuticle cell [32, 35, 36]. Thus, the epicuticle forms part of the resistant membrane system that surrounds all cuticle and cortical cells [10]. These membranes are relatively chemically inert (hence the term 'resistant') and are the last part of the fiber to dissolve when wool is digested in a wide variety of reagents [37, 38]. Stirm and Colle proposed that Allwörden bubbles are formed as a result of an osmotic pressure that arises from the generation of high molecular weight oxidation products formed by reaction of chlorine with protein material near the surface of the cuticle [39]. This suggestion was supported by the observation that the bubbles collapsed when exposed to a concentrated salt solution [40]. Other workers have supported the mechanism that a semi-permeable membrane (the epicuticle) is stretched outwards by the development of an osmotic pressure [31, 35, 36]. The major reaction of wool with chlorine involves the disulphide bonds of the cystine crosslinks, which are oxidized to anionic cysteic acid residues [41]. Some hydrolysis of peptide bonds also occurs and this has been found to be important in the generation of Allwörden bubbles [36, 42–45]. Thus, it appears that oxidation of the high concentration of cystine present in the A-layer of the exocuticle, together with the cleavage of peptide bonds, enables sufficient osmotically active peptides to be produced to generate Allwörden bubbles.

Hydrophobic surface of wool

As shown schematically in Fig. 7.3, the surface of wool consists mainly of the epicuticle. A small part of the surface (approximately 0.05%) is composed of the region between the cuticle cells that extends to the fiber surface [46]. Wool fibers that have been cleaned of wool grease are hydrophobic. This property is difficult to explain if the epicuticle is composed solely of protein. The marked increase in wettability that occurs when wool is treated for a short time with a solution of potassium hydroxide in an alcohol [34] is inconsistent with the chemical resistance of the epicuticle, discussed above [12]. To explain the increase in wettability, Lindberg proposed that alkali removes a thin layer of hydrophobic material, only a few molecules thick [12]. Other workers also suggested the presence of a covalently-bound lipid layer on the surface of wool [47, 48]. Leeder and Rippon used a method involving an anhydrous solution of potassium tertiary-butoxide dissolved in tertiary-butanol to treat dry wool [14]. In the absence of water, the fiber was not swollen and this prevented diffusion of these bulky reagents beyond the fiber surface [49]. Changes were found in a range of properties controlled by the fiber surface, including a dramatic increase in wettability, an increase in dry and wet friction and improved adhesion of polymers. The increased dry friction led to a substantial harshening of handle. There was also a large reduction in felting shrinkage, which can be explained by the increased wet friction. Lubricants, such as cationic softeners, reduced the dry friction and, thus, improved the handle. Wet friction was, however, affected to a much smaller extent by softeners. These effects will be discussed in more detail later. Despite the changes in surface properties, no visible changes could be seen in the electron microscope and Allwörden bubbles could still be raised by immersion in chlorine water [49]. Furthermore, there was no measurable change in the mass of the treated wool. These results suggested that the reaction was confined to the fiber surface and it was proposed that the changes in fiber properties were the result of removal of the postulated lipid layer to expose the clean protein surface of the epicuticle. Leeder and Rippon named the lipid component the 'F-layer', because they considered it to be separate to the proteinaceous epicuticle. In view of the difficulty in removing it, they suggested that the F-layer is chemically bound to the epicuticle. Evans *et al.* confirmed the presence of fatty acids in the anhydrous potassium tertiary-butoxide solutions used to remove the F-layer [50]. They measured the amount removed to be 0.025% on mass of fiber. The major component was shown to be an unusual methyl-branched 21-carbon fatty acid (18-methyleicosanoic acid), which they suggested was linked to the epicuticle via an ester or thioester bond. This fatty acid has since been found in human hair [51] and in the hair of other mammals [52]. The thickness of the lipid layer on the fiber surface has been estimated to be 0.9 nm [53], which is around 40 to 50% of the value calculated from the dimensions of the C_{21} fatty acid [54]. The importance of the F-layer in the mechanism and relative effectiveness of chemical shrink-resist treatments will be discussed later.

7.3 Surface friction of wool

Wool is one of the smoothest textile fibers, with a much lower coefficient of friction than, for example, silk, nylon or viscose rayon [55, 56]. It differs from other textile fibers in that it exhibits a directional frictional effect (DFE). This property, which it shares with other animal fibers, is the result of the surface scales, shown in Fig. 7.2, because a DFE is not shown by de-scaled wool [57]. The scales point from root to tip and the protruding scale edges result in the surface roughness, or friction, being greater against the scales (i.e. when rubbed from tip to root) than with the scales (from root to tip). This difference in roughness can be demonstrated by pulling a human hair slowly between finger and thumb in each direction. The DFE associated with the cuticle cells of wool is believed to assist in expelling dirt and other contaminants from the fleece [22].

A number of techniques have been used to measure the friction of wool [45, 58, 59]. These include the 'violin bow' method [60], the 'twisted fiber' method [61] and the 'capstan' method [62]. In the violin bow method, an assembly of parallel fibers, oriented with the scales in the same direction, is mounted between two wooden blocks with sufficient tension to make the fibers taut. The bow or mat of fibers is then placed on an inclined plane reference surface, the angle of which is increased until slipping occurs. The apparatus can be used in air to measure the coefficient of dry friction, or under water to measure wet friction. Variations of this method have been used in which a 'bow' made from a single fiber is drawn tangentially across a rod made of the reference material [63]. In the twisted fiber method, two fibers are twisted together by a certain number of turns and held under a constant load applied to each end. The tension on the other ends is then increased until the two fibers slip relative to each other. Both dry and wet friction can be measured by this procedure. The twisted fiber method is attractive because it is considered to simulate the interaction between wool fibers in a yarn. The technique most commonly used by workers studying the effect of chemical treatments on friction and shrink-proofing is the capstan method. In this technique, a fiber is suspended over a cylindrical rod with a given weight on one end. The load on the other end is increased until the fiber begins to slip. Several variations of this method have been used, mostly aimed at making it easier to use on wet fibers [45, 58]. In the case of the violin bow method, the reference material was usually a wool fabric but keratin made from horn has also been used. The reference material most often employed in the capstan method has also been animal horn. This material was chosen because it has a similar keratinous composition to wool and, therefore, similar elastic modulus. [59]. Furthermore, even when polished, a horn surface contains tiny cracks that provide asperities with which the scales of wool fibers can interact when they slide over it. Other materials have also been used for the reference or rubbing surface. These include ebonite, glass, polystyrene, PTFE and casein [45, 59]. The friction of wool fibers has been quantified in terms of the with-scale (μ_w) and the against-scale (μ_a) coefficients of friction. Table 7.2 shows values of μ_w and μ_a,

Table 7.2 Values of the coefficients of friction of untreated and treated wool fibers measured in air at 65% RH

Method	Rubbing surface	Wool preparation	μ_a	μ_w	σ	Reference
Violin bow	Wool cloth	Solvent extracted plus water rinse	0.34	0.23	0.11	[60]
Violin bow	Horn polished	Not specified	0.50	0.30	0.20	[64]
Violin bow	Horn polished	Solvent extracted plus aqueous rinse	0.50	0.34	0.16	[63]
Violin bow	Wool fibers	Solvent extracted plus aqueous rinse	0.42	0.25	0.17	[63]
Capstan	Horn polished	Solvent extracted plus aqueous rinse	0.35	0.11	0.24	[62]
Capstan	Horn polished	Solvent extracted	0.54	0.49	0.05	[65]
Capstan	Horn roughened	Solvent extracted	0.53	0.42	0.11	[65]
Twisted fibers	Wool fiber	Solvent extracted plus aqueous rinse	0.15	0.11	0.04	[61]
Twisted fibers	Wool fiber	Solvent extracted plus aqueous rinse	0.16	0.10	0.06	[66]
Twisted fibers	Wool fiber	As above plus KOH in methanol	0.29 (0.16)	0.21 (0.10)	0.08 (0.06)	[66]

Note: values of μ_a, μ_w and σ in brackets are for untreated wool fibers.

measured by various workers on untreated, dry wool fibers and Table 7.3 shows values for untreated wet fibers.

Direct comparison of the data on wet wool cannot be made with the results for dry wool, unless the results are from the same publication. In all cases, however, for both dry and wet untreated wool, the against-scale coefficient has been found to be greater than the with-scale value. The general conclusion can also be made from Tables 7.2 and 7.3 that the frictional coefficients are normally greater when the wool is wet than when dry [59, 66, 69]. This difference has been attributed to fiber swelling [58]. There has been common agreement that the difference in the two coefficients of wet friction is largely responsible for the felting behavior of wool. Various workers have defined the DFE, or frictional difference, in different ways, in an attempt to relate the numerical values to the degree of felting. Speakman and Stott [60] referred to 'scaliness', which they defined as $(\mu_a - \mu_w)/\mu_w$. The term, scaliness is no longer used, as it could be confused with other properties of the cuticle, such as the scale height [70]. Other workers defined 'relative frictional difference' as $(\mu_a - \mu_w)/(\mu_a + \mu_w)$ [63]. Neither of these expressions

Table 7.3 Values of the coefficients of friction of wool fibers measured in aqueous media

Method	Rubbing surface	Wool preparation	Measurement medium	μ_a	μ_w	σ	Reference
Violin bow	Horn	Not specified	Water pH 4	0.6	0.3	0.3	[64]
Violin bow	Horn	Not specified	Water pH 10.8	0.6	0.2	0.4	[64]
Violin bow	Horn	Solvent extracted plus water rinse	Water pH 7	0.9	0.5	0.4	[63]
Capstan	Horn	Solvent extracted plus water rinse	Soap and soda, pH 10.7	0.35	0.11	0.24	[62]
Capstan	Horn	Commercial scour	pH 9 buffer plus non-ionic detergent	0.52	0.23	0.29	[67]
Capstan	Horn	Solvent extracted plus water rinse	Water	0.43	0.16	0.27	[68]
Twisted fibers	Wool fiber	Commercial scour	Water pH 2.1	0.39	0.21	0.18	[61]
Twisted fibers	Wool fiber	Commercial scour	Water pH 6	0.31	0.16	0.15	[61]
Twisted fibers	Wool fiber	Commercial scour	Water pH 9	0.34	0.18	0.16	[61]
Twisted fibers	Wool fiber	Untreated	Water pH 6	0.34	0.14	0.20	[66]
Twisted fibers	Wool fiber	Solvent extracted	Water pH 6	0.33	0.16	0.17	[66]

has a firm theoretical basis, however, and attempts to correlate them with felting have been unsuccessful. This is because, as discussed later, factors other than inter-fiber friction are also involved in wool felting. Bohm [70] proposed that the DFE be quantified simply as the arithmetic difference between the against- and with-scale coefficients of friction ($\mu_a - \mu_w$; referred to as σ). This parameter is now widely used by most workers studying the effect of friction on the felting and shrink-proofing of wool and will be used in this paper.

7.3.1 Friction of dry wool

The friction of dry wool is believed to be a factor in determining the feel or 'handle' of a wool fabric [14, 59]. Treatment of wool with a strongly alkaline solution of potassium hydroxide in methanol increased the mean value of the with- and against-scale dry inter-fiber friction, measured against horn, from 0.42 to 0.65. The increased friction resulted in the fabric feeling harsher [63]. The results in Table 7.2 show that both μ_w and μ_a, measured against a wool fiber, were increased by treatment with alcoholic alkali [66]. It was also found that the mean value of the inter-fiber friction was restored to around the value for untreated wool by application of a softening agent [63]. Alcoholic alkali swells wool fibers and its action is not confined to the fiber surface [46]. It was unclear, therefore, from the early studies involving alcoholic alkali whether the changes in friction and handle were the result of a surface-specific modification, or whether they were the result of deeper attack into the cuticle [66]. As discussed on page 259, specific removal of covalently bound lipids from the surface of the epicuticle also produced fabrics with a very harsh handle. This was the result of a significant increase in (dry) inter-fiber friction from 6.7% to 19.5%, as determined from the frictional component of wrinkling [14]. As was the case in the earlier study [63], it was also found that the dry inter-fiber friction and harsh handle could be reduced to a value close to untreated wool by application of a fiber lubricant or a softener [14, 46]. These findings explain the action of alcoholic alkali on wool and clearly demonstrate the importance of the F-layer on the dry frictional properties of the fibers. As will be discussed later, other treatments, such as plasma, that remove the covalently-bound C_{21} fatty acid also increase the harshness of wool.

7.3.2 Friction of wet wool

In considering felting and shrinkproofing, only the DFE of wet wool fibers is relevant. Table 7.3 summarizes values for μ_a, μ_w and σ obtained for a range of rubbing surfaces and aqueous media. The precise method by which the scales are responsible for the DFE of wet wool has been the subject of considerable debate. This has been comprehensively reviewed by Makinson [45, 59]. The earliest proposed mechanism, which involved the interlocking of scales, was suggested by Monge [71]. In this mechanism, the scales engage with each other so that two fibers move along each other in one direction only. There is no evidence for this, as interlocked fibers have never been found in felted wool [72, 73]. The ratchet mechanism, however, does not require a definite interlocking of the scale edges but only a mechanical ratchet effect between the scales and asperities, or roughness, on another surface. This mechanism can explain why it is easier to draw a fiber across a surface in one direction than in the other. Makinson has used two techniques involving microscopy to provide experimental evidence for this mechanism. The first involved drawing wool fibers under water across the surface of a diffraction grating that had been modified to give a saw tooth profile [69]. The spacing and height of the teeth closely matched those of the scales on wool. When drawn in the against scale direction, it was found that the scales engaged with the teeth on the grating by a ratchet type of contact. The scales were seen to deform as they slid past the teeth and to quickly recover. The second method involved examining two twisted wool fibers under water, as one fiber was pulled against the other [74, 75]. It was observed that the scales on one fiber flipped past the scales on the other fiber in a ratchet action. The ratchet mechanism has been criticized on the grounds that a swelling agent, such as water, would be expected to reduce the frictional effect by softening the scales; whereas the friction is actually increased by wetting the fiber (58, 76). This apparent anomaly was explained when it was shown that although swelling agents do weaken the scales, the softened scales lie flatter against the fiber surface, thereby enabling more of the scales to make steep face contacts [74, 77]. Depending on the degree of softening, the net effect of cuticle softening on the DFE could, therefore, be zero, positive or negative.

A DFE has been observed under circumstances where the ratchet mechanism is unlikely to occur. Thus, a DFE was observed when a wool fiber was slid over a very smooth polytetrafluoroethylene (PTFE) rod [78]. PTFE is much softer than the cuticle of wet wool and, in this case, the DFE has been explained in terms of the scales ploughing through the PTFE when fibers are pulled against the scales. In the case of wool, however, the mechanism would be expected to involve deformation and elastic recovery of the surface, rather than cutting [75]. Unlike the ratchet mechanism, which can operate only when fibers are oppositely orientated, the ploughing mechanism could occur between two fibers with the root and tip ends facing in the same direction. According to the ploughing mechanism, the scale tips of one fiber can dig into the scale surfaces of the other fiber. In this case, interaction between fibers involves no ratchet component of friction. This mechanism is attractive because it could also take place when two fibers move in a transverse direction to each other, such as may occur during the formation of a wool felt.

Several other mechanisms have been suggested to explain the frictional properties of wool. Lincoln, from a consideration of scale geometry, proposed that contacting asperities can slide up the steep face of a scale without deformation of either the scale or the asperity [58, 76]. Another model, also based on fiber geometry, was proposed by Flanagan to explain felting in a mass of loose wool, such as during the manufacture of non-woven felts [79]. Grosberg has suggested that the DFE could arise, not from the scale tips themselves, but from a difference in the area of contact of the flat part of the scales near the tips, during with-scale versus against-scale sliding [80]. He ascribed the difference to anisotropy of the scales. The amorphous structure of the cuticle, discussed above, probably makes this an unlikely possibility [45]. It has also been suggested that the DFE of wool fibers is not due to the geometry and properties of the scales, but to asymmetry of the molecular fields at the surface [81]. Although Martin provided no evidence to support the proposal, this mechanism enjoyed some popular support. It was, however, effectively disproved when it was shown that fibers coated with a thin layer of gold retained the ability to felt [82].

Makinson has concluded that, of the possible mechanisms for the DFE of wool, there is experimental evidence for only the ratchet and the ploughing mechanisms [45]. The former is certainly involved in the felting of untreated wool, while the latter may be responsible for the felting of shrink-resist treated wool. It was also suggested that the Lincoln and Flanagan mechanisms may operate under certain circumstances. For the ratchet mechanism, any treatment that softens the scales, so that they flip over each other more easily, would be expected to reduce the directional frictional effect. In the case of the ploughing mechanism, increasing the inter-fiber friction would be expected to affect the sliding of one fiber over the surface of another. The effect of these factors on felting will be discussed later.

7.4 Shrinkage of wool

Wool yarns, fabrics or garments, can shrink during washing by two mechanisms: relaxation shrinkage and felting shrinkage.

7.4.1 Relaxation shrinkage

Relaxation shrinkage can occur with products made from any textile fiber. It results from the release of strains introduced into the structure during manufacturing processes such as spinning, knitting, dyeing and finishing. In the case of wool, the strains are most likely held by 'cohesive' or 'temporary' set. Unlike 'permanent' set, which primarily occurs via rearrangement of the covalent disulphide bonds, cohesive set is the result of non-covalent interactions [10]. The most important of these is hydrogen bonding, which for wool occurs between suitable donor and acceptor groups in the peptide chains and also the amino acid side chains. Ionic bonds, or 'salt linkages' also exist between basic amino and acidic carboxyl groups of some of the amino acid side chains. Both ionic and hydrogen bonds contribute to the physical properties of dry wool. They are, however, disrupted by water, which then allows the yarn, fabric or garment to contract to its unstrained state. Depending on the structure, not all the relaxation shrinkage in a garment is released by soaking in water. Sometimes mild agitation is required to achieve this, which can make it difficult to separate relaxation shrinkage from the other type of shrinkage that occurs when wool is washed; namely, felting shrinkage.

7.4.2 Felting shrinkage

The directional frictional effect, discussed in Section 7.3, is responsible for wool's unique property of felting [45, 73, 75, 83]. Felting occurs in water when individual fibers, within a loose mass, a yarn or a fabric, move preferentially in one direction. The lower value of with-scale friction compared with friction against the scales causes the fibers to irreversibly migrate towards their root ends. In the case of loose fibers, felting produces entanglement. For fabrics, shrinkage occurs. This results in a decrease in area and an increase in fabric thickness. The phenomenon of felting was probably known in prehistoric times and archaeological finds suggest that the earliest type of fabric made from animal fibers may have been a wool felt [2]. The term 'felting' is generally used when the effect of fiber or fabric consolidation is undesirable, such as when it occurs in laundering. Felting is also carried out in a controlled manner, when it is called 'milling' or 'fulling' [45, 83–85]. Milling is used to close up the structure of fabrics or garments and in the manufacture of a diverse range of products, such as piano hammers, hats, billiard-table cloths, military uniforms and polishing pads.

Several mechanisms have been put forward in an attempt to explain how the DFE causes felting in wool fabrics [45]. A popular mechanism, proposed by Shorter, is illustrated by examining the possible ways in which the DFE can influence the behavior of a single fiber within a textile assembly, such as a fabric [72, 86]. This is shown schematically in Fig. 7.4, in which a wool fiber is considered to interact with other fibers at two points along its length. At one point (B), the interaction is so strong that the fiber is unable to move in any direction. At the other point (A), the fiber is less strongly held and under mechanical action (e.g. laundering) it can move relative to the other fibers. As a result of the DFE it can move, through the point of partial constraint (A), only towards its root end but not towards its tip. When the point at which the fiber is totally constrained is near its tip (Case 1), then movement of the single fiber towards the root will tend to decrease the distance along the fiber between the points of interaction; thus tightening the

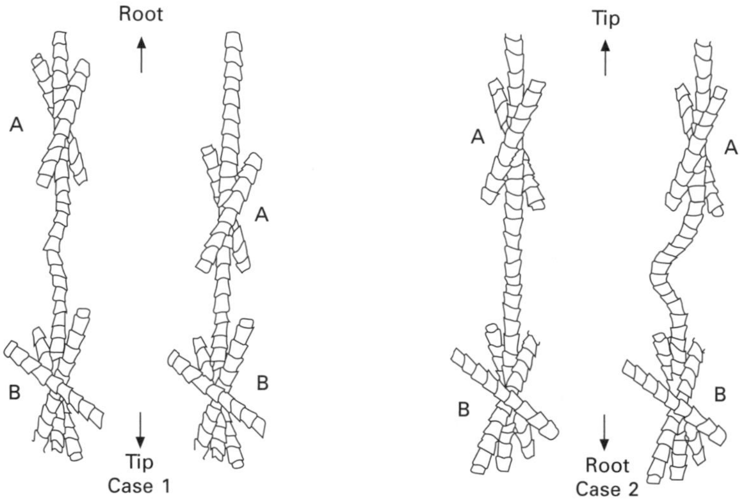

7.4 Schematic diagram of Shorter's mechanism of wool felting.

fiber in the structure. A point will be reached at which the tension generated in the fiber opposes additional movement and, hence, retards further felting. The mechanical action that occurs during washing or milling will, however, produce some slack in the structure as a result of compression of the fabric. This allows the fiber to move further towards its root end, until movement is again constrained by tension. When the distorting force responsible for the compression of the fabric is removed, the DFE prevents the fiber from moving back to its original position in the structure. Under the repeated forces of compression and extension, that occur during laundering or in a milling machine, this self-tightening process is repeated many times to produce a felted structure.

Another situation arises when the point of total constraint (B) is near the root end of the fiber (Case 2). In this situation, movement towards the root end increases the distance along the fiber between the points of interaction. The DFE prevents the fiber from slipping back and this results in it buckling to form a loop. Both Cases 1 and 2 contribute to fabric shrinkage. Shorter's mechanism would be expected to be important in situations where the applied forces are mainly compressive, for example in a washing machine or when woolen-spun knitted garments are milled during wet finishing. Other workers have extended and developed Shorter's mechanism to include the migration of fibers under tension [87]. This was explained in terms of the fiber being pulled through a constraint whilst under tension. Subsequent relaxation and contraction of the fiber causes the fabric structure to close up. The main situation in which tension can develop in a fabric is if the fabric is stretched, for example in piece dyeing. Tensile strain could also develop if a fabric is bent, but it is unlikely that the strain level would be great enough in these circumstances to cause fiber migration.

A mechanism proposed by Martin is similar to that of Shorter, in that he considered the felting process to be driven by externally applied compression forces [81, 88]. Martin's view, however, was that the fibers become locked into a structure by migration of the root ends through loops that are produced in other fibers by compression of the structure. A significant difference between the two theories is that Martin believed that only the root ends are responsible for locking fibers, whereas Shorter's view was that a fiber could be constrained by interaction with other fibers at any point along its length. It has been suggested [89] that Martin's loop-locking mechanism is most likely to operate in fabrics made from very short wool fibers, but that the two mechanisms are not mutually exclusive [90].

7.4.3 Other factors that affect felting shrinkage

In addition to the DFE, a number of other properties also influence the felting shrinkage of untreated wool. These include fiber diameter, length, crimp and elasticity. Of these, it appears that elasticity is probably the most important. The various models to explain the mechanism of felting rely on deformation of the assembly of fibers under the action of external forces. For the fibers to migrate under the influence of the scale structure, it is necessary for them to extend and then recover from the extension. Speakman *et al.* found that felting ability increased with ease of fiber extensibility, when measured in the same aqueous liquor [87, 91]. Although changes in liquor pH do not significantly affect the frictional properties of wool [61], liquor pH was found to influence felting rate by changing the elastic properties of the fiber [87, 91]. The ease of extensibility was found to be at a minimum in the range pH 4–8, which corresponded with the pH range over which felting rate was also at a minimum. Outside this pH region, either acids or alkalis increased fiber extensibility and also felting rate.

A correlation has been found, for a range of wool types, between felting ability and the capacity of fibers to recover after deformation [92, 93]. Wool dyed to a heavy shade showed an increase in the work to stretch the fibers, a decrease in ability to recover and a decrease in felting [94]. This observation led to attempts to make wool shrink-resistant by changing its elastic properties by the deposition of polymers inside fibers [95].

Temperature has been found to also affect fiber elasticity [96, 97]. This is consistent with the increasing in felting rate that has been observed with increase liquor temperature. There has been general agreement that feltability increases with temperature between 20°C and 40–50°C. Above this range the situation is less clear. Some workers reported a maximum in felting rate in the range 40–50°C [87, 98]. Other workers, however, found no maximum, only a limiting value [99, 100]. It appears that whether a maximum in felting rate does occur at a particular temperature depends on the overall conditions employed. Thus, Feldtman and McPhee found that the temperature at which they observed a maximum in felting rate was influenced by liquor pH, washing action of the machine and whether the liquor contained soap or a detergent [98]. The type of detergent used was also important.

In addition to its effect on elasticity, Lindberg and Gralén found that increasing the temperature from 20°C to 70°C decreased the with-scale friction from 0.34 to 0.23, the against-scale value from 0.22 to 0.16 and the frictional difference from 0.12 to 0.07 [61]. They concluded, however, that the increased felting rate of wool at higher temperatures cannot be ascribed to these changes in frictional properties, because the nature of the changed frictional coefficients would be expected to cause a decrease in felting rate. The effect of soap, detergents and neutral salts on the felting behavior of wool have been studied in some detail by several groups of workers. These studies have been comprehensively reviewed elsewhere [45]. There has not been universal agreement on the effects of all of these factors on felting. In general, however, their influence has been found to be relatively small, and also to differ between untreated and shrink-resist treated wool.

Fine wools felt more readily than do coarse wools and this has been considered to be due to the influence of fiber diameter on felting [60]. The situation is more complex, however, because diameter is also associated with differences in crimp and surface friction [87]. Diameter is believed to be important because of its effect on the compressibility of wool [101]. In this respect, crimp has been shown to be more important than diameter.

The structure of a yarn or fabric affects felting shrinkage [102]. Any property that restricts the migration of fibers will reduce the rate of felting. Thus, tightly constructed woven or knitted fabrics felt more slowly than do loose, open structures [103]. Increasing the twist in yarns also decreases felting by increasing fiber security and thereby retarding fiber migration during washing [104].

7.5 Methods of reducing the felting shrinkage of wool

If great care is taken, many garments made from wool can be washed in an aqueous liquor without significant felting taking place. The modern lifestyle has, however, created a consumer demand for wool garments that can be laundered under more severe conditions, such as in a washing machine. The laundering cycle often includes tumble-drying, which can also contribute to excessive felting shrinkage. The propensity for wool garments to shrink by felting under these conditions has led to the development of techniques to make wool 'shrinkproof'. An alternative, and better term is 'shrink-resistant'.

The performance of treatments employed to make wool shrink-resistant is tested by means of a range of procedures that measure the amount of relaxation and felting shrinkage under standard washing conditions. These have been developed by The Woolmark Company (formerly The International Wool Secretariat) [105].

In order to prevent wool from felting, it is necessary to overcome the DFE responsible for unidirectional fiber migration. This can be done by removing the scales by mechanical means, for example with an abrasive [106], chemical means, for example treatment with an oxidizing agent [107], and by procedures involving the use of enzymes [108]. De-scaling methods remove a significant proportion of the fiber and are aimed primarily at increasing luster rather than producing shrink-resistant wool.

Felting shrinkage can be reduced by less drastic methods and over the years several approaches have been used. Commercially important shrink-resist treatments can be divided into three main categories:

- chemical treatments
- polymer-only treatments
- combination of chemical and polymer treatments.

Chemical treatments are often referred to as 'degradative', because they involve chemical attack on the wool fiber. At various times, a number of chemical treatments have been used to reduce the felting shrinkage of wool. These include oxidizing agents, such as chlorine, peracetic acid, permonosulphuric acid and potassium permanganate. Treatments based on the use of reducing agents, solutions of alcoholic alkali and enzymes have also been used [45, 83].

Some of the oxidative treatments discussed in Section 7.5.1 are no longer used commercially. They have been included because of their contribution to the overall understanding of the mechanisms involved in felting and shrink-resist treatments in general. Other techniques that do not involve oxidation, discussed in Section 7.5.2, have also been included for similar reasons. One of these involves treatment with an alkali dissolved in an alcohol. The other one uses a corona or plasma discharge. The latter technique is also important because of current interest in developing new shrink-resist processes that do not involve the use of some type of chlorination.

Although different mechanisms may be involved, all shrink-resist treatments reduce felting shrinkage by retarding or preventing the unidirectional migration of fibers under the action of the directional frictional effect. In the case of the chemical treatments, discussed in detail in Section 7.5.1, this generally occurs by decreasing the frictional difference, usually by a large increase in the with-scale coefficient (Table 7.4).

Depending on the rubbing surface, the against-scale coefficient may also be increased, but usually to a lesser extent. Some treatments, such as those

Table 7.4 Effect of shrinkproofing treatments on the (wet) frictional properties of wool

Treatment	Rubbing surface	Measurement conditions	Untreated			Treated			Ref.
			μ_a	μ_w	σ	μ_a	μ_w	σ	
Acid chlorination	Horn	pH 10.7 buffer plus soap	0.34	0.14	0.20	0.34	0.30	0.04	[62]
Acid chlorination	Horn	pH 9 buffer plus anionic wetting agent	0.39	0.25	0.14	0.48	0.44	0.04	[109]
Acid chlorination	Horn	pH 9 buffer plus nonionic wetting agent	0.42	0.25	0.17	0.49	0.41	0.08	[110]
Acid chlorination	Wool fiber	Water, no wetting agent	0.37	0.21	0.16	0.92	0.91	0.01	[61]
Acid chlorination	Wool fiber	Water plus anionic wetting agent	0.30	0.13	0.17	0.76	0.68	0.08	[111]
Permono-sulphuric acid/bisulphite	Horn	pH 9 buffer plus nonionic wetting agent	0.39	0.25	0.14	0.39	0.32	0.07	[109]
Permanganate/salt	Horn	pH 9 buffer plus nonionic wetting agent	0.39	0.25	0.14	0.40	0.31	0.09	[109]
Ozone	Glass	Water	0.27	0.13	0.14	0.19	0.11	0.08	[112]
Alcoholic KOH	Horn	pH 10.7 buffer plus soap	0.37	0.14	0.23	0.67	0.43	0.24	[62]
Alcoholic KOH	Wool fiber	pH 6 buffer	0.34	0.14	0.20	0.54	0.31	0.23	[66]
Corona	Horn	Water plus nonionic wetting agent	0.43	0.18	0.25	0.70	0.49	0.21	[77]
Plasma	Not known	Water plus wetting agent	0.51	0.24	0.27	0.82	0.49	0.33	[113]

described in Section 7.5.2, reduce fiber migration by markedly increasing both μ_w and μ_a without changing σ.

Treatments based solely on the application of polymers (called 'additive' treatments) prevent fiber migration by forming elastic bonds of crosslinked polymer between fibers. Processes based on a chemical pre-treatment plus the application of a polymer reduce felting by more than one mechanism. Polymer-based shrink-resist processes, either with or without a fiber pre-treatment are commercially the most important. The mechanisms by which they function will, therefore, be discussed in detail in Section 7.6, under Commercial shrink-resist processes.

7.5.1 Mechanisms of oxidative chemical treatments

Chlorination

The early methods of chlorinating wool, involving the use of bleaching powder and hydrochloric acid, were later replaced by methods involving sodium hypochlorite [114]. Yellowing and poor abrasion resistance were found to result from over-treatment and much early research was aimed at minimizing these adverse effects by optimizing the treatment conditions [115]. The next advance in wool chlorination was the development of a method using chlorine gas instead of aqueous hypochlorite [116]. Careful control of the moisture content of the wool minimized fiber swelling and also reduced the rate of reaction between the chlorine and wool. These two factors ensured that the treatment was more confined to the fiber surface than was the case with the earlier aqueous methods.

Speakman *et al.* postulated that chlorine reduces felting shrinkage by changing surface friction through the fission of cystine disulphide bonds on the fiber surface [41]. This mechanism, involving oxidation of cystine in the exocuticle to cysteic acid residues, has been confirmed by other workers; for example by chemical analysis of the cuticle [117] and by examination of the fiber surface using infra red spectrometric techniques [118]. Alexander *et al.* showed that for shrink-resistance to be obtained, some fission of the peptide chains must also occur. They suggested that this takes place primarily at tyrosine residues [42, 43]. The importance of cystine oxidation and peptide bond hydrolysis at tyrosine residues, in the mechanism of shrink-resistance by chlorination, has been discussed by Makinson [119]. From the concentration of tyrosine residues in the exocuticle, an estimate was made of the molecular weight of the peptide fragments expected from a combination of oxidation of cystine and fission of protein chains adjacent to tyrosine residues. Makinson concluded that, although they are water-soluble because of the cysteic acid groups, the fragments are too large to diffuse through the epicuticle membrane. As discussed on page 258, these osmotically-active peptides are also responsible for the formation of Allwörden bubbles when wool is immersed in chlorine water. The absorption of large amounts of water by osmosis swells the scales and makes them softer. This results in a large increase in the with-scale coefficient of friction (Table 7.4). It has been suggested that this is the consequence of the higher area of contact with other fibers produced by the greater deformability of the softened scales [45]. Against-scale friction has also been found to increase, to an extent that depends on whether horn or wool is used as the rubbing surface. In this case, although softening will increase the number of steep face contacts, the increased viscous-elasticity of the softened scales enables them to flip over each other more easily than in untreated wool. The relative importance of these two factors in determining the against-scale coefficient of friction depends on the level of treatment.

The results in Table 7.4 show, however, that there is general agreement that chlorination produces a large decrease in the frictional difference (σ) and that this is a major factor responsible for the reduced felting shrinkage of chlorinated wool. In addition to the oxidation of cystine and hydrolysis of peptide chains, treatment with chlorine also removes covalently-bound fatty acids from the surface of wool [120]. It was discussed on page 259 that removal of surface lipids reduces felting shrinkage by increasing inter-fiber friction [14]. Thus, in addition to the effect of scale softening on the DFE, increased surface friction produced by partial removal of the F-layer also contributes to the shrink-resistance produced by chlorination. The harshness resulting from chlorination has been attributed to large increases in with-scale and against-scale friction and virtual removal of the DFE, as shown in Table 7.4 [61, 102, 121, 122]. This is also consistent with the removal of surface lipids, because harshness is also increased when the lipids of the F-layer are removed by other methods [14].

Other oxidants

Other oxidizing agents have also been used to reduce the felting shrinkage of wool [44, 45, 83]. The most important of these are peracids, in particular permonosulphuric acid, and potassium permanganate. These reagents have often been used in the presence of concentrated solutions of inorganic salts, to suppress fiber swelling and thus confine the reaction to the surface of the cuticle [109, 123].

Peracids

Treatment with peracetic acid produces shrink-resistance only under severe treatment conditions that degrade the fiber [43, 109]. A lower level of peracetic acid can be used, provided the oxidation is followed by treatment with a reducing agent (sodium bisulphite or sulphite) [124]. On its own, peracetic acid produces little change in either shrink-resistance or in the frictional properties of wool. Following a post-treatment with sodium sulphite, however, frictional properties are changed and felting shrinkage is decreased. Similar results are obtained with permonosulphuric acid (PMS; $HOOSO_3H$). As with peracetic acid, oxidation with PMS on its own produces little change in either friction or shrink-resistance [124]. Post-treatment with sodium sulphite or bisulphite lowers the frictional difference and decreases felting shrinkage by increasing the with-scale friction without affecting the against-scale values (Table 7.4).

PMS treatment is much less effective than chlorination in reducing the felting shrinkage of wool [45]. This could be due to two factors. PMS followed by bisulphite produces Bunte salt groups in the cuticle, whereas chlorination generates cysteic acid residues [125–127]. Both species are strongly anionic, which will lead to increased swelling and softening of the cuticle. As discussed on page 273, however, a higher level of cuticle softening is produced by a combination of cystine oxidation and peptide hydrolysis than is produced by cystine oxidation alone [119]. Of the two treatments, the necessary level of peptide hydrolysis only occurs with chlorination. The second factor relates to the relative effectiveness of the two treatments in changing frictional properties by removing the lipids of the F-layer. The wettability and harshness of wool is significantly increased by chlorination; effects that can be attributed to removal of surface lipids [14, 120]. In contrast, wool treated with PMS differs very little in wettability from clean, untreated wool [128]. It also has a less harsh handle than chlorinated wool, which can be explained by a lower level of lipid removal and the unchanged value of the against-scale friction shown in Table 7.4.

Permanganate

The felting shrinkage of wool is decreased by treatment with potassium permanganate. Brown manganese dioxide is deposited in the fiber and this is removed by an after-treatment with sodium bisulphite [129]. The effectiveness of the treatment is improved when carried out in a concentrated solution of sodium chloride. Under neutral conditions, the sodium bisulphite reductive step also markedly improves the level of shrink-resistance [109, 121]. In common with other oxidizing agents, the mechanism is believed to involve oxidation of cystine to cysteic acid [130]. Friction measurements have shown that shrink resistance is increased as a result of a reduction in the DFE caused by an increase in the with-scale friction, with little change in the against-scale value (Table 7.4). This has been claimed to be the reason for the soft handle of wool treated by this method compared with wool treated by chlorination, because, as discussed above, the harsher handle of chlorinated wool has been attributed to a large increase in both the with-scale and against-scale friction [61, 102, 121, 122]. The effectiveness of permanganate oxidation in removing covalently-bound surface lipids does not appear to have been studied. It is possible, however, that permanganate-treated wool feels softer than chlorinated wool because it is less effective in removing the F-layer from the epicuticle.

Ozone

Oxidation with ozone has been investigated for producing shrink-resistant wool [112]. Ozone was found to oxidize cystine to cysteic acid and to reduce felting shrinkage. The data in Table 7.4 show that this occurs as a result of a decrease in both the with- and against-scale friction and also the frictional difference.

7.5.2 Mechanisms of non-oxidative chemical treatments

Some treatments that do not involve strong oxidizing agents also decrease the felting shrinkage of wool. Two of these, namely treatment with a solution of alkali in anhydrous alcohol and treatment with a plasma or corona discharge, do not form the basis of any currently used commercial shrink-resist process. They are, however, worth some discussion in the present review because they involve surface modification and changes in frictional properties.

Alcoholic alkali

The Freney-Lipson process for decreasing felting shrinkage involved treatment of wool with a solution of potassium hydroxide in dry ethanol [131]. This method was used commercially for a number of years [132]. Felting shrinkage was decreased by the treatment but the scale structure, viewed under an optical microscope, was unchanged. Table 7.4 shows that there was a large increase in both with- and against-scale friction and little change in the frictional difference. This was accompanied by a harshening of handle [62, 133]. In view of similar observations made with other alkaline treatments [14], it appears that the main mechanism by which the Freney-Lipson process reduces felting shrinkage is by removing covalently bound lipids from the fiber surface, with a corresponding increase in fiber friction.

Corona and plasma discharges

Kassenbeck reported a range of changes in the properties of wool obtained by means of an electric discharge [134]. The treatment, which was considered to be confined to the fiber surface, made wool easier to wet. It also increased inter-fiber friction, which resulted in better spinning performance and increased resistance to pilling and felting shrinkage, compared with untreated material. The effect on handle was not discussed. Other workers, who used both corona treatment at atmospheric pressure and plasma under reduced pressure, found an increased resistance to felting [113, 135–138], a harsher handle [139] and increased cohesive force of top and tensile strength of yarn [135, 136]. These effects were attributed to large increases in the with- and against-scale frictions as shown in Table 7.4. Plasma treatments have also been carried out in the presence of various gases [140].

Surface analysis of wool treated with a corona or plasma discharge has shown an increased concentration of cysteic acid [137, 138, 141], indicating oxidation of cystine in the exocuticle [142]. Etching of the fiber surface has also been observed [143]. The treatments have been found to produce a large decrease in the amount of the covalently bound C_{21} fatty acid that is the major component of the F-layer [144]. Removal of surface lipid is consistent with the harsh handle produced by electric discharge treatments. The amount of lipid material removed was found to increase with the treatment time and the power input, as also did wettability and resistance to felting shrinkage [145, 146]. Thus, it appears that, in addition to oxidizing cystine in the exocuticle, removal of fatty acid from the epicuticle also plays an important role in the changed fiber properties produced by electric discharge treatments.

7.6 Commercial shrink-resist processes

7.6.1 Processes based on chemical treatment

Of the degradative treatments discussed above, the two most important are chlorination and permonosulphuric acid (PMS). Both can be used for the batch treatment of fabrics, garments and (sometimes) yarns and also for the continuous treatment of wool tops [147–151].

For continuous chlorination treatments, rapid reaction with the wool is required and this is best achieved by the use of acid plus sodium hypochlorite [147], or by the use of chlorine gas dissolved in water [152]. In the batch processing of garments or fabric, an uneven treatment can result if the rate of reaction between the chlorine and wool is too fast. This problem can be minimized by using dichloroisocyanuric acid (DCCA) (Fig. 7.5) as the chlorinating agent [148, 149].

DCCA releases chlorine more slowly than acid hypochlorite. Furthermore, its action can be controlled by temperature and/or pH. Treatment with DCCA is suitable for the batch treatment of fabrics, garments or yarn; and also for the continuous processing of tops. Following chlorination, an 'anti-chlor' step with sodium sulphite is carried out to remove N-chloramines formed by

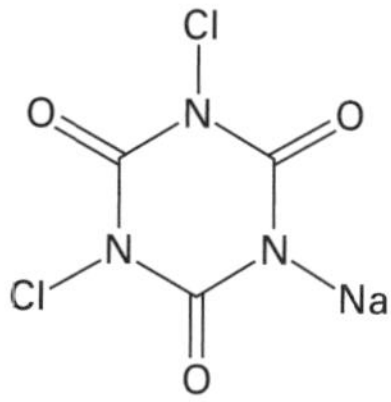

7.5 Structure of the sodium salt of dichloroisocyanuric acid (DCCA).

reaction of chlorine with the amino groups of wool [44]. It has been suggested that the sulphite after-treatment also contributes to the shrink-resist effect by enhancing the softness of the scales, as a result of increased porosity following extraction of proteins rich in cysteic acid [118].

PMS, used in the form of its potassium triple salt, reacts more slowly with wool than chlorine. When used in conjunction with a reductive after-treatment, this makes it particularly suitable for use in batch processes [150].

7.6.2 Processes based on polymer treatment

The felting of products made from wool can be reduced by the application of certain polymers. Over the past 60 years, a large number of polymers have been applied by a variety of techniques. These include melamine formaldehyde resins, silicones, vinyl polymers, polyacrylates and N-methoxymethyl nylon. Monomers, such as styrene, methyl methacrylate and acrylamide have been grafted onto wool by various methods. Interfacial polymerization techniques have been used to form linear polymers across the fiber surface [44, 153, 154].

Spreading and adhesion of polymers on wool

In order to impart shrink resistance with polymers, the surface of the fiber must be wettable. Untreated wool fibers have poor wettability due to the low surface energy of the hydrocarbon F-layer. A liquid will only spread on a solid surface if there is a decrease in the free energy of the system [155]. Thus, a liquid polymer will only spread on the surface of wool if the surface energy of the fiber is greater than that of the polymer [156]. The surface energy of untreated wool is approximately 45 mNm^{-1}. It has been found that treatments that raise the surface energy to around 70 mNm^{-1} enable some polymers to spread on wool. Furthermore, adhesion between the polymer and the fiber surface must be strong enough to resist removal during laundering.

Makinson has described a number of mechanisms by which polymers can reduce the felting shrinkage of wool [45]; the two most relevant being fiber-fiber bonding and scale masking. Both mechanisms decrease felting shrinkage by preventing unidirectional fiber migration during washing.

The processes fall into two categories; polymer-only methods, used on untreated wool, and methods that require a chemical treatment before polymer application.

Polymer-only processes on untreated fabric

The polymer-only methods are used on untreated wool, most commonly in the form of woven or, sometimes, knitted fabric. These methods involve the deposition of an oligomer (pre-polymer) onto the surface of fibers by a pad-dry method. The pre-polymers contain reactive groups that enable them to self-crosslink as the treated fabric is dried [157, 158]. When the polymer is fully cured by crosslinking it forms flexible bonds between fibers and yarns. This mechanism has been referred to as 'spot welding'. With some polymers, such as those based on poly(propylene oxide), the bonds may appear like bridges [157–159]. Silicone polymers are also used. These elastomers, which have a lower surface energy than polymers based on polypropylene backbones, spread readily even on untreated wool and tend to form a continuous layer joining two or more fibers [160].

In general, fiber-fiber bonding increases fabric stiffness because fibers are less able to move independently of each other. This adverse effect on fabric handle is a disadvantage of polymer-based fabric shrink-resist treatments. The problem is alleviated to some extent by using a polymer that is effective at a low treatment level and which forms thinner, flexible and strong bonds [157].

Oxidation plus polymer processes

Two types of processes are used that require the wool to be chemically treated before application of a polymer; namely, batch procedures for garments or fabrics and continuous methods for wool top or sliver.

Batch treatments

Although garments with machine washable performance can be obtained by chlorination, the high treatment level required produces unacceptable fiber damage [149]. Polymer treatments, on the other hand, can produce high levels of shrink-resistance without causing significant fiber damage. This has led to the development of processes for the production of fully machine-washable garments in which a low level oxidative treatment is followed by the exhaust application of a polymer [149,150]. The chlorination pre-treatment is normally carried out with DCCA. Suitable polymers that can be used in conjunction with this treatment are listed in Table 7.5. Although the structures of some of these polymers are not disclosed, it is believed that they all have cationic properties.

Table 7.5 Shrink-resist polymers used on pre-treated wool

Product	Manufacturer	Description	Pre-treatment
Hercosett 125	Hercules	Polyamide/epichlorhydrin (cationic)	DCCA
Basolan SW	BASF	Polyether containing aziridine groups (cationic)	DCCA PMS
Basolan MW	BASF	Polyamino siloxane (cationic)	DCCA PMS
Polymer GE	PPT	Undisclosed mixture	DCCA

Although chlorination alone provides some degree of shrink-resistance, its main purpose is to modify the fiber surface for the polymer application. The treatment increases the surface energy and improves the wettability of wool by removing a high proportion of the F-layer lipids [120] and also by oxidizing cystine in the exocuticle to cysteic acid [41]. The latter groups, near the fiber surface, attract the cationic polymer during the exhaustion step [150]. During curing, the polymer can self-crosslink and also react with thiol and amine groups on the wool surface [154, 161].

Thus, the cationic, water soluble polyamide-epichlorhydrin pre-polymer, Hercosett 125 (Hercules Chemical Co) (Fig. 7.6), which is completely ineffective on untreated wool, decreases the felting shrinkage of chlorinated wool. The polymer is believed to act by both fiber-fiber bonding and scale masking via fiber encapsulation [149,159].

Pre-treatment with PMS can be used as an alternative to chlorination in batch polymer processes [162, 163]. Table 7.5 shows, however, that not all the polymers that are effective on chlorinated wool are suitable for application

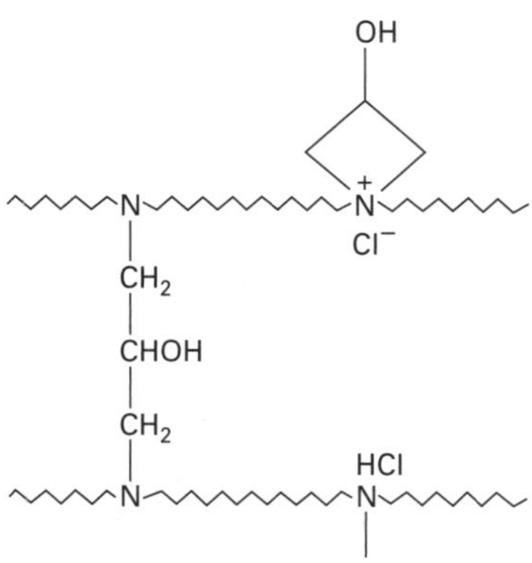

7.6 Structure of Hercosett 125.

to PMS-treated material. This is consistent with chlorination being the more effective of the two treatments in promoting spreading and adhesion of polymers, by increasing the surface energy and wettablity of wool via removal of surface lipids (128).

Continuous treatment of top

The greatest proportion of machine-washable wool is produced by the continuous treatment of wool tops. This process, developed by the Commonwealth Scientific and Industrial Research Organization (CSIRO) and the International Wool Secretariat (IWS), is carried out on specialized equipment in many locations around the world [151,164–166]. A web of parallel wool slivers is treated in a series of steps at a speed of 5–10 m/min and a production rate of 200–500 kg/h. The following sequence of operations is used:

> Acid chlorination → anti-chlor → neutralize → rinse → polymer → softener → dry

A key requirement for a polymer suitable for this application is that it must spread on the surface of the treated wool so that the fibers become encapsulated [167–169]. Very few polymers have been found that have a sufficiently low surface energy to spread on wool, even after the surface energy of the fibers has been raised by chlorination. The product most commonly used in this process is the cationic water soluble pre-polymer, Hercosett 125 (Fig. 7.6), described above [154]. The anionic cysteic acid groups, produced in the exocuticle by chlorine oxidation of cystine, attract the cationic polymer so that it readily exhausts and spreads on the surface of the pre-treated wool, where it cures by self-crosslinking and by reaction with groups on the fiber surface [154,161].

There is agreement between workers that, in sliver treatments, most of the shrink-resistance is produced by the chlorination pre-treatment, via the effect on the DFE produced by softening the scales [168,169]. The polymer, however, raises the level of shrink-resistance to meet the stringent requirement for full machine-washability.

The mechanism involving fiber-fiber bonds, that applies to polymer shrink-resist treatments on fabric, cannot apply to top treatments, as any bonds formed during the treatment of top would be broken during subsequent processing into yarn. The generally accepted mechanism involves decreasing the frictional difference, to a value of around 0.04 or less, by masking the scales [169]. This could occur by fiber encapsulation, preferential deposition of the polymer at the scale edges, or by a combination of both of these.

A scanning electron micrograph of a wool fiber treated in top form by the chlorine/Hercosett process is shown in Fig. 7.7. Following chlorination, the

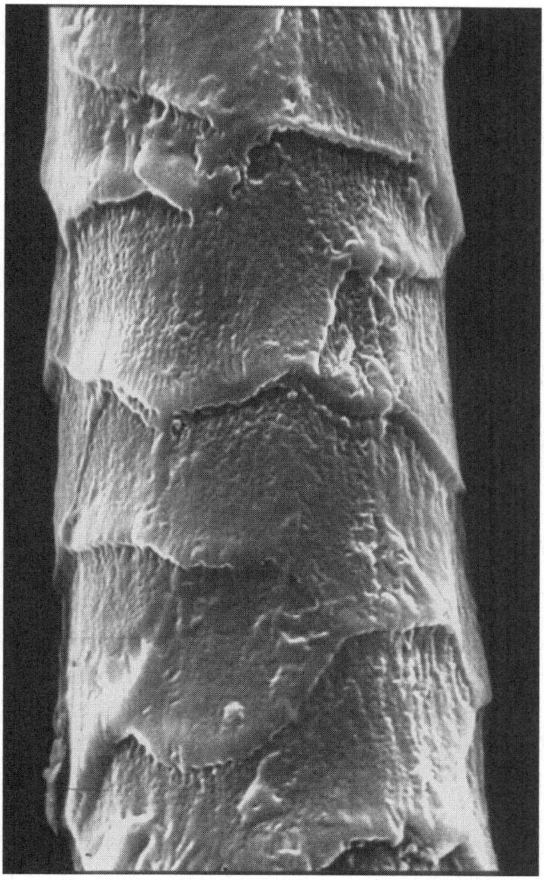

7.7 Scanning electron micrograph of a chlorine/Hercosett-treated wool fiber.

scale height on a 20-micron wool fiber is approximately 0.5 micron. To produce a polymer layer at least this thick would require around 10% Hercosett on mass of wool (omw), whereas in commercial treatments only around 2% omw is used [169]. This apparent anomaly has been explained in terms of the swelling properties of the polymer layer on the treated fibers. Crosslinked films of Hercosett resin swell in water to around five times their air-dry volume. Thus, a level of 2% omw in the dry state produces around 10% of resin gel in water. Consequently, the swollen polymer is capable of masking the scales and Makinson and Lead proposed that this plays an important role in the shrink-resist mechanism of chlorine/Hercosett top-treated wool [169]. These workers also suggested that the swollen polymer increases the adhesion between fibers, thus preventing fiber migration in either direction.

7.7 Future trends in the shrink-resist treatment of wool

The preceding discussion has described a range of methods that have been used commercially over many years in the manufacture of shrink-resistant wool products.

The most effective of these for producing full machine-washability involve chlorination of the wool. During these processes, compounds formed by reaction of chlorine with protein material dissolved from the wool are released into the environment [170]. These compounds are collectively described as Adsorbable Organic Halides (AOX). Concerns about the toxicity of these chemicals and their possible effect on the environment has led to the introduction of legislation controlling their discharge from textile plants [171, 172]. This has resulted in a considerable effort over the past 10–15 years to find a suitable replacement for chlorination in the shrink-resist treatment of wool.

As discussed above, oxidation with permonosulphuric acid can be used successfully for garment treatments, either alone or more commonly with a polymer. PMS treatment is unsatisfactory, however, for use in the continuous treatment of top, which is the route by which the majority of machine-washable wool is produced. A process has been patented, however, in which wool is pre-treated with a mixture of hydrogen peroxide and PMS in the presence of a heavy metal peroxide catalyst. This step is then followed by application of a polymer [173]. Oxidation with potassium permanganate followed by treatment with an enzyme has also been tried [174].

Most recent work has focused on the possibility of using a plasma treatment as an alternative to chlorination. None of the polymers that are effective on chlorinated wool are suitable for application to plasma-treated top [113]. Some new polymers that operate by scale-masking have been claimed to be effective, but none of the methods is used commercially [113, 175, 176].

A process currently used in Japan involves the treatment of wool top with ozone, followed by the application of Hercosett [177]. This method is claimed to be as effective as the chlorine/Hercosett process for producing machine-washable wool, without the AOX problems associated with the use of chlorine.

7.8 Summary

Wool from sheep has been used for the production of cloth for thousands of years. Despite the increasing use of synthetic fibers, it remains an important material used by the textile industry. Wool and other keratin fibers have evolved highly complex chemical and physical structures in order to maximize their primary function, which is to insulate the animal from heat and cold. Wool has a unique surface structure as it is the only textile fiber with a layer of overlapping scales on the surface. Whilst this structure is useful in expelling dirt and other contaminants from the fleece, it is also responsible for wool's directional friction effect. Although wool is one of the smoothest textile fibers, the DFE has both positive and negative implications when it is used to make textile products, because it is responsible for the felting that can occur when wool substrates are agitated in an aqueous medium. This property is useful in the manufacture of items where it is desirable for the fabric density to be increased by closing up the structure; for example wool felts, blankets and overcoat fabrics. Felting is, however, a problem when products are required that have very good dimensional stability when laundered, for example wool knitwear.

The felting shrinkage of wool can be reduced by chemical treatments that change the frictional properties of wet fibers by increasing the swelling of the cuticle cells as a result of the oxidation of cystine to cysteic acid. This produces a decrease in the DFE from values that are typically in the range 0.14–0.20 to values of around 0.01–0.10. Depending on the method used to measure shrinkage in washing, these treatments will reduce area felting shrinkage from very high levels (typically 40–50%) to less than 5%. Felting shrinkage can also be decreased by chemical treatments that do not significantly affect the DFE. These non-oxidative treatments operate by approximately doubling both the with-scale and against-scale frictional values due to the removal of surface lipids. The resulting large frictional values make it difficult for fibers to migrate in either direction during laundering. In general, treatments that involve polymer application are more effective than chemical-only methods and are used when the highest levels of machine washability are required. When applied to fabrics, they reduce felting by forming fiber-fiber bonds that prevent fiber migration during laundering. When applied to top, following an oxidative pre-treatment, the polymer reduces the DFE to very low values, primarily by scale masking.

7.9 References

1. Hyde N, 'Fabric of History – Wool', in *National Geographic*, National Geographic Society, Washington, DC, May 1988, **173**, 552–591.
2. Ryder M L, 'The Evolution of the Fleece', in *Scientific American*, Scientific American Inc., New York, Jan 1987, 100–107.
3. Mercer E H and Maltosy A G, 'Keratin', in Montagna W and Dobson R L, Ed., *Advances in Biology of Skin and Hair Growth*, Oxford, Pergamon Press, 1968, 555–569.
4. Fraser R D B, Macrae T P and Rogers G E, *Keratins – their Composition, Structure and Biosynthesis,* Springfield USA, Thomas, 1972.
5. Lungren H P and Ward W H, 'The keratins', in Borasky R, Ed., *Ultrastructure of Protein Fibers,* Academic Press, New York, 1963, 39–122.
6. Zeigler K, 'Crosslinking and self-crosslinking in keratin fibers', in Asquith R S,

Ed., *Chemistry of Natural Protein Fibers,* New York, Plenum Press, 1977, 267–300.

7. Zahn H and Kusch P, 'Wool as a biological composite system', *Melliand. Textilber.*, Eng. Ed., 1981, **10**, 75–85.
8. Zahn H, 'Wool is not keratin only', *Plenary Lecture, 6th Int. Wool Text. Res. Conf, Pretoria,* **I**, 1980.
9. Baumann H, 'Applied aspects of keratin chemistry', in Parry D A D and Creamers L K, Ed., *Fibrous Proteins: Scientific, Industrial and Medical Aspects* – ***Vol. 1***, London, Academic Press, 1979, 299–391.
10. Rippon J A, 'The structure of wool', in Lewis D M, Ed., *Wool Dyeing,* Bradford UK, Society of Dyers and Colourists, 1992, 1–51.
11. Rivett D E, 'Structural lipids of the wool fiber', *Wool Sci. Rev.*, 1991, **67**, 1–25.
12. Lindberg J, 'Relationship between various surface properties of wool: wettability' *Textile Res. J.*, 1953, **23**, 585–588.
13. Pittman A G, 'Polymer surface coatings for wool and mohair fibers – influence of wetting behavior', *Proc. 4th Int. Wool Text. Res. Conf., San Francisco*, 1970, 593–599.
14. Leeder J D and Rippon J A, 'Changes induced in the properties of wool by specific epicuticle modification', *J. Soc. Dyers and Colour.*, 1985, **101**, 11–16.
15. Bradbury H, 'The structure and chemistry of keratin fibers', in Anfinsen C B Jr., Edsall J T and Richards F M, Ed., *Advances in Protein Chemistry* – ***Vol 27,*** New York, Academic Press, 1973, 111–211.
16. Leeder J D, 'The cell membrane complex and its influence on the properties of the wool', *Wool Sci. Rev.*, 1986, **63**, 3–35.
17. Feldtman H D, Leeder J D and Rippon J A, 'The role of fiber structure in wool fiber and fabric performance', in Postle R, Kawabata S and Niwa M, Ed., *Objective Evaluation of Apparel Fabrics,* Osaka, Text. Mach. Soc. Japan, 1983, 125–134.
18. Evans D J, Denning R J and Church J S, 'Interface of keratin fibers with their environment', in *Encyclopedia of Surface and Colloid Science,* Marcel Dekker, 2002, 2628–2642.
19. Bradbury J H and Leeder J D, 'Keratin fibers: Part IV- Structure of cuticle', *Aust. J. Biol. Sc.*, 1970, **23**, 843–854.
20. Jones L N, Rivett D E and Tucker D J, 'Wool and related mammalian fibers', in Lewin M and Pearce E M, Ed., *Handbook of Fiber Chemistry*, New York, Marcel Dekker, 1998, 355–413.
21. Appleyard H M and Grevelle C M, 'The cuticle of mammalian hair', *Nature*, 1950, **166**, 1031.
22. Fraser R D B, Jones L N, Macrae T P, Suzuki E and Tulloch P A, 'The fine structure of the wool fiber', *Proc. 6th Int. Wool Text. Res. Conf., Pretoria,* **1**, 1980, 1–33.
23. Bradbury J H, Chapman G V and King N L R, 'The chemical composition of wool', *Aust. J. Biol. Sci.*, 1965, **18**, 353–364.
24. Bradbury J H, Chapman G V and King N L R, Chemical composition of the histological components of wool', in Crewther W G, Ed., *Symposium on Fibrous Proteins,* Australia, Butterworths, 1967, 368–372.
25. Leeder J D and Marshall R C, 'Readily-extracted proteins from merino wool', *Textile Res. J.*, 1982, **52**, 245–249.
26. Bradbury J H and Ley K F, 'The chemical composition of wool: Separation and analysis of exocuticle and endocuticle', *Aust. J. Biol. Sci.*, 1972, **25**, 1235–1247.

27. King N L R and Bradbury J H, 'The chemical composition of wool: The epicuticle', *Aust. J. Biol. Sci.*, 1968, **21**, 375–384.
28. Leeder J D, Rippon J A, Rothery F E and Stapleton I W, 'Use of the transmission electron microscope to study dyeing and diffusion processes', *Proc. 7th Int. Wool Textile Res. Conf.*, Tokyo, 1985, **V**, 99–108.
29. Sikorski J and Simpson W S, 'Electron microscopy studies of the chemical reactivity in keratin cuticle', *Nature*, 1958, **182**, 1235–1236.
30. Lehmann, E 'Chemical and histological studies on wool', *Melliand. Textilber.*, 1941, **22**, 145.
31. Lindberg J, 'Allwörden's reaction', *Textile Res. J.*, 1949, **19**, 43–45.
32. Leeder J D and Bradbury J H, 'Conformation of epicuticle on keratin fibers', *Nature*, 1968, **218**, 694–695.
33. Allwörden K Von, 'The action of chlorine water on wool', *Z. angew. Chem.*, 1916, **29**, 77.
34. Lindberg J, Mercer E H, Philip B and Gralén N, 'The fine histology of the keratin fibers', *Textile Res. J.*, 1949, **19**, 673–678.
35. Leeder J D and Bradbury J H, 'The discontinuous nature of the epicuticle on the surface of keratin fibers', *Textile Res. J.*, 1971, **41**, 563–568.
36. Bradbury J H and Leeder J D, 'Keratin fibers: Mechanism of the Allwörden reaction', *Aust. J. Biol. Sci.*, 1972, **25**, 133–138.
37. Peters D E and Bradbury J H, 'The chemical composition of wool: The cell membrane complex', *Aust. J. Biol. Sci.*, 1976, **29**, 43–55.
38. Lindberg J, Philip B and Gralén N, 'Occurrence of thin membranes in the structure of wool', *Nature*, 1948, **162**, 458–459.
39. Stirm K and Colle H, 'On the mechanism of the Allwörden reaction', *Melliand. Textil.*, 1935, **16,** 795–798.
40. Hock C W, Ramsay R C and Harris M, 'Microscopic structure of the wool fiber', *Amer. Dyestuff Reptr.*, 1941, **30**, 449–456 and 469–470.
41. Speakman J B, Nilssen B and Elliott G H, 'Reactivity of the sulphur linkage in wool', *Nature*, 1938, **142**, 1035.
42. Alexander P, Carter D and Earland C, 'The reaction of wool with organic chloramines', *J. Soc. Dyers and Colour.*, 1951, **67**, 17–22.
43. Alexander P, Carter D and Earland C, 'The rôle of the disulphide bond in reactions which render wool non-felting', *J. Soc. Dyers and Colour.*, 1951, **67**, 23–27.
44. Maclaren J A and Milligan B, *Wool Science-the Chemical Reactivity of the Wool Fibre*, NSW, Australia, Science Press, 1981.
45. Makinson K R, *Shrinkproofing of Wool*, New York, Marcel Dekker Inc., 1979.
46. Leeder J D, Rippon J A and Rivett D E, 'Modification of the surface properties of wool by treatment with anhydrous alkali', *Proc. 7th Int. Wool Text. Res. Conf.*, Tokyo, 1985, **IV**, 312–321.
47. Elliott R L and Manogue B, 'An electron microscope study of the surface structure of wool', *J. Soc. Dyers and Colour.*, 1952, **68**, 12–14.
48. Köpke V and Nilssen B, 'Wool surface properties and their influence on dye uptake', *J. Text. Inst.*, 1960, **51**, T1398–1413.
49. Leeder J D and Bradbury J H, 'Effects of shrinkproofing and other chemical treatments on the epicuticle of wool', *Textile Res. J.*, 1971, **41**, 215–218.
50. Evans D E, Leeder J D, Rippon J A and Rivett D E, 'Separation and analysis of the surface lipids of the wool fiber', *Proc. 7th Int. Wool Text. Res. Conf., Tokyo*, 1985, **I**, 135–142.

51. Wertz P W and Downing D T, 'Integral lipids of human hair', *Lipids*, 1988, **23**, 878–881.
52. Logan R I, Rivett D E, Tucker D J and Hudson A H F, 'Analysis of the intercellular and membrane lipids of wool and other animal fibers', *Textile Res. J.*, 1989, **59**, 109–113.
53. Ward R J, Willis H A, George G A, Guise G B, Denning R J, Evans D J and Short R D, 'Surface analysis of wool by X-ray photoelectron spectroscopy and static secondary ion mass spectrometry', *Textile Res. J.*, 1993, **63**, 362–368.
54. Peet D J, Wettenhall R E H and Rivett D E, 'The chemistry of the cuticle surface of keratin fibers', *Textile Res. J.*, 1995, **65**, 58–59.
55. Postle L J and Ingham J, 'The measurement of inter-fiber friction in slivers', *J. Text. Inst.*, 1952, **43**, T77–90.
56. Carter E G, 'Wool science and wool supremacy', *Proc 1st Int. Wool Text. Res. Conf., Australia,* 1955, **E**, 556–563.
57. King A T, 'Unscaled fibers. A new aspect of fiber research', *Biochem. J.*, 1927, **21**, 434–436.
58. Anon, 'Frictional properties of wool fibers', *Wool Sci. Rev.*, 1960, **18**, 38–50.
59. Makinson K R, 'Surface properties of wool fibers', in Schick M J, Ed., *Surface Characteristics of Fibers and Textiles*, New York, Marcel Dekker, 1975, 109–164.
60. Speakman J B and Stott E, 'A contribution to the theory of milling: A method of measuring the scaliness of wool', *J. Text. Inst*, 1931, **22**, T339–348.
61. Lindberg J and Gralén N, 'Frictional properties of wool fibers measured by the fiber-twist method', *Textile Res. J.,* 1948, **18**, 287–301.
62. Lipson M and Howard F J, 'Friction between keratin surfaces as affected by some shrink-proofing treatments', *J. Soc. Dyers and Colour.,* 1946, **62**, 29–32.
63. Mercer E H and Makinson K R, 'The frictional properties of wool and other textile fibers', *J. Text. Inst.*, 1947, **38**, T227–240.
64. Mercer E H, 'Frictional properties of wool fibers', *Nature*, 1945, **155**, 573–575.
65. King G, 'Some frictional properties of wool and nylon fibers', *J. Text. Inst.*, 1950, **41**, T135–144.
66. Lindberg J, 'Relationship between various surface properties of wool fibers: Frictional properties', *Textile Res. J.*, 1953, **23**, 225–236.
67. McPhee J R, 'Rate of felting of untreated and shrink-resistant wool fabrics', *Textile Res. J.*, 1961, **31**, 770–778.
68. Chaudri M A and Whiteley K J, 'Frictional and felting properties of wool fibers treated with benzoquinone', *J. Text. Inst.*, 1969, **60**, 37–45.
69. Makinson K R, 'The use of a diffraction grating as a rubbing surface in the study of the frictional properties of wool fibers', *Textile Res. J.*, 1967, **37**, 763–771.
70. Bohm L, 'The frictional properties of wool fibers in relation to felting', *J. Soc. Dyers and Colour.,* 1945, **61**, 278–283.
71. Monge M, *Annales de Chemie*, 1790, **6**, 300–311 (In French).
72. Shorter S A, 'The moisture content of wool – its relation to scientific theory and commercial practice', *J. Soc. Dyers and Colour.*, 1923, **39**, 270–276.
73. Alexander P and Hudson R F, *Wool – its Chemistry and Physics*, Second Edition, Edited by Earland C, London, Chapman and Hall, 1963.
74. Makinson K R, 'Some observations on the effect of mild shrink-proofing treatments on wool fibers', *Textile Res. J.*, 1968, **38**, 831–842.
75. Anon, 'The role of the scales of wool fibers', *Wool Sci. Rev.*, 1972, **42**, 2–16.

76. Lincoln B, 'The frictional properties of the wool fiber', *J. Text. Inst.*, 1954, **45**, T92–107.
77. Makinson K R, 'Mechanisms involved in shrinkproofing by degradative treatments', *Proc 4th Int. Wool Text. Res. Conf., San Francisco,* 1970, 1083–1096.
78. Makinson K R, 'Effect of $KMnO_4$/ salt shrink-proofing treatment on the friction properties of wool fibers: Dependence on the nature of the rubbing surface', *Textile Res. J.*, 1969, **39**, 1072–1074.
79. Flanagan G F, 'Felting and ratchet action of wool fibers', *Textile Res. J.*, 1966, **36**, 55–65.
80. Grosberg P, 'The differential friction of wool', *J. Text. Inst.*, 1955, **46**, T233–246.
81. Martin A J P, 'Observations on the theory of felting', *J. Soc. Dyers and Colour.,* 1944, **44**, 325–327.
82. Thomson H M S and Speakman J B, 'Frictional properties of wool', *Nature*, 1946, **157**, 804.
83. Moncrieff R W, *Wool Shrinkage and its Prevention*, London, National Trades Press, 1953.
84. Nason R, 'Fulling – past and present', *Amer. Dyestuff Reptr.*, 1965, **54**, 1008–1012.
85. Marsh J T, *Introduction to Textile Finishing,* London, Chapman and Hall, 1966.
86. Shorter S A, 'Observations on the theory of felting' *J. Soc. Dyers and Colour.,*1945, **61**, 172–173.
87. Speakman J B, Stott E and Chang H, 'A contribution to the theory of milling', *J. Text. Inst.*, 1933, **24**, T273–292.
88. Martin A J P, 'Observations on the theory of felting', *J. Soc. Dyers and Colour.*, 1945, **61**, 173–174.
89. Makinson K R, 'Studies on the movement of wool fibers in fabrics during felting; Quantitative relation between duration of milling, shrinkage and fiber movement', *Textile Res. J.*, 1962, **32**, 364–374.
90. Makinson K R, 'Heavier felting in a worsted fabric', *Textile Res. J.*, 1959, **29**, 439–443.
91. Speakman J B and Hirst M C, 'The constitution of the keratin molecule', *Trans. Faraday Soc.*, 1933, **29**, 148–172.
92. Whiteley K J and Speakman J B, 'The plasticity of wool: Assessment of quality in wool', *Textile Res. J.,* 1960, **30**, 46–50.
93. Whiteley K J, 'The stress-strain properties of felting wools', *J. Text. Inst.*, 1963, **54**, T1–8.
94. Liu W T, Speakman J B and King P E, 'The influence of dyes and dyeing processes on the milling, strength and weaving properties of wool fabrics', *J. Soc. Dyers and Colour.*, 1939, **55**, 183–192.
95. Lipson M and Speakman J B, 'The formation of polymers in wool', *J. Soc. Dyers and Colour.,* 1949, **65**, 390–401.
96. Weigmann H-D, Rebenfeld L and Dansizer C, 'A transition temperature in wool fibers under stress in relation to structure', *Textile Res. J.*, 1965, **35**, 604–611.
97. Mitchell T W and Feughelman M, 'The mechanical properties of wool fibers in water at temperatures above 100°C' *Textile Res. J.*, 1967, **37**, 660–668.
98. Feldtman H D and McPhee J R, 'Effect of temperature on the felting of shrink-resistant wool', *Textile Res. J.*, 1964, **34**, 199–206.
99. Schofield J, 'Researches in wool felting', *J. Text. Inst.*, 1942, **33**, T181–194.
100. Bogaty H, Frishman D, Sookne A M and Harris M, 'A simple quantitative test for the felting of wool top', *Textile Res. J.*, 1950, **20**, 270–276.

101. Chaudri M A and Whiteley K J, 'Relationship between loose wool felting and bulk compression properties', *Textile Res. J.*, 1970, **40**, 775–779.
102. Anon, 'Shrink-resist processes for wool: Factors that affect felting shrinkage', *Wool Science Rev.*, 1957, **17**, 16–32.
103. Brown A E and Krasny, J F 'Development of wash and wear wool fabrics for modern home laundering', *J. Text. Inst.*, 1960, **51**, T1414–1437.
104. Feldtman H D and McPhee J R, 'Machine washing and tumble drying of wool fabrics', *Amer. Dyestuff Reptr,* 1964, **53**, 199–209.
105. TM 31 – 'Washing of Wool Textile Products; relaxation and felting shrinkage', The International Wool Secretariat, 1996. TM 254 – 'Tumble Drying Performance after Washing.' The International Wool Secretariat 1997.
106. Speakman J B and Whewell E, 'The use of abrasives to make wool unshrinkable', *J. Text. Inst.*, 1945, **36**, T48–56.
107. Hojo H, 'Improvement of wool fibers by removing exocuticle in the presence of metallic ions', *Proc. 7th Int. Wool Text. Res. Conf.*, Tokyo, 1985, **IV**, 322–331.
108. Levene R and Shakkour G, 'Wool fibers of enhanced lustre obtained by enzymatic descaling', *J. Soc. Dyers and Colour.*, 1995, **111**, 352–359.
109. McPhee J R, 'Reaction of wool with oxidizing agents in concentrated salt solutions', *Textile Res. J.*, 1960, **30**, 349–357.
110. Bradbury J H, 'The theory of shrinkproofing of wool. Part II: Chemical modification of the fiber surface and its effect on felting shrinkage, friction and microscopic appearance', *Textile Res. J.*, 1961, **31**, 735–743.
111. Lindberg J and Gralén N, 'Measurement of friction between single fibers. Part IV: Influence of various oxidizing and reducing agents on the frictional properties of wool fibers', *Textile Res. J.*, 1949, **19**, 183–201.
112. Thorsen W J, Ward W H and Millard M M, 'Wool shrinkage control and surface modification by ozone', *J. Appl. Polymer Sci.*, 1979, **24,** 523–546.
113. Rakowski W, 'Plasma treatment of wool – from lab curiosity to industrial reality', *Proc. 9th Int. Wool Text. Res. Conf., Biella*, 1995, **IV**, 359–368.
114. Davidson A N and Preston R, 'Shrink-resisting wool', *Proc 1st Int. Wool Text. Res. Conf, Australia*, 1955, **E**, 473–488.
115. Trotman S R, 'The chlorination of wool', *J. Soc. Chem. Ind., Lond*, 1922, **41**, T219–224.
116. Wool Ind. Res. Assoc., Chlorination treatment of wool, BP 417719, 1933.
117. Andrews M W, Inglis A S and Williams V A, 'Chemical changes in the cuticle of oxidized wool', *Textile Res. J.*, 1966, **36**, 407–412.
118. Schumacher-Hamedat U, Laurini C and Schneider V, 'The chemistry and physics of the wool surface upon shrink-proofing', *Proc 8th Int. Wool Text. Res. Conf, Christchurch*, 1990, **IV**, 451–460.
119. Makinson K R, 'The role of chlorination in oxidative antifelting treatments of wool', *Textile Res. J.*, 1974, **44**, 856–857.
120. Negri A P, Cornell H J and Rivett D E, 'Effects of processing on the bound and free fatty acid levels in wool', *Textile Res. J.*, 1992, **62**, 381–387.
121. McPhee J R, 'Shrinkproofing of wool with neutral permanganate or acid bromate in concentrated sodium chloride solution', *Textile Res. J.*, 1960, **30**, 358–365.
122. Frishman D and Harris M, 'The nature of the modification of wool by various chlorination processes and their effect on the wear life of shrink-resistant materials' *Amer. Dyestuff Reptr.*, 1954, **43**, 174–179.
123. Williams V A, 'Wool shrink-proofing studies: influence of inorganic salts on the

reaction of potassium permanganate with wool in neutral solutions', *Textile Res. J.*, 1962, **32**, 977–985.
124. McPhee J R, 'Shrinkproofing of wool with peracetic acid followed by sodium sulfite', *Textile Res. J.*, 1963, **33**, 755–757.
125. Sweetman B J and Maclaren J A, 'Chemical reactions involved in the shrinkproofing of wool by oxidation-sulfite processes', *Textile Res. J.*, 1965, **35**, 315–322.
126. Douthwaite F J, Lewis D M and Schumacher-Hamedat U, 'Reaction of cystine residues in wool with peroxy compounds', *Textile Res. J.*, 1993, **63**, 177–183.
127. Douthwaite F J and Lewis D M, 'The formation of cysteine-S-sulphonate groups in wool and the effect on shrink-resistance', *J. Soc. Dyers and Colour.*, 1994, **110**, 304–307.
128. Byrne K M, Ryder A, Jackson J and Lavocah W, 'Non-chlorine shrink-resist treatments for wool top', *Proc 9th Int. Wool Text. Res. Conf, Biella,* 1995, **I**, 234–243.
129. Alexander P, Carter D and Hudson R F, 'The treatment of wool with acid permanganate solutions', *J. Soc. Dyers and Colour.*, 1949, **65**, 152–161.
130. Alexander P, Fox M and Hudson R F, 'The reaction of oxidizing agents with wool', *Biochem J.*, 1951, **49**, 129–138.
131. Freney M R and Lipson M, 'Effect of concentrated aqueous and certain non-aqueous solutions of alkali upon wool', *Nature*, 1940, **145**, 25–26.
132. Lipson M, 'Unshrinkable wools produced by alcoholic alkali', *J. Text. Inst.*, 1947, **38**, 279–285.
133. Gralén N J, 'The cuticle of wool', *J. Soc. Dyers and Colour*, 1950, **66**, 465–469.
134. Kassenbeck P, 'Electric discharge treatment and its application to wool', *Bull Inst., Text. de France*, 1964, **18**, 7–33 (In French).
135. Thorsen W J and Kodani R Y, 'A corona discharge method of producing shrink-resistant wool and mohair', *Textile Res. J.*, 1966, **36**, 651–661.
136. Lee K and Pavlath A E, 'Low temperature plasma treatment of wool: Effects of variables on shrinkage and yarn strength', *J. Polymer Sci., Chem. Ed.*, 1974, **12**, 2087–2090.
137. Thorsen W J, 'Textile modification in a corona field', *Proc 4th Int. Wool Text. Res. Conf., San Francisco*, 1970, 1171–1175.
138. Pavlath A E and Lee K S, 'Wool modification by low temperature glow discharge', *Proc. 5th Int. Wool Text. Res. Conf., Aachen*, 1975, **III**, 263–274.
139. Hesse A, Thomas H and Höcker H, 'Zero-AOX shrink-proofing treatment for wool top and fabric', *Textile Res. J.*, 1995, **65**, 355–361.
140. Pavlath A E and Slater R F, 'Low temperature plasma chemistry: shrink-proofing of wool', *Proc 4th Int. Wool Text. Res. Conf., San Francisco*, 1970, 1317–1324.
141. Millard M M, 'Surface analysis of plasma-treated wool fibers by X-ray photoelectron spectrometry', *Proc. 5th Int. Wool Text. Res. Conf., Aachen*, 1975, **II**, 44–53.
142. Höcker H, Thomas H, Küsters A, and Herrling J, 'Dyeing of plasma treated wool', *Melliand. Textilber,* 1994, **6**, E131–133.
143. Klausen T, Thomas H and Höcker H, 'Influence of oxygen plasma treatment on the chemical and morphological changes of the wool fiber surface', *Proc 9th Int. Wool Text. Res. Conf., Biella*, 1995, **II**, 241–248.
144. Körner A, Schmidt H, Merten T, Peters S, Thomas H and Höcker H, 'Changes in the content of 18-methyleicosanoic acid in wool after UV-irradiation and corona treatment', *Proc 9th Int. Wool Text. Res. Conf., Biella,* 1995, **II**, 414–419.
145. Merten T, Thomas H and Höcker H, 'Comparative investigations on the shrinkage

behavior of corona and glow discharge-treated wool top', *Proc 9th Int. Wool Text. Res. Conf., Biella,* 1995, **IV**, 386–393.
146. Höcker H, 'Plasma treatment of textile fibers', *Pure Appl. Chem.*, 2002, **74**, 423–427.
147. Anon, 'Shrink-resist processes for wool: commercial methods', *Wool Sci. Rev.*, 1960, **18**, 18–37.
148. Anon, 'Shrink-resist treatments based on DCCA', *Wool Sci. Rev.*, 1968, **34**, 1–9.
149. Cockett K R F, 'Production of superwash knitwear by batch processing routes', *Wool Sci. Rev.*, 1979, **56**, 2–44.
150. Byrne K M, 'Machine-washable knitwear – production routes', in Carr C M, Ed., *Chemistry of the Textiles Industry*, London, Blackie Academic and Professional, 1995, 186–209.
151. Connell D L, 'Wool finishes: the control of shrinkage', in Heywood D, Ed., *Textile finishing*, Bradford UK, The Society of Dyers and Colourists, 2003, 372–397.
152. Byrne K M, Smith P, Jackson J and Lewis J, 'Kroy offers control and versatility', *Textile Horizons*, 1984, **4**, 25–28.
153. Anon, 'The application of polymers to wool for the prevention of felting shrinkage', *Wool Sci. Rev.*, 1969, **36**, 2–11.
154. Lewis J, 'A review of Superwash technology', *Wool Sci. Rev.*, 1977, **54**, 2–29.
155. Datyner A, *Surfactants in textile processing*, New York, Marcel Dekker, 1983.
156. Feldtman H D and McPhee J R, 'The spreading and adhesion of polymers on wool', *Textile Res. J.*, 1964, **34**, 634–642.
157. Kilpatrick D J, Rippon J A, Rushforth M A and Shaw T 'Shrink-resist polymers for wool: A comparative study', in Arthur J C (Ed.), *Textile and Paper Chemistry and Technology*, Washington DC, Amer. Chem. Soc., 1977, 176–196.
158. Rippon J A and Rushforth M A, 'The use of the water-soluble bisulphite addition product of a polyisocyanate', *Textiveredlung.*, 1976, **11**, 224–229.
159. Kershaw A and Lewis J, 'The role of polymer treatments in the shrink-proofing of wool', *Textile Month,* 1976, **April**, 40–44.
160. Makinson K R, 'Mechanism of shrink-proofing with silicone resins', *Textile Res. J.*, 1975, **45**, 217–219.
161. Earle R H Jr., Saunders R H and Kangas L R, The chemistry of the chlorination/Hercosett resin shrink-proofing process', *Proc. 4th Int. Wool Text. Res. Conf., San Francisco*, 1970, 707–714.
162. Bereck A, Byrne K M and Rushforth M A, 'Developments in batch shrink-resist treatments for wool knitwear', *Proc. 8th Int. Wool Text. Res. Conf.*, Christchurch, 1990, **4**, 431–440.
163. Byrne K M, Bourne M, Clark M and Duffield P A, 'Zero AOX garment dyeing', *Proc. 9th Int. Wool Text. Res. Conf., Biella*, 1995, **I**, 191–198.
164. Anderson C A, Wood G F and Goldsmith M T, 'The CSIRO chlorine/Hercosett process for shrink-proofing wool sliver', *Text. Manuf.*, 1969, **95**, 184–186.
165. Feldtman H D, McPhee J R and Morgan W V, 'Shrink-proofing wool tops', *Text. Manuf.*, 1967, **93**, 122–124.
166. Lewis J, 'Continuous shrink-proofing processes', *Wool Sci. Rev.*, 1978, **55**, 23–42.
167. Feldtman H D and McPhee J R, 'Application of synthetic polymers to wool', *Proc. 3rd Int. Wool Text. Res. Conf., Paris*, 1965, **III**, 345–357.
168. Anderson C A, Goldsmith M T, Katz H J and Wood G F, 'The roles of chlorine treatment and resin application in the CSIRO chlorine/resin shrink-proofing process', *Proc. 4th Int. Wool Text. Res. Conf., San Francisco,* 1970, 715–726.

169. Makinson K R and Lead J A, 'The nature and function of the resin in the chlorine/resin shrink-proofing treatment of wool tops', *Textile Res. J.*, 1973, **43**, 669–681.
170. De Fries W, Altenhofen V, Föhles J and Zahn H, 'A protein chemical investigation of the chlorine-Hercosett process', *J. Soc. Dyers and Colour.*, 1983, **98**, 13–16.
171. Müller B M, 'Adsorbable organic halogens in textile effluents', *Rev. Prog. Coloration,* 1992, **22**, 14–21.
172. Augustin R F, Crompton J A, Jackson J, Müller B M and Rushforth M A, 'The origins and control of AOX formed in wool shrink-proofing', *Proc. Aachen Textile Conf.*, 1991, **108**, 615–626.
173. Connell D L and Huddlestone K M, 'Method for the treatment of wool', US Patent 5,697,983, Dec 1997.
174. Haefely H, 'Enzyme treatment for wool', *Textilveredlung.*, 1989, **24**, No. 7, 271–276.
175. Hesse A, Thomas H and Höcker H, 'Zero-AOX Shrink-proofing treatment for wool top and fabric', *Textile Res. J.*, 1995, **65**, 371–378.
176. Jansen B, Kümmeler F, Thomas H and Müller H B, 'New resins for shrink-proofing plasma treated wool', *Proc 10th Int. Wool Text. Res. Conf., Aachen*, 2000, **PL–2**, 1–6.
177. Ichimura H, Umehara R, Karakawa T, Oshima K and Nakase K, 'Shrinkproof animal fiber and method for preparation thereof', Eur. Patent 1176245 A1, July 2001.

8
Friction and surface characteristics of synthetic fibers

Y E EL MOGAHZY, Auburn University, USA

8.1 Introduction

The very large number of inherently and categorically different synthetic fibers, used in a very wide range of apparel and industrial products, is a testament to the difficulty of selecting topics for inclusion in a chapter on friction and surface characteristics of synthetic fibers. To make matters additionally complex, a great deal of research in this area represents scattered activities made by different fiber producers that are typically inaccessible due to their proprietary nature [1, 2, 3].

In Chapter 6, we discussed many of the new methods used for measuring cotton fiber friction. Many of these methods were originally developed for testing synthetic fibers (e.g. line and point-contact method, and rotor-ring technique) and they will be referred to in this chapter. In addition, we shall focus on surface characterization, particularly in view of the advanced techniques that can be utilized to manipulate the surface texture of synthetic fibers for the purpose of performing specific end use functions.

Before proceeding with the discussion of friction and surface characteristics, it will be useful to first provide a brief review of the current technological status of synthetic fibers. The purpose will be to make readers aware of the magnitude of development and the design challenges that lie ahead.

8.2 The developments of synthetic fibers

In today's technology, synthetic fibers are produced in an unlimited array of bulk and surface characteristics covering a very wide range of applications from traditional apparel to high-tech industrial products. The first generation of synthetic fibers primarily consisted of fibers that are still available today, i.e. rayon, acetate, nylon, and polyester, introduced in the first half of the 20th century. The focus of development of this group was on producing fibers that could either replace natural fibers in some applications, or merge with natural fibers through blending in other applications. From an end product viewpoint, the first generation aimed largely at the traditional textile market (apparel and household products).

Toward the end of the 20th century, it became clear that there was a global shift in the textile market that made it difficult for industrial countries to compete in commodity and mass-oriented textile products. As a result, synthetic fibers have evolved to a new level of value-added applications competing with natural fibers, e.g. silk and ramie, or joining other non-fibrous materials (metal, wood, soils, etc.) for use in a variety of existing and exciting new products. These developments were stimulated by technological capabilities to produce synthetic fibers of high modulus, high tenacity, light weight, micro-fineness, engineered biodegradability, high thermal resistance, and high electrical conductivity. These fibers have proven to be very attractive in industrial applications, requiring products that are low weight, impact resistant, tear resistant, tough and heat resistant. As a result, countless varieties of polymeric and inorganic fibers have made their way to every conceivable field in human life.

The evolution of synthetic fibers for high-tech industrial use was primarily driven by the need to increase the tenacity and the stiffness for industrial applications [4, 5, 6]. The early generation of fibers exhibited tenacities in the range 3 to 6 g_f/den and modulus in the range 10 to 30 g_f/den. These were suitable for most traditional textile applications. The next generation exhibited values in the ranges 6 to 10 g_f/den for tenacity and 30 to 100 g_f/den for modulus. As demands for new industrial applications such as aerospace, automotive, and composites increased, high performance fibers emerged that have values in the ranges 20 to 40 g_f/den (or 3 to 6 GPa) for tenacity and 400 to over 1000 g_f/den (or 50 to 600 GPa) for modulus.

In addition to the above categories of fibers, a newer generation of fibers described as "smart fibers" have appeared. These are fibers that can sense, interpret and react to dynamically changing environmental conditions or external stimuli including mechanical, thermal, chemical, magnetic or electrical signals [7]. Some smart fibers can function as conductive "wires" and react to signals from electricity, heat or pressure. Others are made of different shapes (oval, square, or triangular) that can be made to contract or expand to loosen and tighten clothing to make the wearer warmer or cooler. Some conductive fibers could change color on command from an electric signal that changes the reflective quality of specially dyed fiber/cloth. Many smart fibers are used as sensors to detect chemical, biological, and toxic substances.

Today, a large number of smart fibers are available commercially. For example, DuPont produces conductive fiber Aracon® that is a type of Kevlar® fiber, which is stronger than steel, more flexible and lighter than copper, and good conductor of electricity. It consists of para aramid structure clad with a variety of metals to yield strong, light weight composite conducting and shielding material that has the strength of Kevlar® but the conductivity of metals. DuPont also produces sleepers that can monitor an infant's vital signs, and garments that feature a weave that, with the press of a button, can open or close to allow greater or lesser airflow, increasing comfort during changes in weather or time of day/night. France's Telecom R&D has designed a prototype for a flexible screen made of woven optical fibers that are capable of downloading and displaying static or animated graphics (such as logos, texts, patterns, scanned images, etc.) directly on clothes. In other words, these are the clothes that can act as a graphical communication interface, displaying visual information in real time, and offering access to all telecom services (Internet, video, e-commerce, and 3G mobiles). Some polyurethane-diacetylene copolymers can be used as thermochromic material for temperature sensor application [8].

In the context of friction and surface characteristics, revolutionary developments have been made through utilization of the friction phenomenon for creating special adhesion effects and through modification of surface texture for achieving special thermal and moisture transfer characteristics. In this chapter, many of these developments will be reviewed.

8.3 Surface characterization of synthetic fibers

Traditionally, the interest in fiber surface has been oriented toward fiber processing performance. In other words, the goal in friction study has been mainly to optimize the fiber surface so that efficient processing and compatible blending with other fiber types could be achieved. Obviously, these objectives will continue to remain as an essential aspect of fiber design. However, the recent developments of synthetic fibers have been aimed at reaching higher levels of functional performance. These levels could only be met through optimization of all characteristics of a fiber including surface.

The increasing trend of incorporating fibrous structures with other types of materials in many industrial products clearly calls for better understanding of the surface characteristics of fibers. These characteristics represent critical design parameters in many products such as the geotextiles, medical textiles, and automotive textiles where fibers come into contact with many materials and under a range of external conditions.

In the following sections, we will discuss the various parameters of fiber surface characterization, summarized in Fig. 8.1, and the roles they play in influencing friction. These include surface area, cross-sectional shape, surface roughness, fiber crimp, and molecular orientation. In addition, we will review some of the surface-related parameters describing the performance of fibers in contact applications and in different media. These include surface wetting, surface finish, and lubrication.

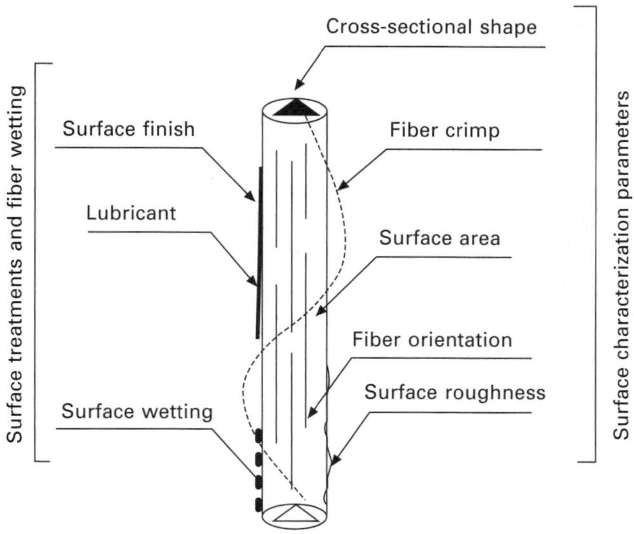

8.1 Surface characterization parameters and surface treatments of synthetic fibers.

8.3.1 Surface area

Surface area represents a key design criterion for synthetic fibers. In practice, fiber diameter or fiber denier is typically used to express surface area on the basis of its definition πdl. Another way to express it is to use the specific surface, which is defined by the surface area per volume $2/r$ or per mass $2/r\sigma$, where r is the radius and ρ is the density. Most commercial fibers have specific surface of about 0.1 m^2/g.

In recent years, surface area has received a great deal of attention after the introduction of polyester and nylon microdenier fibers (0.9 denier, or finer). These fibers exhibit much smaller diameters than the traditional fibers. The surface area of a given mass of microdenier fibers is also very large as compared to the same for regular fibers. These attributes make them suitable for a variety of end products including high-fashion flexible fabrics (with high drape, wickability, and breathability), tents, sleeping bags, track and jogging suits, and industrial filters. The very small diameter and very large surface-to-volume ratio of microdenier fibers imposes a great difficulty in processing, particularly during carding [9, 10].

Currently, increased attention is paid to developing nano or super microdenier fibers. Recently, the Japanese Toray Industries Inc. announced that it has developed a way to synthesize fibers that are mere tens of nanometers in diameter. By using this technology, the company has bundled more than 1.4 million fibers together to develop a nylon textile with a surface area more than 1,000 times greater than regular nylon. This new nylon textile excels at absorbing toxic odors, and it has double to triple the moisture-absorbing ability of regular nylon; in this regard, the competing fiber in mind should be cotton.

Surface area is also a factor that affects flame retardancy. In general, items that have a large surface area for a small amount of material tend to ignite more easily and burn faster; a twig (stick) is easier to ignite than a log. Accordingly, high surface area is considered as a key negative factor in flame-resistant fabrics, particularly if fabrics have raised surface fibers.

In the context of fiber friction, some studies examined the effect of surface area on friction for the purpose of validating the classical laws of friction; i.e. the law that friction force is independent of the apparent area of contact. In an early study, Schick [11] evaluated the effect of surface area (or denier) of polyester and nylon filaments on fiber-to-metal friction. Measurements were made using a Rothschild friction meter. This device operates on the basis of the capstan principle, $T_1/T_o = e^{\mu\theta}$, to measure friction at medium and high speed. Two levels of fiber denier were examined for each fiber type (see Fig. 8.2). Polyester and nylon monofilaments had 1% wof finish on their surfaces.

Figure 8.2 clearly indicates that an increase in surface area or filament diameter results in an increase in frictional force over the range of sliding speeds used. The increase in fiber friction was more or less consistent over the range of speeds used for each fiber type. The average percent increase in

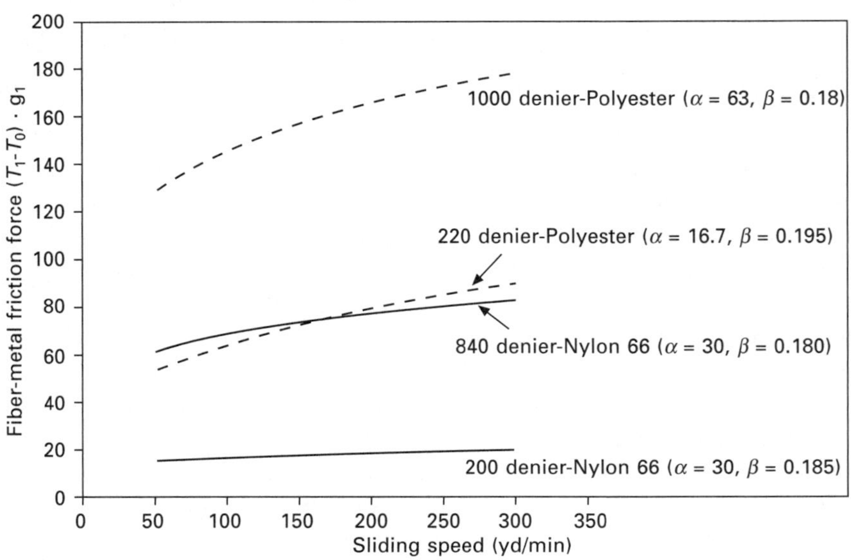

8.2 Effects of fiber denier on fiber/metal friction – T_0 = 0.05g/denier, θ = 360°, chrome capstan pin RMS-60, 1.27 cm diameter (modified after Schick [11]).

friction with denier was higher for nylon fibers (313%) than for polyester fibers (150%).

8.3.2 Fiber cross section

The cross section of a synthetic fiber represents another key design criterion that can also influence fiber processing and functional performance. Most synthetic fibers are made of circular cross sections but different cross sectional shapes can be obtained to have special appearance effects or special performance characteristics. Table 8.1 shows examples of common cross sectional shapes that are commercially available.

In recent years, design engineers of synthetic fibers have taken a leap in realizing and utilizing fiber cross sectional shape as a powerful functional parameter. For example, the C-slit cross sectional shape (Fig. 8.3(a)) was developed to entrap air for thermal insulation, while simultaneously improving elastic behavior [12]. The C-shaped sheath originally has an alkali-soluble polymer core reaching the external surface through a narrow longitudinal slit. After drawing, the filaments can be textured by false twisting or commingling. The core is then removed using alkali finishing, giving a hollow C-shaped cross section with a longitudinal slit. The combination of texturing and cross sectional shape provides void fraction exceeding 30% and a springy feeling. Different fibers including polyester or polyamide based can be used. In another derivative, the sheath can be made of a blend of polyester and hydrophilic polymer [13]. In this case, alkali finishing introduces microcrazes and pores throughout the sheath, which allows liquid sweat absorption and transportation from the skin to the hollow core. This feature allows dryness rather than thermal insulation.

Table 8.1 Examples of common fiber cross sectional shapes

Cross sectional shape	Special Features
(a) Circular: ●	used in most synthetic fibers, reference for other cross sectional shapes (a circular shape factor is one), it has a low surface-to-volume ratio
(b) Hollow: ○	lower density at the same diameter, higher bending resistance, entraps air to provide thermal insulation, light scatter by internal surfaces leading to soil hiding and translucency characteristics
(c) Trilobal:	often used in carpets to provide higher bending stiffness and soil hiding characteristics
(d) Ribbon: ⬬	larger flat surface for sparkling appearance, directional bending characteristics

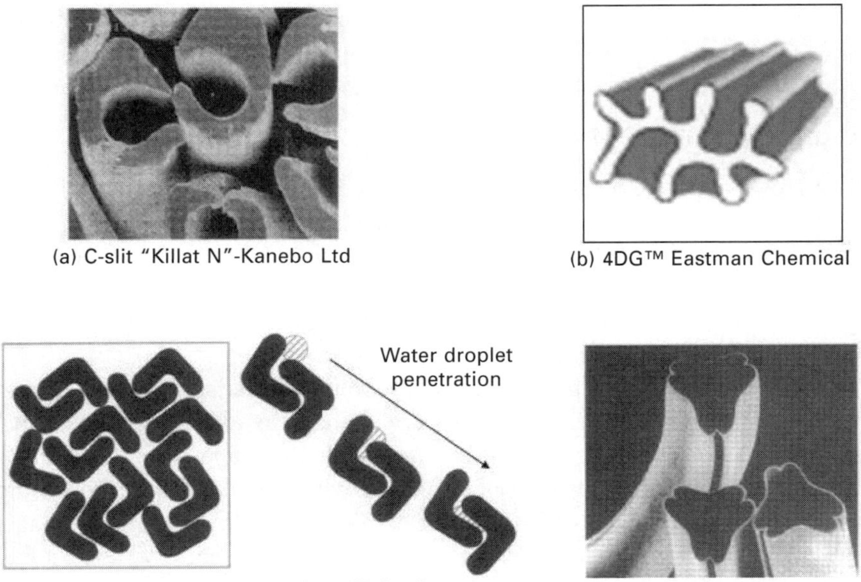

(a) C-slit "Killat N"-Kanebo Ltd

(b) 4DG™ Eastman Chemical

(c) L-shaped nylon cross section "Cleber"–Asahi Chemical Industry Co., Ltd

(d) Trilobal-slit "Silklook Royal S"-Torary Ind.

8.3 Specialty cross-sectional shapes [12–17].

Another cross section introduced during the past decade is the 4DG™ fibers (Fig. 8.3(b)), produced by Eastman Chemical. This fiber has a special, highly modified cross section that provides several deep grooves that run along the length of the fiber [14, 15]. The expanded surface area of this fiber provides about three times the amount of specific surface per denier compared to circular fibers. The geometry of cross section can allow transporting of up to two liters of water per hour per gram of fiber (high capillary wicking). The grooves in a 4DG™ fiber are also good for trapping particles in an air or liquid stream. These grooves provide areas where eddy currents will preferentially deposit particles, and where particles can collect without blocking pores in the fabric. These fibers provide increased filtration efficiency without increase in pressure drop across the fabric. The air entrapping also provides increased thermal insulation.

Figure 8.3(c) shows another interesting cross sectional shape used to produce close packing in conjunction with surface water wettability or wicking effect. The L-shaped cross section creates capillary forces in the inter-fiber volume that are claimed to be sufficient to wick away liquid sweat. These fibers are also claimed to eliminate the need for wettability surface treatment [16].

Another specialty cross section that was actually developed on the basis of frictional concepts is that of the so-called "Silklook Royal S™" polyester introduced by Japanese fiber producers (see Fig. 8.3(d)). The fiber has a trilobal shape with a longitudinal microgroove at the apex of each lobe. It is intended primarily to impart the unique scroop and rustle sounds that are often felt in silk fabrics [16–19]. This configuration acts like a tuning fork when rubbed by an adjacent fiber. The sounds produced are a result of a unique stick-slip pattern in which the difference between static and kinetic friction is largely a function of the microgroove width and depth. This unique cross sectional shape also simulated silk-like luster, depth of shade, and hand.

In light of the above discussion, it follows that most developments of fiber cross sections have aimed at departing from the classic circular cross section to introduce special appearance effects or specific functional performances. In the context of friction, cross sectional shapes deviating from circularity represent surface protuberances or projections that are an order of magnitude (several microns) greater than those of microscopic asperities (fraction of micron). As a result, noncircular cross sections are expected to generate unique stick-slip patterns during the friction process. These patterns can only be obtained when point or line-contact friction modes are considered. When multiple-fiber contacts occur, a key factor that can influence the frictional behavior of synthetic fibers is their potential for inter-fiber contact. Figure 8.4 shows crude mathematical estimates of this

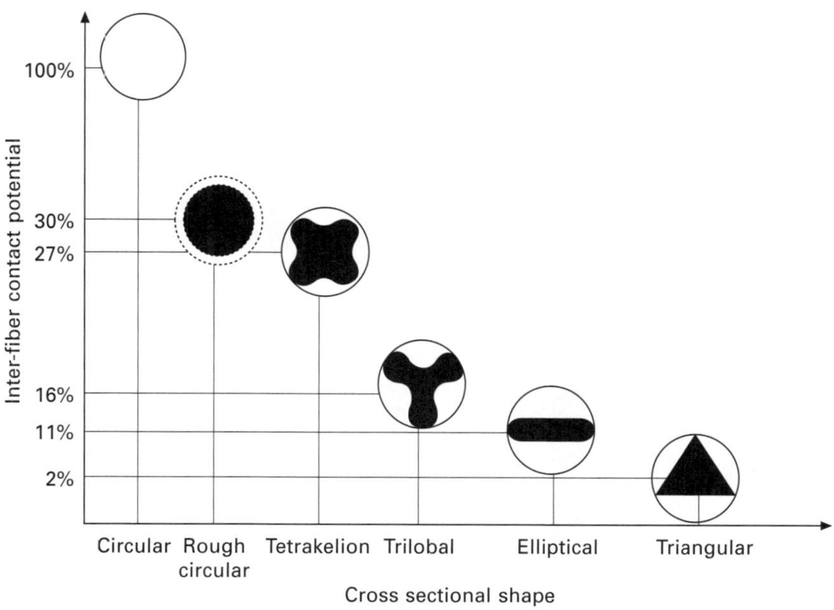

8.4 Estimated potential inter-fiber contact for different cross sectional shapes (based on Scardino estimates [20]).

potential with respect to the circular shape, which is assumed to exhibit a 100% potential contact [20]. It may be pointed out, however, that cross sectional shapes are often distorted after spinning as a result of fiber reactions to deformation forces. Such distortions also lead to other geometrical disturbances such as twisting, spiraling, or curling around their own axes.

In a study by El Mogahzy [21], the effect of cross sectional shape of polypropylene fibers on inter-fiber friction was examined. Polypropylene monofilaments of 18 denier were specially made for this study by Hercules Inc. The monofilaments had no finish and they were made of three different shapes (circular, triangular, and trilobal). Point and line-contact friction methods discussed in Chapter 6 were used.

Figure 8.5 gives the results which show that the coefficient of friction of non-circular fibers was lower than that of the circular. In the point contact test, the coefficient of friction of the circular fiber was about 24% higher than the trilobal fiber and 29% higher than the triangular fiber. In the line contact test, the corresponding differences were 18% and 25%, respectively. The friction constants a and n determined from the empirical equation $F = aN^n$, were also determined for the circular and the trilobal materials. The values of "a" were 0.309 and 0.242, for the circular and the trilobal shapes, respectively. The corresponding values of n were 0.889 and 0.858.

The effect of cross sectional shape on μ was explained on the grounds that noncircular cross sectional shaped fibers had surface protuberances or projections that were inherently stiffer, and thus, more resistant to compression than the bulk material of the fiber or the fiber with a circular cross section.

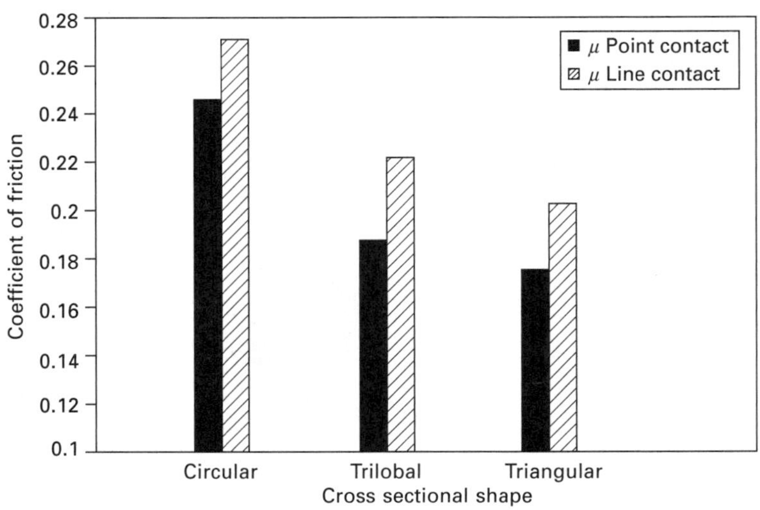

8.5 Coefficient of friction of different cross sectional shapes. Polypropylene monofilament (18 denier), T_0/r = 160g$_f$, sliding speed = 0.5 inch/min (modified after El Mogahzy [21]).

Some research evidences indicate that this is due to the molecules in the lobes of triangular or trilobal fibers being more highly oriented and the structure in these regions being more highly crystalline. This should, at least partially, lead to a relatively lower contact area and therefore to lower friction. Additionally, it is expected that the radius of curvature of the contact region in the noncircular fibers is lower than that in circular fibers [22, 23]. This should also lead to a lower coefficient of friction in noncircular fibers.

8.3.3 Fiber crimp

Fiber crimp is a general term reflecting departure of the longitudinal shape from the normal straight form. The familiar process of texturing can result in a highly crimped filament. When filaments are converted into staple fibers, they are typically crimped using a compacting or gear box process prior to cutting. The crimped configuration of fibers largely simulates the crimp found in natural fibers, which is a necessity for easing fiber manipulation during opening and carding. In general, crimp allows the fiber clusters during processing to conform to the different mechanical components, making them easier to hold together and to separate. Intermediate fiber strands (fiber webs, carded or drawn slivers) are partially held together by the effect of fiber crimp. Furthermore, fiber crimp contributes significantly to fiber resiliency without which fibers are likely to break during processing.

According to Scardino [20], the planar saw-tooth crimp geometry common in synthetic staple is easily detected by visual inspection, especially in unopened stock. What may not be apparent, however, are the prominent nodules and related surface deformations resulting from a typical crimping process. These projections can be easily observed under low power microscope by straightening (unbending) a crimped fiber under low tension. Under such low tension, these crimp nodules, or deformations, tend to project out of the line of the fiber axis for a distance equivalent, at least, to the fiber diameter. Under greater tension (sufficient for removal of crimp), the fiber becomes quite linear and the projections of the crimp nodules are substantially less prominent. The frequency of the nodules depends on fiber crimp frequency. For example, in a synthetic fiber with 5 crimps per cm, one would expect to find a crimp nodule approximately every millimeter along the fiber length.

Earlier in our discussion, we indicated that cross sectional shapes deviating from circularity represent surface protuberances or projections that are an order of magnitude (several microns) greater than those of microscopic asperities or surface roughness (fraction of micron). Fiber crimp typically produces projections an order of magnitude greater than that of fiber cross sectional shape. As a result, crimp is expected to have a relatively more significant effect on fiber friction. Ideally, these points can only be illustrated using single filament friction analysis in which different levels of crimp and different cross sectional shapes are evaluated in a controlled design experiment. Such analysis is not available in the literature; however, one is certainly worth considering by researchers in the field.

Analysis published by Scardino [20] illustrated the relative effects of cross sectional shape and fiber crimp. In it sliver cohesion or drafting force was measured to provide an index of inter-fiber friction at two different stages of processing, namely carding and pin-drafting. The limitation of this study is that different stages of processing are also associated with differences in other surface-related factors such as fiber alignment (a drawn sliver has better fiber alignment than card sliver), and the level of crimp (drafting will remove some of the fiber crimp). Nevertheless, the Scardino study provides useful guidelines.

Values of sliver cohesion (or drafting force) for card sliver and pin-drafted sliver are shown in Figs 8.6 and 8.7, respectively. The crimped fiber gave higher drafting force than did non-crimped fibers, and the circular shape led to higher values than did the trilobal shape. On a relative basis, the percentage increase in card sliver cohesion due to the effect of crimp is substantially higher than that due to the change in cross sectional shapes.

The results of Fig. 8.7 reveal similar trends to those of Fig. 8.6 but with a substantially greater effect of the cross sectional shape, particularly when un-crimped fibers are used. The effect of crimp was also increased as a result of an increase in contacts between the fibers in the pin-drafted sliver. Comparing the results of the carded with those of the pin-drafted slivers, it is seen that the former has higher drafting force. This must be due to greater amount of crimp in and entanglements among fibers in the former than in the latter. The results also reveal the critical importance of crimp in maintaining the integrity

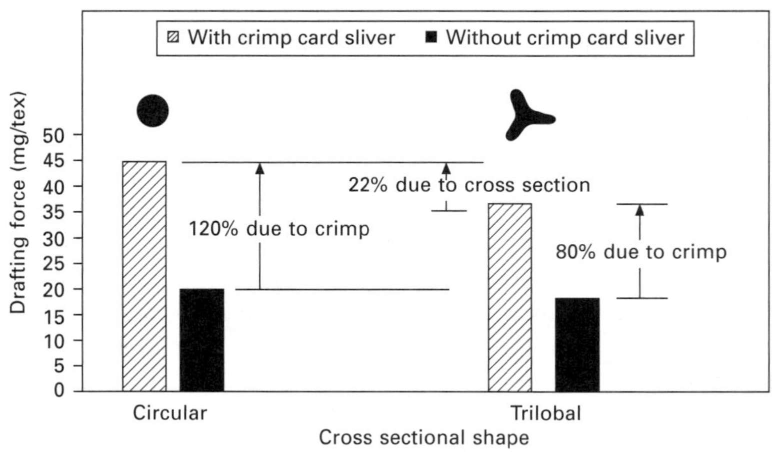

8.6 Effects of cross sectional shape and crimp on drafting force of Polyester card sliver (based on data of Scardino [20]).

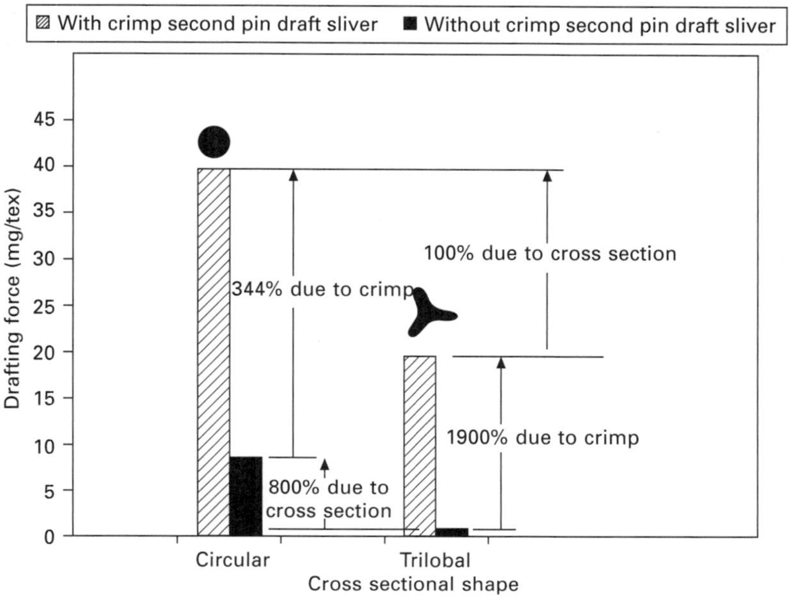

8.7 Effects of cross sectional shape and crimp on drafting force of Polyester second pin-draft sliver (based on data of Scardino [20]).

of the drafted sliver; a drafted sliver with only 1 mg_f/tex strength is likely to fall apart when it is being transported to the next stage of processing.

8.3.4 Fiber orientation

The spinning process used in producing synthetic fibers results in some molecular orientation. However, further orientation is typically achieved by drawing and annealing of filaments after spinning. This is a process that pulls the long polymer chains into alignment along the longitudinal axis of the fiber, which results in a closed packed structure and high inter-molecular cohesion.

As indicated in Chapter 6, fiber molecular orientation is commonly examined using the birefringence index, $\Delta n = n_{\Pi} - n_{\perp}$, in which n_{Π} is the refractive index for light oscillating in a plane parallel to the fiber axis, and $n_{\perp}$ is the refractive index for light oscillating in a plane perpendicular to the fiber axis. Most common textile fibers have a refractive index n_{Π} ranging from 1.5 to more than 1.7, the highest belonging to PET fibers, which have the highest birefringence. It may be noted that birefringence depends on both the degree of symmetry and the degree of orientation of molecules. The former is determined by the polymer type and the latter by the extent of molecular alignment [24]. As will be discussed shortly, there are other techniques for estimating molecular orientation.

In the context of friction, fiber orientation is expected to have a direct effect on inter-fiber friction. El Mogahzy and Gupta [21, 23] evaluated the effect of molecular orientation on the friction of polypropylene yarns. Molecular orientation was varied by the classic stretching procedure. As is usually the case, the stretch resulted in a decrease in diameter. Table 8.2 shows the values of yarn diameter, crystallinity and birefringence of polypropylene yarns at two levels of draw ratio.

The reduction in diameter (or denier) results in a reduction in the surface area and expected reduction in the coefficient of friction. This effect is likely to offset the effect of the increase in molecular orientation associated with drawing on the frictional behavior of fibers. In the analysis made in the above study, the change in the coefficient of friction associated with the change in fiber diameter was estimated by the following equation [21]:

$$\frac{\mu_1}{\mu_2} = \left(\frac{r_1}{r_2}\right)^{2(1-n)}$$

where n is the friction constant from the empirical equation, $F = a\,N^n$, r_1 and r_2 are the values of fiber radius before and after drawing. Note that when n equals 1, μ_1 will be equal μ_2, the classic law of friction); in other words, the change in fiber radius will have no effect on fiber friction. At values of n less than 1.0, an increase in fiber radius leads to an increase in μ.

The above relationship also indicates that the change in μ with fiber radius depends on the value of n and the ratio of fiber radii. In this study, the ratio r_1/r_2 was 0.84, and the average value of n was about 0.873. Accordingly,

$$\frac{\mu(2X)}{\mu(1X)} = \left(\frac{r(2X)}{r(1X)}\right)^{2(1-n)} = (0.84)^{2(1-0.873)} = 0.96$$

This means that at a given normal force N, the coefficient of friction of yarn at draw ratio 2*X* should be about 4% lower than that of yarn at 1*X*, if the only difference was yarn diameter. This reduction was accounted for in a regression analysis relating the coefficient of friction to the initial tension-radius ratio, T_o/r. Figure 8.8(a) illustrates a summary of this analysis, which clearly shows that at any given value of T_o/r, the increase in molecular orientation resulted in an increase in fiber friction.

Table 8.2 Properties of polypropylene yarns at two levels of draw ratio [21]

Draw ratio	Total denier (yarn diameter, mm)	No. of filaments	DSC crystallinity (%)	Birefringence
1*X*	1390 (0.469)	675	53	19×10^{-3}
2*X*	824 (0.395)	675	58	23×10^{-3}

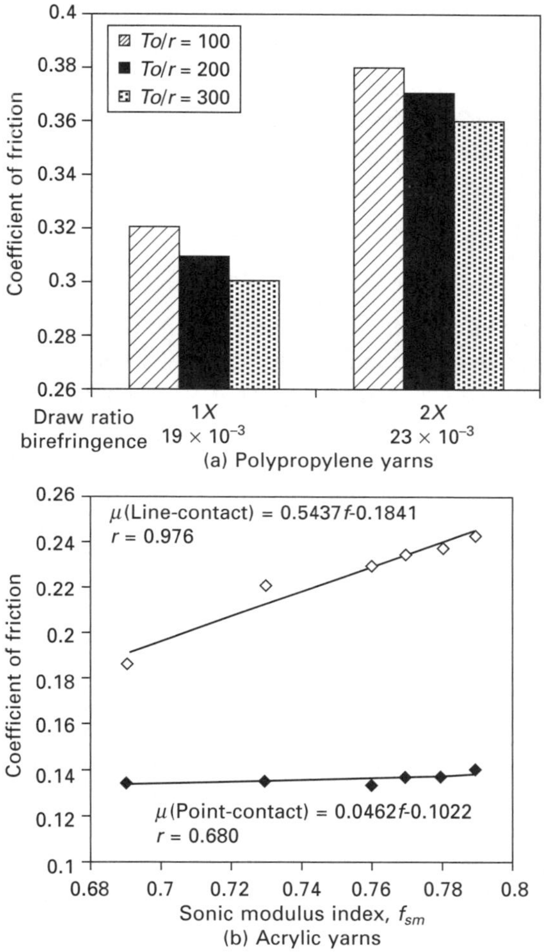

8.8 Effects of molecular orientation on fiber friction (based on data from El Mogahzy and Gupta [23]).

El Mogahzy and Gupta [21, 23] also examined the effect of molecular orientation of acrylic yarns on fiber friction. This experiment was different from the above in that a combination of jet stretch and cascade stretch was used to produce the fibers. The jet stretch did not significantly affect orientation but the cascade stretch did. This combination was unique in that it allowed changes in orientation without a change in the final denier of the filaments.

For acrylic fibers, molecular orientation was measured using the sonic technique in which the dynamic Young's modulus of fiber is determined [25, 26]. The sonic modulus, E_s, of a solid material is calculated from $E_s = \rho C^2$ [27], where ρ is the material density and C is the velocity of the sound wave. In principle, a sonic pulse is transmitted along the length of the fiber, and the flight time between transmitting and receiving transducers is measured. From knowledge of this time and sample length, the sonic velocity C can be computed. An increase in cascade stretch that should result in an increase in orientation gave an increase in sonic modulus.

From the values of sonic modulus an orientation index was calculated as follows [27]:

$$f_{sm} = 1 - \frac{E_0}{E}$$

where E is the sonic modulus of the fiber, and E_0 is the sonic modulus of an equivalent fiber having zero orientation. Morgan [28] found that this index agreed well with birefringence values. Figure 8.8(b) shows the values of coefficient of friction of acrylic yarns measured by point- and line-contact methods at different orientation levels given by the sonic modulus index, f_{sm}.

The results clearly indicate that when the line-contact method is used, a significant linear effect of fiber orientation on friction is observed. When the point-contact method is used, the effect still exists but with a weaker positive correlation coefficient.

8.3.5 Surface roughness

In all engineering fields, surface roughness of solids often receives high attention because of its impact on many performance characteristics, namely friction, abrasion, wear, appearance, water resistance, soil resistance, adhesion, and strength. In engineering applications, surface roughness, given by the ASME B46.1-2002 standard, is used as one of the three main descriptors of surface irregularity or texture. The other two are surface waviness and surface lay (see Fig. 8.9). The standard definition of roughness is the wavelength of the irregularities on a surface. This generally results from a particular production process or material condition. Waviness includes the more widely spaced (longer wavelength) deviations of a surface from its nominal shape. Waviness errors are intermediate in wavelength, i.e. between those of roughness and gross long deviations (typically called form error). Lay refers to the predominant direction of the surface texture, which is ordinarily determined by the particular production method and geometry used.

In relation to fibrous structures, the three engineering parameters should be examined independently, as each provides a different useful dimension of the overall surface characterization. Typically, surface roughness of fibrous structures may be presented in random form, as a result of dimensional variability, or in pre-designed form, as a result of some surface finish treatments that impose special roughness. Similarly, surface waviness can be a result of gross malfunctioning during manufacturing, of medium to long-term variation in fiber or yarn dimensional characteristics, of pre-designed form, or of

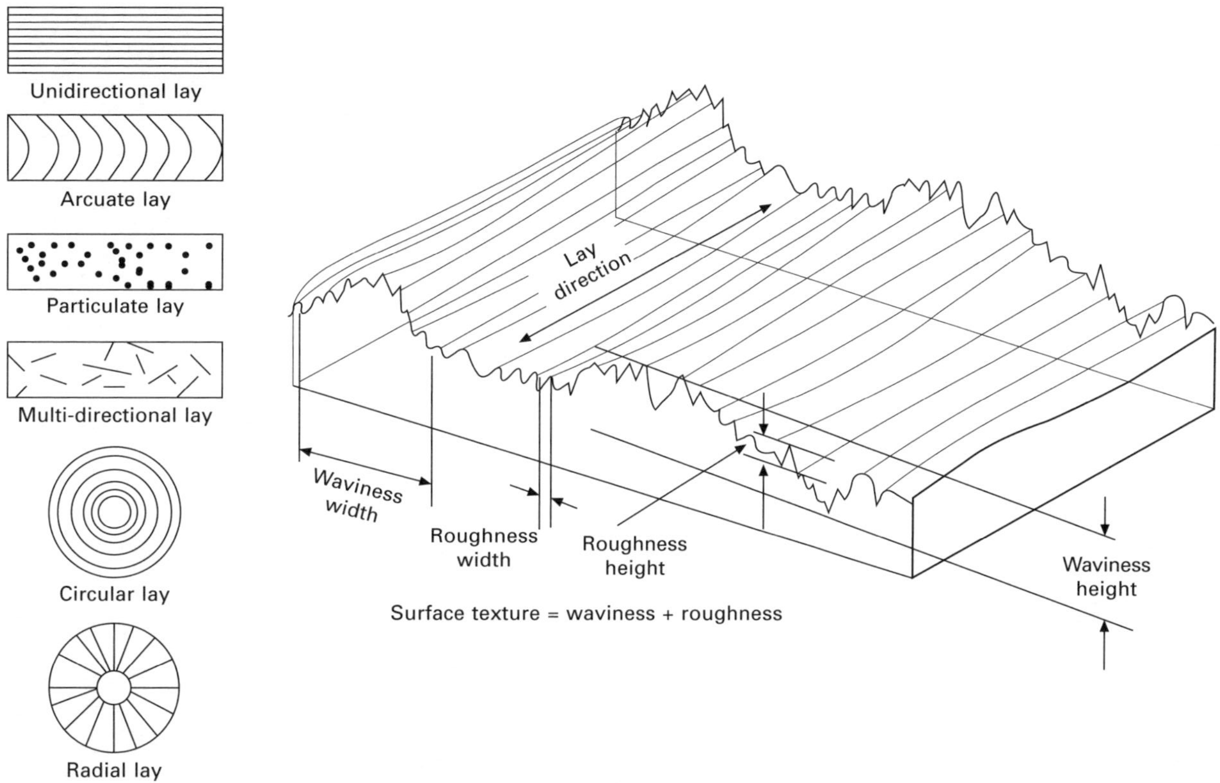

8.9 Surface texture parameters (ASME B46-1-2002).

utilizing special structural features of fabric patterning. Surface lay is largely dependent on the manufacturing method of fibrous assemblies. Figure 8.9 shows different lay possibilities suggested for metal and wood materials. It would be useful that similar categories be identified for fibrous assemblies.

The term surface roughness used to describe fiber surface, often implies subjective description of surface irregularities. Based on microscopic analysis of PET fibers, treated with delusterant such as titanium dioxide (TiO_2), Scardino [20] characterized surface roughness in terms of density of asperities of 0.5 μm in diameter and about 0.25 μm in height. The amount of titanium dioxide not only influences the appearance and dullness of the fiber, it also alters the surface roughness. Based on the weight of the fiber, the inclusion of 1% delusterant can produce a "semidull" fiber, while doubling this amount would produce a "dull" fiber. Microscopic evaluation of PET fiber roughness showed that a 2.0% TiO_2 resulted in about seven to eight times the number of asperities than did a 0.1% TiO_2 [20]. Figure 8.10 shows SEM pictures of surfaces at different TiO_2 levels [20, 29]. Note that polyester treated at 2.0% TiO_2 produced significant roughness as well as arcuate lay. Also note that in case of polypropylene filaments; TiO_2 treatment not only resulted in surface roughness, but also in surface waviness.

Schick added TiO_2 to nylon 66 fiber melt to change the intensity of fiber luster and found that bright fibers (low TiO_2) tended to have a smooth surface whereas dull a rough surface [30]. This change in roughness was not accompanied by a chemical change in fiber surface, which remained essentially as that of the fiber material. He also changed the roughness of the circular

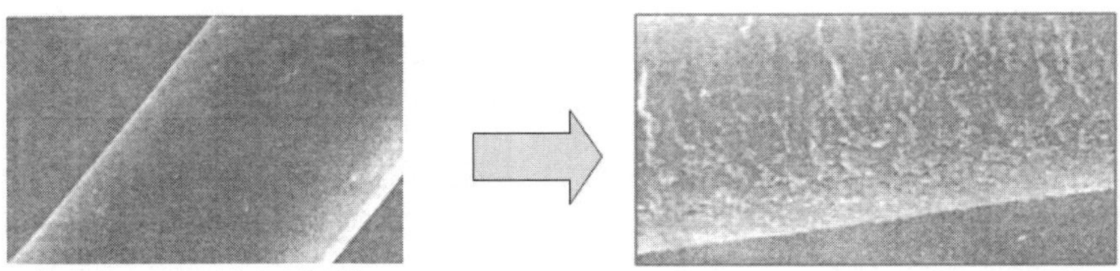

(I) Polyester with 0.1% TiO_2 (II) Same polyester with 2.0% TiO_2

(a) Electron photomicrographs of polyester fibers at different levels of TiO_2 [20]

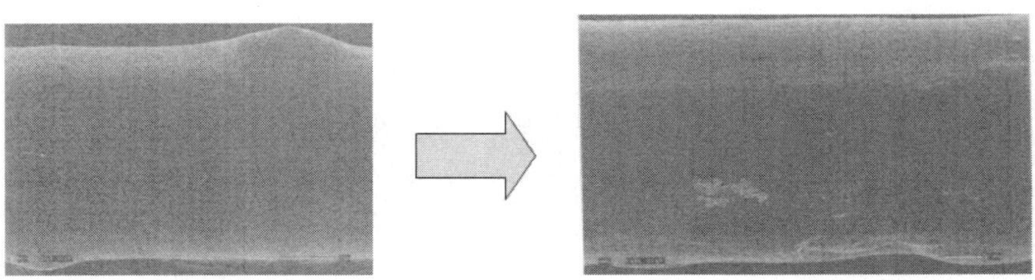

(I) Polypropylene with V_f = 0.05% TiO_2 (II) Same Polypropylene with V_f = 2.0% TiO_2

(b) Electron photomicrographs of polypropylene fibers at different levels of TiO_2 [29]

8.10 Effect of TiO_2 on surface roughness of synthetic fibers.

capstan pin: smooth pin (RMS-4) and rough pin (RMS-60). A summary of his results is given in Fig. 8.11. In general, bright-smooth fibers had higher friction than did dull-rough fibers. This effect was more pronounced with the use of smooth surface pin (RMS-4), and at high speeds (i.e., in the hydrodynamic, region). The low friction in case of rough surface fibers or guides was attributed to the disruption of the continuous fluid film between fiber and guide resulting in a shift toward conditions simulating a semi-boundary region [30].

The titanium dioxide particles, roughly spherical, of about 0.15 μm diameter (e.g., Ti-Pure R102 by DuPont), typically tend to adhere to each other, forming agglomerate. Other types of filler particles of different shapes and dimensions are also used. These include calcium carbonate (parallelepiped, produced by Fisher Chemicals), NYAD 400 wollastonite (needle-like particles by NYCO Minerals Inc.), and talc Jetfil 700C (discotic, Luzenac America, Inc.). Each of these particles produces a unique surface texture [29], and therefore affects friction.

Earlier in this chapter, we discussed how the cross sectional shape of fibers could be manipulated to produce special performance functions such as thermal insulation or wicking effects. Such manipulations represent surface changes of the order of several microns. Surface roughness or surface protuberances or projections can also be manipulated to produce special effects or unique performance characteristics. This type of manipulation is made at a much smaller scale (a fraction of micron).

A unique example of surface roughness manipulation is illustrated by the so-called Microcratered polyester fibers produced by Kuraray Co., Ltd. This

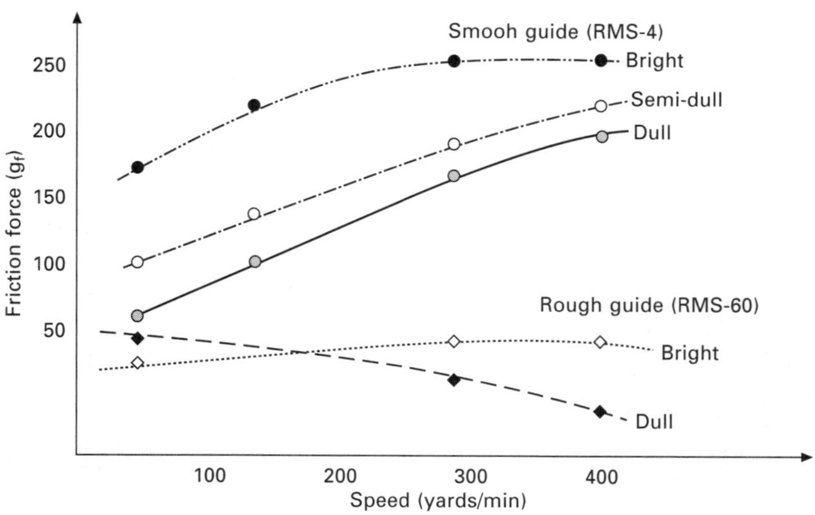

8.11 Effect of fiber luster and guide roughness on fiber friction Nylon 66, 200/34 DuPont (from Data of Schick [30]).

fiber has a unique surface texture of microcraters of critical dimensions that are characterized by depressions and spacings of the order of those of the light. These craters are designed to redirect a major fraction of impinging light rays toward the inside of the fiber rather than toward the observer (see Fig. 8.12). The fiber is produced by melt-spinning polymer compositions containing homogeneous dispersions of silica particles of uniform size, then subjecting the fabric to alkaline etching. Because of their sensitivity to alkali, the silica particles at the surface dissolve more readily than those in the polymer matrices in which they are embedded [31]. The particles inside the fibers remain largely unaffected, so the treatment leaves the tensile properties of the fibers or fabrics essentially unchanged. The ability of black microcratered polyester to confer on formal wear the depth of shade of worsted is a measure of the technology's effectiveness. Besides visual benefits, the hands are crisper and drier because of higher static-to-dynamic friction coefficient ratios. It should be pointed out that these features represent claims by the fiber producer with no considerable published data to support them.

8.4 Surface wettability

Friction being a surface phenomenon, a thorough understanding of a surface is essential. In addition to roughness discussed in previous section, two

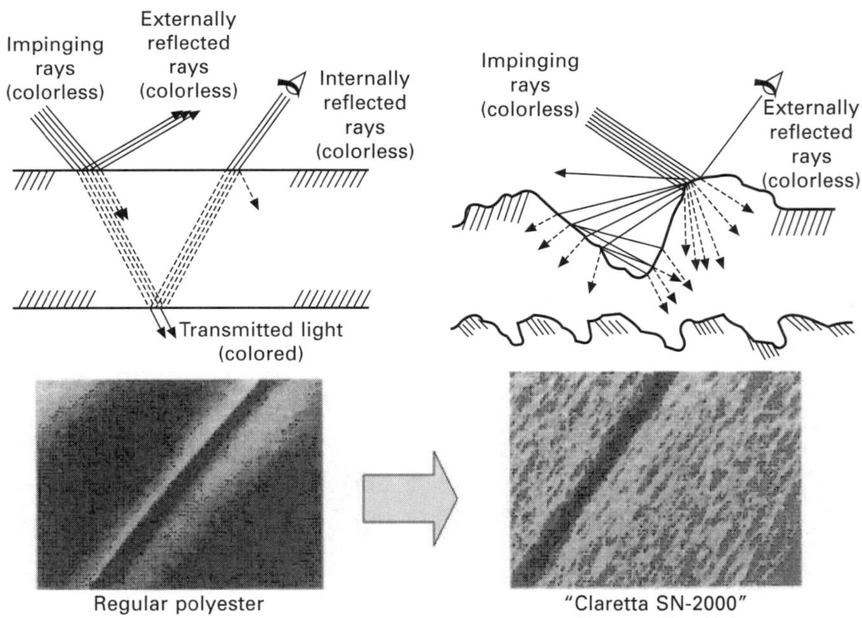

8.12 Comparison of regular and microcratered Polyester fibers [31].

others are surface wettability or energetics, covered in this section, and cleanability and decontamination, covered in the next section. Although no direct results are available to correlate friction with these, an understanding of these will nevertheless provide tools to engineer surfaces that have desirable properties, including friction.

At this point it will be useful to discuss surface wettability due its interesting connection with surface roughness. Surface wettability is an important phenomenon that governs the attraction of one material for another. The two basic ingredients governing surface wettability are chemical composition or structure and surface roughness. In this section, we discuss the critical role surface roughness plays in controlling the wettability of synthetic fibers. From the standpoint of friction, surface wettability or energetics should play an important role in governing the effectiveness of a lubricant in reducing friction.

The equilibrium shape of a liquid drop on a surface is governed by three basic forces existing at three interfaces. These are the solid/liquid, *SL*, liquid/vapor, *LV*, and solid/vapor, *SV*, surface tensions. The balance of these tensions yields the well-known Young's equation (see Fig. 8.13):

$$\gamma_{SV} - \gamma_{SL} = \gamma_{LV} \cos \theta$$

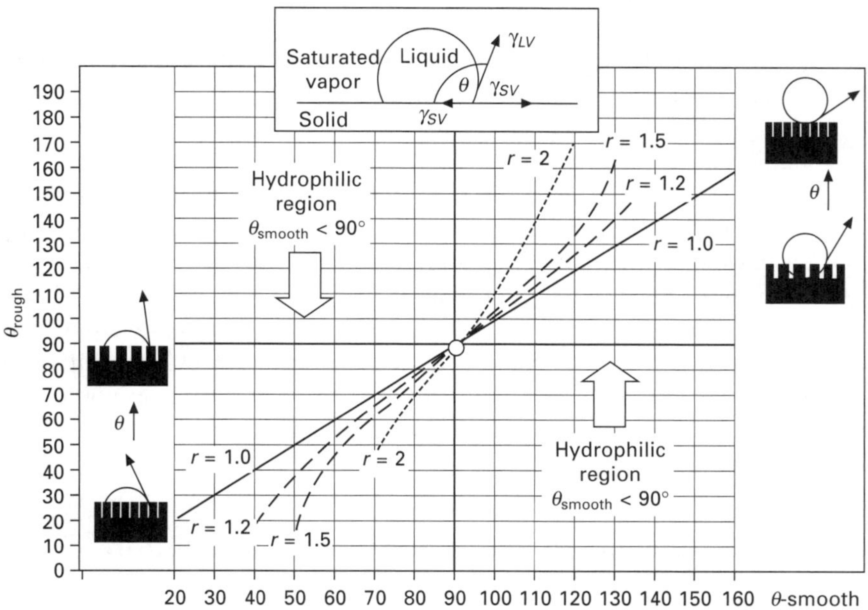

8.13 Demonstration of Young and Wenzel models. In hydrophilic material (left) as *r* increases from 1 to 2, θ decreases. In the hydrophobic material (right) as *r* increases from 1 to 2, θ increases.

where θ is called the intrinsic contact angle of the drop and is a parameter that reflects the degree of wettability of a surface. A surface that is wettable for a given fluid has θ less than 90° and the closer the value to zero the more rapidly wettable the surface.

The role of surface roughness in influencing wettability is given by the Wenzel equation [32, 33] as follows:

$$\cos \theta_{\text{rough}} = r \cos \theta_{\text{smooth}}$$

where r is called the roughness factor defined by the ratio of rough to planar surface areas.

The above equation indicates that for a rough surface ($r > 1$) wettability is improved by roughness of a hydrophilic surface ($\theta_{\text{rough}} < \theta_{\text{smooth}}$ for $\theta < 90°$), and hydrophobicity or dryness is improved by roughness of a hydrophobic surface ($\theta_{\text{rough}} > \theta_{\text{smooth}}$ for $\theta > 90°$). This theoretical concept is in agreement with physical observations; a drop on a rough high-energy surface will likely to sink into the surface. For $\theta > 90°$, the free energy of the dry surface is lower than that of the wet solid, and hence the drop will likely recede from the roughest regions.

In an early study by Cassie and Baxter [34], a rough surface was modeled by a heterogeneous surface composed of air pockets and the solid. They postulated that the cosine of the contact angle of a liquid drop on a heterogeneous surface corresponds to the sum of the cosines of the contact angles of the two homogeneous materials, weighted by the amount of available surface. If one of the surfaces is just air, the cosine of the contact angle on this surface is –1, leading to the following equation:

$$\cos \theta_{\text{rough}} = -1 + \varphi_s(1 + \cos \theta_{\text{smooth}})$$

where φ_s is the surface fraction of the solid. A very rough surface will have a φ_s value approaching zero and, therefore, a rough contact angle approaching 180°, accordingly, the liquid drop will theoretically lift off the solid surface.

In summary, for the equilibrium configurations of liquids on rough surfaces: if the surface has a high interfacial energy, roughness promotes wetting, and the liquid will spread within the corrugation (surface wicking). But for a low-energy surface, roughness promotes repulsion: the drop does not follow the surface corrugations, but achieves its minimum at a position on top of the corrugation [35].

8.5 Surface cleanability and decontamination

Another subject that is directly related to the above phenomenon is the cleanability or decontamination of surfaces. In this regard, the key question is "how can the surface texture be manipulated to enhance the cleanability of hydrophilic and hydrophobic surfaces?" This question is currently a focus of research in many fields, particularly the high-tech field of biotechnology that requires clean surfaces.

For the case of hydrophilic materials, cleaning is basically a process of causing contaminant film to flow. In this case, numerous factors may influence cleanability. The ability of the liquid to dissolve the contaminants (solution/contaminant chemical compatibility), the ability of the liquid to adhere to the contaminants, the pore structure (pore size and its distribution) of the hydrophilic surface, the contaminant particle size and its distribution, the extent to which the capillaries are blocked by unforeseen chemical reactions, and liquid characteristics. These aspects are considered in the design of many fibrous products such as filters and re-usable absorbent medical fabrics.

In the context of surface texture, it was indicated earlier that if a material has a high surface free energy, roughness promotes wetting, and the liquid will flow within the corrugation. When contaminants are present, and assuming no chemical interaction between liquid and contaminants, any removal of contaminant particles will depend largely on the sliding pattern of the liquid against the surface. This is because contaminants should be transported along the surface for efficient cleaning. For this, surface texture can be manipulated to accelerate planar sliding pattern from a liquid. This can be achieved by modification of surface during spinning or by application of special finish.

For the case of hydrophobic surfaces, a fascinating phenomenon was discovered by two botanists, Barthlott and Neinhuis [36, 37] from Bonn, Germany, in the course of their studies of plant leaf structures. They noticed that surface features of the lotus leaf, together with its waxy surface chemistry, rendered the leaf non-wettable. Indeed, the surface of the lotus leaf is one of the nature's most water-repellent surfaces. As seen in Fig. 8.14(a), the surface has countless miniature protrusions coated with water-repellant hydrophobic substance. As a result, the water droplets form spherical globules and roll off the leaves even when they are only slightly inclined. Particles of dirt absorbed by water are removed in the process (see Fig. 8.14(b)). Note that in comparison with human skin, another hydrophobic surface with a contact angle of about 90°, the lotus leaf exhibits a contact angle of 170°. Bird feather, another super-hydrophobic surface, has a contact angle of about 150°.

The lotus leaf exhibits a unique porous surface texture on a micrometer scale. The air trapped in the crevices prevents water from adhering to the solid. Researchers led by H. Yildirim Erbil of Kocaeli University in Turkey [38] re-created such a super-hydrophobic surface by first dissolving polypropylene in a solvent and then adding a precipitating agent and applying the solution to a glass slide. After evaporating the solvent mixture in a vacuum oven, they had a highly porous gel coating with a contact angle of 160° and water-repelling capabilities comparable to those of the lotus leaf.

In a recent study, Luzinov *et al.* [39] examined a number of approaches for mimicking the behavior of the lotus leaf to create synthetic coatings with

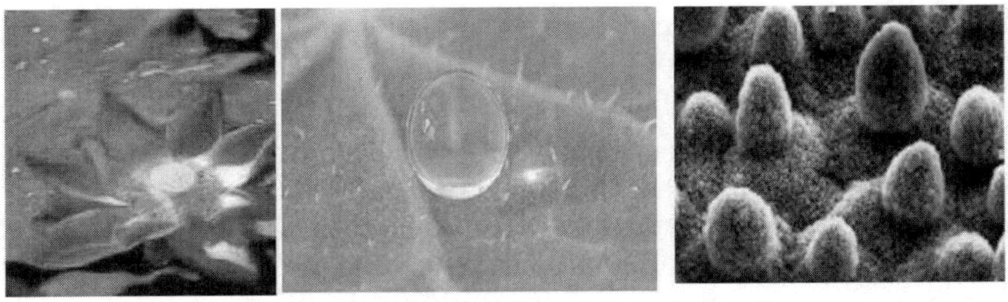

(a) Macroscopic and microscopic lotus leaf surface

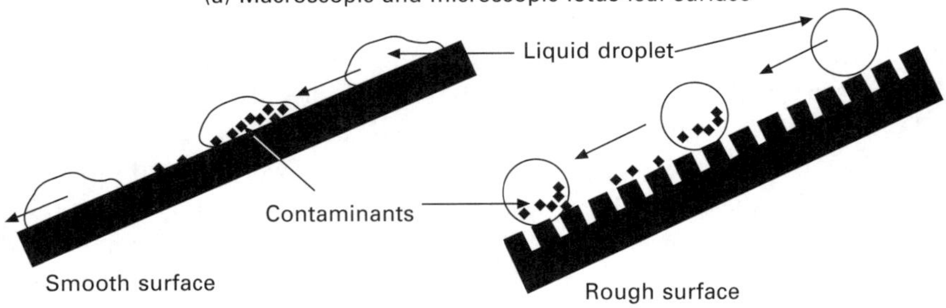

(b) Water movement against smooth and rough hydrophobic surfaces

8.14 The 'Lotus' effect.

exceptional anti-wetting properties. They employed a combination of polystyrene grafted layer (low surface energy component) and nanoparticles (roughness initiation component) and obtained a textile material that demonstrated very low wettability by water. They also attempted to create the "lotus" effect on a fabric surface by evaluating the deposition of both polystyrene (PS) and triblock copolymer polystyrene-b-(ethylene-co-butylene)-b-styrene (SEBS) simultaneously on a model substrate. Polystyrene was then extracted employing selective solvent, ethyl acetate, which acted as a solvent for PS and as a non-solvent for SEBS. The dissolution of PS created a porous (rough) hydrophobic structure on the substrate. The controlled method of surface modification was applied to a polyester fabric which produced a practically non-wettable textile product.

The dynamic behavior of droplets on ultra-hydrophobic surfaces was studied in great detail by David Quéré of the Collège de France and his collaborators [40–45]. Their research indicated that the most important effect of these surfaces on liquid drops concerned the contact line of the drop, that is, the one-dimensional line of intersection of the three interfaces. Because the contact area of the drop shrinks with an increase in contact angle, the contact line can be deformed less easily, and hence the hysteresis in contact angle between the advancing angle (θ_a, or the front angle in direction of droplet motion), and the receding value (θ_r, or the rear angle) is drastically reduced. This hysteresis is expressed by the pinning force per unit length of the drop perimeter:

$$F = \gamma_{LV}(\cos \theta_r - \cos \theta_a)$$

This force has to be overcome by external forces (wind, gravity, etc.) to initiate droplet motion. If the hysteresis is too large, and the driving force is not big enough, the liquid drop will stick or be smeared across the surface. If the contact angles are sufficiently high (> 170°) viscous droplets will roll off the surfaces (not slide).

The dynamic contact angles of a drop moving down a surface are affected by the magnitude of surface roughness [46]. For surface projections of several tens of micrometers, a liquid droplet can still be deformed by them even if it is considerably larger than the projections themselves. Therefore, smaller projections are generally needed for a good ultra-hydrophobic surface.

The second most important factor affecting the dynamics of a drop on an ultra-hydrophobic surface is the velocity of impact. Typically, the impact of the liquid droplet against a surface will result in elastic rebounding with a velocity almost equal to that of impact. This information is useful in designing repellent or drying surfaces.

The self-cleaning mechanism of ultra-hydrophobic surfaces relies on the smallness of the contact area of a drop on a surface. For the ultra-hydrophilic route to self-cleaning, the flow of the liquid film is essential. Ultra-hydrophilic surfaces are wetted easily with very low contact angle fluids: if the surface is inclined, it is the flowing liquid film that carries the contaminants along. The usefulness of this concept thus depends on the rapidity with which a liquid film runs off a surface. For sufficiently thick films (of the order of hundreds of nanometers and above), flow is hydrodynamic. For thinner films, however, the flow of the film will consist of a rapid equilibration by surface diffusion. Not all liquid will move: there will be a stagnant (solidified) layer on microscopic scale.

8.6 Surface lubrication

Lubrication is a common engineering process that involves creating a media between two surfaces for the purpose of controlling friction and wear and protecting the surfaces in contact. The amount and character of the lubricant used have a profound effect on friction. For example, disregarding such related factors as heat and wear but considering friction alone between two oil-film-lubricated surfaces, the friction can be 200 times less than that between the same surfaces with no lubricant. Another critical merit of lubrication is to maintain a uniform noise-free sliding between the surfaces in contact. Uniform sliding is a key aspect for efficient processing and consistent products.

As indicated in Chapter 6, natural fibers have a waxy surface layer that allows them to be readily and smoothly processed. Synthetic or man-made fibers, on the other hand, require an application of finish to make them readily manageable during processing.

It is important to realize that lubrication is used as much to control friction as to reduce it. One of the complex aspects of the friction phenomenon is that its effect may follow different patterns depending on the application or the performance characteristic of interest. As conceptually illustrated in Fig. 8.15, friction can have positive or negative effects on different performance parameters. For example, the strength of synthetic yarns requires an optimum inter-filament cohesion above which strength may reach a plateau, or actually decrease (e.g., Broughton *et al.* [47]). Fiber friction is likely to adversely influence the flexibility of a fiber assembly on the ground of limiting inter-fiber mobility. During processing of fibers into yarns or nonwovens, inter-fiber friction should be at an optimum level, below which control of fiber flow is difficult, and above which fiber clusters and failure of cluster breakdown during opening and drafting can occur (e.g., El Mogahzy *et al.* [48, 49]). Optimum friction is also desirable for blend compatibility of synthetic fibers with natural fibers (e.g., El Mogahzy *et al.* [50, 51]). Fiber friction should also be at an optimum level to achieve acceptable hand characteristics. This level is determined by the extent of surface roughness and the degree of contact between the fabric and the skin (e.g., El Mogahzy *et al.* [52, 53]).

Most lubrication is intended for reducing fiber-to-metal friction. This is perhaps a direct result of the known adverse effect of this friction, not only on fiber surface degradation but also on the wear of machine components (Fig. 8.15). In today's market, there are numerous types of lubricants that

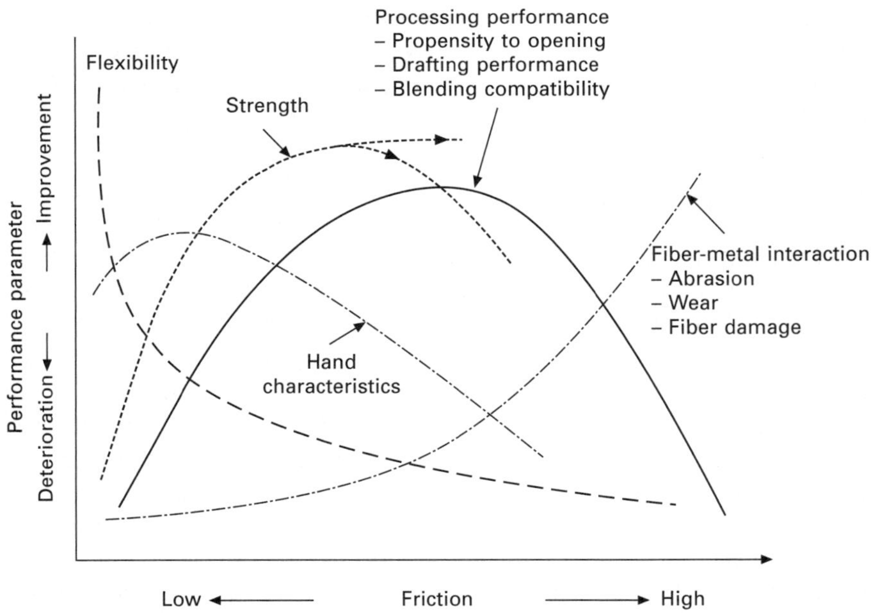

8.15 Conceptual effects of fiber friction (modified from refs [47–53]).

Table 8.3 Basic classes of lubricants and associated general features

Lubricant type	Features
Fluid-film lubrication	• A fluid film that completely separates sliding surfaces. • Examples: (1) intentional – oil in the main bearings of an automobile, (2) unintentional – water between a smooth rubber tire and a wet pavement. • Mostly liquid, but it may also be a gas. The gas most commonly employed is air. • To impart complete surface separation, the pressure within the lubricating film should balance the load on the sliding surfaces. • Hydrostatic lubrication: when the lubricating film's pressure is supplied by an external source. • Hydrodynamic lubrication: when the pressure between the surfaces is generated as a result of the shape and motion of the surfaces themselves. This type of lubrication depends upon the viscous properties of the lubricant. • Under fluid-film conditions, friction is directly proportional to the viscosity of the fluid.
Boundary lubrication	• A lubricating status that lies between un-lubricated sliding and fluid-film lubrication. • In boundary lubrication, the properties of the surfaces and properties of the lubricant, other than viscosity, determine the friction between surfaces. • Boundary lubrication encompasses a significant portion of lubrication phenomena, which commonly occurs during the starting and stopping of machines. • Under boundary lubrication conditions, the effect of viscosity on friction becomes less significant than the chemical nature of the lubricant.
Solid lubrication	• A lubricating process imparted by solid particles. • Solids such as graphite, molybdenum disulfide (Moly) and PTFE (Teflon) are widely used when normal lubricants do not possess sufficient resistance to load or temperature extremes.

can be used. These can be divided into three basic classes: fluid-film lubricants (liquid or gaseous), boundary lubricants (fluid/solid combination), and solid lubricants. Table 8.3 lists these classes and some of their associated general features.

The classification of lubricants into three basic classes (Table 8.3) is driven by the generalized frictional behavior of liquid-lubricated fibers [11, 48, 54, 55], shown in Fig. 8.16. It is characterized by two distinct behaviors: boundary and hydrodynamic, which are largely governed by three factors: the speed of sliding, interfacial viscosity and normal pressure. Boundary lubrication occurs when the three parameters combine to lead to a very thin film – a monolayer of lubricant between the contacting surfaces. Typically,

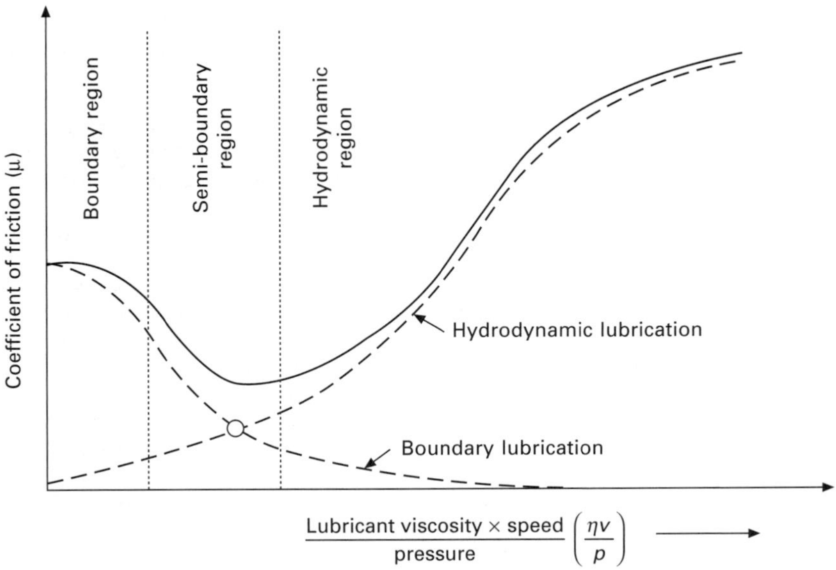

8.16 Generalized frictional behavior of liquid-lubricated fibers [11, 48, 54, 55].

the speed is low and the load is largely carried by the asperities in contact. As the speed increases, the contact decreases and the friction decreases. When the speeds are high, the bodies with a lubricant are separated enough that a relatively thick film of oil develops between the surfaces. Friction is then given not by the shear of junctions but by the viscous resistance to flow of the lubricant. This condition constitutes the hydrodynamic lubrication state. As speed increases, friction also increases.

Another type of behavior is the so-called semi-boundary behavior. This is a transitional behavior that lies between the boundary and the hydrodynamic behaviors. The importance of this region is that it encompasses the minimum friction level of the system, where both the contact and the film thickness are optimally minimal.

Different regions of the general frictional behavior of lubricated surfaces shown in Fig. 8.16 were experimentally confirmed in studies on synthetic fibers. The work by Schick [11], illustrated earlier in Fig. 8.2, represents an example of effects leading to hydrodynamic frictional behavior of polyester and nylon fibers. The constants α and β represented the basic coefficients of the power function of the relationship between frictional force and sliding speed, $F = \alpha v^{\beta}$. The coefficient α is largely sensitive to the fiber dimensions (denier). The coefficient β falls within a range between 0.180 and 0.195. It is essentially the parameter that governs the force-speed relationship, which is expected to vary with the surface area, as confirmed in Fig. 8.2, the range of sliding speeds used, and surface finish, type and amount.

In an earlier study, Lyne [56] determined the effects of speed, viscosity, and pre-tension on the frictional behavior of acetate yarns sliding over a chromium pin of $^1/_4$ inch (6 mm) diameter in which he used mineral oils of known viscosity as lubricants. The study revealed a dominance of the hydrodynamic behavior. Other studies [49, 57] revealed semi-boundary conditions at low sliding speeds for polyester/cotton blends and for pure polyester fibers. These conditions were observed for fiber-fiber friction experiments.

In closing, one may note that the dominance of one frictional behavior or the other will largely depend on many factors that have to collectively come together to contribute to the behavior outcome. Hydrodynamic behavior is sensitive to factors such as speed, lubricant viscosity, temperature, concentration of applied lubricant, yarn denier, pre-tension, yarn surface roughness, and guide material and diameter. On the other hand, boundary conditions are sensitive to factors such as the chemical nature of the lubricant, the physical nature of lubricant (particularly shear behavior), the surface texture of the contacting surfaces, rigidity of substrates, and the level of lateral pressure [11, 48, 54, 55].

8.7 Friction behavior in wet media

In practice, the effect of wetting on friction can be considered under three different sets of wetting conditions: (a) sliding in a highly humid atmosphere, (b) sliding in a soaking-wet environment, and (c) sliding with a lubricant, aqueous or non-aqueous, driven by surface energetics.

Sliding under the first set was examined by Schick [57]. Figure 8.17 shows the results for rayon yarns tested in three different relative humidities. They indicate that fiber-to-metal friction increases with an increase in relative humidity. Under low to medium humidity conditions, the increase in friction with humidity can be largely attributed to the swelling of hydrophilic fibers as a result of moisture absorption. This will result in an increase in the true area of contact and a corresponding increase in friction.

Under relatively higher humidity, the behavior appears equivalent to that given by hydrodynamic lubrication conditions. It is possible that smooth surface and absorbed layers of water led to such behavior.

The second set of conditions represents those that are encountered when surfaces are sliding in a submerged state or soaking-wet environment. The practical interest in this mode is usually realized for hydrophobic fibers. In a previous study by El Mogahzy [21], the effect of wetting on the frictional behavior of hydrophobic fibers was examined at low sliding speed (1.27 cm/min). The fiber types used were acrylic and polypropylene multifilament yarns and the test method used was fiber-to-fiber line-contact twist method (discussed in Chapter 6).

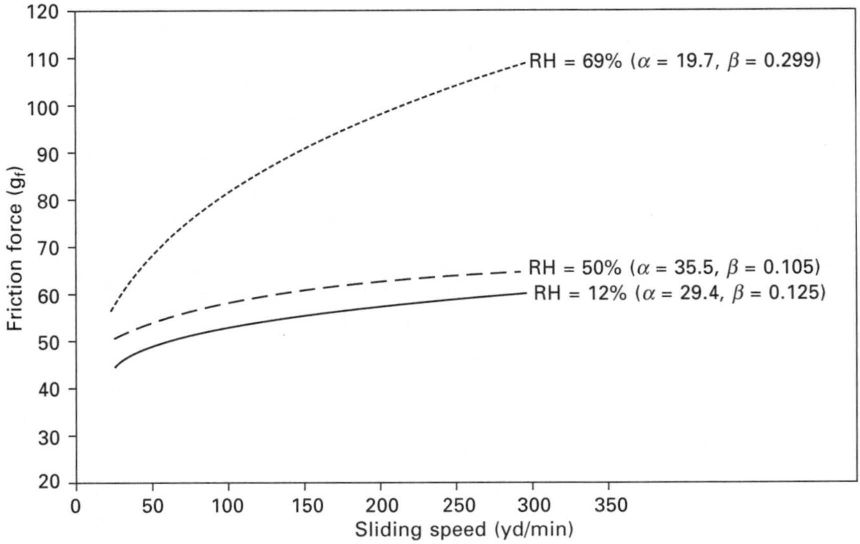

8.17 Effect of relative humidity on friction of rayon yarn 300/44 (AVC) Rothschild friction meter (RMS-60, 0.5-in diameter chrome pin, contact angle 360) (modified after Schick [57]).

The results are given in Fig. 8.18 that show that the coefficient of friction, μ, and the friction index, a, increase with wetting. The change in the value of the friction index n was insignificant.

These results were explained on the basis that since the fibers were hydrophobic, the effect of wetting was merely a result of penetration of water through the interstitial space between filaments that caused the yarn to swell to some extent. This should result in an increase in the area of contact and, therefore, in an increase in friction. An alternative reason could be that the test in the liquid media caused the contaminants to be removed from the fiber surface, which gave an increase in friction.

The third condition or factor is that described by dynamic wettability driven by surface energetics. In this case, the wetting media is the lubricant substance with a behavior largely governed by the hydrodynamic lubrication mechanism. Such behavior was examined for nylon 66, polyester, polypropylene, and Teflon yarns of comparable denier (200–300) using different liquids [58]. The results showed that nylon 66, polyester, and polypropylene yarns had critical surface tension values, γ of 46, 43 and 24 dyn/cm, respectively.

Low friction requires good wetting with a lubricant. Accordingly, for these surfaces, a lubricant should have a low surface tension so that it spreads and adheres to the fiber material. Teflon has low wettability for any liquid because it has low surface energy (γ = 18.5 dyne/cm). Fortunately, it also has very low friction to start with and, therefore, does not require a lubricant to reduce it.

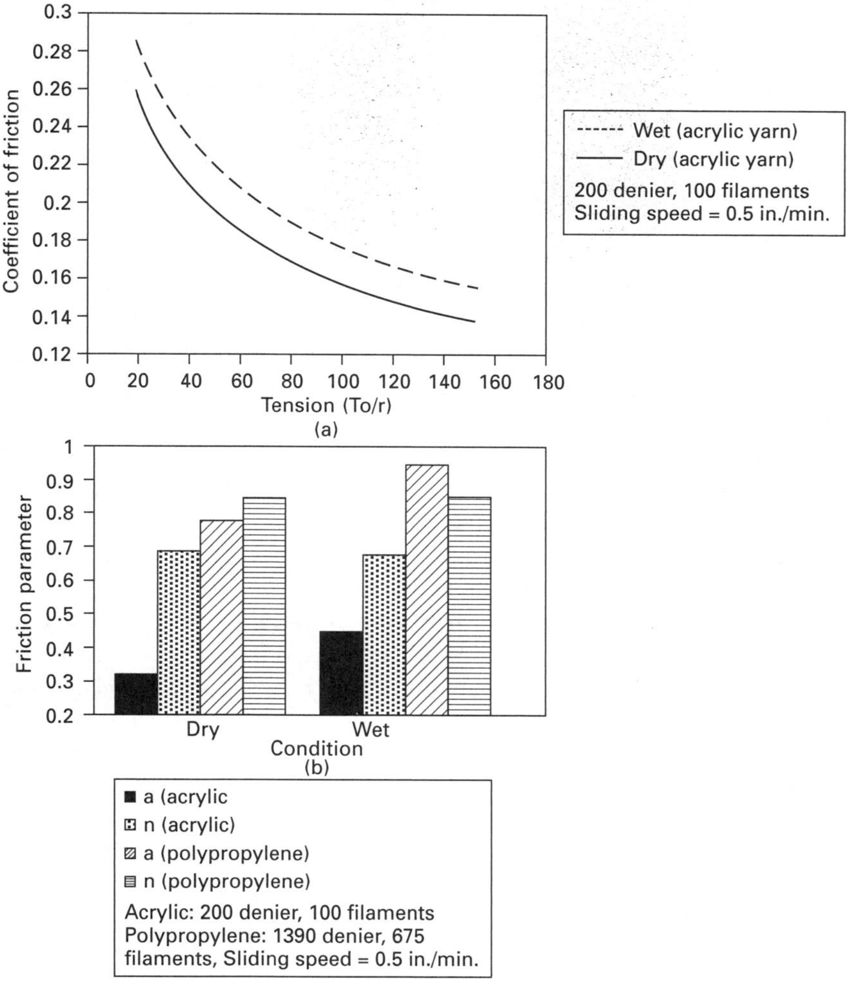

8.18 Effect of wetting on fiber friction-line contact twist method Sliding speed 1.27 cm/min (modified after [21]).

8.8 The stick-slip phenomenon

Subjectively, one may describe a stick-slip pattern as being a reflection of the smoothness or the roughness of contact between two surfaces. This type of description may be supported by the fact that stick-slip patterns are often observed in boundary lubrication conditions and at low sliding speeds. However, stick-slip patterns may also be observed in cases involving semiboundary lubrication or hydrodynamic lubrication. This may lead one to speculate that these patterns may in fact be indicating a lack of uniformity in the friction process, i.e. uniform friction leads to a smooth pattern and irregular friction leads to irregular pattern.

Theoretically, engineers attempted to describe stick-slip patterns using mechanical models consisting of friction blocks, springs and dashpots [59, 60]. Two of these models providing different stick-slip patterns are shown in Fig. 8.19. The first (Fig. 8.19(a)) consists of a friction block loaded against a surface which moves at velocity V_o. The block is connected to a spring of certain stiffness, k. The equation describing this system is as follows:

$$\mu W - kx = Ma_o$$

where μ is the coefficient of friction, W is the normal load, k is the spring constant, x is the displacement of the spring, and a_o is the acceleration.

If μ was constant, the system would be stable when, $\mu W = kx$, and the spring would have deflected by $x = \mu W/k$. A stretch-contract pattern in the movement of the spring may, therefore, be considered as a mechanical instability in the system that results in an oscillating movement of the spring. In this regard, two types of coefficient of friction should be introduced: the static coefficient of friction, μ_s, which indicates the friction involved in initiating the movement, and the kinetic or dynamic coefficient of friction, μ_k, which indicates the friction during the actual movement.

When the system starts to move with velocity V_o, the spring force is zero and the spring deflection, x is also zero. The mass M moves with velocity V_o until the spring force kx just equals the static friction force $\mu_s W$ (i.e., the displacement is $\mu_s W/k$). At this point, the system (block and spring) starts to slide and a dynamic state of a lower friction, μ_k dominates the sliding condition. Accordingly, M will accelerate until x reduces to $\mu_k W/k$. M will then continue to move under its stored kinetic energy until it comes to rest when static

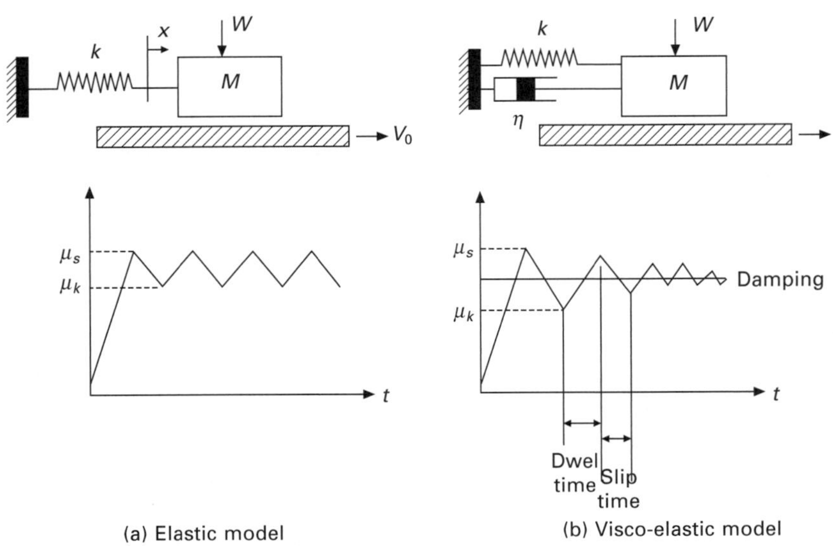

8.19 Stick-slip simulating models [59, 60].

friction dominates again and the cycle is repeated. When a dashpot is added to the mechanical model, the stick-slip pattern will involve a damping effect with time such as that shown in Fig. 8.19(b).

The models discussed above only describe the mechanical system leading to a stick-slip pattern. In other words, they relate more directly to the friction process and its associated parameters (i.e., speed, stiffness, and viscosity). In practice, stick-slip patterns are often irregular and are difficult to interpret. However, there are three common observations that should be expected in rationalizing these patterns:

- As the sliding speed increases, the amplitude of the stick-slip pattern is likely to decrease. A critical speed above which stick-slip pattern completely disappears should be expected.
- As the system stiffness increases, the magnitude of the stick-slip pattern should decrease. This point is extremely important in designing a friction device involving sliding of fibers. Different devices may provide different amplitudes of stick-slip patterns depending on the stiffness of the sliding mechanism
- Certain boundary lubricants do reduce static friction to a level below kinetic value and, therefore, entirely eliminate stick-slip [59].

Accordingly, any consideration of using stick-slip patterns to rationalize friction results should be based on an understanding of the friction process and the external parameters involved. Comparison between stick-slip patterns produced by different friction testing techniques could certainly be misleading.

8.9 Surface finish

Throughout this chapter, the issue of surface finish was repeatedly brought up in relation to surface-related parameters such as molecular orientation, surface roughness and lubrication. In practice, surface finish has been a key design parameter in numerous traditional and industrial textile applications. In the general sense, surface finish reflects all types of surface treatments that are intended to compensate for inherent deficiencies in the fiber surface, as fibers are spun and yarns are made, or to enhance existing characteristics. On this ground, lubricants represent a special category of surface finish.

In today's technology, numerous types of surface finish are being used for a wide range of applications and we encourage the reader to refer to the many references dealing specifically with surface finish treatments for synthetic fibers [e.g., 24, 29, 61]. In the context of fiber friction, lubrication has been the primary treatment used for controlling this complex phenomenon. Other finishes that are intended to enhance various performance characteristics may also directly or indirectly influence frictional behavior of synthetic fibers. These include: finishes used for improving fiber hydrophilicity, abrasion resistance, pilling resistance, and antistatic properties. A brief discussion on these types of finishes follows.

Abrasion resistance is primarily the phenomenon that characterizes the energy-to-rupture under severe rubbing conditions; the greater the energy-to-rupture, the greater the abrasion resistance. Finishes that lubricate or coat the fiber improve its ability to withstand some or all types of wear or abrasion. Lubrication with substances such as polysiloxanes, is usually only beneficial for improving flex abrasion. Polymeric coatings such as Teflon, poly(vinyl chloride) and polyurethanes, applied to fabrics typically result in overall improvement in abrasion resistance. These coatings have an inherently high resistance to abrasion and thus absorb wear and friction stresses rather than such mechanical stresses being absorbed by the fiber.

Pilling is the entanglement of loose fiber ends that form balls or pills on the surface of a fabric. It is particularly problematic with high tenacity fibers such as polyester, since the fibers do not wear off due to abrasion, but instead remain on the surface and become entangled. During processing, pilling may be overcome by singeing or by using tightly woven fabric constructions. At the fiber level, finishes such as hydrolysis, alcoholysis, and ammonolysis are used to alter the surface of polyester fibers. Alternatively, film-forming polymers may be used to reduce the mobility of the fibers by providing interfiber adhesion and thus minimize entanglement and subsequent pilling.

Antistatic finishes represent a common category of surface treatment used for most synthetic fibers. Like most materials, fibers have the ability to acquire a static charge and transfer it from one material to another. The moisture content of fibers can significantly influence their ability to dissipate such charges. For synthetic fibers, this may not be an option in many situations. Finishes yielding hydrophilic surfaces can make fibers more able to dissipate static charge. Special finishes are also used as antistatic agents. Some agents act by increasing the rate of charge dissipation and others act by reducing the generation of static charge on the fiber. Most antistatic finishes were developed for synthetic fibers with low to moderate moisture content such as polyamides, acrylics and polyesters.

Earlier in this chapter, we discussed the relationship between surface texture and fiber wettability. Poor wettability or hydrophobicity of synthetic fibers contributes a great deal to other less desirable properties such as poor soil-release behavior in aqueous liquids, attraction to oily soils, low adhesion to rubber and plastics, and the tendency for static electricity buildup. Surface energy or polarity of a polymer can be increased by incorporating hydrophilic compounds by means of copolymerization/addition/blending and surface modification by coating, grafting, and other reactions [24].

The basic condition for efficient wetting is that the surface tension of the substance, $\gamma_{substance}$ must be equal to or greater than that of the liquid, γ_{liquid}. At room temperature, the surface tension of water, γ_{water} is 72.8 mJ/m^2. This value is higher than the value of surface tension of most common synthetic fibers (polyester, nylon, and polypropylene). This necessitates the use of hydrophilic surface finish to enhance fiber wetting. One of the common methods is the use of alkaline hydrolysis finish. In general, surface hydrolysis causes chain scission to form polar hydroxyl and carboxylic groups and increases surface polarity and hydrogen-bonding capacity with water molecules, thus increasing water wettability. Another finish type is enzyme treatment. This provides improved water wettability and it is often accompanied by full-strength retention as compared to the significantly reduced strength and mass loss from alkaline hydrolysis [24].

8.10 Summary

The chapter dealt with friction and surface characteristics of synthetic fibers. The emphasis was placed on highlighting key design parameters that can assist in modifying the surface characteristics and the frictional behavior of fibers. These parameters, which could be changed during the spinning process of most fibers, included: surface area, fiber cross section, fiber crimp, fiber orientation, and surface roughness. In addition, the impressive development in nanotechnology has made it possible to alter fiber surface at the nano level without influencing key physical characteristics of fiber. The chapter also dealt with friction in wet media by discussing the effect of wetting on fiber friction and that of surface texture on the extent of wettability of fibers. This is an interesting area as it relates to many important applications of textiles, such as those requiring surfaces that are soil-resistant, or can be easily cleaned, stain-resistant, and absorbent and comfortable but quick drying. Although it is not explicitly discussed, the chapter calls for the development of standard test methods for fiber friction and other surface-related properties. These methods will assist in unifying the terminologies associated with critical measures and in better understanding the nature of fiber friction.

8.11 References

1. Howell H G, Mieszkis K W and Tabor D, *Friction in textiles*, Butterworth, London, 1959.
2. Schick M J, *Surface characteristics of fibers and textiles*, Marcel Dekker, Inc., New York and Basel, 1975.
3. Hong J and Jayaraman S, 'Friction in textiles', *Textile Progress*, **34**(1/2), The Textile Institute, Manchester, UK, 2003.
4. Hearle J W S, *High-performance fibers*, Woodhead Publishing Limited, Cambridge, 2001.
5. Miraftab M, 'Technical fibers', in Horrocks A R and Anand S C, Ed., *Handbook of technical textiles*, Woodhead Publishing Limited, Cambridge, 2000.
6. Mukhopadhyay S K, 'High-performance fibers', *Textile progress*, 1993, **25**, 1–85.

7. Tao Xiaoming, *Smart fibers, fabrics and clothing*, Woodhead Publishing Limited, Cambridge, 2001.
8. Shishoo R, 'Safety and protective textiles: The opportunities and challenges ahead', *Safety and protective fabrics proceedings*, IFAI, Roseville, MN, 3–17, 2004.
9. El Mogahzy Y E, 'Fiber/machine interaction', *Proceedings Beltwide Cotton Conference*, 14 January 1994.
10. El Mogahzy Y E, *Utilization of cotton/micro-denier polyester blends in air-jet spinning*, Paper presented in Textile World Microdenier Fibers Conference, Greenville, SC, June 1–2, 1994.
11. Schick M J, 'Friction and lubrication of synthetic fibers, fiber science series', in M J Schick, Ed., *Surface characteristics of fibers and textiles*, Part I, Marcel Dekker, Inc., NY and Basel, 1975, 1–63.
12. Industrial Report, 'New hollow filaments from Kanebo', *JTN*, October 1992, **455**, 45.
13. Industrial Report, 'Highly perspiration-absorbing quick drying polyester from Mitsubishi rayon', *JTN*, May 1992, **450**, 32.
14. Pramanick A K and Crouse B W, 'Application of deep grooved polyester fiber in composite high absorbent paper', TAPPI Nonwovens Conference, St. Petersburg, Florida, *IPST Technical Paper Series* Number **693**, 1998.
15. Haile W A, 'Deep grooved polyester fiber for wet lay applications', *Tappi J.*, 1995, **78**(8), 139.
16. Berkowitch J, Yoshida P G, *Trends in Japanese textile technology*, Department of Commerce, Office of Technology Policy, December 1996.
17. Hongu T and Philips G O, *New fibers*, Ellis Horwood Limited, a division of Simon & Schuster, West Sussex, England, 1990.
18. Fujimoto M and Hayakawa K, Toray Ind. Inc., 'Approach to silk-like polyester fiber which can produce rustling and scrooping sounds', *The 14th Textile Research Symposium* at Mt. Fuji, 1985.
19. Morooka H, 'Frictional sound of fiber assemblies: Silk scroop and shari sound', *J. Text. Mach. Soc.*, Japan [Sen-I Kikai Gakkaishi], 1991, **44**(10), 29–34.
20. Scardino F L, 'Surface geometry of synthetic fibers', in M J Schick, Ed., *Fiber science series*, *Surface characteristics of fibers and textiles*, Part I, Marcel Dekker, Inc., NY and Basel, 1975, 165–191.
21. El Mogahzy Y E, *A study of the nature of fiber friction*, Ph.D. dissertation, North Carolina State University, Raleigh, NC, 1987.
22. Gupta B S and El Mogahzy Y, 'Friction in fibrous materials. Part I: Structural model', *Textile Res. J.*, 1991, **61**(9), 547–555.
23. El Mogahzy Y E and Gupta B S, 'Friction in fibrous materials, Part II: Experimental study of the effects of structural and morphological factors', *Textile Res. J.*, 1993, **63**(4), 219–230.
24. Hsieh, You-Lo, 'Surface characteristics of polyester fibers', in Pastore C M and Kiekens, Ed., *Surface characteristics of fibers and textiles*, Surfactant Science Series, Marcel Dekker, NY, Basel, **94**, Chapter 2, 2001, 33–57.
25. Gupta B S, El Mogahzy Y E and Selivansky D, 'The effect of hot-wet draw ratio on the coefficient of friction of wet-spun acrylic yarns', *J. Appl. Polymer Sci.*, 1989, **38**, 899–905.
26. Gupta B S, 'Frictional properties of textile materials', in Pastore C M and Kiekens, Ed., *Surface characteristics of fibers and textiles*, Surfactant Science Series, Marcel Dekker, NY, Basel, **94**, Chapter 3, 2001, 59–92.

27. Moseley W W Jr., 'The measurement of molecular orientation in fibers by acoustic methods', *J. Appl. Polymer Sci.*, 1960, **3**, 266.
28. Morgan H M, 'Correlation of molecular orientation measurements in fibers by optical birefringence', *Textile Res. J.*, 1962, **32**, 866–870.
29. George B, Hudson S and McCord M G, 'Surface features of mineral-filled polypropylene filaments', in Pastore C M and Kiekens, Ed., *Surface characteristics of fibers and textiles*, Surfactant Science Series, Marcel Dekker, NY, Basel, **94**, Chapter 6, 2001, 139–160.
30. Schick M J, 'Friction and lubrication of synthetic fibers, Part I: Effect of guide surface roughness and speed on fiber friction', *Textile Res. J.*, 1973, **43**, 111–117.
31. Akagi T, Maeda K, Kawamoto M and Yamaguchi S, 'Fibers and fabric properties of microcrater polyester fibers', *Proceedings of the Japan-Australia joint symposium on objective measurements: Applications to product design and process control*, Kyoto, Japan, 5 September 1985.
32. Shafrin E G and Zisman W A, 'Upper limits to the contact angles of liquids on solids', in Gould R F, Ed., *Contact angle, wettability and adhesion*, Advances in chemistry series, Vol. **43**, American Chemical Society, Washington D. C., 145–157, 1964.
33. Wenzel R N, 'Resistance of solid surfaces to wetting by water', *Ind. Eng. Chem.*, 1936, **28**, 988.
34. Cassie A B D and Baxter S, 'Wettability of porous surfaces', *Trans. Faraday Soc.*, 1944, **40**, 546–551.
35. Blossey R, 'Self-cleaning surfaces – virtual realities', *Nature materials*, May 2003, **2**, 301–306, (www.nature.com/naturematerials).
36. Barthlott W, Neinhuis C, 'Purity of the sacred lotus, or escape from contamination in biological surfaces', *Planta*, 1997, **202**, 1–8.
37. Von Baeyer H C, 'The lotus effect', *The Sciences*, January/February, 2000.
38. Yildirim Erbil H, Levent Demirel A, Yonca Avci and Olcay Mert, 'Transformation of a simple plastic into a super-hydrophobic surface', *Science*, February 2003, **299**, Issue 5611, 1377–1380.
39. Luzinov I, Brown P, Chumanov G, Minko S, '*Ultrahydrophobic fibers: Lotus approach*', Annual report, Project No. C04-CL06, National Textile Center, 2004.
40. Bico J, Marzolin C and Quéré D, 'Pearl drops', *Europhys. Lett.*, 1999, **47**, 220–226.
41. Richard D and Quéré D, 'Viscous drops rolling on a tilted non-wettable solid', *Europhys. Lett.*, 1999, **48**, 286–291.
42. Richard D and Quéré D, 'Bouncing water drops', *Europhys. Lett.*, 2000, **50**, 769–775.
43. Aussillous P and Quéré D, 'Liquid marbles', *Nature*, 2001, **411**, 924–927.
44. Richard D, Clanet C and Quéré D, 'Contact time of a bouncing drop', *Nature*, 2002, **417**, 811.
45. Bico J, Tordeux C and Quéré D, 'Rough wetting', *Europhys. Lett.*, 2001, **55**, 214–220.
46. Öner D and McCarthy T J, 'Ultra-hydrophobic surfaces. Effects of topography length scales on wettability', *Langmuir*, 2000, **16**, 7777–7782.
47. Broughton R M, El Mogahzy Y E and Hall D M, 'Mechanism of yarn failure', *Textile Res. J.*, 1992, **62**(3), 131–134.
48. El Mogahzy Y E, Broughton R and Wang Qin, 'The friction profile of cotton fibers and its importance in determining fiber performance in the nonwoven process. Part I: Fundamental aspects of fiber friction and lubrication', *International Nonwoven Journal* (*INJ*), 1995, **6**(4), 35–42.

49. EL Mogahzy Y E, Broughton R, Guo H and Rollins C, 'Evaluating staple fiber processing propensity. Part II: Processing propensity of cotton/polyester blends', *Textile Res. J.*, 1998, **68**(12), 907–912.
50. Yehia Elmogahzy, 'An integrated approach to analyzing the nature of multi-component fiber blending. Part I: Analytical aspects', *Textile Res. J.*, 2004, **74**(8), 701–712.
51. El Mogahzy Y E, Farag R, Abdelhady F and Mohamed A, 'An integrated approach to the analysis of the nature of multi-component fiber blending. Part II: Evaluation of structural and attributive blending of cotton/polyester blends', *Textile Res. J.*, 2004, **74**(9), 767–775.
52. El Mogahzy Y E, 'The theory and the practice of understanding fabric comfort', *Proceedings of Textile Science*, 98-International Conference, Technical University of Liberec, Czech Republic, 1998.
53. El Mogahzy, Y E, Kilinc F S, Hassan M, Farag R and Kamel A, 'Protective clothing: The unresolved ultimate trade-off between protection and comfort', *Proceedings of the 4th International Conference on Safety and Protective Fabrics*, IFAI Books, Roseville, MN, 36–60, 2004.
54. Hansen W W and Tabor D, 'Hydrodynamic factors in the friction of fibers and yarns', *Textile Res. J.*, 1957, **27**, 300.
55. Olsen J S, 'Frictional behavior of textile yarns', *Textile Res. J.*, 1969, **39**, 31.
56. Lyne D G, 'The dynamic friction between cellulose acetate yarn and cylindrical metal surface', *J. Textile Inst.*, 1955, **46**, 112.
57. Schick M J, 'Frictional and lubrication of synthetic fibers. Part I: Effect of guide surface roughness and speed on fiber friction', *Textile Res. J.*, 1973, **43**, 103.
58. Schick M J, 'Frictional and lubrication of synthetic fibers. Part III: Effect of guide temperature, loop size, pretension, denier and moisture regain on fiber friction', *Textile Res. J.*, 1973, **43**, 254.
59. Cameron A, *Basic lubrication theory*, 3*rd* edition, Ellis Horwood Series in Engineering Science, Ellis Horwood Publishers, Chichester, 1981.
60. Armstrong-Helouvry B, *Control of machines with friction,* Kluwer Academic Publishers, Boston, Dordrecht, London, 1991.
61. Vigo T L, *Textile processing and properties*, Elsevier, Amsterdam, London, NY, Tokyo, 1994.

9
The role of fiber friction in staple fiber processing

V K KOTHARI and A DAS, Indian Institute of Technology, India

9.1 Introduction

A primary step that leads to the formation of a staple fiber yarn is the drafting of a continuous strand of fibers from a card web, a sliver or a roving. Such drafting or drawing of a fiber strand is only possible if fibers can slide over each other. Some degree of control over this sliding is essential for good spinning.

The contribution of fiber-to-fiber or fiber to other engineering surfaces (metal, rubber, plastic, ceramic, etc.) friction in controlling fiber flow during the staple fiber spinning process is well established. This is because fiber friction is a unique index that expresses interactive behavior. In the traditional staple fiber to yarn conversion process, the fibers are converted into yarn through a series of opening and drafting steps involving interaction between different groups of fibers and between fibers and other materials (solids or fluids). These interactions determine relative fiber position and fiber movement during processing. Friction is an important physical characteristic of staple fibers from the standpoint of the role it plays in governing fiber behavior during spinning and in determining the mechanical properties of the resultant yarn.

9.2 Friction in fibers

Friction, in general, is a force, which opposes relative motion between two surfaces in contact. As discussed in Chapter 2, fibers are visco-elastic in nature; their behavior differs from that given by the Amontons' law, in which μ is a material property. Its behavior is more exactly given by the model, $F = a\,N^n$, in which a and n are constants [1–4]. The value of n has been shown to vary with the deformational behavior of the material. The parameter a is analogous to coefficient of friction but is not dimensionless: it has the units of force^{1-n}. It becomes dimensionless and equal to μ only when $n = 1$. If μ is calculated by the classical equation F/N, its value varies with N.

It has been discussed elsewhere in the book that a plot of F against distance or time yields a stick-slip profile. From this, one can obtain the values of the static (F_S) and the kinetic (F_k) frictional forces, which can be used to calculate the values of the coefficients of friction, μ_s and μ_k, respectively. In staple fiber spinning, both static and dynamic frictional coefficients, between fibers themselves and between fibers and other engineering surfaces (metal, rubber, plastic, ceramics) play important roles. Figure 9.1 shows a typical profile obtained with prominent stick-slip character when friction is measured between two cotton fiber beards [4].

Subramaniam *et al.* [5, 6] found that for cotton of a variety of types, the fiber characteristics length (2.5% span length), fineness and strength have significant correlations with inter-fiber friction, with the correlation coefficients being: –0.73, +0.73 and –0.69, respectively (Table 9.1). The other fiber characteristics (e.g., maturity, convolution angle, and convolutions/cm) show low linear correlations with friction.

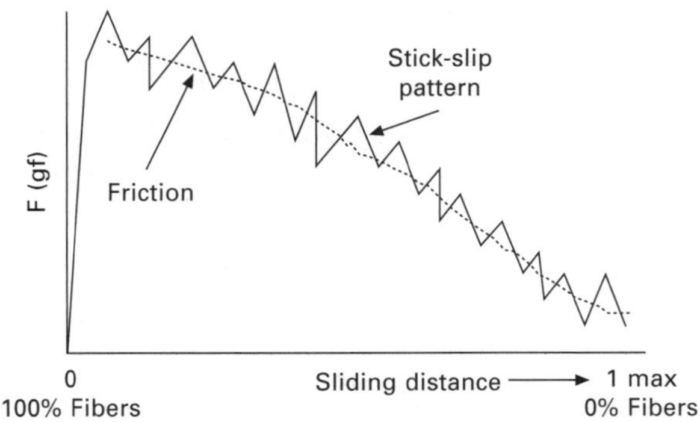

9.1 Friction profile of cotton fiber fringe moving relative to a metal block [4].

Table 9.1 Correlation between friction and other fiber characteristics of cotton [5]

Variables correlated		Correlation coefficient	Coefficient of multiple correlation
Dependent	Independent		
Coefficient of friction at 30 gf	Maturity	–0.245	0.814
	Strength at 1/8 in gauge, g/tex	–0.695	
	Convolution angle	0.326	
	Convolution/cm	0.239	
	2.5% Span length, mm	–0.733	
	Micronaire value	0.732	

9.3 The importance of fiber friction in staple yarn formation

In trying to correlate the behavior of staple fibers during spinning with their frictional properties, it should be kept in mind that the latter control only a part of the behavior. The spinnability of staple fibers depends on both the geometrical structure and the surface properties. The most important geometrical properties of a fiber affecting spinning are the linear density, length, crimp, cross-sectional shape (circular, flat, serrated, trilobal, etc.) and surface structure (i.e. the presence of scales, asperities or cavities on the surface). These surface properties define frictional behavior which may be modified by the presence of lubricating agents. Apart from these factors, the spinnability is also affected by the hygroscopic nature and static electrical behavior, which depend on the ambient conditions (humidity and temperature). Accordingly, fiber friction is an important parameter but a simple correlation should not be expected between this parameter and spinning performance of a fiber.

Coefficient of friction plays a dual role in spinning. Very low static or dynamic frictional coefficient of fiber may cause easier slippage during drafting but this may also create difficulties due to uncontrolled movement of fibers. High fiber-to-fiber friction is required for more uniform formation of card web, for development of compact slivers and rovings, i.e. less sliver and roving breaks during processing, and for production of stronger yarn. It also results in a better control of yarn withdrawal during rotor and other types of open-end spinning. However, too high inter-fiber friction could result in difficulty in separating fibers during processing and high drafting forces. On the other hand, low fiber-to-fiber friction is required for better opening at blow room and carding as well as easy and smooth drafting at different stages of processing. In spinning, dual requirements also apply to friction between fibers and machine part surfaces. Higher value is desired for efficient carding and transport of fibers by card clothing, for effectively false twisting in rotor spinning, and for imparting real twist in friction spinning. However, low friction is needed for reducing the chances of card over loading and fiber damage, and for easy withdrawal of fibers from any surface.

The handle of fibers and their behavior during yarn formation is found to correlate well with the difference between the static (μ_S) and dynamic (μ_k) coefficients of friction [7]. If the difference is large, the material will have a coarse, crunchy, feel and will lead to jerky movement of fibers during opening and drafting processes. This tendency can be minimized by adding a lubricating agent at the mixing stage that reduces the static friction and thus the frictional difference ($\mu_S - \mu_k$). Also the handle of fibers will then be softer.

In the two primary operations of lap making and carding, emphasis should be laid on the properties of fiber that promote mutual coherence. In subsequent operations, including opening actions at blow room and carding, slipping of fibers becomes the more important requirement. The amount and type of oil present on fibers, and the geometrical and surface characteristics of the latter govern the static and dynamic coefficients of friction. Roder [8] investigated the carding behavior of rayon staple fibers in relation to the difference between μ_s and μ_k. His findings are summarized in Table 9.2. According to his results, a high value of ($\mu_s - \mu_k$) led to generally a more compact and higher quality sliver.

On heating, moisture content of cotton decreases and hence the friction between the fibers also decreases [9]. When the staple fibers are subjected to different mechanical actions in the spinning preparatory machines, a substantial amount of heat is generated; this results in drying up of fibers and thus in a reduction of friction in fibers. Such change in friction due to change in moisture is largely recovered before the next process if the fibers are allowed time to equilibrate with the environment. Thus, allowing time between processes involving strong mechanical actions assists fibers to stabilize and regain their original frictional characteristics. This point is particularly critical in view of the higher operational speeds and the more condensed processing lines used in today's technology.

9.4 Changes in fiber friction during spinning

It has been observed that there is a gradual increase in the coefficient of friction as the material passes from the original untreated and unprocessed stage to the card sliver stage, with the increase being as high as 24%. With further processing, i.e. through drawing and roving stages, the frictional coefficient is reported to fall consistently, with the fall being greater for the treated than for the untreated cottons (Figs 9.2 and 9.3) [5, 6]. PEG, Waxol PA and Syton (Fig. 9.3) are the commercial spin finishes used to control fibre friction during spinning.

The initial increase in friction coefficient is, to some extent, due to the abrasive effect of beaters and grid bars in the blow room and wire points in the card, which result in damage to fiber surface. The drop in friction may be

Table 9.2 Fiber friction and type of card sliver [8]

Difference in friction, $\mu_s - \mu_d$	Observation during carding
+0.018	Compact sliver
+0.015	Normal sliver
+0.009	Normal sliver
+0.001	Reasonable sliver
<0.0	Flabby card fleece and voluminous sliver

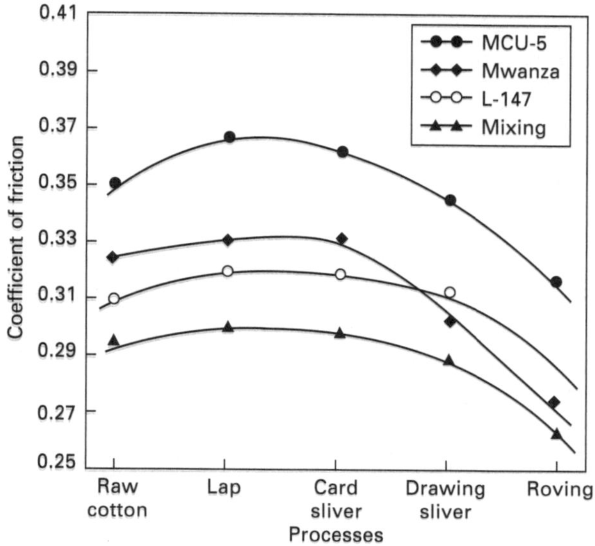

9.2 Changes in the coefficient of friction due to mechanical treatments of different varieties of untreated cottons having differences in friction only [6].

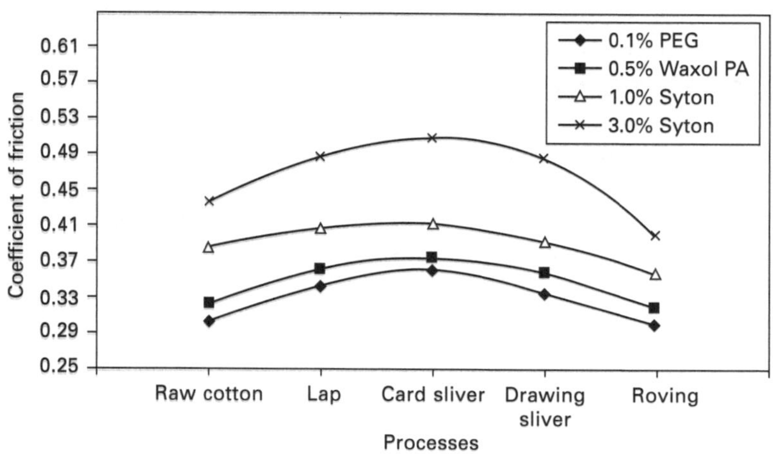

9.3 Changes in the coefficient of friction due to mechanical treatments in treated cottons (*L*-147) having different types and levels of finishes [6].

due to ironing of fibers by drafting rollers and consequent decrease in asperity heights. This effect may be expected to be greater when the number of fibers in the cross section is smaller. In the case of 'dull' polyester staple fibers, it has been observed that the average number of asperities on the fibers decrease during processing due to rubbing of polymer and the removal of the pigment particles. The repeated draw frame passages also resulted in reduction in fiber friction, which was mainly due to removal of crimp and straightening of fibers.

9.5 Control of fiber friction by spin finish

An important function of spin finish is to modify fiber-fiber and fiber-other surface friction during spinning. A defect during drafting appears as drafting waves that are characterized by thick and thin places. The amplitude of such waves during a drawing process may be minimized by increasing control over movement of short fibers, which may be accomplished by the use of a suitable finish that increases inter-fiber friction. Such control may also be achieved by adjusting friction between floating fibers and machine components, such as aprons in roving frame and ring frame or pins in gill drawing system, which move at some speed along with the floating fibers.

Natural fibers usually contain lubricating compounds (e.g., natural wax or grease) that assist in efficient spinning. Manufactured fibers, with their smooth surface and usually high coefficient of friction, must be treated with a spin finish to assist in processing [9]. The parameters that play a role are the finish type, viscosity, and the quantity applied. Uniform application over the entire surface of fibers ensures effective running performance. Figure 9.4 shows that in actual spinning of staple fibers with spin finish, there are two types of friction encountered, i.e. solid to solid and solid to fluid, and both types depend on the nature and the add-on of finish. These frictions arise together but with different magnitudes depending on the processing conditions. The solid-body friction occurs where more or less direct contact exists between two fibers or between a fiber and a machine component. Generally, the solid-fluid friction occurs in a zone adjacent to the solid-solid friction zone. The latter friction is of concern in that it determines the wear on fiber and on machine parts while the former controls the drafting properties of fiber

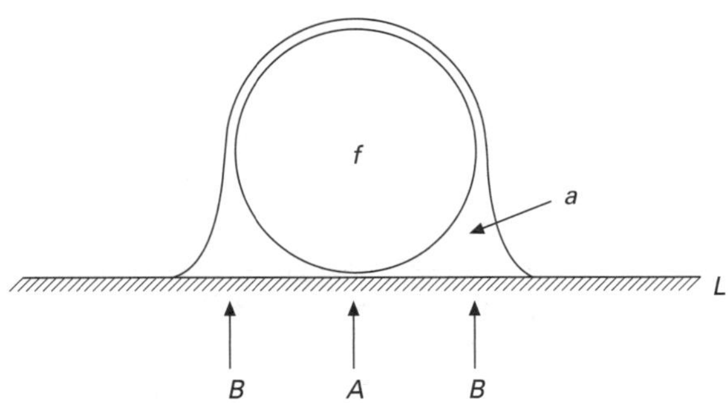

9.4 Solid-body friction (*A*) and fluid friction (*B*) when a fiber (*f*) surrounded by spin finish (*a*) glides over a solid body (*L*) [10].

strand and fiber flow characteristics over different machine components. The phenomenon of solid-body friction occurs when either the application of spin finish is not uniform or the quantity and the type are not adequate. Ideally, the finish should uniformly surround the fiber and be bound to the surface during the entire process, and it should not leave and smear on the machine parts. Usually, however, on account of surface energetics, a spin finish has a high affinity for wetting a metal surface and, therefore, tends to smear on it. This can be detrimental to the ideal functioning of spin finish as it can lead to an increase in solid-body friction. This tendency can be minimized if the surface tension of the spin finish is held below that of the fiber, assuring that the finish will remain on the fiber surface. This is achieved by the addition of surface-active agents to the finish.

9.6 Fiber friction at different stages of spinning

9.6.1 Opening and carding

In a recent study [11] the impact of fiber friction on opening behavior of fiber mass was reported. The authors showed that when the fiber to fiber coefficient of friction was low, the fibers separated easily and the process resulted in a more open fiber mass. With low inter-fiber friction, therefore, since the fibers opened easily, the impact of licker-in speed on opening performance was low. On the other hand, the fibers with higher fiber-fiber friction required higher opening force to separate them from each other and thus higher licker-in speed was needed to perform the operation.

In the two primary operations of lap formation and carding, involved in staple spinning, emphasis must be given to an increase in mutual coherence between fibers, i.e. higher fiber-to-fiber friction. In subsequent processes, however, involving drafting, the slipping of fibers, i.e. low friction, becomes desirable for smooth operation.

The fiber-to-fiber and fiber-to-card wire-point frictions have great impact on the relative movements of fiber and wire-point when one end of a fiber is considered fixed and the other is pulled by a wire-point [12]. Figure 9.5 shows a fiber held between two wire-points AC and BD, where BD is moving to the right of AC as indicated by the arrow. As wire-point BD moves, it will straighten the section of fiber between C and D, and a force T will be produced in this section. This force will have two components, $T \sin \alpha$ and $T \cos \alpha$, at point D, where α is the angle of the card wire to the surface. If the fiber, which is part of a tuft, is to move down the wire towards its base, then $T \cos \alpha$ must be greater than the frictional force F between the fiber and the wire-point. This force is approximately given by $F = \mu P$, where $P = T \sin \alpha$, and μ is the coefficient of friction between fiber and wire-point. Hence, for a fiber to move to the wire base:

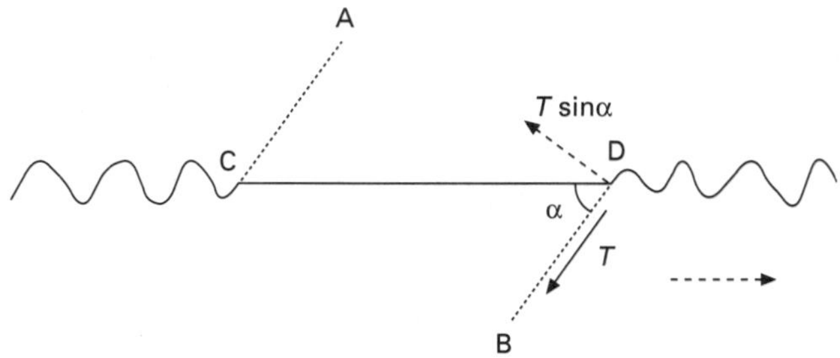

9.5 Fiber held between two pins of a card moving relative to each other [12].

$$T \cos \alpha > F \ (= \mu P = \mu \ T \sin \alpha)$$

Accordingly, if the fiber tuft is to move down the wire-point, the condition $\cot \alpha > \mu$, should be satisfied. During the initial stages of carding, it is desirable to compress the fiber tufts among the wire-points. As a result, the tufts already in the pin bed will increase the force needed to press the new tuft into the bed of pins. Hence, the angle of inclination of the wire-points and the coefficient of friction between fiber and wire-point should be selected in such a way that $\cot \alpha$ is greater than μ.

The purpose of carding is to produce a relatively uniform and continuous strand or web of fiber material. The product is an assembly of pseudo-oriented fibers in the form of a web, which is a weak and fragile network with low density. It should be as uniform as possible. The inter-fiber friction plays an important role in controlling the cohesion among the fibers and, therefore, in imparting a degree of elasticity to the web [13]. The stability of card web is important in producing a uniform yarn. A typical load-elongation curve of card web is shown in Fig. 9.6. Three distinct sections can be recognized. Initial small region (A) is due to the take-up of slackness. The next straight elastic zone (B) is due to the resistance offered by the inter-fiber friction. After this, the curve loses its smoothness and begins to undergo large local deformations due to slippage of fibers over each other. The frictional resistance offered is mainly by the mechanical interlocking of crimped fibers. The magnitude of frictional resistance (μN) is very small in card web, as N at each contact is small, determined largely by fiber weight and position in web.

It has been observed that the card waste increases with increase in inter-fiber friction (Table 9.3), and the amount of licker-in and the flat strip wastes are also quite sensitive to friction. A high coefficient of friction requires high force to open and separate fibers, increasing the possibility that more fibers will remain in clusters in the licker-in region and be removed as waste. An increase in inter-fiber friction also leads to cylinder loading. It is expected that fiber transfer from cylinder to doffer is easier and more efficient if the

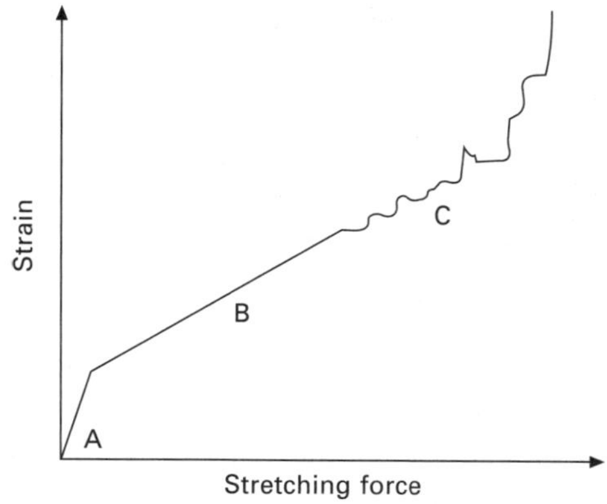

9.6 Load–elongation curve of a carded web [13].

Table 9.3 Effect of fiber friction on waste and nep count during carding of cotton [5]

Treatment	Coefficient of friction at 30 gf	Total card waste, %	Nep count, Neps/100 cm^2
Untreated	0.315	4.19	3.0
PEG, 0.125%	0.253	3.98	2.8
Waxol PA, 0.5%	0.285	4.27	2.8
Syton, 1%	0.338	5.04	3.4
Syton, 2%	0.394	5.19	4.0
Syton, 3%	0.524	6.10	7.0
Mercerized	0.386	8.41	10.6
Scoured and mercerized	0.410	8.96	10.9
Scoured, mercerized and dyed	0.383	9.94	11.7

fibers have low rather than high friction. It has been observed that high inter-fiber friction leads to the formation of neps due to poor fiber separation during processing and therefore to increase in the number of clusters in card web (Table 9.3). When the fibers are greatly matted into wires, the incidence of neps in the web becomes high. This is because the loading of cylinder and the saturation of flats not only makes them incapable of removing neps present in the feed, but also creates conditions favoring the formation of additional neps during the process. Hence, a lower static coefficient of friction should in general result in improved carding performance as observed in carding of cotton.

9.6.2 Drawing

Drafting involves the attenuation of a fiber strand caused by the difference in speeds of the front and the backset of rollers. The separation of rollers must be greater than the length of fibers to avoid the breakage of fibers; this creates a situation where there will be fibers that are not gripped by either set of rollers. These are the uncontrolled or floating fibers whose dynamics are not well understood, but they are acted on by frictional forces of surrounding fibers. These fibers move at a speed different from those of either the front or the back rollers. Knowledge of the frictional characteristics of fibers can help in understanding the dynamic behavior of these floating fibers and, therefore, in developing a control over them.

According to the theory of drafting, the dependence of inter-fiber friction on relative velocity plays an important part, since the friction-velocity relationship affects the movement of floating fibers. In the drawing operation, the fiber-to-fiber or fiber to other surface (e.g. aprons in drafting zone) friction is of importance. Too high a dynamic coefficient of friction will yield a thin, compact sliver which does not draft satisfactorily. Too high a friction between fiber and apron or difference in friction between fibers and top apron and fibers and bottom apron may result in uncontrolled fiber movement in the drafting zone.

During drafting, two sets of forces act on the fiber web at the front roller nip. The first causes the web to be pressed into the surface of the roller which tends to carry it around. The second opposes the tendency of the web to lap around the roller by withdrawing or pulling the web from the front roller nip. The web is withdrawn due to the pull from the calendar roller which is transmitted through the fiber matrix to the front nip. An adequate inter-fiber cohesion is a prerequisite for transmitting this pull to the nip without undue stretching or rupture of web. Fibers with low crimp have poor cohesion and are therefore more prone to lapping. Such fibers can be processed effectively if the fiber-to-fiber coefficient of friction is increased without increasing the friction of fiber to machine parts.

The process of attenuation of linear assemblies by roller drafting causes a tension to be generated in the fibers in the drafting zone; this is known as drafting force. The way a fiber behaves during drafting depends on the variation of frictional forces acting on it in this zone. Single-fiber properties and dimensions, assembly size and compactness, and drafting parameters cause variations in frictional forces acting on short fibers. These variations cause fibers to move in an irregular manner in the drafting zone, which in turn lead to a variation in the mass per unit length of the drafted material. Drafting force and its variability determine the irregularity added during drafting, the number of faults generated and the drafting failures. The drafting force of any fiber strand depends mainly on two parameters, i.e. the number of fibers in the cross-section and fiber-to-fiber friction. These in turn can be said to be affected by the fiber length, fiber fineness, fiber crimp, strand size, fiber orientation, packing factor, twist and fiber frictional coefficient. Table 9.4 shows the results of drafting force in sliver for different level of fiber-to-

Table 9.4 Effect of fiber friction on drafting force of sliver at different level of drafting parameters [14]

				Drafting force (watts)		
S. No	Draft	Drafting speed (m/min)	Roller setting (mm)	Frictional coefficient, (μ = 0.3)	Frictional coefficient, (μ = 0.4)	Frictional coefficient, (μ =0.5)
1	1.2	12	40	1.684	2.061	2.278
2	1.3	12	40	2.075	2.428	2.694
3	1.4	12	40	1.527	2.572	2.802
4	1.5	12	40	0.879	1.235	2.773
5	1.4	6	40	0.955	0.999	1.040
6	1.4	9	40	1.258	1.794	2.098
7	1.4	12	40	1.527	2.572	2.802
8	1.4	12	40	1.527	2.572	2.802
9	1.4	12	45	1.149	2.354	2.743
10	1.4	12	50	0.984	2.206	2.621
11	1.4	12	55	0.962	1.998	2.558

fiber coefficient of friction and process variables, i.e. the draft, the drafting speed and the roller setting [14]. It is observed, as expected, that keeping all other parameters constant, the drafting force increases with the increase in the level of fiber friction. For a given value of fiber friction, one can expect that the number of fibers in the cross section will determine the magnitude of the drafting force.

Owing to the absence of short fibers and generally higher inter-fiber friction in man-made fibers, the drafting force associated with these fibers is higher than that expected in cotton. To have problem free drafting, higher roller weighting, lighter sliver and wider roller settings are used in draw frame for processing these materials.

The generation of drafting wave, which is a periodic variation in thickness of drafted sliver, is also influenced by the frictional characteristics of fibers. The cause is considered to be the sudden increase in velocity, which occurs in the drafting zone when a number of fibers are gripped by the front pair of rollers. This snatch by the front rollers leaves behind a thinned-out area populated by a comparatively small number of fibers not gripped by either of the rollers. These in turn reach the front rollers, but their small numbers are not able to drag forward many other fibers with them. However, the slower back rollers are sending more fibers in the drafting zone, and when these reach the grip of the front rollers they repeat the snatching process. The frequency of snatching depends on the fraction of fibers under the grip of front rollers, the fraction loose, and the static and the dynamic frictional coefficients of the fibers.

Fiber configurations in carding and drawframes

Fiber friction plays a role in the formation and removal of hooks during carding and drawing. As friction increases, the number of hooks in both card and drawframe slivers increases. Low friction is expected to facilitate transfer of fibers from cylinder to doffer, thus reducing cylinder loading and improving carding action. Subramaniam *et al.* [6] studied the effect of fiber coefficient of friction on the arrangement of fibers in card and draw frame slivers by using Lindsley's [7] method. This method is used for estimating the extent of fiber orientation in lap, sliver, etc., thus making possible a quantitative evaluation of the effectiveness of carding and drawing operations in aligning the fibers. In principle, the method involves clamping the sample in a special device, and discarding the fibers extending beyond a fixed length ($^1/_2''$) by cutting the tuft with the assistance of a straight edge. The clamped tuft is combed out carefully to remove and discard loose fibers and straighten those held on one end in the grip. The fibers now extending beyond the fixed distance ($^1/_2''$), considered to have been out of alignment in some way, are cut out and weighed (E). The fiber tuft remaining within the fixed length is also carefully cut and weighed (N). Lindsley termed the ratio E/N as an index characterizing the lack of orientation. Referring to this ratio as the disorientation index, the results of Subramaniam *et al.* on the effect of coefficient of friction on fiber configuration in card and drawing sliver are illustrated in Table 9.5. The results show that the lower the fiber friction, the higher is the removal of hooks (lower disorientation indicates lower number of hooks).

9.6.3 Combing

The efficiency of combing or the extraction of comber noil depends on the fiber-to-fiber and fiber-to-combing pin friction. If the fiber-combing pin friction is high and that of fiber-fiber low, the level of noil extraction will be high. Very high fiber-fiber and fiber-pin friction will cause high combing force and high fiber damage.

In combing of wool, an optimum quantity of lubricant, giving optimum friction, is needed for obtaining minimum noil extraction. Under ideal conditions of scouring, the merino wool may have residual grease content of

Table 9.5 Effect of fiber friction on fiber configuration given by disorientation index (DI), in slivers [6]

Characteristics	Coefficient of friction				
	0.306	0.327	0.353	0.383	0.438
Card sliver DI ($\times 10^3$)	84	103	107	133	137
Drawing sliver DI ($\times 10^3$)	22	36	59	55	64

about 0.8%. However, a lower content than this would be preferred for reducing nep formation in carding. With low residual grease content, i.e. higher friction, a lubricant is needed (2% add on) to obtain minimum noil removal in rectilinear wool combing [15]. Accordingly, all these considerations are taken into account in order to control fiber-fiber and fiber-other machine component friction.

9.6.4 Yarn formation

In unconventional spinning technologies such as rotor, air-jet and friction spinning, there is a lack of traditional mechanical control, which is inherent in ring spinning. One common feature of these systems is the existence of a gap in which fibers flow in an air-stream, as shown in Fig. 9.7 (a, b, and c). The only mechanism of control in these is a combination of fiber-fiber, fiber-metal and fiber-air friction. Thus the performance of fibers in these systems depends largely on their frictional characteristics.

Ring spinning

The friction between yarn and traveler is responsible for a large component of the winding tension. The latter is given by the forces acting on the traveler, governed by the traveler weight, the reaction of ring, the friction between traveler and ring, and the friction between yarn and traveler. Keeping all other parameters constant, the higher the coefficient of friction between yarn and traveler, the higher is the winding tension. It has also been found that with a reduction in yarn to traveler friction the balloon size increases, which sometimes creates problem in spinning.

In the ring spinning system, the fiber-traveler and traveler-ring friction not only influence running performance but also yarn quality. The friction between ring and traveler affects the speed. At very high speed, due to friction between fiber and traveler and ring and traveler, burn out damage may occur to fiber or traveler. This is due to abrasion and generation of heat at the interacting points. Suitable selection of frictional coefficients of ring, traveler and fibers is necessary to alleviate these problems in ring spinning. Additionally, the fiber-lappet guide and fiber-balloon control ring frictions are important. If these frictions are high, the spinning tension will be high and so chances of ends down and yarn surface damage will also be high.

Crank and Whitmore [17] reported a detailed study on the impact of coefficient of friction between ring and traveler and between yarn and traveler on spinning tension and balloon formation. They examined bobbin tension by measuring the forces acting on the traveler by changing the traveler weight and the frictional coefficient between ring and traveler (μ_1). The yarn tension itself was varied by changing the friction between yarn and traveler

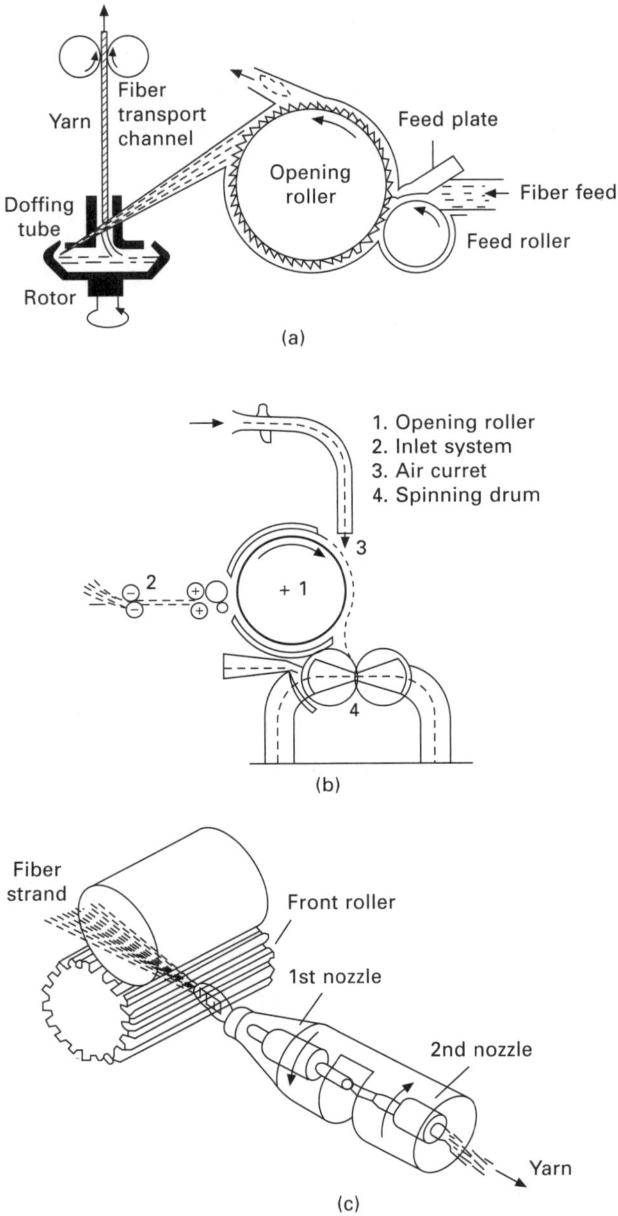

9.7 General schematics of non-ring type spinning processes (a – rotor, b – friction, c – air-jet) [16].

(μ_2). Figure 9.8 shows the relationship found between bobbin tension and traveler weight at different levels of friction between ring and traveler (μ_1) with very low friction between traveler and yarn (μ_2). The yarn tension increased with an increase in both the traveler weight and μ_1. They also

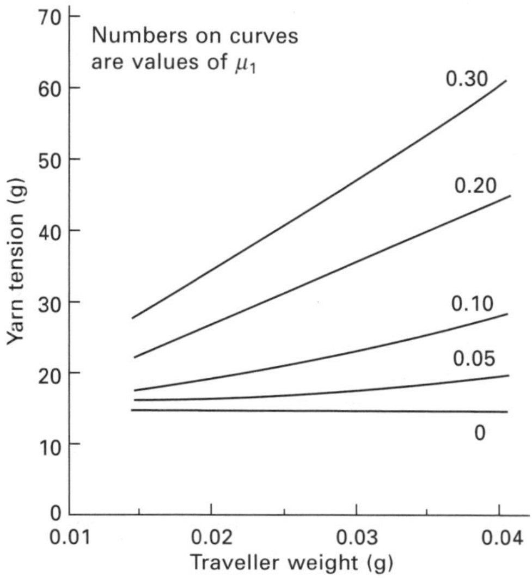

9.8 Effect of traveler weight and traveler-yarn friction on yarn tension [17].

observed that there was an upper limit to μ_1, beyond which no ballooning occurred and spinning was not possible. On the impact of frictional parameters on balloon size, they observed that for high friction between ring and traveler (μ_1) and low friction between yarn and traveler (μ_2) large balloon was obtained with a light traveler. Alternatively, for a given traveler weight, with low friction between yarn and traveler (μ_2), the balloon size increased as μ_1 decreased.

Rotor spinning

In rotor spinning, from sliver opening to yarn winding, the fiber-fiber and fiber-other material frictions play important roles in determining running performance and product quality. The action of the opening roller is dependent on the frictional interaction between fibers and wire points. Keeping all other parameters constant (opening roller speed, density and dimension of pins) the opening performance and fiber damage largely depend on fiber-fiber and fiber-pin friction. The transport behavior of fibers from opening roller to rotor surface and the nature of fiber deposition on the latter are influenced by friction between the fibers and the inner surface of the transport tube. As is often the case in situations involving friction, the requirements of friction vary among steps within a process. This is true for rotor spinning. The fiber-fiber friction should be low for opening of strand in opening roller so that the opening is efficient and gentle and does not generate fiber packets.

On the other hand, the inter-fiber friction should be high for fibers to get caught and be twisted together in the rotor groove. After yarn is made, low friction is required for it to be lifted off easily from the rotor wall without damage to fiber. A high friction between fibers and navel is required for generation of false twist that will result in lower end breakage [10].

Friction spinning

Among all yarn formation technologies, the running performance and yarn quality of friction spinning depends largely on the fiber-fiber and fiber-other surface frictions. Unlike ring spinning, friction spinning lacks positive control over fibers during assemblage and twisting. A common feature of the new systems is the existence of a gap in which fibers flow with the aid of an air stream. The mechanism of control over fibers is, thus, a combination of fiber-fiber, fiber-metal and fiber-air frictions. The twisting rate and the ability of fibers to remain together without slipping in the system are largely dependent on friction between the fibers and the surface of the perforated drums. Also, the fiber-fiber friction governs the ability of parallel fibers or filament core to be properly bound to each other and resist slippage. Therefore, a higher fiber-fiber and fiber-metal friction is desirable for the process. However, as with rotor spinning, high fiber-fiber and fiber-metal friction can be a disadvantage at the opening roller where improper opening and damage of fibers may occur. In core-sheath type yarn, the anchorage of sheath fibers within the structure and the friction between the core and the sheath fibers has a decisive effect on the integrity of yarn and the latter's tensile properties. In yarns composed of filament core and staple sheath, it has been observed [18] that the magnitude of twist or wrapper turns of sheath fibers is higher with higher staple fiber-filament friction, which in turn is responsible for higher tensile strength of yarn. The contribution of sheath fibers to strength also increases with an increase in friction of sheath fibers. In case of yarns with staple core and sheath, increase in friction of core fiber, sheath fiber, or both, results in a significant increase of false twist in core and of real twist and of wrapping of sheath over the core. Table 9.6 shows the effect of fiber-fiber friction on tensile behavior of 50 tex friction spun yarn [18]. The 'friction-tenacity' plot in Fig. 9.9 shows a good correlation, with R^2 of 0.87.

Air-jet spinning

As in other systems, friction plays an important role in air-jet spinning in controlling the formation of yarn and determining its quality. The outer layer fibers wrap around the core with the help of high-pressure air jet; the friction between fiber surface and air affects the wrapping efficiency. The friction between the fibers and the inner surface of the nozzle controls the movement

Table 9.6 Effect of fiber friction on properties of friction spun yarn [18]

Sample no.	Fiber/fiber frictional force (cN)	Tenacity, (cN/tex)	Breaking extension (%)	Initial modulus (cN/tex)	Energy to break ($\times 10^{-2}$ J)
1	20.0	11.1	18.3	223.5	32.7
2	20.3	11.6	17.8	252.6	32.8
3	20.6	12.4	18.3	259.6	36.0
4	20.3	11.5	18.6	232.1	33.6
5	20.8	12.3	18.1	259.3	35.6
6	21.2	12.6	18.2	260.5	37.3
7	20.6	12.4	18.5	254.1	36.5
8	21.3	12.9	18.4	263.4	37.6
9	21.8	13.1	18.5	264.3	39.4

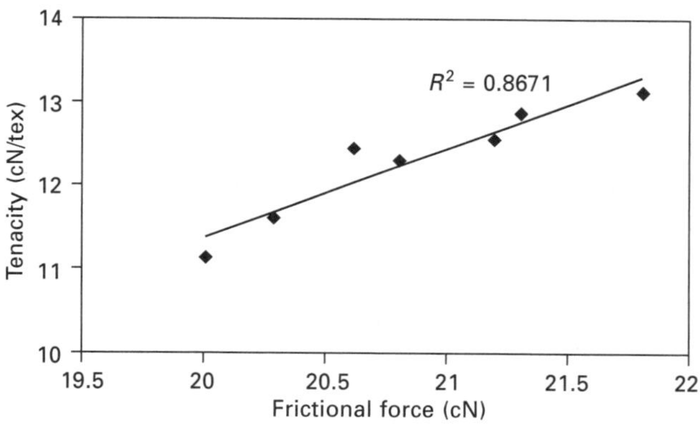

9.9 Effect of fiber-fiber friction on tenacity of friction spun yarn [18].

of fibers through the latter. A very high value of this friction disturbs the flow of fibers by restricting the movement. In one study, conducted on air-jet spun yarns, it was observed [19] that the tenacity and the flexural rigidity of the yarn increased considerably as the add-on finish level increased, which increased the fiber-fiber and fiber-metal coefficients of friction. The increase in flexural rigidity could be expected to be due to a decrease in the freedom of movement of fibers with increase in friction.

9.7 Product quality

In the cotton system, the fibers are short and fine and hence the coherence of fibers is particularly a decisive factor affecting processing. Increase of cohesion by application of finishing agents, which increase static friction, is desirable for improving the evenness of yarn, but, at the same time, it has been seen to produce an increase in neps in yarns, as seen in Table 9.7 [8]. Although the

Table 9.7 Effect of change in fiber friction of polyamide fiber with addition of colloidal silica solution, Syton WX, on neppiness and yarn evenness [8]

% Syton added	Coefficient of friction of fiber		Yarn evenness (%)	Neppiness*
	Static	Dynamic		
0	0.183	0.169	24.0	1
0.1	0.206	0.184	23.5	2
0.2	0.210	0.193	23.4	3
0.3	0.212	0.196	21.3	4
0.4	0.213	0.196	20.4	5
0.5	0.221	0.200	19.9	6
0.6	0.221	0.200	19.0	7

*1 means best and 7 means worst

evenness of yarn increased with increase in friction, the neppiness also increased greatly.

In spinning chemically treated fibers, e.g. scoured, bleached, dyed, and mercerized, which have high inter-fiber friction, a lubricant finish must be added in order to reduce friction and improve processability at the opening and carding stages. Since different chemical treatments given and lubricants added to enhance processability produce different frictional results, accurate assessment of the effects of these on friction should be made in order to determine the conditions optimum for lubrication treatment.

Short-term variations in uniformity of card sliver are influenced by fiber friction. In general, low inter-fiber friction facilitates orderly movement of fibers during carding and drafting and improves the regularity of sliver, whereas high inter-fiber friction shows the opposite effect. When cotton fibers are scoured to remove surface wax and other impurities, inter-fiber friction goes up and the tendency of fibers to cluster increases [10]. This causes an increase in sliver unevenness.

The importance of inter-fiber friction in influencing yarn strength is acknowledged widely. Broughton and El Mogahzy [20] showed that inter-fiber friction could be a dominant factor in determining tensile properties of a ring spun yarn; an increase of 59% in inter-fiber friction can produce a large increase (900%) in yarn strength, keeping other parameters constant. In a normal yarn, the frictional forces should exceed yarn strength so that it fails by rupture. If, on the other hand, friction is low and the tension exceeds it, the fibers will have a tendency to pull apart and lead to yarn rupture by fiber slippage. For a fiber with a particular value of friction, as yarn twist increases the frictional resistance between the fibers increases; this increase will continue until a point is reached at which the tendency of the fibers to slip is overcome [15], as shown in Fig. 9.10. As hypothetically illustrated in Fig. 9.11, keeping all other fiber properties constant, a higher frictional coefficient will enable higher maximum yarn strength (S_3) to be obtained at

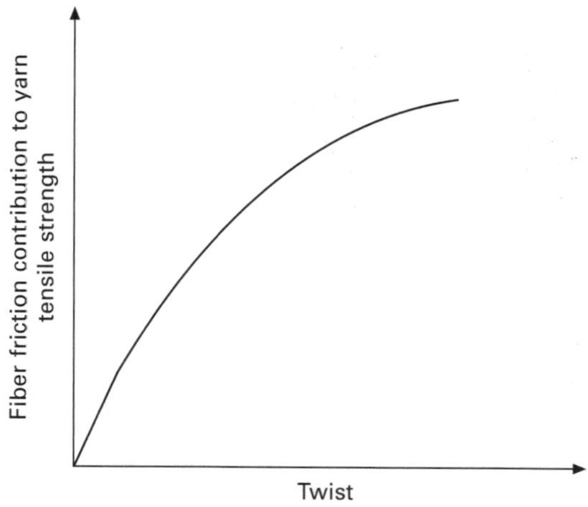

9.10 The influence of yarn twist on fiber friction-yarn strength relationship [15].

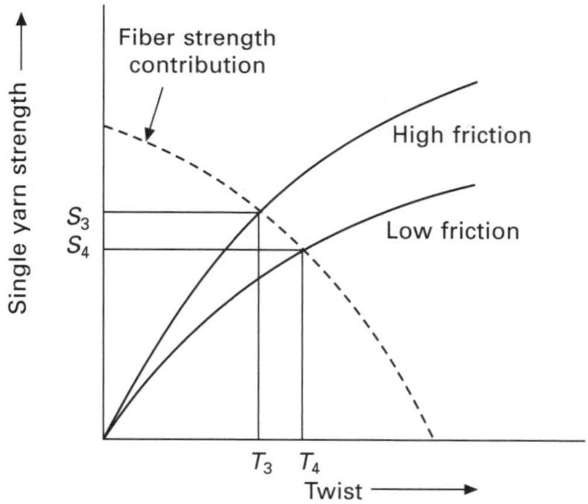

9.11 The influence of fiber friction on the twist-yarn strength relationship [15].

lower twist (T_3) than would be possible if the same fiber had lower frictional coefficient (S_4, T_4). Clearly, therefore, the requirements on friction are not simple; high is required for mechanical integrity of yarn but low is required for proper functioning during many stages of the staple process. Viswanathan [21] reported a negative correlation coefficient of –0.60 between yarn strength and coefficient of friction measured at low normal load. This is because yarn irregularity is influenced directly by the coefficient of friction.

In a recent study [22], the combined effects of fiber-to-fiber friction, roving hank, and roving twist multiplier, on properties of yarn have been reported (Figs 9.12 and 9.13). The increase in yarn tenacity with increase in

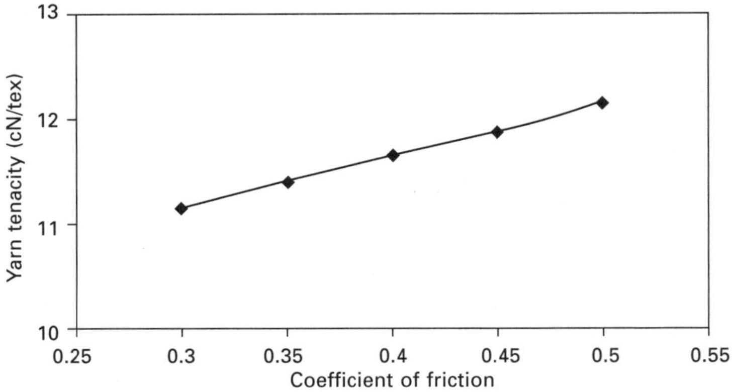

9.12 Effect of fiber-fiber friction on tenacity of ring-spun yarn [22].

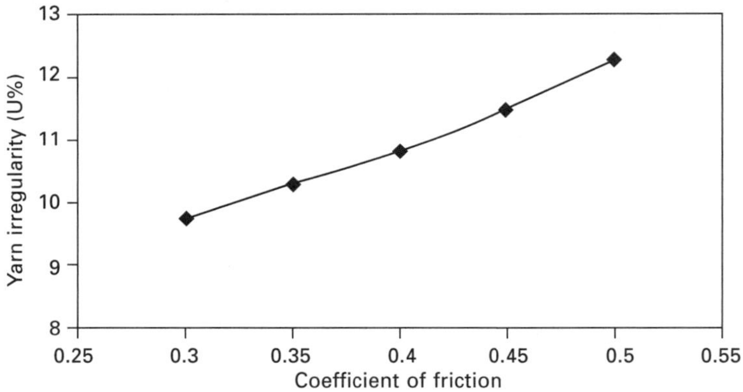

9.13 Effect of fiber-fiber friction on irregularity of ring-spun yarn [22].

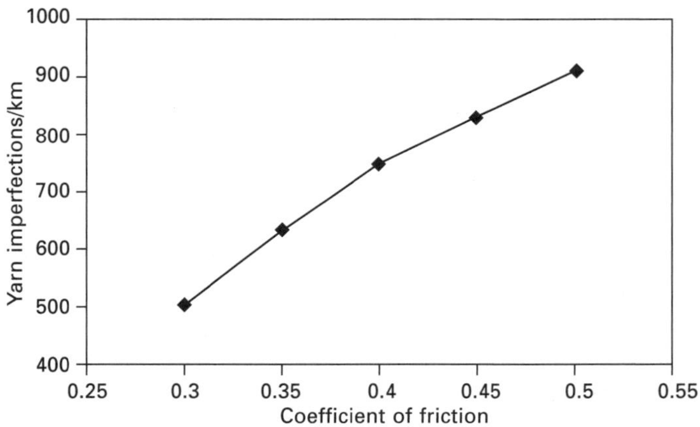

9.14 Effect of fiber-fiber friction on imperfections of ring-spun yarn [22].

friction is attributed to increase in inter-fiber cohesion. Increase in friction, however, significantly increases yarn irregularity, which is due to improper drafting of rovings. Other defects also arose. Figure 9.14 shows that an increase in the fiber-to-fiber friction caused an increase in total imperfections, i.e. thick places, thin places and neps, in a unit length of yarn.

9.8 References

1. Bowden F P and Tobar D, *The friction and lubrication of solids*, Oxford, London, 1950.
2. Howell H G and Mazur J, 'Amontons' law and fibre friction', *J. Text. Inst.*, 1953, **44**, T59–T69.
3. Gupta B S and El Mogahzy Y E, 'Friction in fibrous materials-I: Structural model', *Textile Res. J.*, 1991, **61**(9), 547–555.
4. El Mogahzy Y E and Broughton R M, 'A new approach for evaluating the frictional behavior of cotton fibers. Part I: Fundamental aspects and measuring techniques', *Textile Res. J.*, 1993, **63**(8), 465–475.
5. Subramaniam V, Sreenivasan K and Pillay K P R, 'Studies in fiber friction. Part I: Effect of friction of fiber properties and processing performance of cotton', *Indian J. Text. Res.*, 1981, **6**, 9–15.
6. Subramaniam V, Sreenivasan K and Pillay K P R, 'Studies in fiber friction. Part II: Influence of fiber friction on cohesion, configuration and evenness of fiber assemblies', *Indian J. Text. Res.*, 1981, **6**, 16–21.
7. Lindsley C H, 'Measurement of fiber orientation', *Textile Res. J.*, 1951, **21**, 39–46.
8. Roder H L, 'The relation between fiber friction and the behavior of fibers and yarns during processing', *J. Text. Inst.*, 1955, **46**, P84–P100.
9. Morton W E and Hearle J W S, *Physical properties of textile fibers*, The Textile Institute, Manchester, 2nd edition, 1975.
10. Klein W, *Short-Staple Spinning Series: Man-made Fibers and their Processing*, Vol. **6**, The Textile Institute, Manchester, 1994.
11. Ishtiaque S M, Salhotra K R, Das A and Sukhadeva N S, 'Study on fiber openness: Its impact on roving drafting force and quality of yarn', *Indian J. Fiber Text. Res.*, 2005, **30**, 26–31.
12. Grosberg P and Iype C, *Yarn production: Theoretical aspects*, The Textile Institute, Manchester, 1999.
13. Howell H G, Mieszkis K W and Tabor D, *Friction in textiles*, The Textile Institute Publication, Butterworths Scientific Publications, London, 1959.
14. Ishtiaque S M, Das A and Kumar R, 'Study on drafting force of roving: Part III: Effect of process parameters and roving irregularity on drafting force variability', *Indian J. Fiber Text. Res.*, 2004, **29**, 308–312.
15. Oxtoby E, *Spun yarn technology,* Butterworth & Co. (Publishers Ltd.), London, 1987.
16. Lawrence C A, *Fundamentals of spun yarn technology*, CRC Press, 2003.
17. Crank J and Whitmore D D, 'The influence of friction and traveller weight in ring spinning', *Textile Res. J.*, 1954, **24**, 1006.
18. Gowda M R V, Ph.D. Thesis entitled 'Influence of frictional characteristics of core and sheath on structure, twist, and tensile properties of DREF-III friction spun yarns', IIT Delhi, 2001.

19. Tyagi G K, Singh A, Gupta A, Goyal A and Salhotra K R, 'Effect of add-on finish and process variables on properties of air-jet spun polyester yarns', *Indian J. Tex. Res.*, June 2003, **28**, 163–169.
20. Broughton R M, El-Mogahzy Y E and Hall D M, 'Mechanism of yarn failure', *Textile Res. J.*, 1992, **62**, 131–134.
21. Viswanathan A, 'Some experiments on the friction of cotton fibres', *J. Text. Inst.*, 1973, **64**, 553–557.
22. Ishtiaque S M, Das A and Kumar R, 'Study on drafting force of roving: Part IV: Correlation between drafting force, roving strength and yarn quality', *Indian J. Tex. Res.*, 2004, **29**, 313–317.

10
Friction in woven fabrics

J O AJAYI, Abubakar University, Nigeria

10.1 Introduction

In addition to having systematic thread interlacing such as those found in plain, twill and rib fabrics, woven fabrics might be smooth (continuous filament yarns) or hairy (staple fiber yarns). Any protrusions may be expected to alter the resistance to motion detected when a fabric surface is either stroked tactually by a human subject or rubbed mechanically by another fabric. The frictional properties of fabrics have been accepted for a long time in the evaluation of the degree of smoothness and roughness, and perhaps taken for granted. This chapter examines the multiplicity of factors that may possibly affect fabric friction.

Many fabrics contain fibers, which may be natural, synthetic, or a blend of these, and many contain yarns in form of staple or filament fibers. The frictional properties of fabrics can be expected to depend upon the properties of their component fibers and yarns. The constructional and structural features also affect the surface morphology of fabrics and hence the frictional properties. For example, the number of stick-slip pulses and their amplitudes are closely associated with fabric sett [1, 2] and height of surface protrusions [1, 3–5], respectively. Thus, the higher the yarn sett, the greater the number of pulses, and the higher the protrusion on fabric surface, the higher the amplitude of frictional resistance. The protrusions or obtrusions of yarn crowns on fabric surfaces are calculable from the geometry of fabric structure. The effects of process variables, such as finishing treatments, fabric structure and yarn construction, on fabric friction are discussed in this chapter. The influence of experimental variables such as normal pressure, velocity of sliding, time of loading and nature of sled surface on fabric friction are also considered. These factors have been proven beyond reasonable doubt that they decide the magnitudes of the frictional parameters in fabrics. In view of the diverse nature of fabric surfaces and the fact that polymeric materials do not obey the linear relation between frictional resistance and normal reaction, a single-valued function such as coefficient of friction has been considered inadequate for surface characterization. For these reasons, some new frictional parameters have been proposed that characterize a stick-slip trace and more fully describe the frictional behavior of a fabric.

10.2 Definition of fabric friction

Fabric friction has been defined as the resistance to motion, which is detected when a fabric is rubbed mechanically against itself or tactually between fingers and thumb. This property determines quality features such as handle, for example smoothness or roughness, and performance features such as strength, wear and shrinkage.

The evaluation of friction has been frequently based on the simple linear relationship between frictional resistance (F) and normal load (N), known as Amontons' law, or

$$F = \mu N \tag{10.1}$$

In this, μ is the coefficient of friction. As discussed elsewhere in the book, this equation has not been found valid on visco-elastic materials such as fibers and fibrous products. Equation 10.2 gives a model that fits effectively on the F-N data in such materials.

$$F = aN^n \tag{10.2}$$

In this equation, n is the friction index and a is a constant which becomes equal to μ when n is 1. This power relation between the frictional resistance (F) and normal load (N), with the units of either force or stress, has been found to be valid on fibrous materials and textile assemblies in general. Clearly, the friction constant a has the dimensions of $(load)^{1-n}$, so that its value would depend on the units of normal force or pressure used. For polymeric materials, the value of n has been found to lie between 0.67 and 1.0, being the limits of elastic and plastic deformations, respectively [6].

A negative correlation between the values of n and a has been reported [1, 2, 7, 8]. That is, lower (higher) values of a were associated with higher (lower) values of n. Since the static frictional resistance is usually greater than the kinetic, the difference between the two and the influence of speed on them are primarily the cause of the intermittent (stick-slip) motion.

10.3 Stick-slip motion in fabrics

Such motion frequently accompanies the sliding of one body over another under a steady force. The intermittent motion has been classified into two forms, namely the regular and the irregular stick-slip traces and both types have been illustrated for fabrics [9–13] and metal surfaces [14]. A detailed discussion of the phenomena applying to fabrics has been included in Chapter 4.

It is important to recapitulate, however, that in addition to the material properties, the surface morphology and fabric compressional properties can also influence the stick-slip pattern. It should, therefore, be possible to correlate the stick-slip phenomenon with surface morphology and tactual properties. Results show that smooth surfaces generally yield low frictional resistance and amplitude of stick-slip pulses. The number of stick-slip pulses is usually high in a regular trace and low in an irregular trace. Rough surfaces usually lead to larger frictional resistance and lower pulse frequency. The parameter $F_s - F_k$ (differential frictional resistance) depends on the deformability of the surface.

Since textiles are visco-elastic materials, a longer time of contact should result in a larger deformation [6]. This should cause a steady junction growth (adhesion) and consequently an increase in the stick-slip amplitude. Recent findings [5] also show that the amplitude of the stick-slip trace is proportional to fabric compression, height of yarn crowns in woven fabrics, size of ribs in knitted structures and projection of fiber piles in weft pile fabrics. Accordingly, the topography of a fabric surface can be correlated with the nature of the stick-slip profile. The following parameters are used to characterize a friction profile (see Section 5.3.3, Chapter 5): static friction resistance F_s, kinetic friction resistance F_k, amplitude of peaks F_a, number of peaks/unit length F_n, and the difference $F_s - F_k$. These parameters describe the complete surface topography and are well correlated with the tactile sensations of smoothness, scroopiness, softness, roughness and ridginess normally felt on fabric surfaces. In parallel with the characterization of a stick slip profile, it will be useful to also characterize the topography of a fabric using a geometric model.

10.4 Geometric considerations in woven fabrics

The frictional properties of woven fabrics may be predicted in relation to surface texture from the geometric considerations of the structure [10, 11]. In accordance with the adhesion theory [14], the overall frictional resistance between two bodies, such as a fabric against a fabric, is composed of two terms, namely:

(a) The ploughing term, which is related to the relative height of surface asperities.
(b) The adhesion term, which is related to the true area of contact between the mating fabrics.

In a plain weave fabric, the ratio of the surface areas of the warp and weft yarns is expressible by the fabric balance, i.e. the ratio of the cover factors of the warp to that of the weft yarns. The height of the surface asperities can be worked out as shown in Section 10.4.1, which follows.

10.4.1 Crown height

The models of fabric-on-fabric (woven) contact for warp-over-warp and weft-over-weft motion are shown in Figs 10.1(a) and (b), respectively. A tangent to the weft yarn crown (XX^1) was taken as an arbitrary plane of the fabric surface. The hatched portion depicts the protrusions of the yarn crowns (warp) from the plane of the fabric (see Fig. 10.1 (c)).

Following a procedure similar to that used by Ohsawa *et al.* [9], Ohsawa and Namiki [10] and Zurek *et al.* [11], the relative height of the protrusion of yarn crown is calculable from the geometrical configuration. According to Pierce [12], the relationship between the yarn crimp (C), spacing (P) (i.e., 1/ threads per cm), and the crimp amplitude (h), is given by the following equation:

$$\frac{h_P}{P_T} = \frac{4}{3}\sqrt{C_P} \qquad 10.3$$

The subscripts P and T refer to warp and weft yarns, respectively. The

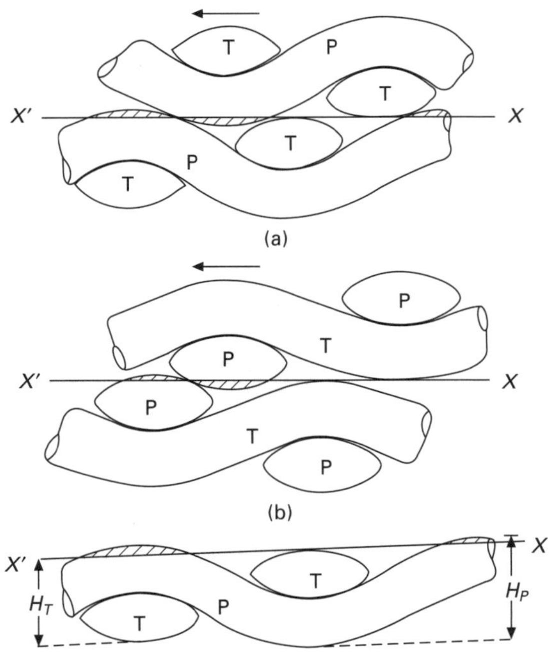

10.1 Model of fabric on fabric contact: (a) warp-on-warp motion, (b) weft-on-weft motion (c) A unit cell of a woven fabric. XX^1 = arbitrary plane of fabric surface, P – warp yarn, T – weft yarn. Arrows indicate the direction of sliding.

distance between the planes tangent to the warp and weft yarn crowns XX[1] (Fig. 10.1(c)) are given by Equations 10.4 and 10.5, respectively:

$$H_P = h_P + b_P \quad 10.4$$

$$H_T = h_T + b_T \quad 10.5$$

where, H is the distance between the planes tangent to the yarn crowns, and b is the minor axis diameter of the flattened yarns.

In a balanced yarn system, where the warp and weft sett and linear density are similar, it will be evident that:

$$h_P + h_T = b_P + b_T \quad 10.6$$

In a yarn system, in which linear densities are different, the minor axis diameters of the two yarns will be different. The ratio of the minor axis diameter of warp to weft is denoted by δ, or,

$$\delta = \frac{b_P}{b_T} \quad 10.7$$

Substitution of Equation 10.7 into 10.6 yields Equations 10.8 and 10.9:

$$b_P = \frac{\delta}{(1 + \delta)}(h_P + h_T) \quad 10.8$$

$$b_T = \frac{(h_P + h_T)}{(1 + \delta)} \quad 10.9$$

Combining Equation 10.8 with 10.4, and 10.9 with 10.5, we get the equations for H_P and H_T, respectively, as shown below:

$$H_P = h_P + \frac{\delta}{(1 + \delta)}(h_P + h_T) \quad 10.10$$

$$H_T = h_T + \frac{(h_P + h_T)}{(1 + \delta)} \quad 10.11$$

The mean height of the yarn crown (H_C) can now be given as:

$$H_C = {}^1\!/_2\,(H_P - H_T) \quad 10.12$$

$$H_C = {}^1\!/_2\,\{(h_P - h_T) + (h_P + h_T)\,(\delta - 1)/(\delta + 1)\} \quad 10.13$$

The factor of (${}^1\!/_2$) represents the protrusion of warp yarn crown from the plane of weft yarns on only one side, i.e. face. The magnitude of H_C may be positive or negative. A positive value would indicate a warp yarn projection, a negative value a weft yarn projection.

10.4.2 Fabric balance

Generally, fabric balance is defined as the ratio of the cover factors of warp to weft yarns [9]. However, the amount of crimp in yarns is not taken into consideration in this definition. Yarn crimp may not be ignored in considering fabric balance if the surface texture of a fabric is important. In order to compute the apparent area of contact between two sliding fabrics, advantage is taken of the ratio of warp and weft yarn surface areas.

The ratio of the surface area made by warp and weft yarns within a unit cell (Fig. 10.1(c)) is expressible by the following equation [9]:

$$\beta = \frac{l_T \cdot d_T}{l_P \cdot d_P} = \frac{l_T \cdot \sqrt{T_T}}{l_P \cdot \sqrt{T_P}} \qquad 10.14$$

where, l is the modular length of yarn, d is the diameter of yarn (circular section assumed), and T is the yarn linear density (Tex). Assuming the crimp in the yarns is denoted by C, then according to Pierce [12]:

$$C_P = \frac{l_P}{P_T} - 1 \qquad 10.15$$

$$C_T = \frac{l_T}{P_P} - 1 \qquad 10.16$$

Accordingly, the fabric balance, β, is given as follows:

$$\beta = \frac{l_T \cdot \sqrt{T_T}}{l_P \cdot \sqrt{T_P}} = \frac{P_P(1 + C_T)\sqrt{T_T}}{P_T(1 + C_P)\sqrt{T_P}} \qquad 10.17$$

P represents yarn spacing i.e. 1/threads per cm, and the subscripts $_P$ and $_T$ refer to warp and weft yarns, respectively. The last equation, i.e. 10.17, takes into account the ratio of the surface areas of the crown parts (protruding parts) of the yarns making up the surface of a fabric, and can easily be related to the surface smoothness and friction. Thus, the larger the surface area of rubbing yarns, the larger will be the fabric balance and, therefore, the larger will be the frictional resistance.

10.5 The effects of fabric structure on friction

10.5.1 Fabrics considered

Structurally, the protrusion of yarn crowns and fiber tufts from the plane of fabric surface are two factors that influence fabric smoothness and frictional properties. In an investigation [13], changes in fabric quality were predicted from a systematic variation in woven fabric structure. This entailed a systematic increase in the yarn sett, with the yarn count maintained constant (Table 10.1). This would be expected to alter yarn crimp (surface boundaries), which would affect fabric smoothness and, presumably, the frictional properties. This hypothesis is considered. Another case involving a series of fabrics with piles (weft pile) is also considered.

Table 10.1 Details of plain weave cotton fabrics [3]

Fabric code	Fiber content	Fabric structure	Sett,[a] (threads/cm)		Yarn count,[b] (tex)		Yarn crimp[c] (%)		Cover factor[d]		Mass[e] (g/cm^2)	Thickness[f] (mm)
			Warp	Weft	Warp	Weft	Warp	Weft	Warp	Weft		
C_6	Cotton	Plain	36	10	21	55	5	1.0	1.7	0.7	126	0.47
C_7	Cotton	Plain	36	13	22	55	8	1.5	1.7	1.0	144	0.47
C_8	Cotton	Plain	37	15	23	55	10	2.1	1.8	1.1	154	0.47
C_9	Cotton	Plain	36	18	23	56	12	2.5	1.7	1.4	171	0.45
C_{10}	Cotton	Plain	36	21	23	56	14	3.0	1.7	1.6	204	0.47

[a] Test method BS2862 1974
[b] Test method BS2865 1974
[c] Test method BS2863 1974
[d] Cover factor = threads/cm $\times 10^{-2} \times \sqrt{\text{tex}}$
[e] Test method BS2471 1974
[f] Test method BS2544 1967 (pressure 36 cN/cm^2)

Some details of the cotton fabrics used in this work are given in Tables 10.1, and 10.2.

10.5.2 Yarn crimp

Figures 10.2 and 10.3 illustrate the effects of increasing weft yarn sett on yarn crimp. The increase in crimp is greater in the warp than in the weft yarns [1, 3]. These results are due, firstly to the greater tension imposed on the weft yarns during weaving, and secondly to their greater stiffness because of higher linear density. The photomicrographs of weft and warp sections are shown in Figs 10.2 (a) and (b), respectively.

These results can be explained with a model proposed by Zurek *et al.* [11]. Figures 10.4 (a) and (b) represent the hypothetical sketches of warp-on-warp and weft-on-weft motions, respectively. The fabric specimens were placed between parallel flat plates, and relative motion was made along the warp. In accordance with Zurek *et al.* [11], as demonstrated in Fig. 10.4, the contact points occur along the warp yarn axes and traversing the upper fabric over the lower will involve small forces (Fig. 10.4 (a)). Consequently, one can assume that the frictional resistance will be low.

The motion along the weft yarn is depicted in Fig. 10.4 (b). The warp yarns of the upper fabric easily engage with the warp yarns of the lower fabric mainly because of lower crimp amplitude of the weft yarns.

The ascending distance of the weft yarn will be shorter and a higher force will be required to raise warp over warp, or in accordance with the adhesion theory, higher force will be required to shear the mechanical interlocking of these yarns. Such a theoretical model is, of course, an approximation for woven fabrics. With fabrics woven from spun yarns, other extraneous factors such as surface hairiness, warp size, finish and color applied to the fabric will also influence the results.

10.5.3 Yarn spacing

Yarn spacing is the reciprocal of the yarn sett or threads per cm. Changing the yarn sett, yarn linear density, or both, will alter yarn spacing. As shown in Fig. 10.5, the influence of a systematic increase in weft yarn sett on yarn spacing is significant; the weft yarn spacing decreases, as expected. The warp yarn spacing, that might be expected to remain constant, actually increased. This latter effect was ascribed to the effects of denting and lifting plan during the weaving operation. For example, insertion of four ends per dent frequently caused the yarns to group together, particularly at lower sett. As the density of weft yarns increased, some separation of warp yarns occurred and the spacing between them increased.

An increase in the linear density, and, therefore, in yarn diameter of weft

Table 10.2 Details of weft pile cotton fabrics [3]

Fabric code	Fabric structure	Sett[a] (threads/cm)		Number[b] of cords per 5 cm	Cord width[c] (mm)	Cord height[d] (mm)	Mass (g/m^2)	Thickness (mm)
		Warp	Weft					
C_{12}	Velveteen	35	39	–	–	–	319	1.23
C_{13}	Corduroy	18	17	32	1.0	0.67	228	0.85
C_{14}	Corduroy	25	21	20	2.0	0.95	257	1.14
C_{15}	Corduroy	35	39	16	2.5	1.06	316	1.30
C_{16}	Corduroy	18	23	8 (8)	4.5 (2.1)	1.18 (0.84)	382	1.42

[a] Sett of base weave (plain)
[b] Test method: microscopic (essentially a traversing microscope with eye piece)
[c] Test method: projection microscope
[d] Two types of cords were used; values in parentheses refer to smaller cord

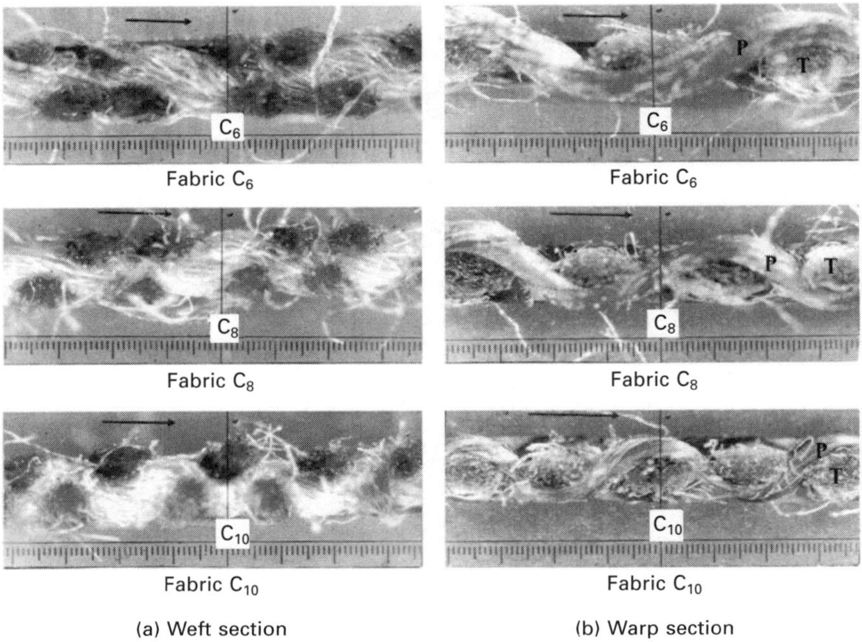

10.2 Effects of increasing weft yarn sett on yarn crimp of fabrics [1].

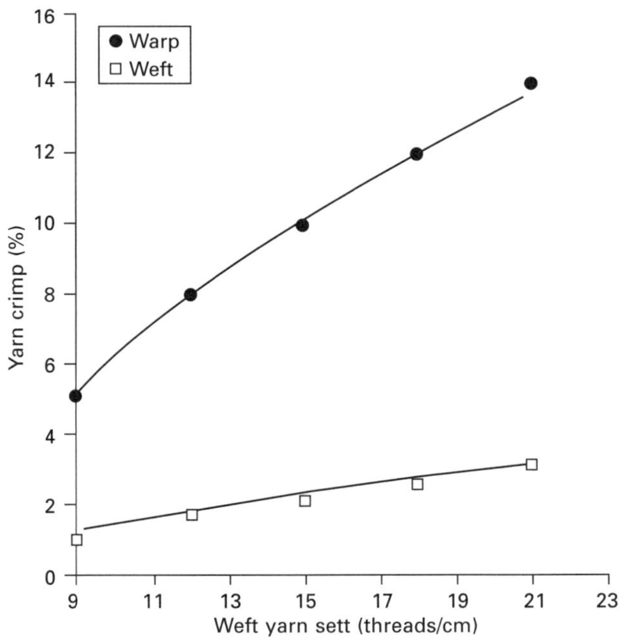

10.3 The effects of increasing yarn sett on yarn crimp [1, 3].

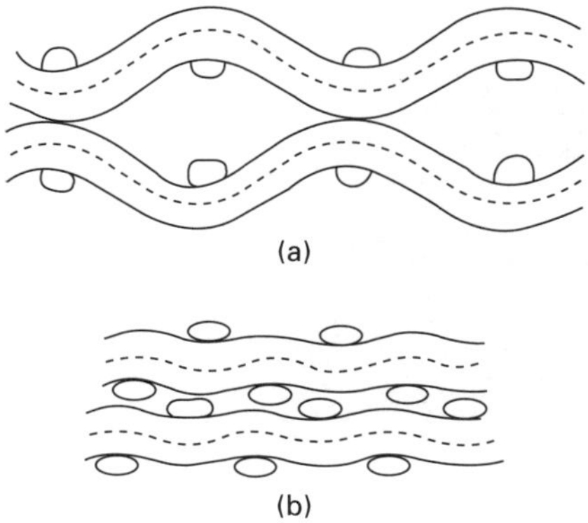

10.4 Hypothetical sketches of (a) warp-on-warp yarn and (b) weft-on-weft yarn motions [11].

yarns will cause an increase in warp yarn crimp; the spaces between weft yarns will also diminish. These factors will increase the frictional resistance along the weft.

10.5.4 Crown height

The projection of yarn knuckles above the plane of the fabric surface is referred to as the crown height. The relation between the crown height and other geometric parameters in a woven fabric has been worked out in Section 10.4.1. The experimental and calculated values of warp and weft projections and calculated values of crown height are shown in Table 10.3 for fabrics C_6 through C_{10}. The results of the geometric parameters assessed from measurements on a projection microscope are also displayed [15].

In spite of an increase in yarn crimp as the density of consolidation of weft yarns increased, the magnitude of the yarn crown height decreased consistently. The surface of the fabrics also felt smoother and more regular. The diminishing values of yarn crown height may be due to a decrease in the modular length of warp yarns [1, 3, 5, 15] or, perhaps more likely, due to the diminishing differences between the crimp balance of the two sets of yarns. In this assertion, the values of H_P derived from Equation 10.11 decreased from C_6 through C_{10}, while that of H_T derived from Equation 10.12 remained roughly constant or varied only slightly. The net effect of these changes when included in Equation 10.13 was a decrease in H_C for both theoretical and experimental values.

10.5.5 Fabric balance

As stated earlier, the relative area of contact between two fabrics is expressible as fabric balance. Theoretically, the frictional resistance to motion should

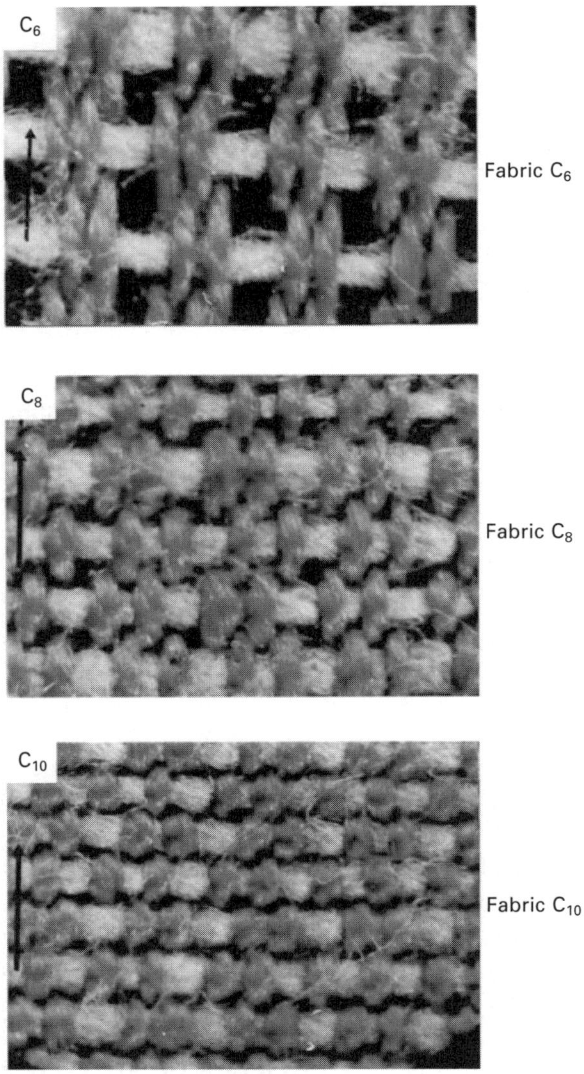

10.5 Effects of increasing weft yarn sett on yarn spacing.

increase if the relative area of contact (fabric balance) is increased. The frictional properties of the five sets of fabrics under consideration in this section are plotted in Fig. 10.6 against fabric balance, estimated from Equation 10.17. An increase in fabric balance increased the frictional resistance as expected.

It is interesting to note that an increase in yarn sett decreased the crown height (Table 10.3). This resulted in a more regular, compact, and smoother fabric surface. On the other hand, an increase in yarn linear density increased

Table 10.3 Geometric parameters of plain weave cotton fabrics

Fabric code	Calculated values (mm)*				Experimental values (mm)**			
	H_P	H_T	H_C	δ	H_P	H_T	H_C	Crimp (%)
C_6	0.54	0.30	0.12	0.38	0.53	0.22	0.16	5 (1.0)
C_7	0.53	0.31	0.11	0.40	0.53	0.28	0.13	8 (1.5)
C_8	0.52	0.33	0.10	0.42	0.47	0.33	0.07	10 (2.1)
C_9	0.50	0.33	0.09	0.41	0.46	0.33	0.06	12 (2.5)
C_{10}	0.43	0.31	0.06	0.41	0.42	0.32	0.05	14 (3.0)

H_P – height of warp yarn projection, H_T – height of weft yarn projection, H_C – crown height, δ – ratio of the minor diameters of warp to weft yarns.
* Calculated from the equations given in Section 10.4.1.
** Measured on projection microscope; values in parenthesis are for weft crimp.

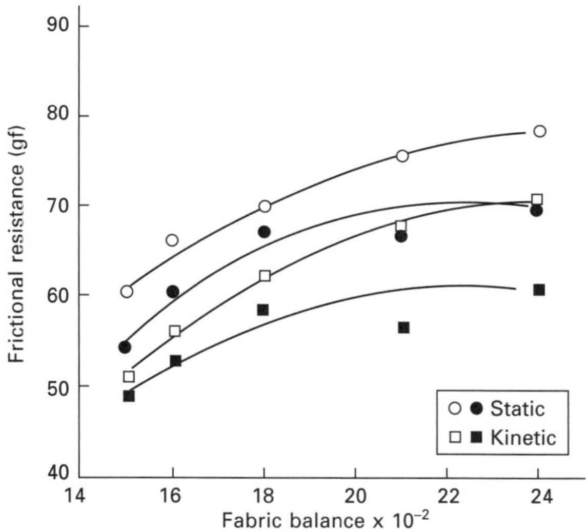

10.6 Effects of fabric balance on frictional resistance of fabrics (filled signs – warp-over-warp motion; empty signs – weft-over-weft motion) [1, 3].

the crown height [15]. This led to a coarser and rougher surface. Accordingly, surface roughness was more likely due to the natural propensity for the surface interlocking of yarn crowns with each other during relative motion [1, 11, 15]. Tactually, such roughness should be felt as fingers move across hard yarn knuckles.

10.5.6 Surface protrusions

The effects of surface protrusions such as of cords, fiber piles and ribs, on frictional resistance are very significant [1, 10]. The results on weft pile fabrics with distinctive surface cords are used as a case study here. The cord fabrics show some distinctive anisotropy in their frictional behavior, that is, the frictional properties along and across the cords as well as on the face and back are different. For example, an inspection of the results in Table 10.4 shows that the frictional resistance (static, F_s, and kinetic, F_k,), and the values of $F_s - F_k$ are usually greater, but the number of peaks, F_n, is usually lower for motion of cord-across-cord (see the results under face-to-face).

A comparison of the results of corded fabrics with those of the plain-woven fabrics will show that the magnitude of frictional resistance of the former is approximately twice that of the latter. However, unlike the plain woven fabrics, in which the frictional properties are very similar on both sides [1, 2], i.e. face-to-face and back-to-back, the resistance of the corded fabrics is significantly higher on the face than on the back.

The amplitude of the stick-slip motion was plotted against the cord height as measured on the projection microscope (Fig. 10.7). As expected, the amplitude increased for cord-across-cord motion but remained constant for cord-along-cord motion, indicating that the amplitude of frictional resistance was related to the height of protuberances on the fabric surface.

Figure 10.8 shows a linear relationship between the number of peaks of the stick-slip friction trace and the number of cords (for cord-across-cord motion) on weft pile fabrics and the yarn sett for plain weave fabrics. Thus, the frequency of spacing, and the size of the cord, which enhance mechanical

Table 10.4 The influence of fabric structure on the frictional properties of weft pile cotton fabrics

Fabric code		Frictional parameters (gf)							
		Face-to-face				Back-to-back			
		F_s	F_k	F_n	$F_s - F_k$	F_s	F_k	F_n	$F_s - F_k$
C_{12}	1	132	126	64	6	66	52	44	14
	2	134	132	63	2	60	50	51	10
C_{13}	1	74	66	40	8	66	58	43	8
	2	100	84	32	16	65	56	40	9
C_{14}	1	97	78	41	9	78	60	42	18
	2	122	108	20	14	72	58	40	14
C_{15}	1	80	78	40	2	72	63	41	9
	2	142	116	16	26	76	60	40	16
C_{16}	1	102	78	40	24	60	52	41	8
	2	158	122	8 (8)	36	80	50	40	30

1– cord along cord, 2 – cord across cord
F_s – static frictional force; F_k – kinetic frictional force; F_n – number of peaks per 5 cm displacement (see section on fabric friction measurement in Chapter 5 for further details).

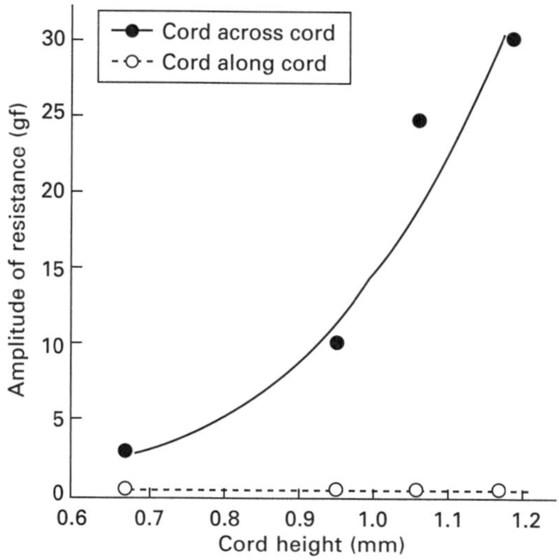

10.7 The influence of cord height on the amplitude of frictional resistance [1, 3].

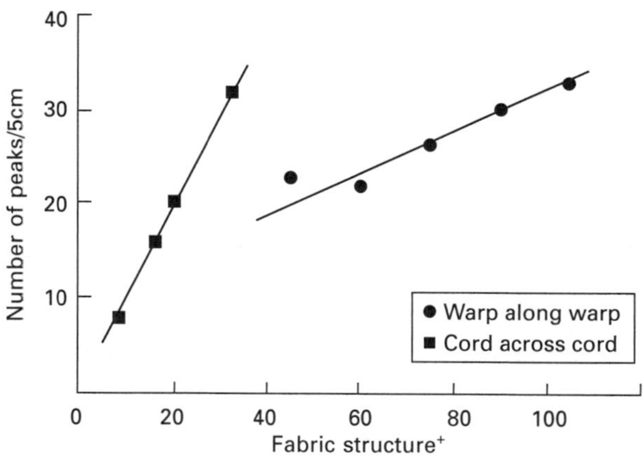

10.8 Relationship between the number of stick-slip peaks and fabric structure. +(For weft pile fabrics, number of cords/5 cm.)

interlocking and intermeshing of threads, influence the nature of the stick-slip profile.

10.5.7 Resolution of friction traces

An important characteristic of corduroys in general is the 'ridginess' of the fabric surface. An examination of friction traces such as those shown in Fig. 10.9 enables such features to be characterized objectively and related to

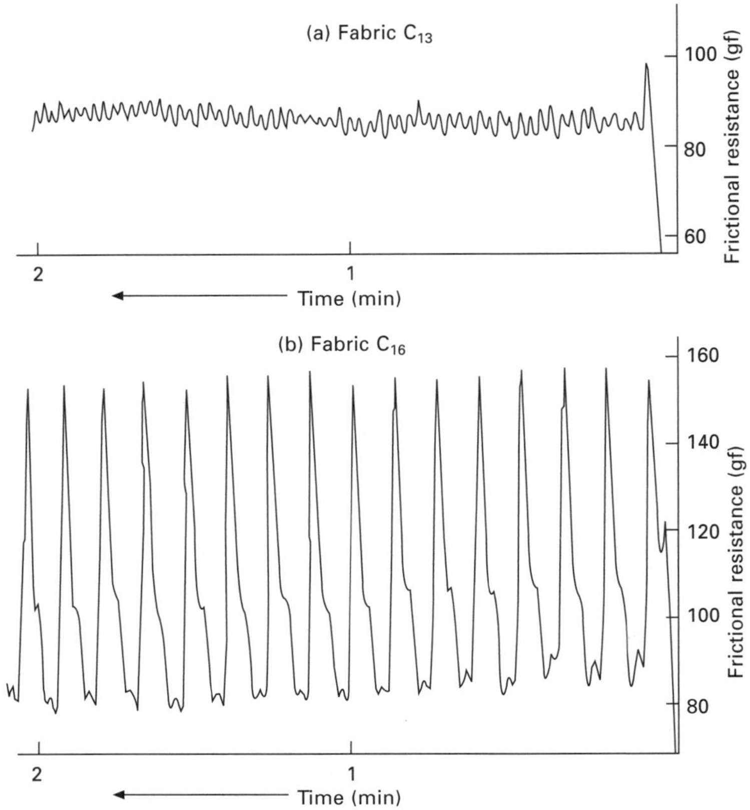

10.9 Frictional traces of weft pile cotton fabrics (C_{13} and C_{16}) for cord-across-cord motion. (Arrows indicate direction of motion) [1, 16].

fabric handle and texture [16]. These figures give examples of regular stick-slip traces.

The friction profiles of two cord fabrics are shown in Fig. 10.9. Details of these fabrics are given in Table 10.2. Fabric C_{13} is smoother than fabric C_{16}, accordingly, the number of peaks of the former is relatively greater, but the frictional resistance, amplitude and magnitude ($F_s - F_k$) of C_{13} are lower than those of fabric C_{16}. This agrees with the findings of Hearle and Husain [17], who showed that on smoother fabrics, the magnitude of $F_s - F_k$ was relatively lower and the number of peaks relatively larger than on rougher fabrics.

10.6 The effects of yarn friction

Fabrics are composed of fibers and yarns, which are assembled in forms of non-woven, knitted or woven fabrics and their frictional properties are expected to influence the frictional properties of fabrics. In a study, Ajayi and Elder [18] measured the frictional properties of wool, cotton, acrylic and viscose rayon yarns by the capstan and by the WIRA linear friction methods. The results obtained are shown in Table 10.5. The results obtained on the capstan device are much higher than the values obtained with the linear method. The yarns were then knitted into plain weft knitted structures, and the friction of the resultant fabrics determined in the dry relaxed states. A plot of the coefficient of friction of fabric on fabric against yarn on yarn, measured on the linear friction meter, shows good linear correlation (Fig. 10.10) and thus lends credence to the fact that the fabric and yarn friction are correlated.

10.7 The effects of finishing treatments on fabric friction

Finishing treatments are applied to textile materials for various reasons. This may be to improve quality features such as handle, i.e. smoothness, stiffness, softness and stretchiness, or to enhance performance features such as crease recovery, dimensional stability, wear and strength. The works of Röder [19, 20] suggested that the difference ($F_s - F_k$) was associated with the tactile sensation of smoothness of the material. A greater difference gave a material with crunchy and scroopy handle; such fabrics rustled like silk owing to a marked stick-slip motion. When the difference was moderate (0–0.015), a softer handle resulted. A negative difference matched a very soft and slippery handle. Röder's line of investigation was enterprising, unfortunately however, his results and conclusions were based on values small enough to be within experimental variations. Later works by Ajayi [1], Hearle and Husain [17], Ajayi and Elder [21], and Ajayi *et al.* [22], supported Röder's findings.

Table 10.5 Coefficient of friction of yarn on yarn [1, 18]

Fiber content	Coefficient of friction			
	Capstan[1]		Linear[2]	
	Static	Kinetic	Static	Kinetic
Wool	0.88	0.78	0.47	0.43
Cotton	0.64	0.60	0.26	0.23
Acrylic	0.43	0.43	0.22	0.20
Viscose Rayon	0.78	0.73	0.39	0.33

1 Capstan method
Capstan radius = 5.6 cm
Input tension = 10 ± 0.5 gf
Angle of lap = 3.4 rad.
Yarn velocity = 5 cm/min

2 WIRA linear method
Normal load = 50 gf
Yarn velocity = 5 cm/min

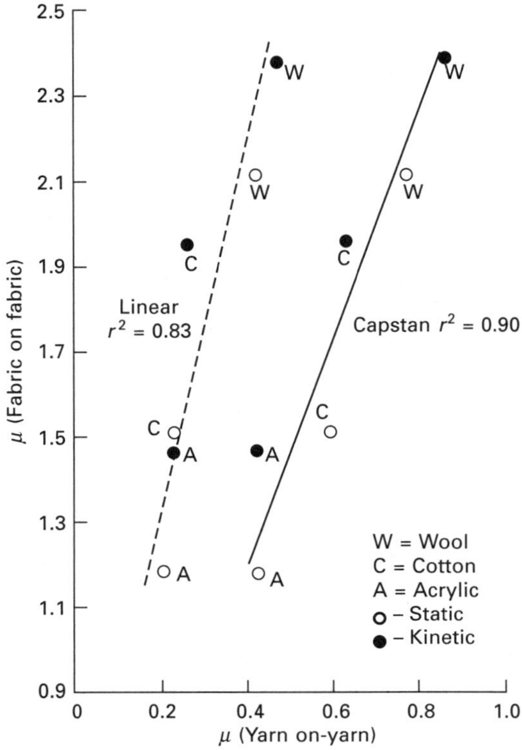

10.10 A plot of the coefficient of friction of fabric on fabric against yarn on yarn. (Filled signs – static coefficient of friction, empty signs – kinetic coefficient of friction) [18].

It is possible to simulate finishing treatments that would either increase or decrease the frictional resistance of fabrics, thereby altering their handle and other physical properties such as abrasion, wear resistance, degree of consolidation and shrinkage. The following finishing treatments are examples that affect the surface frictional properties of fabrics.

10.7.1 Scouring

Relaxation of stresses imposed upon fibers, yarns, and fabrics occurs during scouring and other wet treatments. This along with mechanical agitation causes structural consolidation, i.e. shrinkage, which in turn may be expected to cause changes in frictional resistance. It has been shown [21] that scouring produced small but consistent changes in fabric frictional properties.

The values of F_s, F_k, F_n and the difference ($F_s - F_k$) increased as the number of washes increased (Table 10.6). The scoured fabrics became harder and rougher with washing despite the inherent tendency for the fabrics to develop surface fuzziness due to wet abrasion.

Table 10.6 Effects of scouring on the frictional properties of woven cotton fabrics

No. of washes		Frictional parameters (gf)*			
		F_s	F_k	F_n	$F_s - F_k$
0	P	208	138	50	70
	T	160	115	77	45
1	P	240	155	54	85
	T	170	120	82	50
10	P	265	155	66	110
	T	240	150	98	90

P – warp direction, T – weft direction, normal pressure = 470 N/m^2
*see footnote of Table 10.4 for explanation of friction parameters.

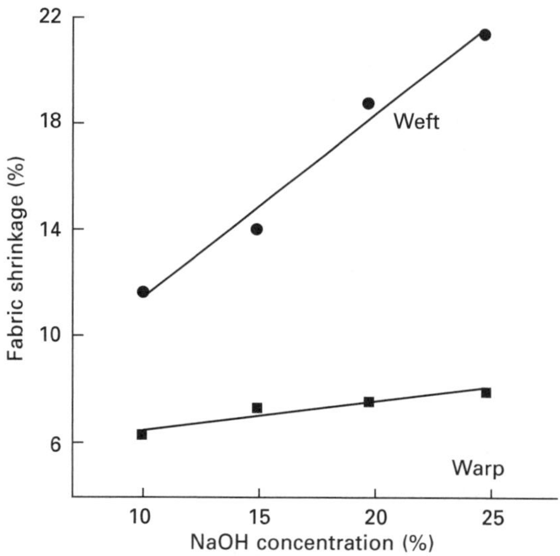

10.11 Effects of concentration of sodium hydroxide mercerization on fabric shrinkage [1].

10.7.2 Mercerization

The mercerization treatment applied industrially to cotton fabrics to improve dye uptake, absorbability, stretchiness and strength, is usually accompanied by an increase in fabric shrinkage (see Fig. 10.11) and frictional properties. This process entails the treatment of cotton fabrics with sodium hydroxide, 20–25% concentration at 0–5°C, under tension. The treatment led to increases of friction (Table 10.7) [1, 21, 23]. The increase in fabric shrinkage is also accompanied by increase in yarn crimp as shown in Fig. 10.12.

The frictional resistance was greater for the plain knitted fabrics than for the plain weave fabrics studied (Table 10.7). These results were ascribed to

Table 10.7 Effects of increasing the concentration of mercerizing solution on the frictional resistance (gf) of plain knit and plain weave fabrics [1]*

Concentration of sodium hydroxide (%)	Plain knitted fabric			Plain weave fabric		
	F_s	F_k	$F_s - F_k$	F_s	F_k	$F_s - F_k$
0	80	63	17	69	52	17
5	83	70	13	71	55	16
10	88	74	14	71	57	14
15	90	78	12	74	60	14
20	95	81	14	76	62	14
25	105	85	20	79	62	17

*See legend under Table 10.4 for explanation of frictional parameters.

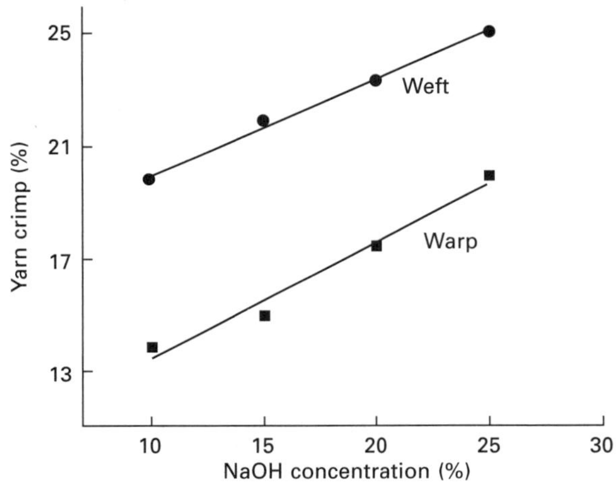

10.12 Effects of concentration of sodium hydroxide mercerization on yarn crimp [1].

structural consolidation as a result of increase in shrinkage and consequent increase in yarn crimp. The treatment led to greater shrinkage and, therefore, to greater crimp in the weft than in the warp yarns (Fig. 10.12). Zurek *et al.* [11], and Ajayi [1–3] have reported that a greater frictional resistance to motion was found in the direction perpendicular to the axes of yarns with greater crimp. Thus a higher crimp in the weft yarns gave a greater resistance along the warp. It is obvious that the barriers to motion when rubbing along the warp (weft) are the weft (warp) yarns. Accordingly, the knuckles of the rubbing and the rubbed fabrics engaged thereby restraining relative motion. We examined this concept fully when the effects of fabric structure on frictional properties were considered in Section 10.5.2.

10.7.3 Softening

Commercially, fabric softeners are applied to goods to make them softer and more pleasing to touch. Softeners are usually made up of emulsions of fatty acid amides as well as silicone. The agents frequently increase fabric smoothness and softness, but, contrary to expectations, this treatment may produce a slight increase in the frictional resistance and in the magnitude of friction difference ($F_s - F_k$) (Table 10.8).

10.7.4 Lubrication

The objective of lubrication treatment is to reduce both the static and kinetic frictional resistances with the intention of producing a smoother fabric handle. A typical result is shown in Table 10.8, in which cotton fabrics were given different treatments [21]. It can be seen that polyethylene glycol lubricants produced large reductions in the static and the kinetic frictional resistances and in the frictional difference, but there was a progressive decrease in the magnitude of change as the molecular mass of the lubricant increased from 1000 to 6000.

The possible mechanism responsible for these observations is hydrodynamic lubrication. Treated fabrics therefore may have a thick lubricant layer on their surface (depending upon the number of carbon atoms involved in the micelles formation), these lubricants or softeners may orientate themselves either parallel or perpendicular to the surface as discrete particles. Increase in friction due to increase in molecular weight is because of an increase in the viscosity of the lubricant. An additional reason for which polyethylene glycol lubricant is used is that it may also serve as an effective antistatic agent by promoting the formation of a condensed film of moisture on the

Table 10.8 Effects of chemical finishing treatments on frictional resistance (gf)* of knitted and woven cotton fabrics [1, 21]

Finishing treatments	Plain knitted fabric			Plain weave fabric		
	F_s	F_k	$F_s - F_k$	F_s	F_k	$F_s - F_k$
Original	70	55	15	65	49	16
Scoured (Teepol)	80	63	17	69	52	17
Mercerized (NaOH)	105	85	20	79	62	17
Softening (Softlan)	75	56	19	60	46	14
Lubrication (PEG 1000)	41	30	11	38	32	6
Lubrication (PEG 4000)	50	32	18	45	37	8
Lubrication (PEG 6000)	56	38	18	60	44	16
Roughening (Syton)	150	120	30	100	80	20

NaOH – Sodium hydroxide.
*See legend under Table 10.4 for explanation of frictional parameters.

fiber surface. In a study by Olsen [24], concerned with the application of five Newtonian oils to Nylon fiber substrates, it was shown that a plot of frictional resistance against the logarithm of viscosity of 2% solution of lubricant was linear. Hearle and Husain [17] have reported a reduction in the coefficient of friction of rayon card webs treated with polyethylene glycol lubricants. These authors also found that the treated sample yielded a lower value of $(F_s - F_k)$. In a study of the effect of finishing treatments of cotton fabrics, Ajayi and Elder [21] reported that the smaller the difference between the static and the kinetic frictional forces the smoother the fabric surface felt.

10.7.5 Roughening

It is not only desirable to reduce friction, for example by the application of a lubricant, it is also sometimes desirable to increase friction in order to provide excellent cohesion and firmness, for example between fibers in nonwoven fabrics and fibers in yarns. Increase in inter-fiber friction also enhances yarn and fabric strength.

The frictional properties of textile materials are frequently modified by the deposition of colloidal silica or starch in order to enhance inter-fiber cohesion during spinning or to impart transverse strength to nonwoven articles. Various workers have used these finishing treatments to achieve different objectives. Hearle and Husain [17] used the application of Syton or silica deposits to strengthen nonwoven fabrics, and Ajayi [1] and Ajayi and Elder [21] used it to increase both the static and the kinetic frictional resistances in order to simulate a rough handle of fabrics. The deposition of colloidal silica (Syton) produced a large increase (>100%) in frictional resistance and in friction difference (Table 10.8). This effect was ascribed to micro interlocking of silica deposits on the surface of the substrate.

10.7.6 Raising

The raising of fabric surfaces by mechanical method, for example by brushing, is well known. In a study by Ajayi and Elder [25], a fabric was subjected to an abrasive action by rubbing with a rotary emery roller, shown in Figs 10.13. Briefly, the apparatus consisted of a wooden roller (30 cm diameter) covered with a grade 150 carborundum sandpaper. The test specimen was lapped around the roller at an angle of about 180° and gripped at both ends with clamps. The latter were movable (upward or downward) to increase or decrease the pressure as desired between the roller and the test specimen.

The air permeability of the treated fabric, the loss in mass of fabric and the mass of detritus were determined after the raising procedure. Some typical results for a cotton canvas fabric are given in Table 10.9 [1].

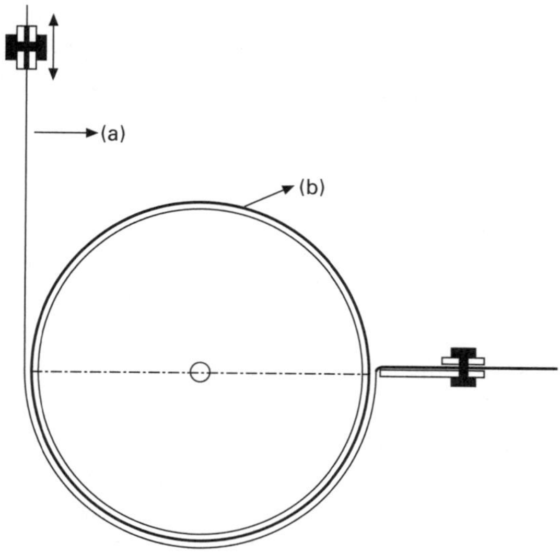

10.13 Schematic of the fabric surface raising apparatus: (a) fabric, (b) roller covered with emery paper.

Table 10.9 Effects of raising on the physical properties of plain weave cotton canvass fabric C_{11}. (Fabric details – Sett (thd/cm): warp 18; weft 12. Yarn count (Tex): warp 160; weft 180. Fabric mass (g/m^2): 500; Fabric thickness: 1.01 mm.) [1]

Time of rubbing (min)	Mass of detritus (g)	Loss in fabric mass (g)	Air permeability (cm^3/cm^2/cm/s)
0	0	0	3.0
1	0.26	30	4.0
3	0.38	43	4.5
5	0.44	50	5.8

Table 10.10 Frictional resistance (gf)* of raised (abraded) fabrics [1]

Rubbing time	Against hair				With hair	
(min)	F_s	F_k	$F_s - F_k$	F_s	F_k	$F_s - F_k$
0	28	23	5	28	23	5
1	35	30	5	33	28	5
3	38	34	4	35	30	5
5	42	37	5	37	33	4

*See legend under Table 10.4 for explanation of frictional parameters.

The frictional properties of the raised fabrics are displayed in Table 10.10. The frictional resistance increased as the time of abrasion increased. The increase is relatively higher when rubbed against the direction of hairs. The amount of detritus also increased. The difference between the loss in mass of fabric and the mass collected after rubbing presumably represented the mass of airborne dust. As the surface is raised, the air permeability increased. While the term roughening has been used to describe the treated material in the frictional context, the raised fabrics actually felt softer because of the ease of compression of the raised fibers compared to the original hard yarn knuckles. The frictional resistance measured by traversing a sled over the surface of partly raised fabric is shown in Fig. 10.14. Portions A, D, E are not raised while portions B and C are raised. The initial portion indicates the static frictional force, which decreases to a constant value (kinetic frictional force). The frictional resistance measured against the direction of hair (filled circles) is consistently greater than that measured along the direction of hair (open circles).

It is interesting to note that as the sled reached the raised portion B, the resistance to motion increased gradually over a displacement of 4 cm, corresponding to the length of the sled. The resistance to motion in the raised portion then remained steady but higher than that of the unraised portion. At the end of the raised portion D, the resistance gradually decreased to the original unraised value over a displacement of 4 cm. The high value in the raised portion is expected to be due to mechanical interlocking of surface hairs during sliding.

10.7.7 Compressibility of surface

The effects of fabric compression on frictional properties have been investigated by using chemically modified fabrics [5]. The finishing treatments consisted

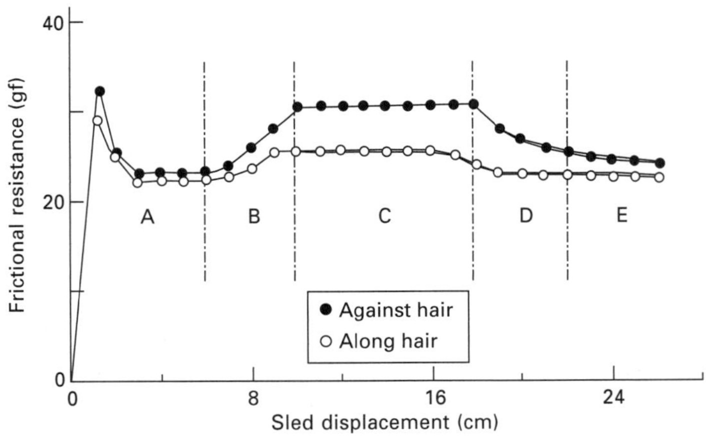

10.14 Effects of the traversing of a sled over the raised and the unraised portions of a fabric. (ADE – unraised portion of fabric surface, BC – raised portion of fabric; filled sign – frictional resistance against direction of hairs, empty sign – frictional resistance along direction of hairs) [1].

of treating a fabric with starch paste and then calendaring (R_1), scouring a fabric with soap solution of Teepol (R_2), finishing a fabric with a lubricant (polyethylene glycol) (R_3), and softening the latter with commercial softening agent (Softlan) (R_4). Compression measurements were made on a tensile tester with appropriate compression load cell [5, 21, 26]. Typical results are given in Table 10.11.

The results of load compression curves are shown in Fig. 10.15. As expected, the fabric treated with the softener was the most compressible, followed by the lubricated, scoured, and starched samples. The load-compression parameters, such as compression under a given pressure, compressive moduli,

Table 10.11 Compressive properties of fabrics

Fabric code	Compression (mm)		Compressive moduli (N/m^2)		Compressive energy (kJ)
	A	B	A′	B′	
R_1	0.0995 (8)	0.292(26)	3.74	31.5	4.56
R_2	0.125 (11)	0.314(28)	3.60	29.9	4.66
R_3	0.134 (12)	0.334(30)	3.28	27.4	4.86
R_4	0.133 (12)	0.357(32)	2.97	25.1	5.23

Values in parentheses refer to compressibility (%)
Pressure (N/m^2): A 0.196, B 1.96, A′ 200–400, B′ 1600–2000

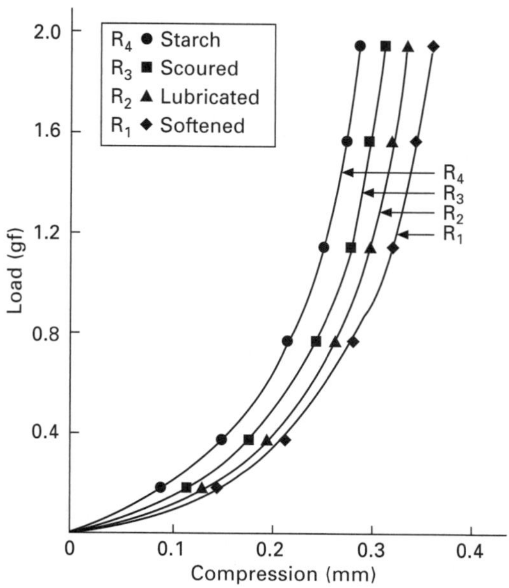

10.15 Typical load-compression curves of differently treated fabrics.

and compressive energy (Table 10.11), are useful parameters for evaluation of fabric handle, of which friction is an important component. The compressive moduli at low pressure (0.2–0.4 kPa) as well as at high pressure (1.6–2.0 kPa) diminished with increase in compressibility; this matches with the increase in tactile sensation of softness of the fabrics. The compressive energy, however, increased with increase in fabric softness.

The responses of fabrics to compression and frictional resistance tests under the same pressure were evaluated and the results are displayed in Fig. 10.16. Both the frictional resistance and the fabric compression increased with increase in normal pressure. Frictional resistance increased with compression, as expected due to increase in the area of contact. At any given value of compression, the fabric treated with starch had higher frictional resistance than fabrics treated with softening agents.

The increase in frictional resistance of starched fabric was ascribed to a possible micro interlocking of the starch deposits in the inter-stitches between yarns, and also on the peripheral boundaries of the fabric surface. Among the lubricated and the scoured fabrics, the former as compared to the latter had lower resistance and softer feel.

10.8 The effects of testing variables

In addition to the influence of fiber content, and yarn and fabric structures, and chemical finishing treatments, fabric friction is also affected by the

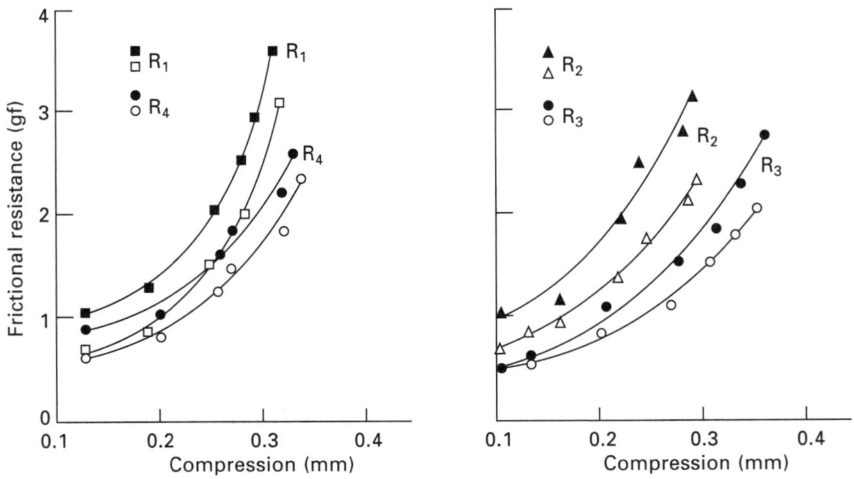

10.16 Effects of fabric compression on frictional resistance under identical pressures (R_1 – starch and calendared, R_2 – scoured with Teepol, R_3 – lubricated with polyethylene glycol, R_4 – softened with Softlan). Filled signs are static frictional resistance, empty signs are kinetic frictional resistance [5].

experimental conditions used in testing. These include normal pressure [2, 7, 10, 27], velocity of sliding [1, 22, 27, 28] number of traverses [2, 29], nature of sled surface [1, 2, 30], and temperature and relative humidity [27]. The results show that the effects of these factors on frictional properties are quite significant. They suggest that either testing conditions must be exactly noted when specifying results or, better, standard test methods should be available so that tests are conducted under identical conditions.

10.8.1 Normal pressure

The effect of increasing the normal pressure on the frictional properties of fabrics is shown in Fig. 10.17. The data fits well on a relation of the form $F = aN^n$, where a and n are friction constants. There is an excellent linear correlation ($R^2 = 0.97$) between the logarithms of frictional resistance and normal force. By means of a regression analysis shown in Chapter 5, the values of a and n were calculated. Typical results by Ajayi [2] are shown in Table 10.12.

The values of n ranged between 0.51 and 0.70, which were lower than the ones found by Wilson [31] (0.57–1.06), Ohsawa and Namiki [10] (0.85–1.07), and by Carr *et al.* [7] (0.70–0.94).

The values of a are generally negatively correlated with those of the exponent n [1, 6, 11, 31]. This was shown by Clapp *et al.* [8] as being generally expected based on a structural model presented in Chapter 3. According to the authors, a and n are related as follows (see Section 3.6):

$$a = S\, C_M\, K^{-n}\, m^{1-n}$$

Table 10.12 Frictional parameters of cotton fabrics calculated from the relation $F = aN^n$

Fabric code		Frictional parameters			
		Static		Kinetic	
		a	n	a	n
C_2	P	0.31	0.55	0.28	0.52
	T	0.26	0.61	0.25	0.55
C_3	P	0.16	0.52	0.06	0.58
	T	0.13	0.66	0.01	0.70
KC_2	C	0.49	0.55	0.45	0.53
	W	0.55	0.51	0.50	0.53

P – warp, T – weft, C – course, W – wale, unit of a is $(gf/cm^2)^{1-n}$
Fabric details – sett (threads/cm): warp C_2 28, C_3 40; weft C_2 39, C_3 20.
KC_2: wales/cm 14.1, courses/cm 15.2.
Yarn count (tex): warp C_2 10, C_3 38; weft C_2 15, C_3 58, KC_2 14.
Fabric mass (g/m^2): C_2 83, C_3 275, KC_2 185.

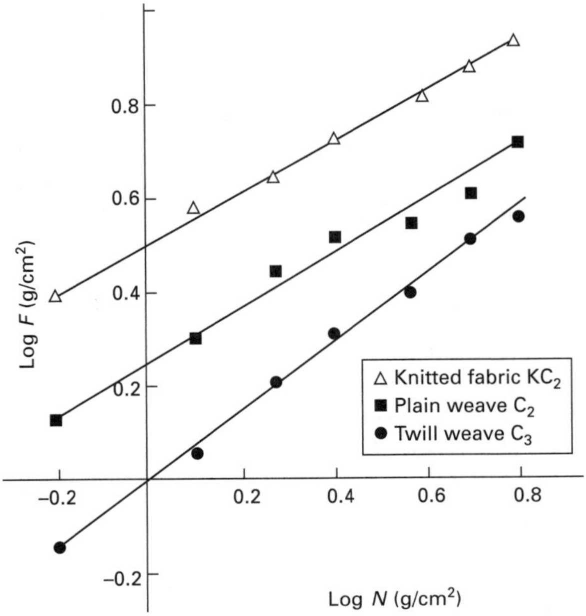

10.17 Plot of log *N* against log *F* for knitted and woven fabrics.

In this *S* is the specific shear strength of junctions, C_M is a model constant whose value is less than but near 1, *K* is the hardness factor (equivalent to compressional modulus), and *m* is the number of asperities making contact. Thus, in going from one fabric set to another, if the values of *S*, C_M, *K* and *m* either did not vary much or they changed in a manner that their combined effect was negligible, the inverse relationship between *a* and *n* would hold.

Apart from the influence of normal pressure on the frictional resistance, it is pertinent to examine the effects of normal pressure on other frictional parameters. The coefficient of friction decreased with increase in normal pressure, which indicates that the simple linear relation between frictional resistance and normal pressure was not valid on fabrics. Also, with increase in normal pressure, the amplitude of the stick-slip motion increased and the number of stick-slip peaks decreased [1–3, 5]. In some cases, the stick-slip motion disappeared altogether at higher pressure, which may be ascribed to a smoothening of fabric surface due to compression and structural flattening.

10.8.2 Velocity of sliding

An increase in the velocity of sliding implies a decrease in the time of contact between sliding members [14]. In visco-elastic materials, therefore, one could expect that the speed of sliding would affect the frictional properties. Nishimatsu and Sawaki [28, 32] studied the frictional properties and handle of some warp pile (terry towel) cotton fabrics. They investigated the effects of normal load, area of contact, relative humidity and velocity of sliding and found that the frictional resistance increased as the normal load and area of contact increased. With speed, the frictional resistance decreased first (0.5–5.0 m/s) and then increased. As implied above, with increase in velocity, the deformation is expected to decrease because of decrease in time of contact and this should result in a decrease in frictional resistance. Nishimatsu and Sawaki [28] also examined the frictional traces of fabrics at different speeds. Their results showed that the amplitude of the stick-slip pulses gradually decreased as the velocity increased.

Using a series of weft pile fabrics, Ajayi, [2] and Ajayi *et al.* [22] evaluated the effects of sled velocity on friction parameters. The results are shown in Table 10.13. The static frictional resistance, as expected, i.e. being the first peak, shows no consistent change. The number of stick-slip peaks in a given length of fabric also did not change with the velocity of sliding. The relationship between the sled velocity and kinetic frictional resistance is illustrated in Fig. 10.18. With the possible exception of fabric P_1, in which there was a consistent increase in kinetic frictional resistance, all other fabrics show an initial decline in resistance, followed by an increase. These results agree with those of Nishimatsu and Sawaki [28] for fabrics, and of several other workers [19, 24] for yarns. The initial decrease is consistent with the general hypothesis in which it is wholly assumed that the kinetic frictional resistance is deformation controlled, any increase in velocity should bring about a

Table 10.13 Effects of sled velocity on frictional properties of fabrics [22]. (Normal pressure = 400 N/m^2, time of loading = 3 min, friction – fabric on fabric)

Frictional parameters	Velocity (mm/min)	Fabrics				
		P_1	P_2	P_3	P_4	P_5
F_s (gf)	50	245	350	447	407	360
	80	315	375	467	420	373
	100	280	390	420	305	395
	200	280	405	450	350	440
F_k (gf)	50	200	340	370	297	305
	80	205	300	328	290	295
	100	230	310	327	252	225
	200	250	325	330	270	335
F_n (per/5cm)	50	10	11	26	9	9
	80	10	11	26	8	9
	100	11	11	30	9	9
	200	9	11	28	8	9

Number of cords/5 cm: P_1 8, P_2 17, P_3 35, P_4 12, P_5 13;
Cord width (mm): P_1 6, P_2 3.1, P_3 1.2, P_4 4.0, P_5 3.5;
Cord height (mm): P_1 1.56, P_2 1.16, P_3 1.68, P_4 0.48, P_5 1.27;
Fabric weight (g/m^2): P_1 340, P_2 370, P_3 220, P_4 204, P_5 369.

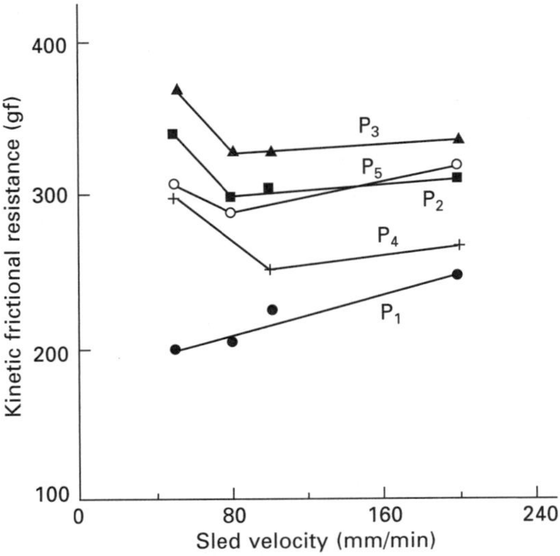

10.18 Effects of sled velocity on the kinetic frictional resistance of fabrics [22] (see fabric details in Table 10.2).

decrease in contact time of the sliding members, which in turn will reduce the resistance to motion.

It is likely that another mechanism operates at higher velocity where contact time is shorter. Under such conditions, the surface asperities may be acting as rigid bodies so that mechanical interlocking takes place that leads to an increase in resistance to motion.

10.8.3 Effects of number of traverses

Successive traverses of a sled over a fabric surface caused a decrease in the magnitude of the frictional resistance (2–34%), number of peaks (0–56%), and the amplitude of resistance (0–100%) [1, 2]. These results were attributed to surface polishing, i.e. compacting and aligning of surface hairs and other asperities in the direction of motion. This was in fact confirmed [2, 33, 34] by turning the fabrics though 180° on both the sled and platform and measuring friction. The ensuing intermeshing of hairs or asperities increased the resistance to motion. It is, therefore, important in friction measurement to employ a uniform set of conditions when comparing behavior of materials, for example by using only one traverse and a fresh sample for each test.

10.8.4 Effects of the nature of the sled surface

There are various types of materials suggested for use as the sled surface. These include cowhide [34] (in order to simulate human skin), rubber, Perspex and fabrics [2, 33, 34]. The choice was generally based on reproducibility of results, simulation of tactile assessment and availability of materials. Ajayi [1, 2] employed three different sled surfaces, i.e. fabric/fabric, rubber/fabric and Perspex/fabric, and a normal pressure of 470 N/m^2, recommended for testing of plastics and used by other investigators [1, 10]. In order to permit comparison of results, only coefficient of friction measured under identical conditions was reported [1, 2]. Results are given in Table 10.14.

Results show that the coefficient of friction for Perspex/fabric was significantly lower than for fabric/fabric and rubber/fabric. With the rubber/fabric test, the coefficient of friction was usually, but not always, greater than with the fabric/fabric test. This agrees with the findings by Yoon *et al.* [30] for knitted cotton and polyester cotton blended fabrics.

It is clear that different surfaces provide different results. The rankings of specimens with fabric/fabric and rubber/fabric were found to be generally similar. The ranking was changed if Perspex/fabric combination was used.

10.8.5 Effects of environmental conditions

Very little has been published on the effects of temperature on frictional properties of fabrics; most studies have been performed under conditions of room temperature. On the effects of moisture, Nishimatsu and Sawaki [28] measured the frictional resistance of cotton terry towels (warp pile fabrics) at 55, 65, and 90% relative humidities. The frictional resistance consistently

Table 10.14 Effects of sled surface on coefficient of friction of fabrics (normal pressure = 470 Pa, velocity of sled = 5 cm/min, number of traverse = 1, time of loading = 3 min, environmental conditions: 20°C, 65% R.H., [1])

Fabric code		Coefficient of friction					
		Static			Kinetic		
		Fabric	Rubber	Perspex	Fabric	Rubber	Perspex
C_2	P	0.74	0.88	0.19	0.57	0.83	0.17
	T	0.82	0.81	0.15	0.66	0.78	0.15
C_3	P	0.53	0.81	0.27	0.37	0.78	0.25
	T	0.59	0.76	0.21	0.44	0.72	0.16
W_1	P	0.81	0.83	0.25	0.68	0.80	0.25
	T	0.85	0.83	0.17	0.70	0.79	0.16
W_2	P	1.09	0.94	0.22	0.92	0.91	0.22
	T	1.14	0.94	0.23	0.96	0.93	0.23

Fabric details – sett (thd/cm): warp C_2 28, C_3 40, W_1 21, W_2 10; weft C_2 39, C_3 20, W_1 19, W_2 11.
Yarn count (Tex): warp C_2 10, C_3 38, W_1 30, W_2 85; weft C_2 15, C_3 58, W_1 33, W_2 100.
Fabric mass (g/m^2): C_2 83, C_3 275, W_1 121, W_2 179.

increased as the humidity increased. These results were attributed to fiber swelling and perhaps ease of compression and increase in area of contact, which, in turn favored increased resistance to motion.

10.9 Surface classification

The friction traces found on fabrics may be classified as shown in Fig. 10.19. Fabric surfaces may be regular and smooth, such as expected in nylon filament fabric, which will yield a regular stick-slip trace. The coefficient of friction, frictional resistance, and amplitude, are usually small but the number of peaks for such surfaces are large. It is also possible to have regular stick-slip motion for undulating surfaces, such as those made up of cords. The characteristics of such surfaces are large coefficients of friction and friction amplitude. Compared with smooth surfaces, the numbers of peaks are relatively fewer.

In the case of surfaces that yield irregular stick-slip motion, it is possible to have smooth and rough textures. In those having smooth textures, such as velveteens, the coefficient of friction, frictional resistance, and $F_s - F_k$ are small. In these, the friction trace is almost straight and the number of peaks

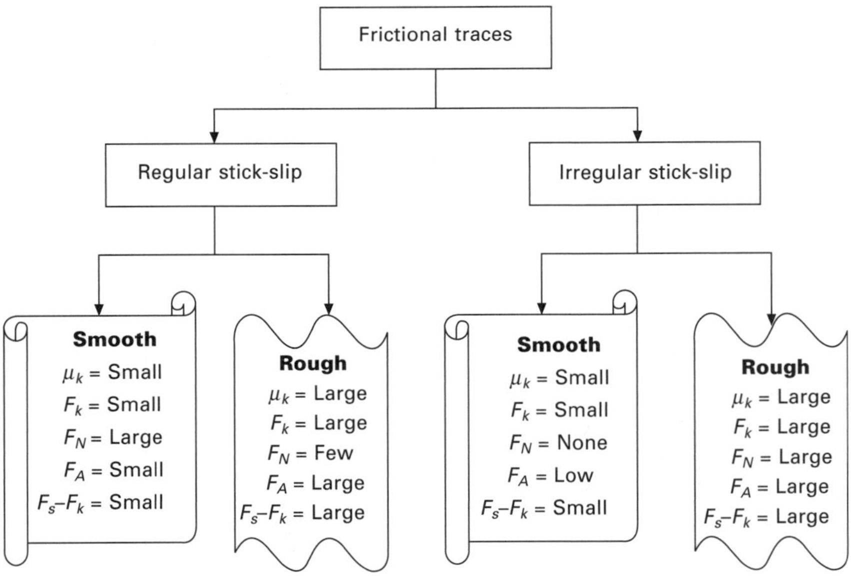

10.19 Classification of friction traces of fabrics
μ_k – Coefficient of friction,
F_k – Kinetic frictional resistance,
F_A – Amplitude of frictional resistance,
F_N – Number of peaks,
F_s–F_k – Difference between static and kinetic frictional forces.

indistinct. For irregular stick-slip motion, emanating from rough surfaces, such as canvas fabrics, the coefficient of friction, frictional resistance, and frictional amplitude are large.

The results found in this chapter broadly support this classification of fabric surfaces.

10.10 Summary

The frictional properties of fabrics are important in the determination of degrees of roughness, smoothness and other surface characteristics. Fabric friction is an important factor in determining the abrasion resistance, shrinkage and esthetic properties. The frictional properties of fabrics are sensitive to variations in fiber content, yarn mechanical properties, fabric structure, type of finishing treatments used in each case, as well as to the conditions used during testing.

Fabric friction has been accepted for a long time for objective measurement of the surface attribute of degree of smoothness. The use of the traditional parameter, the coefficient of friction, based on the classical Amonton relation, cannot fully and effectively characterize surface behavior. It is, therefore, necessary to evaluate friction in terms of additional parameters based on friction profile; the suggested parameters being frictional resistance, amplitude of friction, frequency of stick-slip motion, and the difference between the static and kinetic frictional forces.

Since friction is a surface phenomenon, fabric topography and testing system in addition to fiber content can have a large impact on the behavior. Accordingly, a fuller description of fabric frictional behavior and its understanding can be obtained by examining topography using different types of probes and microscopical examination, and combining these with subjective assessment of handle. A discussion of the probes that have been used in assessing friction in fabrics is included in Chapter 5.

10.11 Glossary

a	Friction constant
b	Minor axis diameter of the flattened yarns
C	Yarn crimp (%)
F	Frictional resistance
F_a	Amplitude of peaks
F_k	Kinetic frictional resistance
F_n	Number of peaks/unit length
F_s	Static frictional resistance
$F_s - F_k$	The differences between static frictional resistance and kinetic frictional resistance (differential frictional resistance)

H	Distance between the planes tangent to the yarn crows
h	Crimp amplitude
H_C	Crown height (mm)
H_P	Height of warp yarn projection
H_T	Height of weft yarn projection
N	Normal load
n	Friction index
P	Yarn spacing
$_P$	Warp yarn
$_T$	Weft yarn
$X X^1$	Arbitrary plane of fabric surface
β	Fabric balance
δ	Ratio of the minor diameter of warp to weft yarns
μ	Coefficient of friction

10.12 References

1. Ajayi J O, *Some studies of frictional properties of fabrics*, Ph.D. Thesis, University of Strathclyde, Glasgow, 1988.
2. Ajayi J O, 'Fabric smoothness, friction and handle', *Textile Res. J.*, 1992, **62**, 53–59.
3. Ajayi J O, 'Effects of fabric structure on frictional properties', *Textile Res. J.*, 1992, **62**, 87–93.
4. Ajayi J O and Elder H M, 'Tactile and instrumental evaluation of surface contour of fabrics', *Journal of Textile Machinery Society of Japan*, 1996, **49**, 98–107.
5. Ajayi J O, Garba E J D, Bello K A and Elder H M, 'Theoretical consideration of fabric friction', *Spectrum Journal*, 2000, **7**, 130–140.
6. Morton W E and Hearle J W S, *Physical properties of textile fibers*, Textile Institute and Heinneman, London, 1975, 611–641.
7. Carr W W, Possey J E and Ticher W C, 'Frictional characteristics of apparel fabrics', *Textile Res. J.*, 1988, **58**, 129–136.
8. Clapp T G, Timble N B and Gupta B S, 'The frictional behavior of textile fabrics', *Journal of Applied Polymer Science: Applied Polymer Symposium*, 1991, **47**, 373–382.
9. Ohsawa M, Namiki S and Kodaka H, 'Relationship between fabric balance and surface friction in plain-woven fabric', *Journal of Textile Machinery Society of Japan*, 1969, **15**(3), 98–105.
10. Ohsawa M and Namiki S, 'Anisotropy of the static friction of plain woven filament fabrics', *Journal of Textile Machinery Society of Japan*, 1966, **12**(5), 197–203.
11. Zurek W, Jankowiak D and Frydrych I, 'Surface frictional resistance of fabrics woven from filament yarns', *Textile Res. J.*, 1985, **55**, 113–121.
12. Pierce F T, 'The handle of cloth as a measurable quantity', *J. Text. Inst.*, 1930, **21**, T377.
13. Ukponmwan J O, 'Appraisal of quality of woven fabrics', *Textile Res. J.*, 1987, **57**, 283.
14. Bowden F P and Tabor D, *Friction and lubrication of solids*, Methuen, London, 1964.

15. Ajayi J O and Elder H M, 'Effects of surface geometry on fabric friction', *Journal of Testing and Evaluation*, 1996, **25**, 190–196.
16. Ajayi J O and Elder H M, 'Objective evaluation of surface characteristics to determine smoothness of file fabrics', *Journal of Testing and Evaluation*, 1994, **22**, 388–396.
17. Hearle J W S and Husain A K M M, 'Studies in needled fabrics. Part VIII: The effect of friction on the processing and properties of needle-bonded fabrics', *J. Text. Inst.*, 1971, **62**, 83–107.
18. Ajayi J O and Elder H M, 'Comparative studies of yarn and fabric friction', *Journal of Testing and Evaluation*, 1994, **22**(5), 463–467.
19. Röder H L, 'Measurement of the influence of finishing agents on the friction of fibers', *J. Text. Inst.*, 1953, **44**, T247.
20. Röder H L, 'The relation between fibre friction and the behaviour of fibres and yarns during processing', *J. Text. Inst.*, 1955, **46**, 84.
21. Ajayi J O and Elder H M, 'Effects of finishing treatments on fabric friction', *Journal of Testing and Evaluation*, 1995, **23**, 55–58.
22. Ajayi J O, Elder H M, Bello K A, Kolawole E G and Darma M U, 'Effects of sled velocity on fabric friction', *Journal of Textile Machinery Society of Japan*, 1998, **51**, 64–71.
23. Ajayi J O, Elder H M, Kolawole E G, Bello K A and Darma M U, 'Resolution of the stick-slip friction traces of fabrics', *J. Text. Inst.*, 1995, **86**, 600–609.
24. Olsen J S, 'Frictional behavior of textile yarns', *Textile Res. J.*, 1969, **39**, 31–37.
25. Ajayi J O and Elder H M, 'Effects of surface hairiness on fabric friction.' Unpublished results.
26. Ajayi J O and Elder H M, 'Fabric friction, handle, and compression', *J. Text. Inst.*, 1997, **88**, 232–241.
27. Nishimatsu T and Sawaki T, 'Study on pile fabrics. Part I: Hand of pile fabrics', *Journal of Textile Machinery Society of Japan*, 1983, **29**(4), 84.
28. Nishimatsu T and Sawaki T, 'Study on pile fabrics. Part IV: Investigation of factors affecting frictional properties of pile fabrics', *Journal of Textile Machinery Society of Japan*, 1984, **30**(4), 100.
29. Dreby E C, 'Physical methods for evaluating the hand of fabrics and for determining the effects of certain textile finishing processes', *American Dyestuff Reporters*, 1942, **31**, 497.
30. Yoon H N, Sawyer L C and Buckley A, 'Improved comfort polyester. Part II: Mechanical and surface properties', *Textile Res. J.*, 1984, **54**, 357–365.
31. Wilson D, 'Study of fabric-on-fabric dynamic friction', *J. Text. Inst.*, 1963, **54**, T143.
32. Nishimatsu T and Sawaki T, 'Study on pile fabrics. Part III: Frictional properties of pile fabrics', *Journal of Textile Machinery Society of Japan*, 1984, **30**(3), 67.
33. Thorndike G H and Varley L, 'Measurement of the coefficient of friction between samples of the same cloth', *J. Text. Inst.*, 1961, **52**, 255–271.
34. Dreby E C, 'A friction meter for determining the coefficient of kinetic friction of fabrics', *Journal of National Bureau for Standards*, 1943, **31**, 237.

11

Finishes and treatments to control friction in textile fibers

M KUTSENKO and T W THEYSON,
Goulston Technologies, USA

11.1 Introduction

11.1.1 The development of friction as a concept

Friction and its related phenomena, i.e. lubrication, cohesion, adhesion, slip-stick behavior and wear, all belong to a class of properties that span the fields of physics, chemistry and engineering. Like many fields of science, these most fundamental of concepts have resisted our efforts to move beyond simple Edisonian formulations to more complete understanding that are anchored at the molecular and atomic levels. Although the knowledge remains incomplete, we live in a world that is dominated by these phenomena: the performance of everything that moves, animate or inanimate, while in contact with another body is greatly influenced by friction.

Given friction's universality and breadth of impact, it is not surprising that it has been the subject of intense, long-term research. Ranging from the early studies of Leonardo da Vinci and Guillaume Amontons in the 16th and 17th centuries to the most recent studies in nano-tribology, the intricacies and counterintuitive behavior of friction has been widely investigated. Over time, researchers in this area have proposed and discarded a number of theoretical models. The difficulties here are well documented, along with our attempts to explain and apply the "laws of friction" to events in daily lives [1, 2].

Friction in fibers and fiber assemblies presents an additional layer of complexity relative to the inherent problems of friction phenomena:

- Issues of friction between high energy surfaces and low energy surfaces.
- The impact of fluid material characteristics (fiber finishes) at the interface between these surfaces (wetting, fluid film strength, uniformity, etc.).
- Issues related to the fact that the low energy surface of the fiber is relatively deformable and can be reversibly or irreversibly stretched during the process.
- Issues of friction, cohesion and stick-slip within the fiber bundle, itself.

At various stages, through the range of fiber processes that are practiced and the variety of products that are produced, we can find almost any combination of frictional behaviors imaginable. It is the effort to control these behaviors, in the context of high speed processes and enormous production that has led to over 50 years of technical efforts at developing suitable finishes and surface treatments. To support this work, the industry has developed equipment and protocols to measure various aspects of fiber friction and lubrication. It is to the area of friction phenomena in general and its specific application to textile products and process that we now turn.

11.1.2 Friction at work

Amontons's and Coulomb's work in the 17th and 18th centuries led to what constitutes the classical laws of friction relating to dry, sliding surfaces:

- The "friction force" that resists sliding is proportional to the normal load that is forcing the surfaces together.
- The "friction force" does not depend on the surface area (a very counterintuitive concept).
- The "friction force" is independent of speed once the surfaces are moving against one another [1].

These laws have long outlived a variety of failed efforts to explain their fundamental basis in terms of surfaces and their interactions. By the 1950s, roughness had been ruled out as a possible explanation with the observation that rough surfaces often seem to have lower frictions than less rough surfaces [2]. In the 1960s, it was the work of Tabor, Bowen and others [3] that suggested that the origins of friction might lie in the area of molecular adhesion. Their work showed that while friction was independent of apparent contact area, it was nevertheless proportional to the microscopic or "true" contact area. This seemed to suggest that friction was a phenomenon that had its basis in some type of "adhesive bonding" at the true contact points. But this theory also did not apply consistently to all materials and under all sets of conditions.

In the 1970s, Tabor's graduate student, Jacob Israelachvili showed that there was substantial friction, even in the absence of wear [4]. This put a question mark on the surface adhesion mechanism for the explanation of friction, especially in some instances, e.g. when a surface has a finish. It was up to the researchers in the 1980s and 1990s to develop a clearer picture of the true source of frictional forces:

- Gary McClelland (IBM) who co-developed the expanded concept that friction arises from atomic lattice force vibrations and developed the atomic force microscope (AFM) to study friction at the atomic scale [5].

- Jacqueline Krim (North Carolina State University) whose work on atomic scale sliding friction has confirmed that Coulomb's third law is incorrect (in systems with constant contact area, friction increased in direct proportion to speed) [6].
- Mark Robbins (Johns Hopkins University) who demonstrated that the behavior of surfaces in contact lubricated with molecular films was largely explainable within a model based on lattice vibrations and that proposed electronic effects are small in comparison [7].
- Steve Granick (University of Illinois) who demonstrated that slip-stick behavior exists even in a lubricated, molecularly flat, friction regime. The full implication of this work is still being developed [8].

Friction and the tools to study it are still areas of very active research. It is expected that early in the 21st century, we will develop a much clearer view of both the source of frictional forces and the mechanism of lubrication and wear under a variety of friction regimes.

11.1.3 Friction in fibers and textiles

Studies of the friction of fibers and fabrics have been extensive due to the importance of friction and wear in determining the design of equipment for the processing of fiber and the quality of resulting yarn and fabrics. They range from the Tabor – Howell – Mieszkis book (*Friction in Textiles*) published in 1959 [3] to the extensive studies of Martin Schick carried out in the early 1960s [9] to more recent works that are largely published as academic [10–12] or industrial articles [13, 14]. These investigations have been largely carried out in the context of traditional thinking about friction (Coulomb's laws – hydrodynamic models for high-speed lubricated friction and adhesion models for friction, abrasion and slip-stick behavior).

More recent work carried out over the last 20 years has largely centered on designing advanced equipment for conducting frictional measurements [15] or developing theoretical models for explaining the phenomenon [16]. The entire area is complex because of the involvement of a number of variables (polymer surface, fiber cross-section, yarn bundle structure, fiber finish composition, etc.) that come together, in affecting friction. The general tendency, therefore, has been to describe the behavior in terms of the force values obtained and the trends based on these values. If the normal force, or the initial tension in the case of the Capstan method, is maintained constant, then often the values of the coefficient of friction are also given (see Section 5.5.4). An enormous amount of work has been done in the area of tribology as applied to friction and lubrication of metals but there has been little effort in applying these to the area of friction and lubrication of fibers and fabrics.

The current work at companies producing lubricants for industrial application has focused on developing sensitive friction measuring equipment and methods for obtaining consistent and reproducible results. These provide effective comparative data that can be used to control the composition of a given type of lubricant or develop new lubricants for the textile industry.

11.1.4 Finishes and treatments for synthetic fiber products

Fiber finish technology was developed in conjunction with the emerging synthetic fiber industry in the 1940s to 1960s. The foundations of the technology have their base in the processing oils developed for wool in the 1920s and 1930s. The finishes that emerged in the 1950s and 1960s utilized components from a range of newly developed chemistries:

- High temperature ester lubricants developed as base lubricants for jet engines.
- Nonionic emulsifiers developed for a range of cleaning and personal care applications.
- Phosphate esters developed as hard surface cleaners.
- Quaternary salts developed as fabric softeners.
- Ethoxylated/propoxylated products developed as specialty lubricants.

The lubricants are combined with a variety of additives (antioxidants, wetting agents, varnish control agents, etc.) and applied to the fiber at point of extrusion or during the processing. Broadly, finishes provide friction control, cohesion and static protection under the range of thermal and processing conditions that are experienced by a fiber.

These finishes are designed variously, i.e. to serve as a processing aid during the early stages of fiber production, or to provide processing support for conversion of fibers into a final product or to even go beyond a processing aid by imparting desirable performance properties to the fiber/fabric structure within a final product. There had been a consistent effort, in recent years, towards developing finishes that can generate value in the final product into which they are incorporated.

While there are ever increasing demands on the performance of finishes (end use properties, high thermal stability, controlled rheology, etc.), their primary function is to control friction in its variety of forms:

- Fiber to metal friction in processing.
- Fiber to fiber friction and cohesion within a fiber bundle.
- Slip-stick behavior between fibers (scroop).
- Abrasion damage to fiber and metal surfaces.

The ultimate goal in constituting a finish is that it leads to a fiber that can be extruded, drawn, cut, bailed, made into a stable package, shipped, textured, carded, bonded, spun, twisted, woven, and converted into a textile product in an efficient, cost effective manner.

11.2 Using the Stribeck lubrication curve to analyze friction in fibers

11.2.1 General laws of friction

The three classical laws of friction have been reiterated earlier (Section 11.1.2). They have worked quite well for many macroscopic friction systems. However, it has been known for some time that textile materials do not behave according to these laws. The Capstan equation, which is commonly used to characterize friction of yarns, is derived using certain assumptions of classical behavior [3]. It describes the resistance to sliding of a flexible but inextensible cord wrapped around a cylinder. The equation is as follows: $T_1 = T_0 e^{\mu\theta}$. Where T_1 is the output tension and T_0 is the input tension, or pretension, μ is the coefficient of friction and θ is the contact (wrap) angle in radians. The Capstan equation makes the implicit assumption that the friction force does not depend on the size of the cylinder, which is practically a rewording of the second classical law of friction. A given contact angle around a larger cylinder results in a larger apparent area of contact, and if textile is used as the string, it has been found to lead to larger friction. Additionally, the dependence of friction on contact angle is shown to be closer to linear than exponential predicted by the equation (Howell, Miezikis, Tabor) [3]. Others have attempted to remedy part of this situation by modifying the Capstan equation in which the more general relation between frictional force, F, and normal force, N, is used as the base, namely, $F = aN^n$. The new Capstan relation yields $\mu = a(r/T_0)^{1-n}$ (see Section 5.5.4). In this a and n are constants, r is cylinder radius, T_0 is pretension. As discussed elsewhere (see Sections 2.14, 2.17 and 3.7), $n = 0.67$ for perfectly elastic materials and 1 for perfectly plastic materials and it lies between 0.67 and 1 for visco-elastic materials.

In work involving the application of a finish and determining its impact on friction, the Capstan method has usually been used. It is convenient in that it can be set up to measure friction on a running yarn, the Capstan pin can be changed to accommodate different types of rubbing surfaces (metals, ceramics, Teflon, etc.), and the speed and the pretension can be varied as needed. Additionally, antistatic properties of a finish can be assessed by introducing an on-line sensor during friction measurement.

11.2.2 Journal bearing theory and its applicability to fiber processes

A characteristic curve that describes the effect of lubrication on friction is given in Fig. 11.1, which relates the observed value of the coefficient of friction with three process parameters: namely speed, pressure, and lubricant viscosity. A good explanation of the curve is found in Howell *et al.*'s 1959

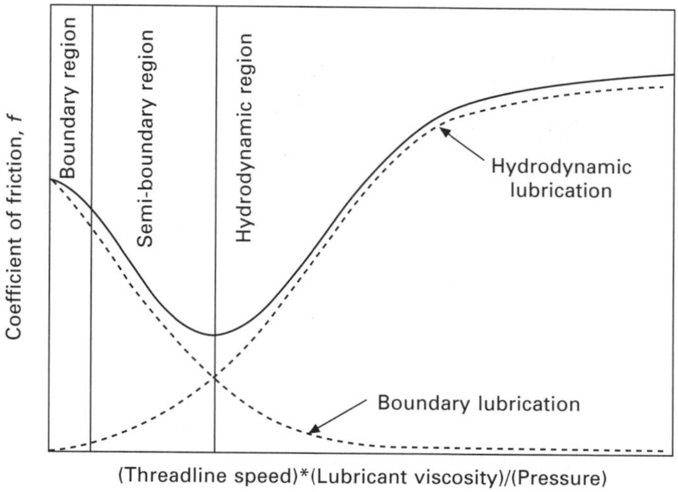

11.1 Stribeck lubrication curve characteristic of a journal bearing adapted to textile applications.

Friction in Textiles [3]. Essentially, there are three regions in the curve that have been well-characterized: the low-speed region characterized by a flat or negative slope, the higher speed region characterized by a positive slope, and the transition zone between these two. The changes in slope from negative to essentially zero and, finally, to positive correspond to a change in the mechanism of lubrication.

In the low speed/high pressure region, the lubrication is boundary. This behavior in fibers is typically found in the range of speeds 0 to 0.1 m/min. According to Schick [9], boundary friction occurs when the pressure between the two surfaces is supported by direct contact of asperities or irregularities on the solid surfaces. The lubricant that is present forms a thin layer on the remainder of the surfaces but is displaced from the actual contact area. Sliding of the two surfaces involves shearing of the solid-solid junctions and sliding over the lubricant film, with the former being the more significant process in terms of generating the friction force. Wear and surface damage can be quite severe under these conditions as there is physical interaction between the surfaces. The phenomenon described by Schick is actually the breakdown of boundary lubrication. As described by Howell, Mieszkis and Taylor [3], boundary lubrication occurs when the entire surface is covered by a molecular-scale thin film of "boundary lubricant". Wear is still acknowledged to exist, but is considerably less severe than that described by Schick. A good "boundary lubricant" has a high affinity for at least one of the surfaces, the ability to form a tightly held layer which can support the applied load.

Hydrodynamic lubrication in fibers is believed to be the controlling mechanism in high-speed/low-pressure range, typically above 4 m/min. The model for this mechanism is based on a type of machine bearing, a journal bearing [3, 9]. A journal bearing is essentially a cylindrical shaft in a cylindrical cavity of larger diameter and the space between them contains a liquid lubricant. When the shaft rotates at or above a certain speed, it forces lubricant between itself and the cavity wall. The weight of the shaft is borne by the pressurized lubricant and the apparent friction is solely the result of hydrodynamic drag. This behavior is highly dependent on the liquid's viscosity; a high viscosity lubricant will lubricate hydrodynamically at a lower shaft speed than a low viscosity lubricant, but will have a greater drag at a given speed.

The semi-boundary region is found in the transition zone between the boundary and the hydrodynamic regions. This is the region that yields minimum friction in a running lubricated system and in which sliding is desired to occur.

11.2.3 Friction profiles of lubricated and non-lubricated filament yarns

Although the Stribeck lubrication curve (Fig. 11.1) seems to be widely accepted by the textile community for explanation of frictional behavior in fibers, the somewhat controversial character of the curve has to be addressed. First of all, it is hard to accept that the journal-bearing model is directly transferable to the filament yarn-to-solid body friction process. While journal bearing represents a closed pressurized cavity, moving yarn wrapped around a solid counter body represents an open system, where leakage of the lubricant from the contact zone, especially at high pressure, is virtually unavoidable. Since the surface has a lubricant, it is hard to imagine that high friction value will be obtained at low speeds/high pressures/low viscosity conditions predicted by the Stribeck curve. In fact, the friction observed at these conditions is low, although it is accompanied by a large variation or "noise" in measured friction force. To understand and further investigate this phenomenon, we performed a series of experiments with lubricated and non-lubricated yarns.

Three of the most commonly used multifilament yarns were given in the lab a medium viscosity (around 100 cps) lubricant at 1% by weight of finish-on-yarn (FOY). The yarn-to-metal friction tests were run using a broad range of speeds (1 to 1,000 m/min) in order to generate the so-called complete friction profile. Examples of these profiles are presented in Figs 11.2–11.4.

The data in these figures suggests an alternative interpretation of the Stribeck lubrication curve. First of all, friction of a non-lubricated yarn is higher at low sliding speeds compared to its lubricated counterpart, which fits well with the general concept of lubrication. Also, in strict accordance

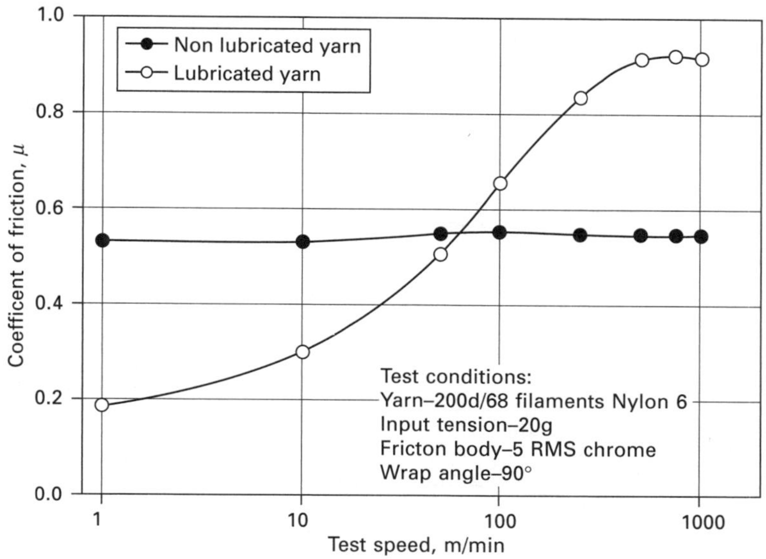

11.2 Complete friction profile for nylon yarn.

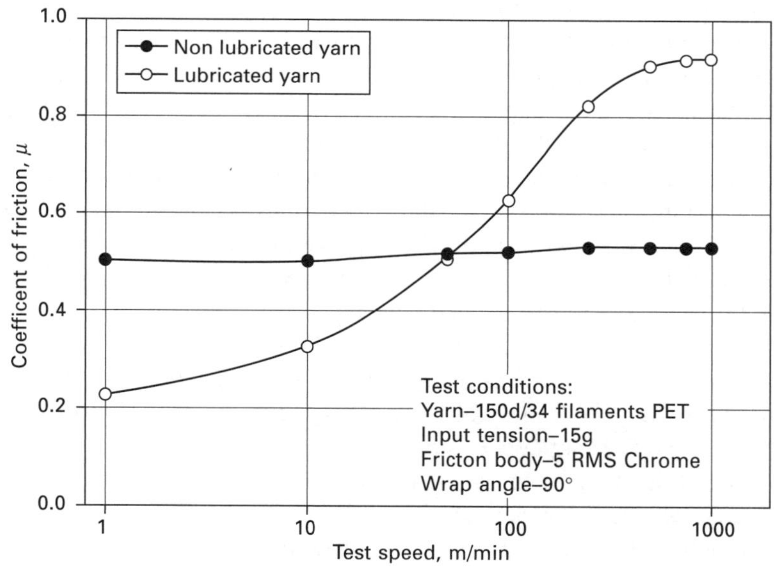

11.3 Complete friction profile for polyester (PET) yarn.

with the classic friction laws, friction of non-lubricated or dry yarn does not change with the sliding speed and remains constant over the whole experimental range. For lubricated yarns, friction always obeys the laws of hydrodynamic lubrication, where a higher speed results in a higher hydrodynamic drag and

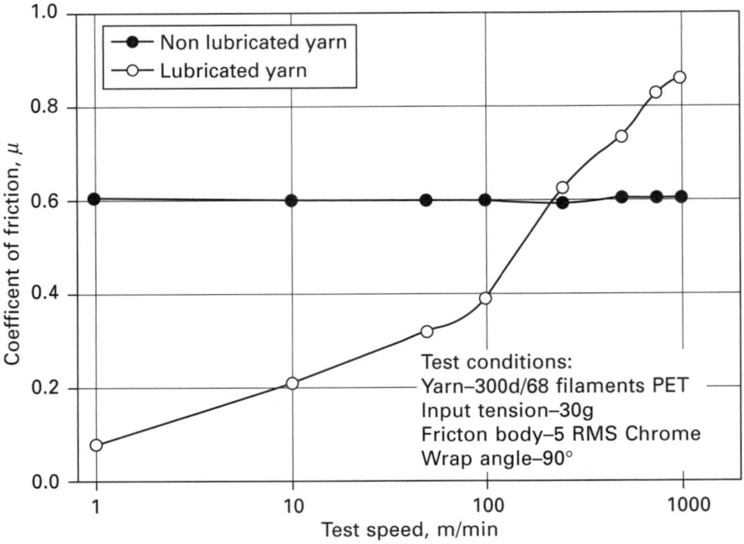

11.4 Complete friction profile for polypropylene (PP) yarn.

correspondingly in a higher coefficient of friction. Thus, the phenomenon observed in Figs 11.2–11.4 is hydrodynamic lubrication. At very low speeds, however, we encounter the "boundary region" where stick-slip friction occurs. What we believe is happening in this region is not boundary lubrication, but the intermittent breakdown of hydrodynamic lubrication. The slipping portion of the stick-slip behavior would be hydrodynamic sliding, and the sticking portion indicates physical contact or *boundary (dry) friction* as illustrated in Fig. 11.5. This also explains the large variations in friction or noise usually observed in this region. The noise is actually caused by oscillation between lubricated and non-lubricated states.

The data in Figs 11.2–11.4 can also be used to explain an apparent well-known paradox, i.e. that the friction of a lubricated yarn at high sliding speed is higher than the friction of a non-lubricated yarn. Thus, we are dealing with two very different phenomena – dry friction, which is independent of speed, and hydrodynamic lubrication, or, more precisely, hydrodynamic drag, which increases with increasing sliding speed. Accordingly, we may conclude that a function of lubrication is not primarily to decrease friction but to control it and decrease wear.

11.2.4 Elasto-hydrodynamic and hydrodynamic friction in textiles

A number of researchers have noted a phenomenon in which with fiber lubricants the friction actually decreases in the higher speed range of the

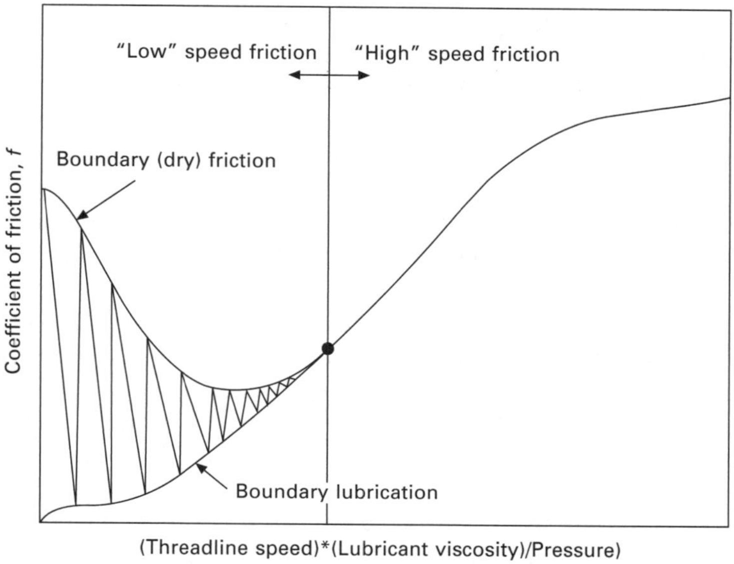

11.5 Revised lubrication curve.

hydrodynamic region. Schick [9] observed this effect for high viscosity finishes, as have we [17]. Schick attributes the decrease in friction to a reduction in viscosity caused by the heat generated in the friction process. Alternatively, Hansen and Tabor have done work that suggests that heat-induced viscosity changes were not responsible for the flattening of the curve at high speeds [3]. They performed friction experiments with silicone oils (which presumably had very minor changes in viscosity with temperature) and saw the same phenomenon.

We believe that several possible hypotheses should be examined that could also possibly account for this behavior. If viscosity reduction is the mechanism behind this high-speed reduction of friction, it could also be caused by shear thinning. The high shear rates present in the friction zone could be orienting the lubricant molecules and breaking up any structure that might have existed. This phenomenon would be more prominent with increasing viscosity, which is consistent with observations of Schick and us [9, 17]. Lower-viscosity lubricants will tend to contain smaller, shorter molecules in which the possible amount of anisotropy due to orientation is lower. Thus, shear thinning tends to occur in high-viscosity and more complex lubricating systems.

Another hypothesis involves the characteristic reaction time of liquids. A liquid is a state of matter characterized by the ability of the chemical units (atoms or molecules) to move in three dimensions while remaining in close physical proximity (unlike a gas). Molecular motion is a kinetic, time- and energy-dependent process. At a given temperature, a molecule will have energy available for kinetic processes proportional to kT, where k is the Boltzmann constant ($k = 1.381*10^{-23}$ J/K). The rate of the process will be proportional to $e^{-(\Delta E/kT)}$. A finite amount of time is necessary for the process to occur. If an event occurs in less time than it takes for the molecules to react, then the molecules will not react, and the liquid will begin to show solid-like properties. In the study of fluid dynamics this relationship is characterized by the Deborah number ($D = \lambda/t$), where λ is the characteristic reaction time of the liquid under given conditions, and t is the time in which the process or event occurs. A small Deborah number corresponds to liquid-like behavior, and a large number to solid-like behavior. The total time that a yarn in our apparatus is in contact with the friction bodies is 3 milliseconds at a 90° wrap angle and 300 m/min yarn speed. If this time is of the same order of magnitude as λ, then we would expect to see a change in the physical behavior of the system compared to the behavior at lower speeds.

A third hypothesis to explain the mechanism of friction reduction at high speeds uses the concept of surface interactions. Generally, metals have high-energy surfaces and polymers low-energy surfaces. The lubricant, usually a low surface tension liquid, would tend to have a much stronger affinity to the high-energy metal surface than to the low-energy polymer surface. At some speed, the shear stress in the lubricant layer may exceed the cohesive strength of the lubricant-polymer interface. At this point, the system would begin to shear at the polymer surface, rather than within the lubricant layer. As a result, the yarn would glide over the surface of the lubricant, with an apparent lowering of friction.

An important criterion for lubrication is that the flow behavior is laminar. To estimate the applicability of rheological approach to the friction phenomenon (which assumes a laminar flow), we analyzed the conditions at which we measure friction in terms of the Reynolds number (R_e). R_e is a dimensionless number which indicates when fluid flow in a system is laminar or turbulent. The calculation of R_e is shown in Fig. 11.6 along with the criteria for transition from laminar to turbulent flow. Hydrodynamic flow is laminar; in order for our assumptions to be valid, our system must be in laminar flow. We had to estimate the value of certain parameters, notably the value of the gap between the yarn and the pin and the width of the contact. We believe that the values are of the order of approximately 1 μm gap and 1 mm contact width. Making this assumption, one sees that the figure shows that the system is in laminar flow conditions over a wide range of speeds and viscosities. The conditions under which turbulence will possibly occur are high speeds and low viscosities.

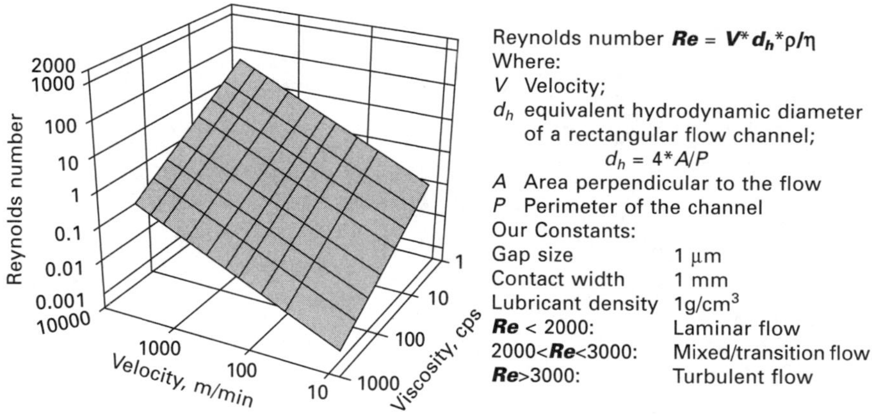

11.6 Reynolds number of friction measurement process.

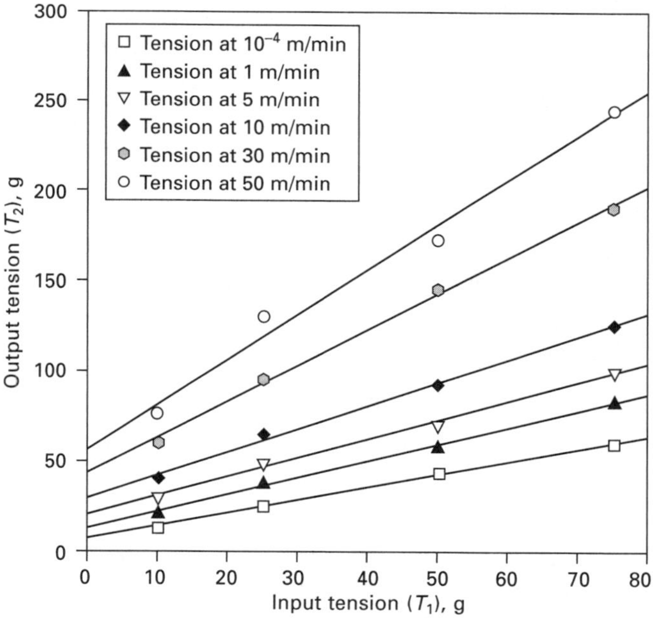

11.7 Effect of pretension on friction.

11.3 Fabric finishes and high speed yarn-to-solid body friction

11.3.1 Effect of input tension

Schick [9] observed (Fig. 11.7) and we have confirmed (Figs 11.8 and 11.9) the linear dependence of friction force on pretension at a constant yarn speed. This result means that the coefficient of friction (the slope of the line),

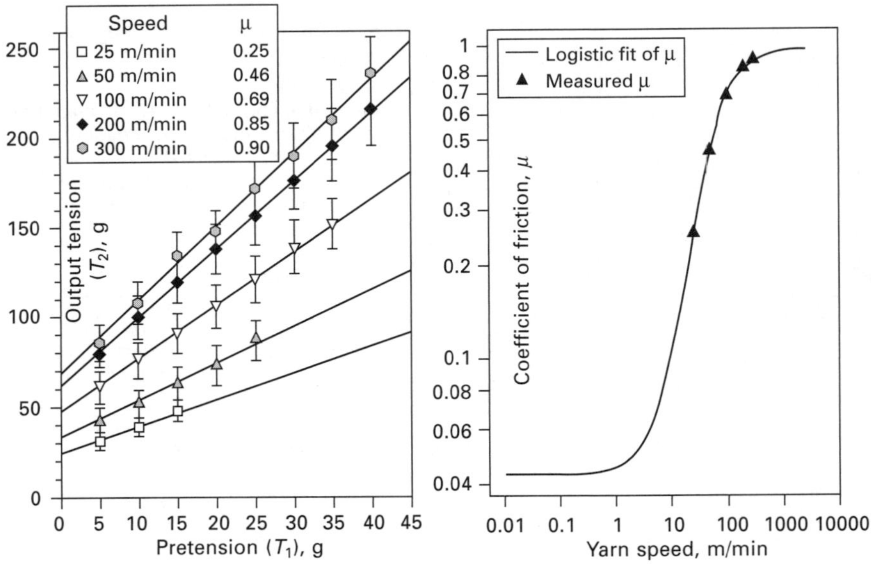

11.8 Effect of pretension on friction of 150 denier PET yarn.

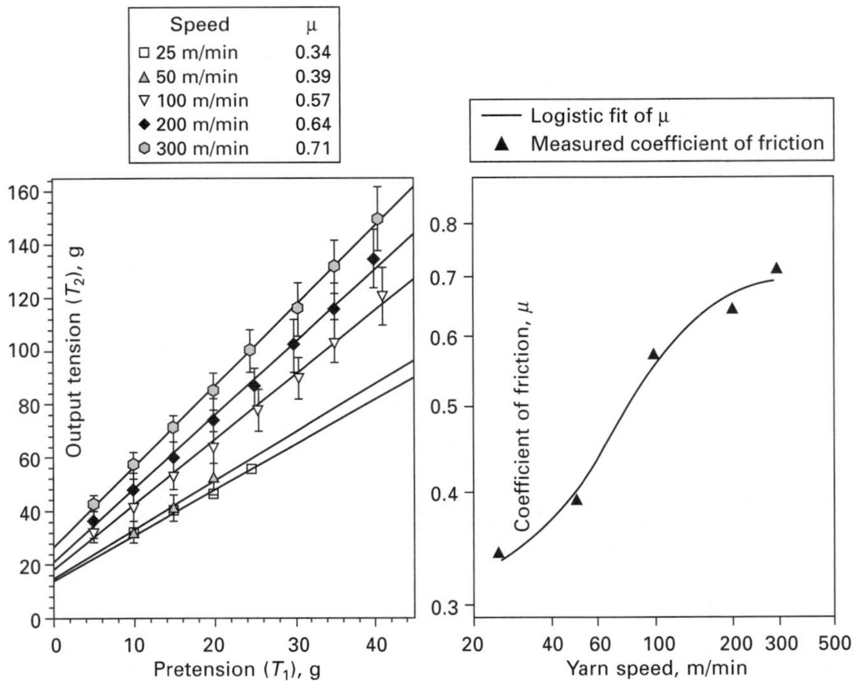

11.9 Effect of pretension on friction of 200 denier Nylon yarn.

is dependent on the speed but not on the pretension. A reason given for this is that as the pretension increases, so does the true area of contact [Schick]. This assertion, however, neglects other possible factors. The film thickness may change with increasing pretension and the rheological properties of the lubricant may change as discussed previously. This suggests that a more complex set of changes may underlie this apparently simple linear relationship.

Figures 11.8 and 11.9 correspond to 150 denier polyester (PET) and 200 denier Nylon yarns, respectively. The data were plotted as output force versus pretension in the left-hand graph of the figures. Regression lines were drawn through the points corresponding to a given yarn speed. The lines appeared to meet at a point near the X-axis at a value below zero. We believe that this corresponds to the residual friction inherent to the particular friction meter.

The right-hand graphs are a plot of μ against speed. Looking at these, we see that the change of μ with yarn speed is not linear. The points fit well to a logistic function. The latter is a four-parameter function, describing an s-shaped curve having flat plateaus at low and high values of the independent variable and a steep transition region between them. This curve is recognizable as part of the 'characteristic' friction curve of a lubricated yarn (Fig. 11.1). The transition region then corresponds to the domain of hydrodynamic friction. It is easy to see that the proper way to characterize differences in friction of two yarns is by the coordinates associated with the plateau regions. The curves of two yarns may intersect in the transition region, but they are predicted to be parallel in the plateau regions.

11.3.2 Effect of lubricant viscosity

As we would expect from data in Figs 11.1 and 11.5, we see increasing friction force with increasing speed up to a plateau region. We would also expect to see an increase in friction with an increase in lubricant viscosity.

In the past, a significant amount of data was generated to investigate this phenomenon. The compiled results of that past investigation are shown in Fig. 11.10 [17]. This figure shows that for the range of viscosities and speeds investigated, the trend is such that on a semi-log plot the relationship between friction force (in the hydrodynamic region) and viscosity is linear. In this case, $F = b_0 + b_1 \, log(\eta)$, where b_0 is the intercept and b_1 is the constant of proportionality (the slope of the line) which is dependent on the yarn substrate. The dependence of friction on yarn substrate will be discussed in more detail in a later section.

In an investigation focused on examining the effects of speed, temperature and viscosity, a series of lubricating systems covering a range of viscosities were used. These finishes were applied to finish-free nylon, polyester, and polypropylene (PP) at a level of 1% by weight. The treated and untreated yarns were tested on the friction meter at room temperature (25°C), using

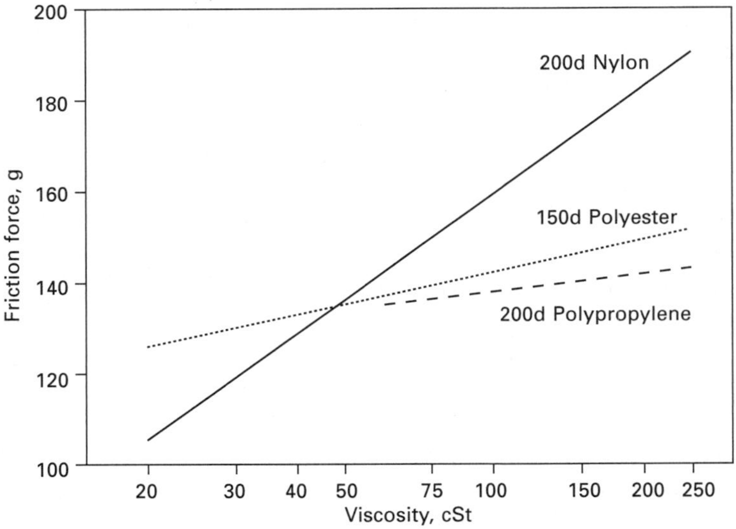

11.10 Effect of viscosity on friction.

both the one-pin and three-pin systems. They were also tested on the one-pin system at 40°C and 60°C.

Figures 11.11 and 11.12 compare the results at room temperature for the one-pin and three-pin methods. These friction response surfaces show how the measured friction force varied with both test speed and lubricant viscosity. The surfaces have very similar shapes, which means that, for both the one pin and the three pin methods, the same mechanisms govern the friction behavior in the same regions and we can generalize the one-pin results to the three-pin setup. The main advantage of the three-pin setup is that the absolute value of friction is lower, thus yarn and pin abrasion is lessened.

A common feature of these surfaces is that they become flat at higher viscosities. This demonstrates that the test has no resolution at the speeds at which it is currently run.

These surfaces are somewhat more complex than the ideal case, illustrated in Fig. 11.13 [17, 18]. In this model, we assumed that Stribeck's curve (derived for the journal bearing) applied equally to test speed and viscosity and plotted the theoretical friction response surface. A test that produced this response would have very good resolution over a wide range of speeds and viscosities.

Real tests as seen in Figs 11.11 and 11.12 and in the following Figs 11.14, 11.15, and 11.16 show behaviors that considerably depart from the ideal. Figure 11.14 shows friction response for 150 denier PET with temperature. Figures 11.15 and 11.16 show the same for 200 denier Nylon 6 and 200 denier PP respectively. For each yarn, it is found that the differences between

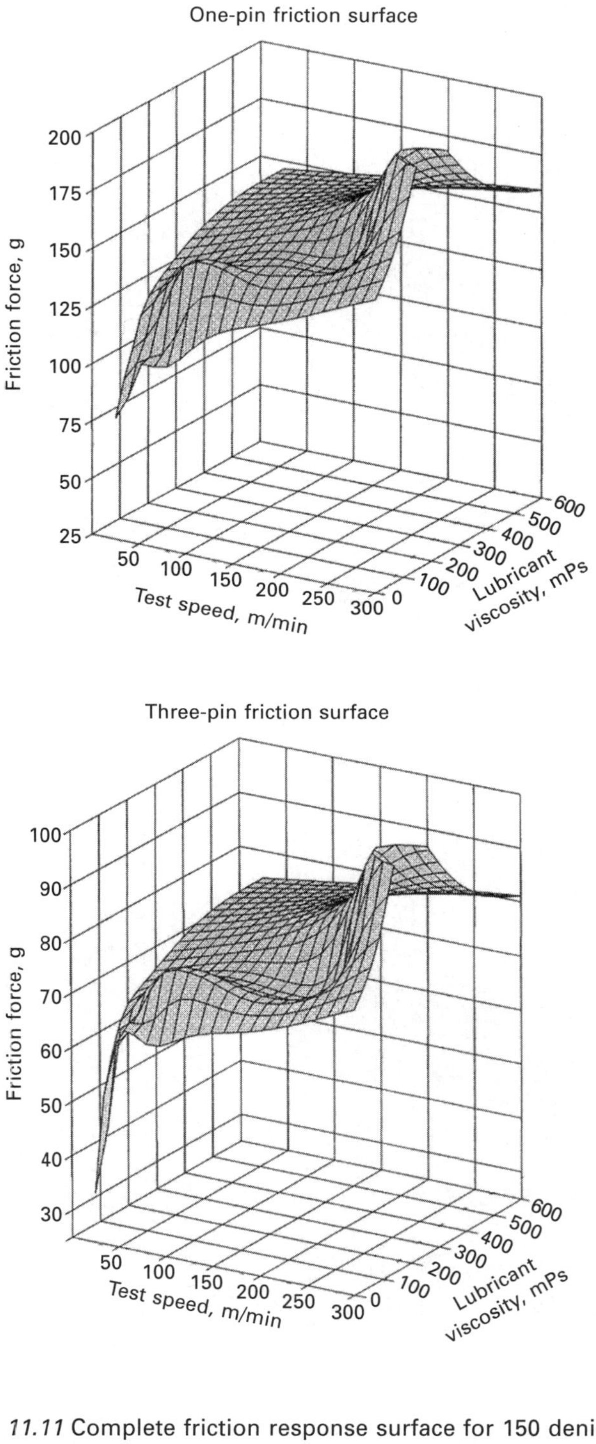

11.11 Complete friction response surface for 150 denier PET yarn.

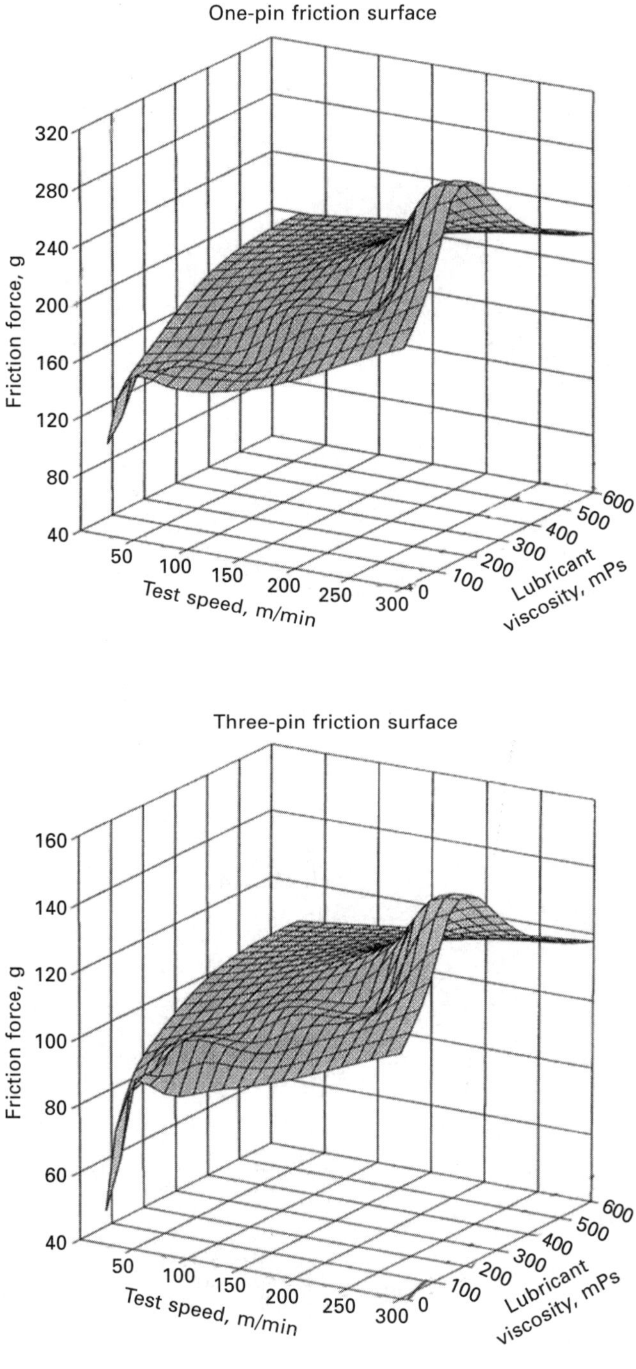

11.12 Complete friction response surface for 200 denier Nylon yarn.

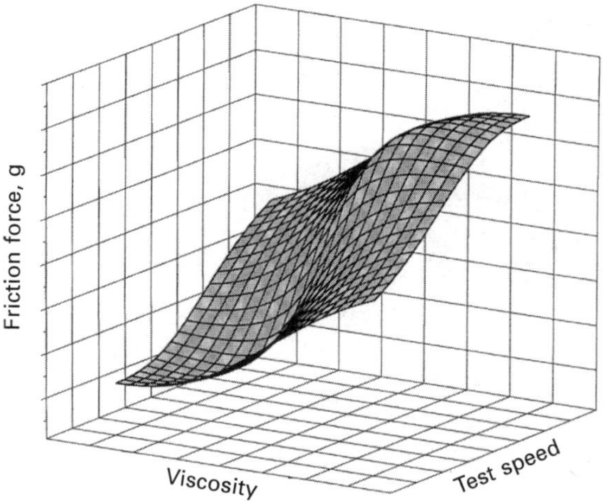

11.13 Model friction surface.

lubricants decrease with increasing temperature. This is because as the temperature is increased, the range of viscosities becomes smaller, and their absolute values are lower. In effect, when we raise the temperature, we are examining portions of the low viscosity end of the room temperature response.

11.3.3 Effect of yarn nature

In one study, Schick examined the effect of yarn material on friction [9]. He used three materials (Nylon 66, polyester (PET) and polypropylene (PP)) and applied a common finish at 1% level. His results are shown in Fig. 11.17. The critical surface energy of these materials decreases in the order: nylon 66 > PET > PP; and he suggested that the wetting behavior of the lubricants used could be correlated with the friction behavior. This is logical, as a lubricant with poor wetting properties would exhibit boundary-type friction behavior at higher speeds and lower pressures than would a lubricant with good wetting properties.

If the yarn is lubricated and one accepts the hydrodynamic model, then it would be expected that the composition of the yarn should not affect friction in the hydrodynamic region. Some departure from this concept has been noted, however.

In a study by the current authors, on the same three materials, PP was found to have lower friction than either PET or Nylon [17, 18]. In another study, polyester yarns of similar denier and finish but dyed to different colors, were used. The friction was found to vary with the color (Table 11.1). In both instances, one suspects that the surface energetics played a role. In the first example, it is possible that the finish used was more compatible with

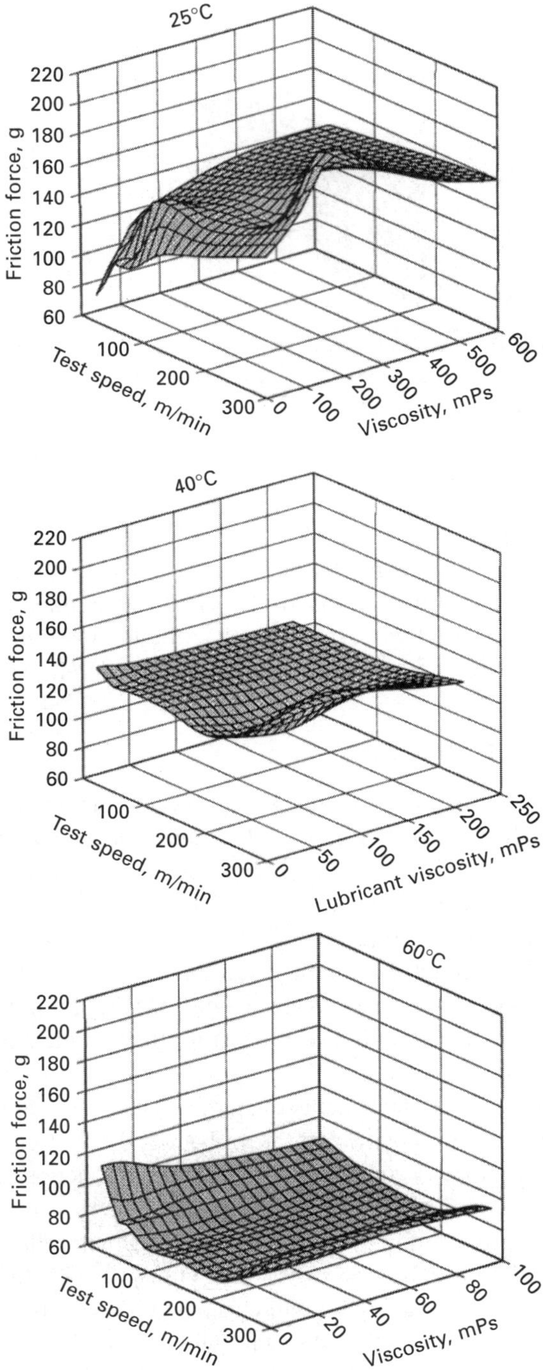

11.14 Complete frictional response surfaces for 150 denier PET yarn at different temperatures.

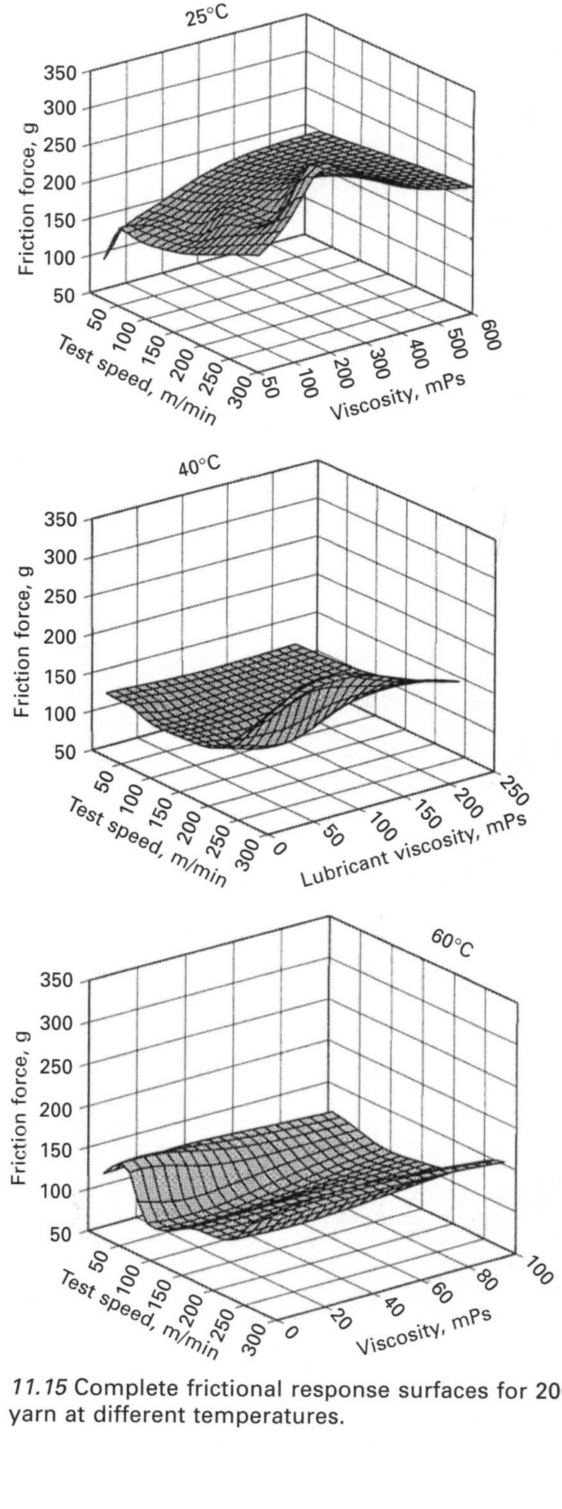

11.15 Complete frictional response surfaces for 200 denier Nylon yarn at different temperatures.

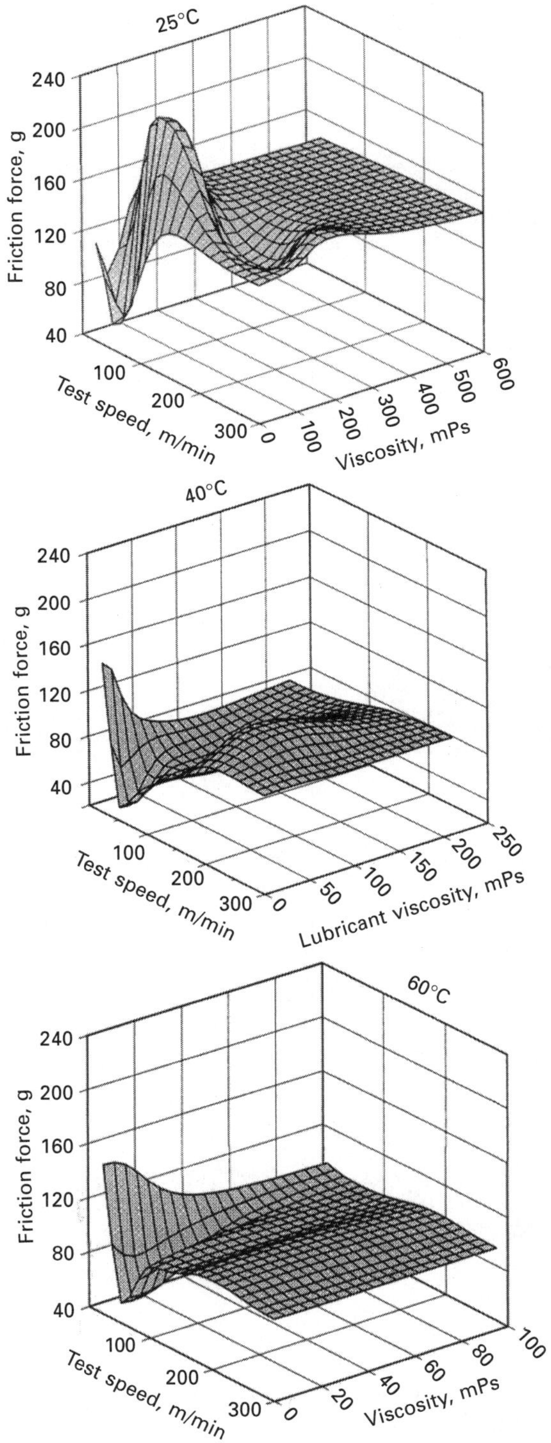

11.16 Complete frictional response surfaces for 200 denier PP yarn at different temperatures.

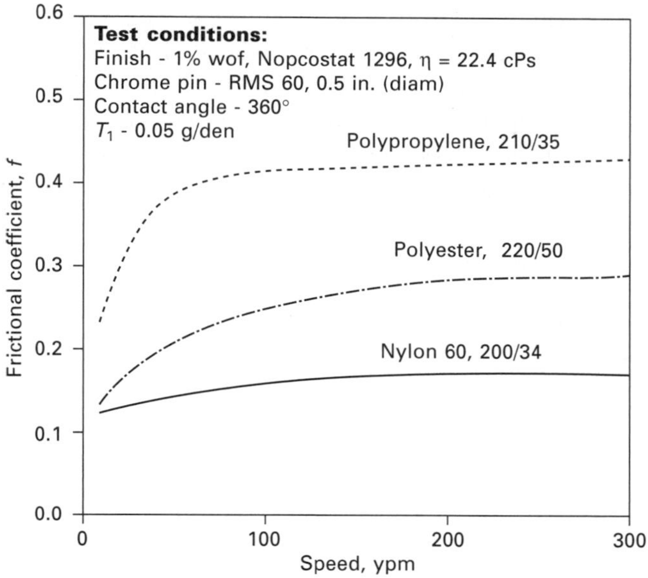

11.17 Dependence of friction on yarn nature.

Table 11.1 Effect of yarn's color on friction behavior (FOY: finish on yarn by weight)

Yarn color	Friction force, grams			FOY,%
	Test speed, m/min			
	100	200	300	
Black	212 ± 20	216 ± 12	216 ± 12	0.51
Purple	300 ± 40	336 ± 38	336 ± 26	0.79
Aqua	168 ± 20	214 ± 22	228 ± 20	0.82
Pink	234 ± 30	268 ± 32	288 ± 24	0.80

PP than with the other two, and in the second example, it is likely that dyeing with different colors led to different values of the critical surface energy.

11.3.4 Effect of lubricant uniformity

It is logical to suggest that uniformity of lubricant on the fiber surface should affect frictional response. Moreover, there were claims [19, 20] that enhanced uniformity of lubricant distribution greatly improves efficiency of lubrication and provides lower friction at comparable bulk lubricant's levels. To investigate this phenomenon we prepared two yarns, one of which was treated by means of a metering pump using standard application procedure, while the other yarn was treated by means of a cam that periodically lifted the yarn from the applicator guide. Both yarns were treated with 0.5% by weight of the same lubricant system. Using the dynamic fluorometer [21] we confirmed that the yarn treated by the cam method did have a square wave pattern of lubricant on its surface (Fig. 11.18). Nevertheless, the frictional profiles of both yarns (Fig. 11.19) were practically identical. This suggests that the effect of lubricant

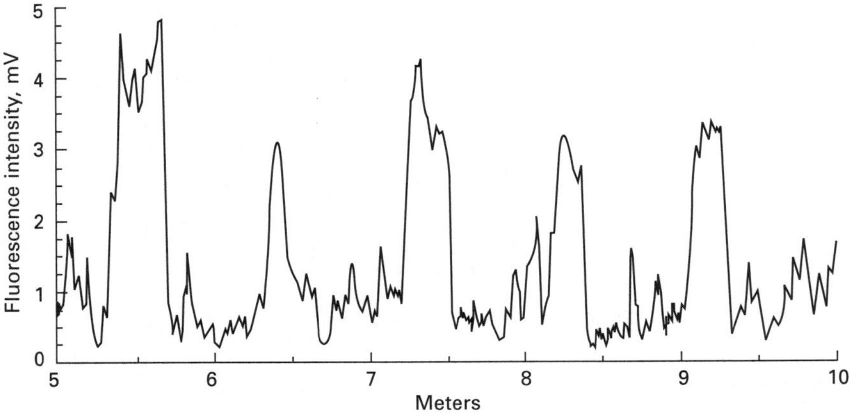

11.18 Square-wave lubricant distribution on PP yarn.

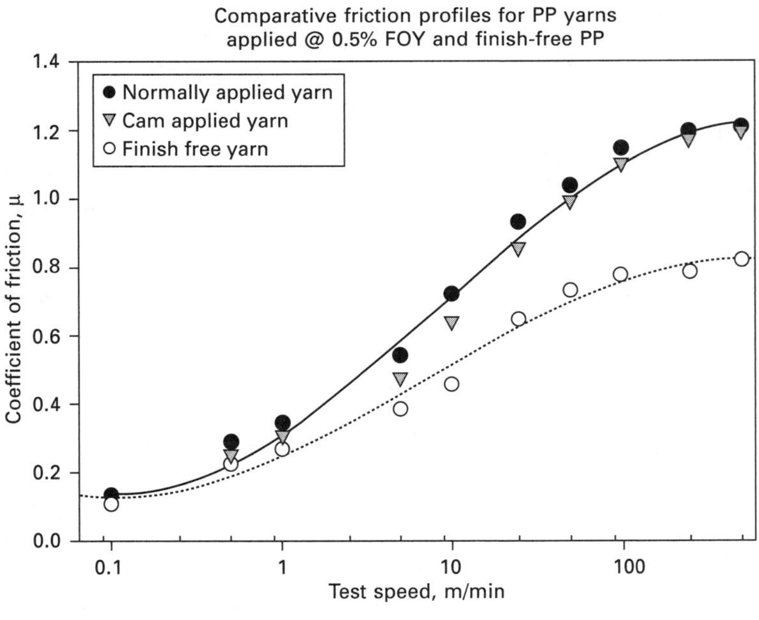

11.19 Effect of lubricant's distribution on friction.

uniformity on the high-speed frictional response is negligibly small [21]. It is suspected that this is due to lubricant flowing laterally on the surface, thereby reducing any irregularity in the level of finish application.

11.4 Fabric finishes and low speed fiber-to-fiber friction

This is the type of friction that plays a role in determining interfilament cohesion and hand or tactile feel of fibrous structures [22]. Traditionally, this friction is measured at extremely low sliding speeds (<1 mm/sec), thus occupying the very left quadrant of the lubrication curve (Figs 11.1 and 11.5). The friction test used to assess this behavior is normally run in either of two standard configurations (Figs 11.20 and 11.21). Although the capstan apparatus, shown in Fig. 11.20, is not produced commercially, the capstan test method is definitely the preferred procedure for measuring slow speed fiber-to-fiber (F/F) friction. The major advantage of this test compared to the twist method (Fig. 10.21) is the absence of the additional pulleys that may significantly alter test data. Regardless of these differences both tests generate data that very clearly differentiates between the fibers given soft, harsh, or the so-called scroopy type topical treatments [23]. The data presented in Figs 11.22 and 11.23 was generated at a test speed of 0.04 mm/sec. A soft hand finish on the fiber led to a frictional response (Fig. 11.22) that showed a smooth curve without any appreciable oscillation in measured friction force. Friction results of this test are traditionally presented either in terms of friction force, or in terms of the coefficient of friction calculated from the capstan equation. In both cases, the value of the dead weight should be

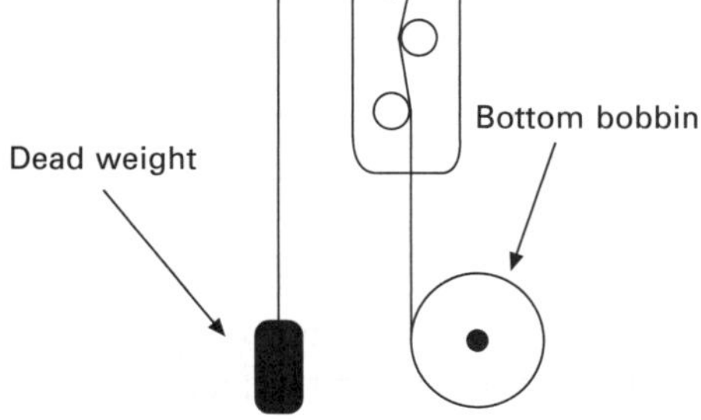

11.20 Capstan tester.

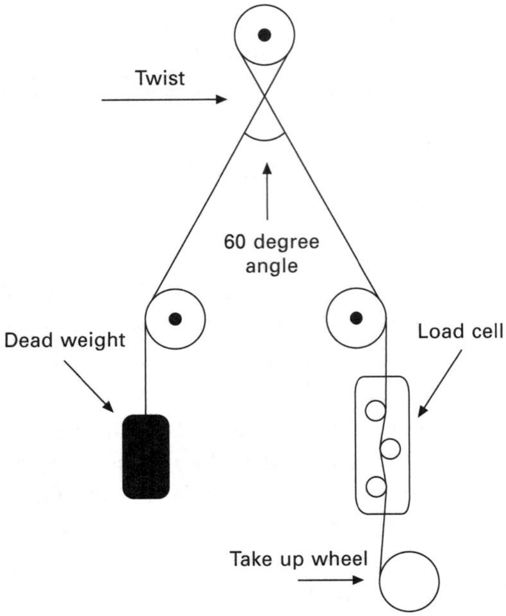

11.21 Twist tester.

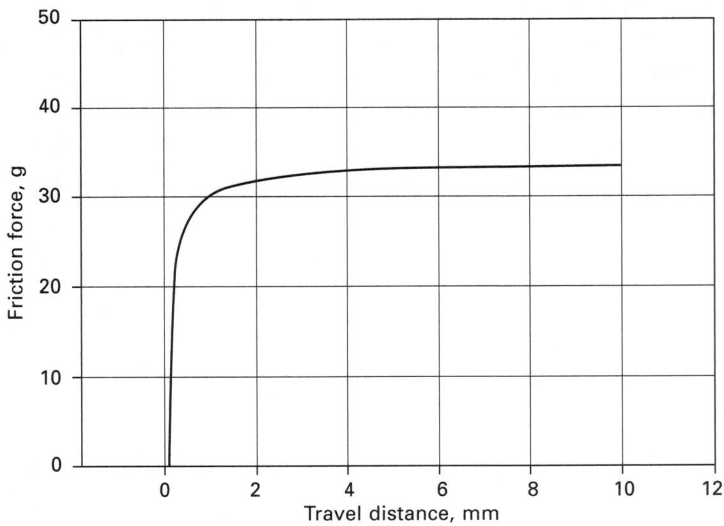

11.22 Soft-hand fiber.

specified. In the case of harsh hand or scroopy surfaces, the frictional graph shows a characteristic saw tooth pattern known as stick-slip (Fig. 11.23). In this case, the frictional response is not described adequately by a value of the coefficient of friction but by those of the following four parameters (see Section 5.3.3 for further details):

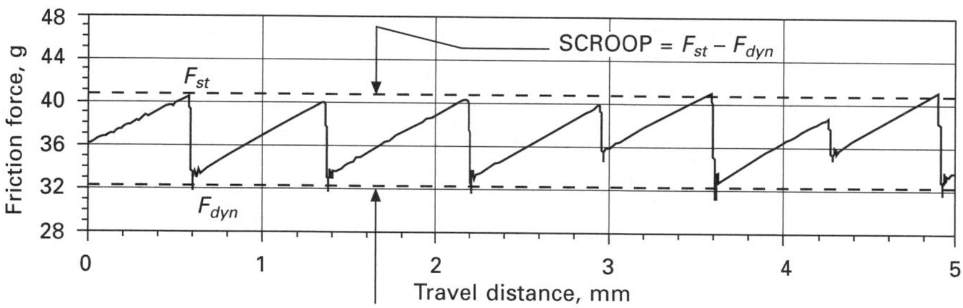

11.23 Harsh-hand fiber.

- F_s – Static friction force. This force represents the average of the peaks.
- F_k – Dynamic friction force. This force represents the average of the troughs.
- F_{av} – Average friction force. This force is an average of F_s and F_k.
- F_a (Scroop) – The difference between F_s and F_k.

This defines the terminology we will be using to describe experimental work in the field of low speed F/F friction below.

11.4.1 Inter-filament cohesion in continuous filament yarns

Inter-filament cohesion is a property that each yarn manufacturer must control. Lack of cohesion manifests itself in flaring of filaments, instability or "walking" of yarn on godet rolls and poor package buildup. On the other hand, excessive inter-filament cohesion leads to low bulk yarns, packages that are too tight and an unacceptable performance during weaving on air jet looms. Fiber producers use two major approaches to control inter-filament cohesion. They can use periodic mechanical (usually air) intermingling of filaments to produce nodes of loosely interlocked filaments that provide required yarn integrity and compactness. Alternatively, they can modify F/F friction by applying a topical treatment or a lubricant system during spinning. Usually, both techniques are applied simultaneously for better control of inter-filament cohesion.

In the case of high-speed fiber to solid surface friction, however, where correlations between friction coefficient and lubricant properties are reasonably well established, the effects of lubricant properties, such as viscosity, are not well understood. There is a simple rule of thumb that claims that high inter-filament cohesion, or high F/F friction, is accompanied by low fiber-to-metal (F/M) friction measured at high (above 100 m/min) speed. Unfortunately, this rule does not work well for all fiber types and lubrication systems. A typical exception to this rule is the frictional response of solid lubricating systems or waxes. Thus, application of these solid lubricating systems to yarn surfaces, results in both low F/F and low F/M frictions.

Our experimental approach to the problem was similar to the one described above for high speed F/M friction (Section 11.3). A range of lubricants was applied at 1% level by weight onto the surface of polyester (265 denier), nylon 66 (100 denier) and polypropylene (300 denier) multifilament yarns. Lubricants of two different chemical origins were used; one was based on predominantly hydrophobic ester compounds and another based on mostly hydrophilic highly ethoxylated compounds. All of these products are used commercially, and we marked them as high, moderate or low cohesive materials based on pure practical industrial experience. The properties of these lubricating systems are summarized in Table 11.2. The empirical cohesive rankings, viscosities measured on a Rheostress® rotational rheometer using 40 mm 2° cone/plate geometry, and measured cohesiveness values are listed along with the lubricants, separated into groups based on their nature.

Analysis of the data in Table 11.2 is presented in Figs 11.24, 11.25 and 11.26. It is generally noted that the exact same lubrication system applied on different yarns results in very different cohesiveness values of the final product. For example lubricant HB-B provides very low cohesiveness for polyester and nylon yarns but very high one for the polypropylene yarn. On the other hand, the lubricant HP-A provided a moderate level of cohesiveness for the polyester and the polypropylene yarns but low cohesiveness for the nylon yarn. This again allows one to conclude that the effect of the lubricating system strongly depends on the substrate fiber material. The data for PET is presented in Fig. 11.24 and that for nylon in Fig. 11.25 in two forms. In figures marked as "A" all cohesiveness numbers are plotted against viscosity regardless of the lubricant's nature. On the other hand, data in "B" figures is

Table 11.2 Lubricating systems of known cohesiveness

Lubricating system	Empirical cohesiveness rating	Viscosity, cps	Measured cohesiveness[1], g		
			265 denier PET yarn	100 denier Nylon yarn	300 denier PP yarn
Hydrophobic ester type					
HB-A	Low	235	1.30	1.69	18.5
HB-B	Low	182	2.86	1.30	18.5
HB-C	Medium–Low	147	1.00	1.00	25.0
HB-D	Medium–High	100	0.64	1.00	25.0
HB-E	Medium–High	68	0.64	0.64	0.64
Hydrophilic highly ethoxylated					
HP-A	Low	285	10.8	4.83	8.16
HP-B	Medium	142	15.9	14.7	17.1
HP-C	High	81	30	17.1	14.7

[1]We used values of static friction force (Fs) for cohesiveness characterization. Dead weight was 10g for all yarns. Experimental setup was Capstan test (F/M).

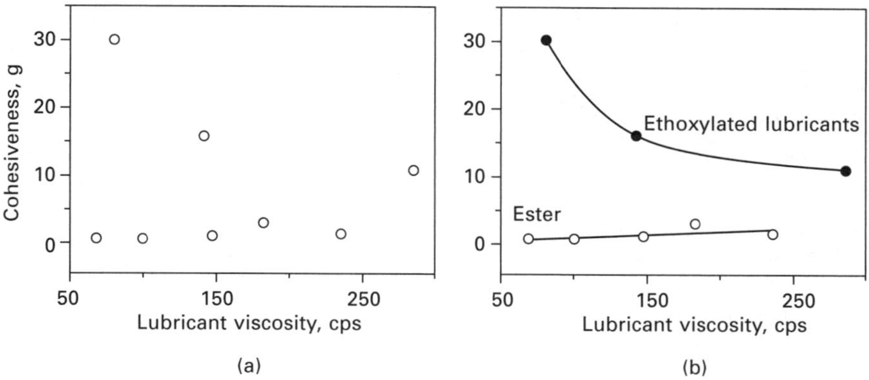

11.24 265 denier polyester yarn. Effect of lubricant viscosity on cohesiveness.

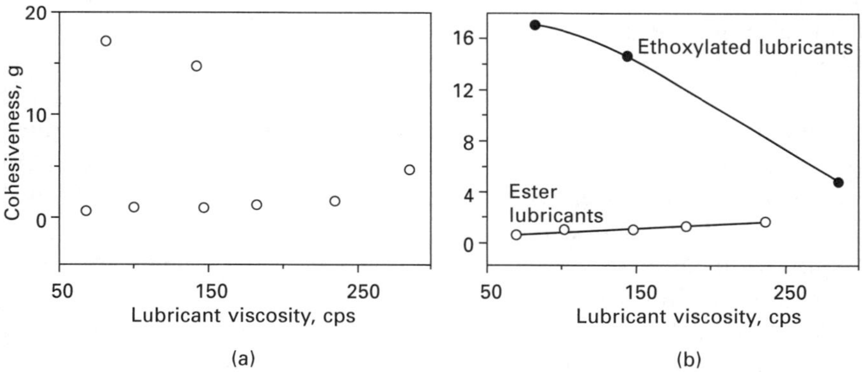

11.25 100 denier Nylon yarn. Effect of lubricant viscosity on cohesiveness.

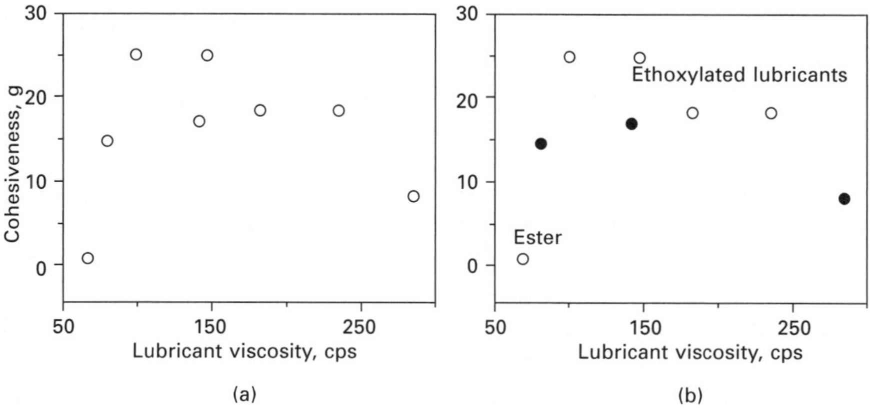

11.26 300 denier PP yarn. Effect of lubricant viscosity on cohesiveness.

separated according to the lubricant's nature. It is clear that if the lubricant's nature is omitted from consideration, then there is no dependence of cohesiveness on lubricant viscosity. However, if one takes the lubricant's nature into consideration, definitive effects are then seen. Thus, for both polyester and nylon yarns, increasing the viscosity of ester-based lubricants leads to an increase in cohesiveness. On the other hand, increasing the viscosity of ethoxylated-based lubricating systems leads to a decrease in cohesiveness. In the case of PP, the results are irregular and clearcut trends are not obvious.

To conclude analysis of data in Table 11.2, one may say that it is possible to predict cohesive effects of lubricating systems only if the same types of lubricants applied on the same fiber are compared.

11.4.2 Cohesion and "hand" of staple fiber

The process of converting staple fiber into yarn involves converting a somewhat random arrangement of fibers (baled fiber) into an assembly of parallel fibers, which can then be twisted to form a yarn. This is accomplished by carding and/or combing the opened staple fibers to initiate the formation of parallel fibers. The sliver is then drawn to increase the percentage of parallel fibers and to improve sliver uniformity. During each of these steps, the individual fibers slide against one another. Control of dry fiber-to-fiber friction is, therefore, very important, and in order to control it with some degree of confidence there must first be a method for measuring it under controlled laboratory conditions.

As mentioned above, F/F friction is a critical parameter for processing fine denier staple fibers into spun yarns or nonwoven fabrics. In the case of the former, during each stage of mill processing (opening, carding, combing, drafting and spinning), F/F friction plays a role and thus must be carefully controlled by the fiber producer. Typically, this is done by controlling the crimp and by using an appropriate fiber lubricant system. In the case of carded fibers for nonwoven products, the fiber must perform well during opening and carding, especially during high-speed operations. Again this is typically done by the use of a suitable combination of crimp and fiber finish.

In order to be able to control F/F friction, one must be able to measure it. Unfortunately, test methods applicable to filament yarns, described in the previous section, are not applicable to staple fibers. The mostly acceptable method of measuring the F/F friction of staple fibers is the staple pad friction test [22]. This is shown in Fig. 11.27.

The most widely used configuration for this test is based on the universal tensile testers. Two small pads of well-carded staple fiber webs are placed over each other, and this fiber "sandwich" is then positioned on the top of the test stage, which is covered with sandpaper. A weighted sled is then carefully put on the top of the fiber "sandwich". The bottom of the sled is also covered

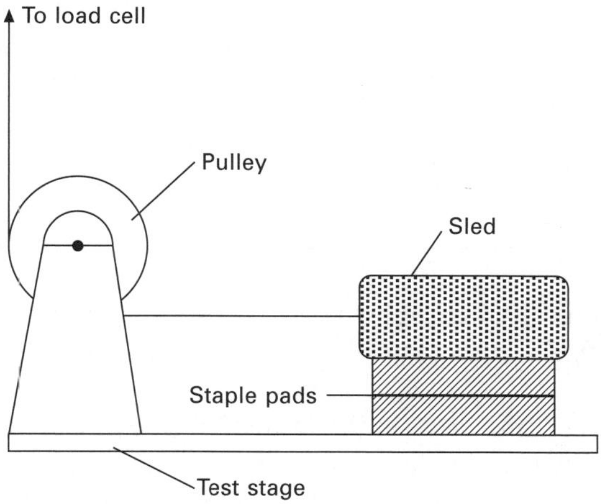

11.27 Staple pad friction (SPF) test.

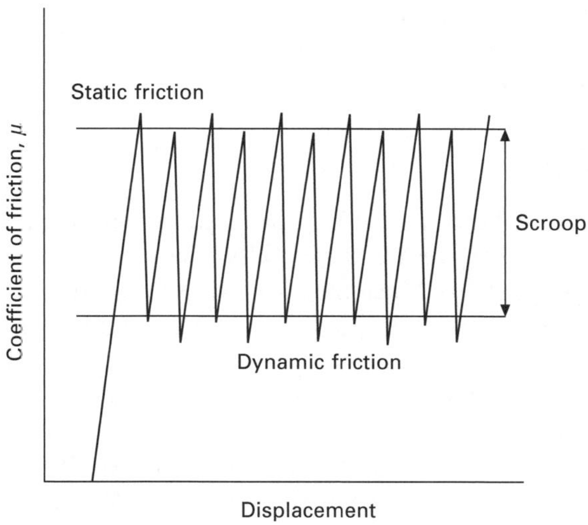

11.28 Typical SPF chart.

with the same sandpaper. The sled is dragged along the test stage at a constant speed. Sliding between test stage and fiber pad and sled and fiber pad is prevented by the grip provided by the sandpaper surfaces. The only sliding movement that occurs is between the two staple fiber pads. The friction force is recorded and plotted against the displacement. A characteristic chart obtained in this test is shown in Fig. 11.28. As with filament yarns, the output from this test is reported as: static friction, dynamic friction, scroop (difference between static and dynamic), average friction and coefficient of friction based on the average friction value. Table 11.3 contains SPF results for two

Table 11.3 Typical staple pad friction results

Finish	Static friction F_s, g	Dynamic friction F_k, g	Scroop F_a, g	Average friction, g	μ
Textile type	701	701	0	701	0.504
Nonwovens type	815	646	169	733	0.528

different finishes – one used in nonwovens and the other in spun yarn applications. Both finishes were topically applied on the exact same staple fiber.

There are some clear differences in the results given by these two finish systems. The textile finish yields lower static friction and scroop, and this makes fiber easier to open, and easier to draft. This results in better sliver uniformity, which is critical in spinning operations. The nonwoven finish, on the other hand, yielded higher static friction and scroop. This leads to stronger carded webs, which enables a mill to increase carding speeds while maintaining web integrity. Both of these are critical parameters in the manufacture of staple-based nonwoven fabrics.

11.5 Summary

The necessity of modifying friction in man-made fibers was identified soon after the development of the first commercially successful synthetic fiber (Nylon 66 by DuPont in 1939). Since then, spin finishes or, more correctly, fiber lubricating systems have been used to facilitate spinning of all man-made fibers and yarns. It is well known that without proper lubrication, the efficient production of man-made fiber-based products would be impossible. Nowadays, contemporary fiber lubricating systems are designed not only to provide high efficiency in fiber spinning, drawing and texturing processes, but also to impart a range of secondary properties appreciated by consumer, such as acceptable hand (tactile feel) and antistatic characteristics. Most surprisingly, the modern lubricating systems work at minute add-ons, normally not exceeding 1% by weight of fiber. The development and formulation of these lubricating systems (spin finishes) has always started with an analysis of the frictional behavior of fibers and yarns.

The measurement and characterization of friction in fibrous systems was an area of active research in the 1950s, 1960s and 1970s. During that time, the complexities involved in measuring and characterizing friction were realized and tests were developed to support the development of fiber lubricant products within the context of expanding processing conditions in the fiber and fabric industries. Such work was successful in providing a testing methodology that supported the industry through a period of rapid development.

However, theories regarding friction as molecular surface phenomena have developed over the last 20 years and the application of these contemporary approaches to the specific problems of fiber friction is just beginning to occur. Major, directly relevant questions that still remain unanswered include:

- What are the effective film thickness and corresponding shear rates encountered under "typical" fiber processing conditions? What variables in terms of fiber and finish control these factors?
- What molecular surface mechanism produces the shape of the Stribeck curve?
- What are the molecular or surface origins of slip-stick and cohesion in fiber systems?
- What is the mechanism by which the nature of a fiber surface affects friction in the hydrodynamic region?

With an expanded view of the basis of friction itself, there is now renewed interest in revisiting these questions in fiber friction phenomena. The hope is that advances in testing and related technologies should allow us to measure variables that could only be guessed in the past. If successful, this effort should lead to an enhanced understanding in this area and, hopefully, provide the basis for future developments in the field of finishes for fibrous materials.

11.6 References

1. Krim J, 'Friction at the atomic scale', *J. Scientific American*, 1996, **275**, 74–80.
2. Salmeron M, 'Observing friction at work', *Chemtech.*, 1998, **9**, 17–23.
3. Howell H G, Mieszkis K W and Tabor D, *Friction in Textiles*, London; Butterworths Scientific Publications, 1959.
4. Israelachvili J N and Tabor D, 'The measurement of van der Waals dispersion forces in the range 1.5 to 130 nm', *Proc. R. Soc. London A*, 1972, **A331**, 19.
5. Salmeron M, Liu G and Ogletree D F, *Forces in scanning probe method*, Kluwer Academic Publishers, 1995 (and references therein).
6. Dayo A, Alnasrallah W and Krim J, 'Superconductivity – Dependent sliding friction', *Physics Rev. Lett.*, 1998, **80**(8), 1690.
7. Cieplak M, Smith E D and Robbins M O, 'Molecular origins of friction – The force on adsorbed layers', *Science*, 1994, **265**(5176), 1209.
8. Granick S, 'Motions and relaxations of confined liquids', *Science*, 1991, **253**(5026), 1347.
9. Schick M J, *Surface characteristics of fibers and textiles: Part I*, New York, Marcel Dekker, Inc., 1975.
10. Gupta B S and El-Molgahzy Y, 'Friction in fibrous materials. Part 1: Structural model', *Textile Res. J.*, 1991, **61**(9), 547–555.
11. Brisco B J, 'Role of adhesion in the friction, wear and lubrication of polymers', *Adhesion-5*, Applied Science Publishers, 1981, 49–80.
12. Persson B N J and Tosatti E, *Physics of sliding friction*, Kluwer, 1996.

13. Virto L and Naik A, 'Frictional behavior of textile fabrics. Part I: Sliding phenomena of fabrics on metallic and polymeric solid surfaces', *Textile Res. J.*, 1997, **67**(11), 793–802.
14. Virto L and Naik A, 'Frictional behavior of textile fabrics. Part II: Dynamic response for sliding friction', *Textile Res. J.*, 2000, **70**(3), 256–260 (and reference therein).
15. Myer E and Heinzelmann H, 'Scanning force microscopy', *Scanning tunneling microscopy II: Further applications and related scanning techniques*, Springer-Verlag, 1992.
16. Krim J, 'Friction at the atomic scale', *Scientific American*, 1996, **275**(4), 74–80 (and references therein).
17. Kutsenko M, 'High-speed fiber-to-metal friction in textiles', *Chemical Fibers International*, 2001, **51**(2), 63–69.
18. Kokenes S and Kutsenko M, 'New look at fiber-to-metal friction', *Fiber Producer Conference*, Greenville, SC, 1998.
19. Williams M, 'Electro-static control for fibers in fabric production', *Institute of Physics (Static Electrification Group)*, London, 1997.
20. M. Williams, 'Making a little go a long way. Enhanced surface coverage spin finishes', *UITEX*, Oudenaarde, Belgium, 1997.
21. Kutsenko M and Theyson T, 'Effect of finish uniformity on friction and antistatic protection of Polypropylene yarn', *International Nonwovens Technical Conference*, Baltimore, MD, USA, 2001.
22. Theyson T W and Kutsenko M, 'Fiber-to-fiber interactions in textile processing', *Fiber Producer Conference*, Greenville, SC, 1992.
23. Kutsenko M and Brooks N, 'Controlling tactile feel in BCF yarn', *International Fiber Journal*, 2004, **19**(3), 60–61.

12

Boundary lubrication phenomena in coated textile surfaces

Y LI and J P HINESTROZA, Cornell University, USA

12.1 Introduction

A major part of friction in textiles, given a finish, is governed by the boundary lubrication phenomena. This includes friction during processing of fibers into yarns and fabrics, conversion of fabrics into products, and day-to-day handling of textiles by consumers. The lubrication phenomenon is considered boundary because the finish applied is usually quite low (~0.1%) and, in most applications, the speed involved is also small. In this type of lubrication, the finish, ideally a monolayer in thickness, is not considered to totally mask the morphology and the chemical potential of a fiber surface, but to modify them. The chemical interaction involved during a friction process is considered to be at the molecular level. Accordingly, in order to fully understand the boundary lubrication phenomena and determine the role of a finish composition, the interaction between finish and fiber surface must be explored at this level.

The availability of lateral force microscopy (LFM) has made it possible to explore friction and wear at the molecular level and to examine the effectiveness of a finishing treatment in modifying a specific behavior of the textile substrate. The technique is based on scanning probe microscopy, and is one of the few experimental methods capable of assessing forces at the single contact or atomic level.

In this chapter, we start out by providing an introduction to the LFM technique, the principle by which it works, the calibration aspects of the forces measured, and the friction phenomenon assessed by it at the atomic level. In the next section, friction and lubrication phenomena are discussed from a molecular perspective and relationships between lubrication performance and lubricant's properties such as its molecular weight, chain configuration, chemical composition, and viscoelastic properties are established. This is followed by a description of the chemical structures and properties of lubricants frequently used in textile processing. Finally, boundary lubrication phenomena is addressed by presenting observations of self-assembled molecular structures over polymeric substrates and describing how these assemblies influence the friction behavior of the coated surfaces. It is visualized that the discoveries in the field of boundary lubrication will continue with further research and advancement in instrumentation. A fuller understanding of the behavior at molecular level is expected to enable the industry in developing new and better finishing treatments.

12.2 Using lateral force microscopy to analyze friction behaviour

12.2.1 Background

In 1986, the atomic forces between a tip and a sample's surface were measured by Binning *et al.* using the first atomic force microscope [1]. One year later, Mate *et al.* observed that the atomic structure of a surface manifests itself directly in the frictional properties of the interface [2]. This observation led to the development of the lateral force microscopy (LFM) technique. This technique, also known as friction force microscopy (FFM), translates the lateral force acting on a sliding tip to the frictional properties of the surface. The tips used in LFM can be customized so the contact area between the tip and the probed surface is of molecular dimensions. This allows one to probe single asperity contacts and study behavior with nanoscale resolution. When a lubricant is applied onto a sliding surface, strong interaction forces arise between the molecules. These forces cause particular molecules to aggregate in the interface and to form self-assembled structures. The conformations and behaviors of these aggregates influence the friction and wear behavior of the coated materials. Recently, the tips used in LFM have also been functionalized by including specific chemical groups, such as – COOH and –OH. Using chemically functionalized tips, the friction behavior could be directly correlated with the lubrication potential of finishes containing these chemistries.

Following its invention, LFM has been used extensively to study molecular lubrication phenomena on hard surfaces, such as mica, silica, and graphite. Studies on polymer surfaces, relevant to textile applications, however, have been limited, primarily due to the fact that polymer surfaces deform easily, which adds complexity to the experiment and to the interpretation of the data.

12.2.2 Principle involved

When an atomic force microscopy (AFM) tip slides on a surface, it will be deformed both in the vertical and the horizontal directions (Fig. 12.1). The force F_n, which is normal to the surface of the sample, results in vertical

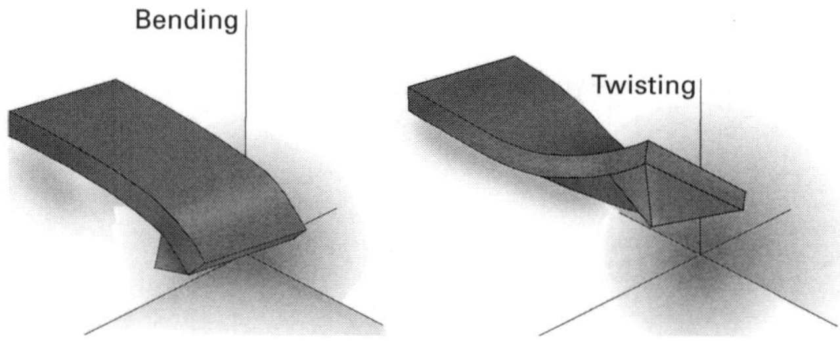

12.1 Twisting and bending motions of a rectangular cantilever.

bending of the free end of the cantilever. By contrast, the force F_l, which is parallel to the probed surface and is in the opposite direction to the sliding direction, leads the cantilever into a twisting motion. Both motions, driven by F_n and F_l, are orthogonal and this orthogonality is what enables the simultaneous, yet independent, acquisition of topographic images and lateral force images.

In order to precisely detect the forces between the tip and the surface, a laser beam is reflected off the back of the cantilever onto a quadrant photodiode detector. The output of the quadrant detector is used to determine the degrees of bending and twisting of the cantilever. The laser beam method is the most commonly used monitoring technique as it can achieve a resolution comparable to that of an interferometer while it is also inexpensive and easy to use.

12.2.3 Calibration, and understanding of friction behavior at atomic level

An accurate quantitative determination of the surface forces is critical for decoupling the LFM signal and obtaining the values of the normal and lateral forces. AFM normal force curves shown in Fig. 12.2 are usually known as force-distance plots. These curves are constructed by monitoring the deflection of an AFM cantilever as its tip approaches or retracts from a sample. The movements of approaching and retracting tip are perpendicular to the sample's surface, so the interaction forces measured are indeed normal to the sample's surface.

In a force curve, two terms, "steric barrier height" and "steric barrier thickness," are usually defined to interpret the nature of the lubricant layer adsorbed. The steric barrier height is the maximum repulsion force and the steric barrier thickness is the separation distance at a force of 0.2nN. These parameters were initially used by Wang *et al.* to study the absorption behavior of silicon and hydrocarbon-based lubricants on silica substrates and provided an indication of the lubricity and wear resistance properties of an additive [3].

Figure 12.3 shows a spring-ball model commonly used to mimic the behavior of the tip-cantilever assembly in an AFM [4]. This simplification is

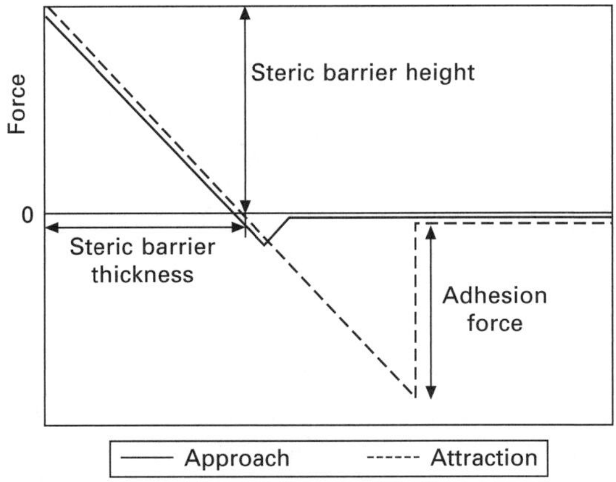

12.2 A schematic representation of force curves commonly observed during AFM analysis. The solid line represents the approach curve and the dashed line represents the attraction curve.

a useful tool to carry out analytical and numerical modeling of the interactions between a tip-cantilever assembly and a sample's surface. The ball represents the tip and a weak spring represents the cantilever used to detect the forces between a sample's surface and the tip. This simple model highlights the central features of AFM as a mechanical sensor for probing interaction forces. The interaction forces between ball and sample depend on many factors, including the tip contact geometry, the tip surface chemistry, and the properties of the sample's surface.

The interaction between ball and surface includes two components: the internal atomic forces between the tip and the surface, and the externally applied force on the spring-ball. The sum of these interactions correlates with the friction behavior of the surface. Therefore, it is essential to know the normal forces in order to study the friction phenomena in LFM.

When a tip initially approaches a surface, no interactions between the tip and sample are detected, resulting in a flat signal. As the tip-sample distance decreases, short-range forces could be detected. In Fig. 12.3(a), a weak attractive force is explained as van der Waals forces between the molecules of the tip and those of the sample. As the tip gets even closer to the sample's surface, it bends, and the force becomes strongly repulsive. In Fig. 12.3(b), a polymer brush attached to the sample's surface and representing lubricant, depresses as the tip approaches. For a thermally diffused motion of the brush, the entropy of confining these dangling polymer chains results in a repulsive entropic force [5]. In Fig. 12.3(c) the relatively hard tip causes an indent in the elastic sample. The true contact is near the point at which the cantilever begins to deflect [6]. Figures 12.3(d–f) show retraction curves. In Fig. 12.3(d), the tip is pulled from the sample and exhibits an adhesion force,

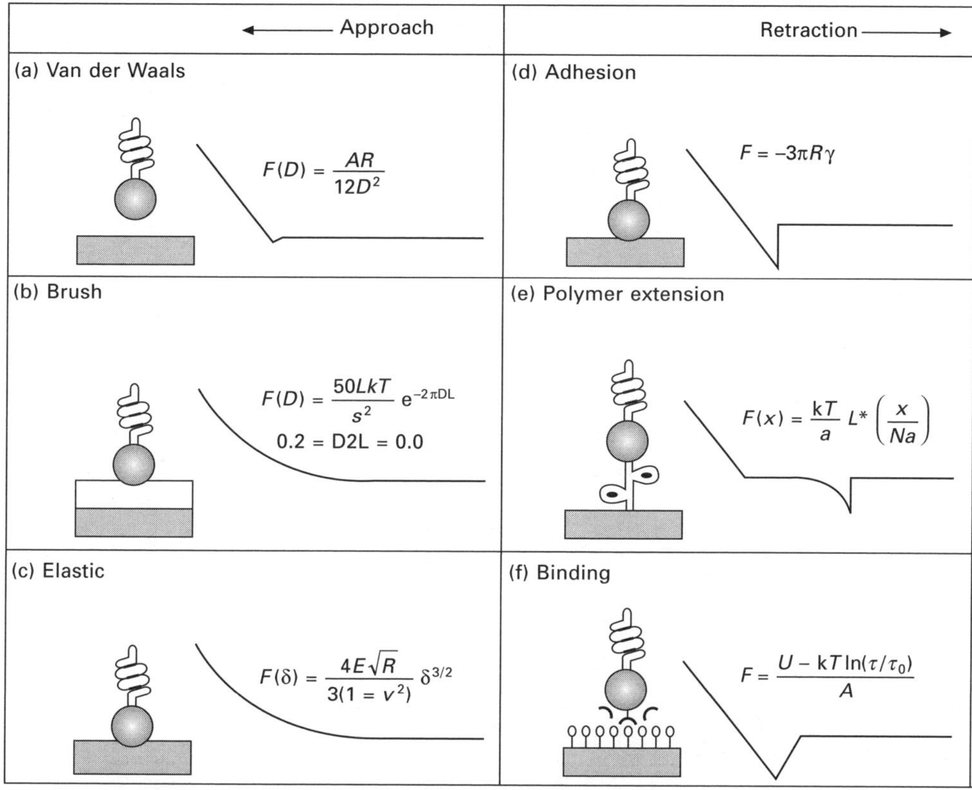

Definitions

- A Hamaker constant
- a Monomer length
- D Probe–sample separation distance
- E Elastic modulus
- k Boltzmann's constant
- L Brush thickness in a good solvent
- L^* Inverse Langevin function
- N Number of units in polymer
- R Radius of probe sphere
- T Absolute temperature
- U Bond energy
- x Elongation of polymer
- δ Indentation depth
- v Poisson rato
- γ Surface energy between tip and sample
- γ_L Surface energy of the liquid
- τ Period over which the bond will rupture
- τ_0 Reciprocal of the natural bond frequency
- s Mean distance between polymers

12.3 Six types of force curves; (a–c) show approach behavior. (d–f) show retraction behavior. (a): van der Waals (attractive) force; (b): an exponential repulsive force on surface coated with a polymer brush (lubricant); (c): an exponential repulsive force on an elastic surface; (d): a general adhesion force; (e): a retarded adhesion force from an unbroken interlink derived with a polymer extension; (f): multiple adhesion points from multiple unbindings.

which is attractive in nature and it retards the separation between the tip and sample [5]. In Fig. 12.3(e), a polymer chain from the surface is adhered to the tip and stretched until the interlink breaks. The force curve shows a characteristic attraction far from the surface and then a jump back to zero [7, 8]. In Fig. 12.3(f), there are multiple adhesion points, so when the tip is moved away, multiple binding points between the tip and the surface's molecules are broken in succession. This is shown by multiple unbinding points in the force curve [9–13]. These details simulate to some extent the frictional forces generated when an asperity on one body slides over the surface of another body.

A force signal obtained by LFM will be the result of the attractive and repulsive forces between the tip and the molecules of the surface as well as the force involved in the tip deforming the surface. These forces will be decoupled to get a reliable quantitative measurement of the lateral forces. A number of calibration methods have been developed to interpret the signals. The details of these methods are adequately described in the literature [14–18].

12.3 Effect of lubricant properties on friction behavior

12.3.1 Molecular weight

Barrena *et al.* [19] studied the friction behavior of alkylsilanols (CH_3-$(CH_2)_{n-1}$-$Si(OH)_3$) with varying chain length ($n > 10$) using LFM. They observed that the height signal decreased in discrete amounts at critical loads, while the friction signal increased in a stepwise manner which was correlated with the height decreases (Fig. 12.4). Generally, this behavior is assumed to be due to the fact that the upright chains are compressed and they initiate a tilting process under load. The tilting motion is discrete in terms of the packing and the interlocking of the molecules following the zig-zag skeletal structure of the carbon chains. The increase in friction is then consistent with the estimated increase in the activation energy required for the tilting motion.

The behavior of shorter chains ($n < 10$) might be distinguished from those of longer ones ($n > 10$) in terms of their mechanical response under load. Lio *et al.* [20] discovered that a short chain ($n = 6$) yielded higher adhesion and higher friction forces for several alkanethiols (CH_3-$(CH_2)_{n-1}$-SH) supported on gold. The initial monolayer may be assumed to be in a close-packed state, so adding more covalent carbon-carbon bonds to the molecular chain provides additional stabilization energy. Therefore, longer chains ($n = 8$, 11, 12, 18) appear to be more stable while maintaining a densely packed configuration. A friction image of a long chain dodecanethiol illustrating hexagonally ordered

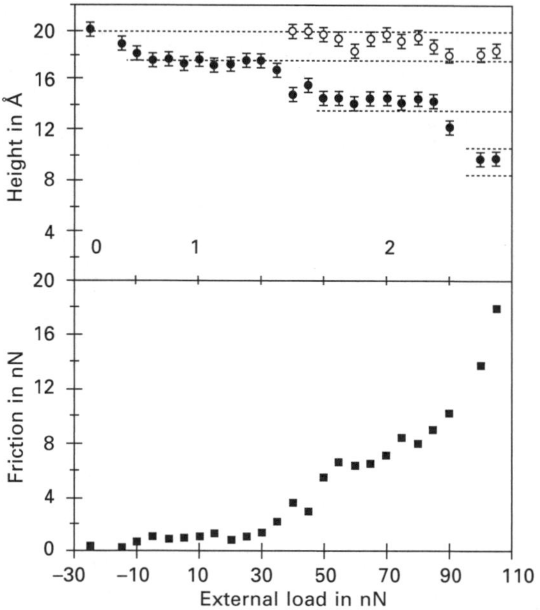

12.4 (top) Height of a C_{12} alkylsilane island vs. load (filled circles). Open circles correspond to the height of the film after the decrease in the load to the lowest practical value for imaging (-25nN). Dashed lines correspond to heights in discrete steps. (Bottom) A step-like behavior correlated with the height decreases is observed. Transition regions are shaded [19].

structures is shown in (Fig. 12.5 left). A disordered structure (not shown) was observed for hexanethiol monolayers ($n = 6$), because a poorly packed structure increased the possibilities of numerous defects and energy dissipation. For long chain alkylsilanes on mica, no long-range order could be detected by the friction image, although residual short-range order was observed (Figure 12.5 right). A possible explanation for this difference in behavior could be based on the cross-linking of the thiol head groups. In thiols, the disulfide bonds formed between pairs of atoms probably produced a superstructure with C hexagonal periodicity relative to the basic structure. It is important to note that the distance between Si atoms in adjacent molecules forming Si-O-Si bonds (2.5 Å) was larger than their chain radius (2.25 Å). This observation implies the presence of considerable chain distortions near the interface. Although the Si-O-Si networks were observed to form across the surface, the size of their cross-linking network was indeed limited. This behavior was probably at the origin of the lack of long-range order in silane based lubricants.

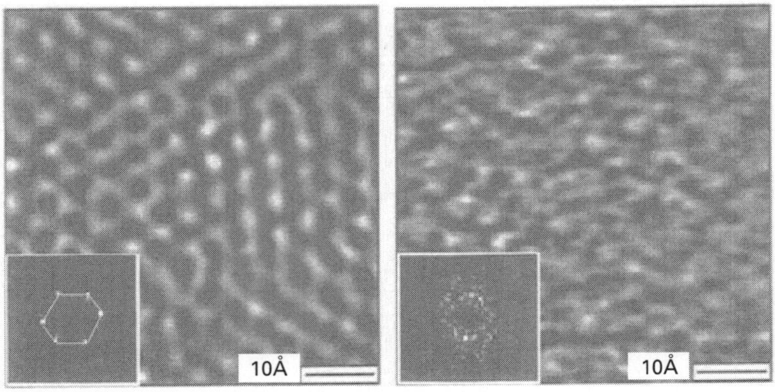

12.5 Frictional force images for dodecanethiol C_{12} on gold and octadecyltriethoxysilane C_{18} on mica. The corresponding 2D Fourier transform power spectra are shown on the bottom left. Alkanethiol monolayers exhibit a hexagonal ordered structure. Residual short-range order is observed for alkylsilane monolayers [20].

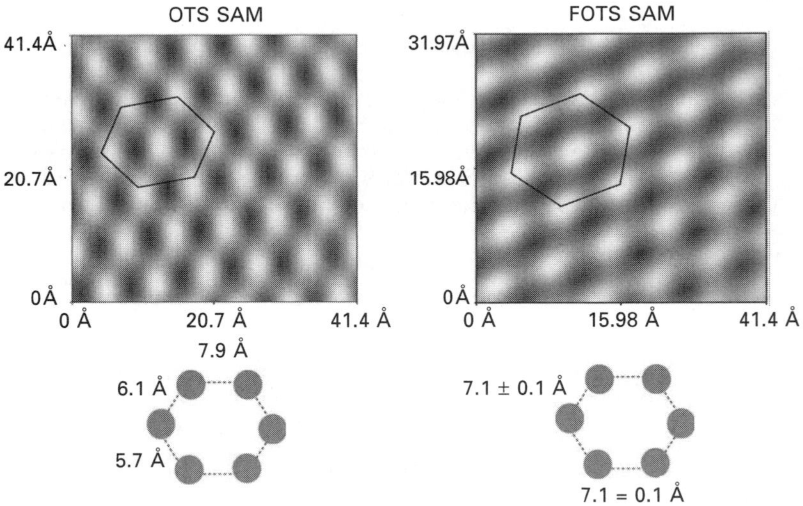

12.6 Lateral force images for OTS and FOTS self-assembled monolayers on mica. Both have hexagonal structures. However, there is lack of order in the hexagonal structures of OTS SAM, while FOTS SAM exhibits a regular hexagonal lattice with a lattice constant = 7.1±0.1 Å [21].

12.3.2 Chain configuration

Devaprakasam *et al.* [21] used the LFM technique to characterize the properties of self-assembled perfluorooctyltrichlorosilane (FOTS) and octadecyltrichlorosilane (OTS) monolayers. FOTS has a rigid rod-like helical structure and OTS has a zigzag backbone configuration. Figure 12.6 illustrates lateral force images of FOTS and OTS self-assembled monlayers on mica. It can be observed that FOTS displays a regular hexagonal lattice and more symmetrical structures in contrast to OTS, indicating that the helical structures were more ordered than zig-zag backbone molecular configuration.

12.3.3 Friction anisotropy

Carpick *et al.* [22] performed LFM measurements on polydiacetylene coated as monolayer films on mica to correlate the molecular structure of the lubricant with friction anisotropy and friction asymmetry behaviors.

Friction anisotropy refers to the variation of friction relative to the orientation angle between sliding surfaces, or with the sliding direction itself. It was observed that the friction force varied substantially from one domain to the next, but was nearly uniform within each domain. Figure 12.7 illustrates the angular orientational dependence of the friction force, where θ represents the domain orientation measured independently via topographic AFM imaging

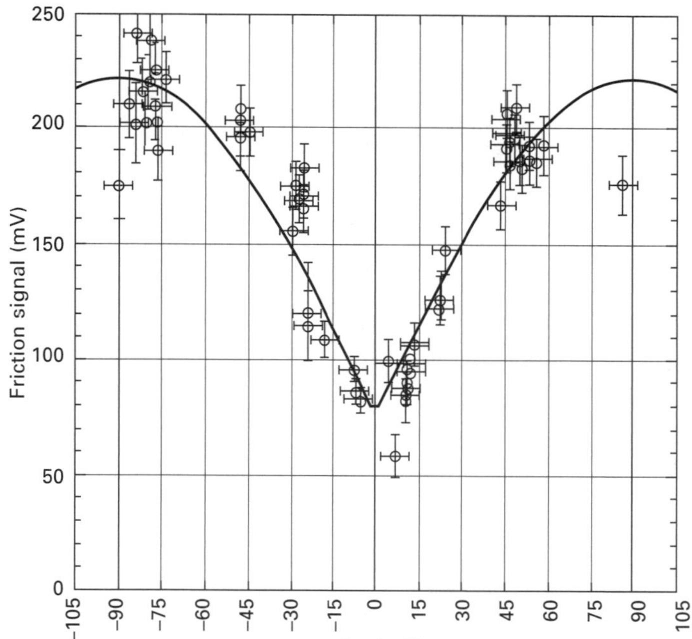

12.7 Friction raw signal (relative friction force) vs. orientation angle. 0° indicates sliding parallel to the backbone direction. Friction is calculated by taking the difference between forward and backward traverse images on single domains. The standard deviation is used for the friction error bar. The solid line represents the fit of equation (12) to the data [22].

and polarized fluorescence microscopy. A simple equation, given below, was used to model the angular dependence of friction force:

$$F_f = F_1 + F_2 |\sin\theta| \qquad 12.1$$

In this equation, F_1 represents the angle-independent contribution to friction and can be thought of as an intrinsic friction force between the tip and the film. F_2 represents an additional contribution to friction that arises when there is a component of sliding motion perpendicular to the polymer backbone direction. It is generally explained that anisotropic packing and/or ordering of the alkyl side chains and the anisotropic stiffness of the polymer backbone itself cause the anisotropic lateral film stiffness, hence resulting in anisotropic friction.

12.3.4 Friction asymmetry

Friction asymmetry refers to a change in friction behavior when the sliding direction is flipped by 180°. This effect is possibly induced by molecular groups that tilt from the surface's normal direction. This phenomenon was not clearly observed with a polydiacetylene monolayer film in Fig. 12.8(a),

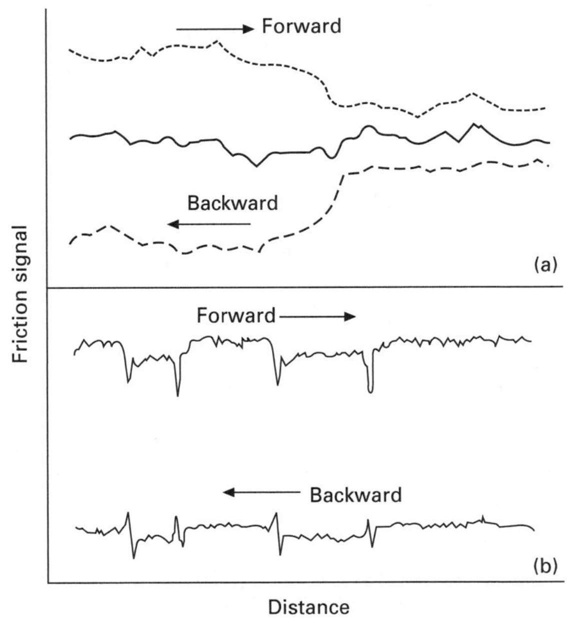

12.8 Friction vs. distance in a friction loop. In (a) there is no observable friction asymmetry on domains in Carpick's experiments. In contrast, (b) shows this effect in Bluhm's LFM testings.

probably because the spatial fluctuations of the friction forces may have obscured the effect. However, Bluhm *et al.* [23] observed friction asymmetry with a ferroelectric triglycine sulfate (TGS) (Fig. 12.8(b)). The friction contrast occurred not only between domains with different orientation, but also inside the domains. It was proposed that the crystallography of TGS, which features asymmetric molecular arrangement on the surface, caused an asymmetric surface potential detected by the tip.

12.3.5 Superlubricity

A superlubricity phenomenon was firstly observed by Hirano *et al.* [24] when a monocrystalline tungsten tip slid on a Si (001) surface. This phenomenon was explained on the basis that the sliding direction of the tip was parallel with the orientation of molecules on the surface. Lateral force microscopy techniques are appropriate for these types of measurements as LFM could magnify molecular anisotropy and asymmetry detecting super-lubricity behavior [25–27].

12.3.6 Chemical composition

The work by Overney *et al.* [28] using LFM confirmed that friction and elasticity measurements were indeed correlated. A monolayer composed of a mixture of hydrocarbon acid and fluorocarbon acid was examined simultaneously for the local elastic compliance and lateral forces. An island-sea structure was noted in both elasticity and friction images. The hydrocarbon island-like domains with the higher elastic modulus and lower friction were found to be surrounded by the fluorocarbon sea-like areas exhibiting lower elastic modulus and higher friction.

Energy dissipation at the molecular level was examined by Burns *et al.* [29, 30] on alkanethiol monolayers with –COOH and $–CH_3$ end groups. They found that under identical experimental conditions, the friction forces of monolayers containing the –COOH groups were higher than those containing the $–CH_3$ groups. A significant fraction of the thiol–COOH chains that attached to the substrate with a 30° tilt appeared to have contacted the tip [31]. The molecules seemed to stand up with one end anchoring to the surface and the other end interacting with the tip (Fig. 12.9). For the thiol–CH_3 chains, a similar tilting arrangement possibly occurred since they have similar backbone molecular chains. However, the effective number of standing thiol–CH_3 chains is not significant enough to prevent the tip from moving under relatively smaller forces. This behavior is likely caused by the relatively weak atomic interaction forces between the tip and the thiol–CH_3 chains.

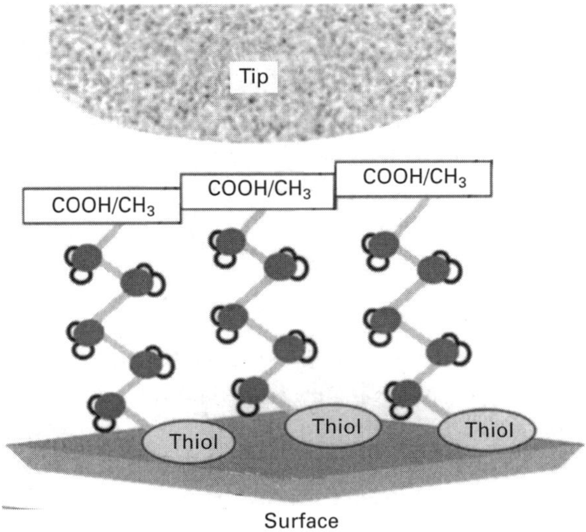

12.9 A scheme (not to scale) of a tip interacting with self-assembled head-tail molecular chains that are anchored to a surface. In every chain, the end with the thiol groups is attached on the surface, while the other with the – COOH/-CH_3 groups interacts with the tip.

12.3.7 Visco-elastic properties

Early in the 18th century, it was determined that if two surfaces adhere to each other, a finite friction force can arise, even in the absence of any externally applied load. This effective additional force is due to the molecular attraction between the two surfaces. Therefore, the general friction force might be composed of two parts. One is a constant force due to adhesion and the other is a force determined by the friction coefficient and is due to the externally applied load [32, 33]. However, a general relationship between adhesion and friction forces is not established, although in specific situations they may be correlated [34, 35].

Yoshizawa *et al.* [36] used a conventional surface force apparatus to study friction and adhesion in surfactant monolayers between two mica surfaces. They noticed that the adhesion property, which qualitatively correlates with the friction force, might be more likely given by adhesion hysteresis than adhesion force itself. The term adhesion hysteresis characterizes the energy dissipation when two surfaces approach and then retract from each other. In other words, it is the difference between the energy necessary for separating two surfaces and that restored when the surfaces are brought back together. The dissipated energy in adhesion hysteresis is calculated as the area of a closed loop during an approaching-retracting cycle (Fig. 12.10).

In order to clarify the correlation between adhesion hysteresis and friction force, the monolayers were classified into three categories: solid-like, liquid-

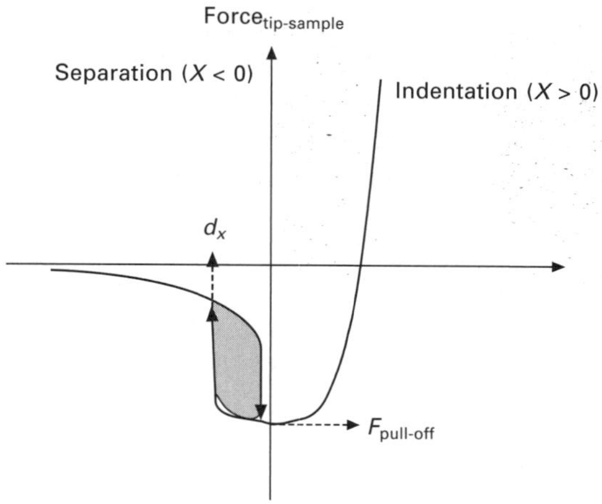

12.10 Theoretical force-distance curve presents the total interactions between a tip and a sample during indentation (x<0) or separation ($x > 0$). The adhesion hysteresis (dotted area) is approximated by the product of critical separation δ_x and adhesion forces $F_{pull-off}$.

Table 12.1 Friction and adhesion hysteresis of dihexadecyl dimethyl ammonium acetate (DHDAA) monolayers on mica at the temperatures: $T < T_m$, $T \approx T_m$, and $T > T_m$

Temperature (°C)	Status	Friction coefficient	Adhesion hysteresis
15	Solid like	Low	Low
25	Amorphous	High	High
35	Liquid like	Low	Low

like, and amorphous [37]. The classification was based on the difference between the testing temperature T and the material melting point T_m. A surfactant monolayer of dihexadecyl dimethyl ammonium acetate (DHDAA) was studied in dry air in the temperature range 15~35°C.

Table 12.1 shows that the friction force and adhesion increased and then decreased, with a maximum value being attained at 20~25°C. It was found in the literature that the melting temperature of DHDAA may be expected to be somewhere between 20 and 40°C [38]. If $T < T_m$, a monolayer might be considered to be in a solid-like state. During an approaching-retracting cycle or a back-forth sliding, there is little interdigitation of the chains or rearrangements of the surface molecular groups. The adhesion hysteresis and friction force are both small because the surfaces do not deform significantly.

At $T \approx T_m$, the monolayer chains become more flexible and mobile, but are still not in a truly melted or liquid-like state. This is the amorphous regime. As two surfaces come into contact or slide past each other, there is now significant interdigitation of the chains across the interface, as well as local rearrangements or reorientation of surface molecular groups [39, 40]. This interdigitation enhances the effective number of molecular contacts or "bonds" across the interface and results in an increased adhesion hysteresis as the surfaces have to break more bonds on separating than are recovered when coming into contact. Similarly, the friction force is relatively high, because a small degree of interpenetration causes a great increase in friction force. Analogous trends were observed in the profile of friction force between bulk polymer and rubber surfaces at temperatures close to their glass transition temperature [23].

At $T > T_m$, the interlocks among chains break down and the monolayer becomes truly liquid-like. The surface layer now behaves like a liquid so that while interdigitation readily occurs as soon as two surfaces come into contact, they can relax or disentangle just as rapidly and easily so that the system is always at equilibrium. There is therefore no adhesion hysteresis, just as one finds for a liquid surface where the surface tension is the same on expansion or contraction. For similar reasons, the friction force is expected to be low.

Furthermore, Marti *et al.* [41] applied the Johnson-Kendal-Roberts (JKR) model [42, 43] and calculated the dissipated energy in the interface between a silicon nitride tip and an oxidized silicon surface under aqueous electrolyte solutions. It was shown that a linear relationship between friction force and adhesion hysteresis existed. The relationship was built upon the dissipated energy in the system. The authors also found that the dissipated energy was a function of the pH of the solution hence implying that improved LFM resolution can be achieved by adjusting the electrolytic nature of the solution.

12.4 Types of textile lubricant

In general, textile lubricants are a mixture of formulated compounds rather a single component. The use of mixtures allows the optimization of the lubricity performance as well as minimizing the cost. Low molecular weight polymers have a natural affinity for textile fibers. In most cases, these polymers create self-assembled molecular structures on a fiber's surface. These polymers are fluids at process temperatures, and associatively slide with the fiber surface during the friction process. High molecular weight polymers are usually formulated as additives, which significantly reduce abrasion and wear to fibers. This is particularly important in dynamic and high-speed operations. Waxes, such as hydrocarbons, polyethylene, and amides, are traditional boundary lubricants that function in both the low-speed (fiber to fiber) and the high-speed (fiber to metal, or fiber to ceramic) processes. Waxes tend to act as solid layers on the fiber's surface enhancing its lubricity. Silicon oils are also commonly chosen as lubricants for their low viscosity, low volatility, and good affinity for most textile fibers. All lubricants are usually applied to textile materials with other finishing agents to assist in processing. The lubricants most generally used in the textiles are fatty acids, mineral oils, as well as synthetic compounds, such as the ethoxylated alcohols, the ethoexylated acids, and the silicone oils. Table 12.2 lists their chemical structures, properties, and common applications in textile processing.

A fatty acid is a carboxylic acid, often coupled to a long unbranched aliphatic chain (C_4~C_{22}), which can be either saturated or unsaturated. There are only carbon and hydrogen atoms in each molecule, except for the carboxylic acid group at the end. The unbranched aliphatic chain can be easily packed together tightly due to a linear structure. This characteristic allows fatty acid molecules to form self-assembled structures in some situations. Fatty acids and their derivatives can function as effective lubricants in many applications. Erucoy glycolates are usually found as lubricants for cotton yarn [44]. A patent reported that diesters of glycolic acid are also effective as yarn lubricants, which additionally inhibit the growth of bacteria and fungi on the material [45].

Mineral oil is often used in a number of applications for its relatively low cost. Solvent-refined mineral oil is found in a commercial textile lubricant blend, which also contain methyl oleate and solvent-refined peanut oil [46]. Mixture of low viscosity shear reducing mineral oil and high molecular weight oil-soluble polymers have been also reported as a fiber lubricant [47].

Since the 1950s synthetic lubricants have become competitive enough in cost and functionality to be selected for textile applications. For example, the polyglycols, high molecular weight polymers of ethylene or propylene oxide, are currently available in a wide range of viscosities. The chemical structures for a polyglycol are shown in Fig. 12.11. Polyglycols have high flash points, good viscosity-temperature properties, low wax-free pour points, and acceptable shear stability. Water-soluble polyglycols are used primarily as base lubricants for spin finishing of textured fibers, which also can reduce fire hazards in processing and improve oxidation resistance [48]. Polyglycols and other similar compounds, such as ethylene glycols and propylene glycols, have been commercialized as effective textile processing lubricants.

Silicon oils are silicon analogs of carbon-based organic compounds, which can form relatively long and complex molecules based on silicon rather than carbon atoms. The most common silicon oil is polydimethylsiloxane (PDMS), which is exclusively used as an additive in the textile lubricant market. A model of this molecule is shown in Fig. 12.12 with the orientation of the PDMS molecules on a fiber surface. PDMS has oxygen atoms that attach to the polymeric fiber surfaces while the methyl groups orient away from the surfaces. The methyl groups are known to reduce friction and enhance abrasion resistance.

Table 12.2 Characteristics and properties of frequently used chemicals in textile lubrications

Lubricant			Chemical structure	Characteristics	Textile lubricant application
Fatty acids	Saturated Unsaturated		$CH_3(CH_2)_nCOOH$, n: 2~20 $CH_3(CH_2)_n(CH{=}CHCH_2)q(CH_2)_mCOOH$, n: 1, 2, 3, 4, 5, 7; m: 2, 6, 10; q:1 ~ 4	Easy packing of molecules on an interface	Cotton yarn lubricant
	Mineral oils		$CH_3(CH_2)_nCH_3$, n:13~38	Low viscosity Low cost Large production	Base component Fiber lubricant
Synthetic lubricants	Polyalkylene glycols		$RO(EO)_m(PO)_n(EO)_mH$ $RO(PO)_{m'}(EO)_{n'}(PO)_{m'}H$ EO: $-CH_2CHO-$; PO: $-CH_2CO(CH_3)-$ R: alkyl; m/m′≥0; n/n′≥0;	Affinity to synthetic fibers Self-assembled structures formation; Resistance to bacteria and contaminants	Spin finish for synthetic fiber lubricant
	Silicon oils	Polydimethyl-siloxane Silicon polyether	$CH_3[SiO(CH_3)_2]_nSi(CH_3)_3$ $-(EO)_m(PO)_n[SiO(CH_3)2](PO)_n(EO)_m-$ EO: $-CH_2CHO-$; PO:$-CH_2CO(CH_3)-$ $m \geq 0$; $n \geq 0$;	Molecular orientation on fibers Softness	Lubricant additives Fabric softener

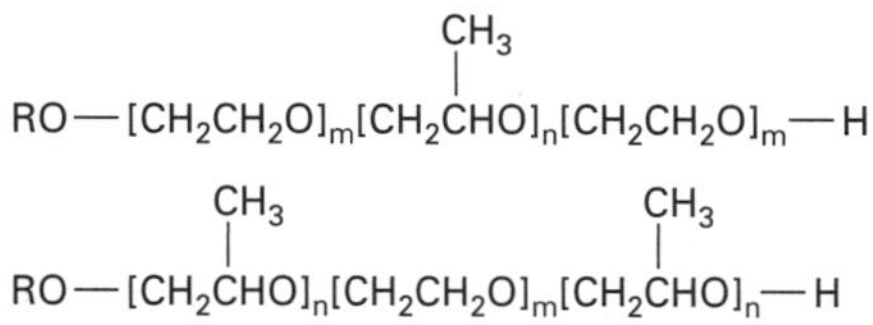

12.11 Commercial polyglycol triblock copolymers with alkyl groups R at the end.

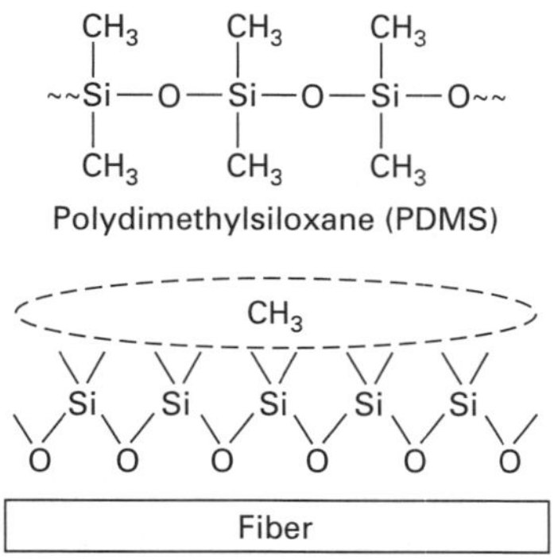

12.12 PDMS molecule orientation on a fiber surface provides the lubricity and softness to the fibers.

12.5 Boundary lubrication phenomena in textile materials

12.5.1 Materials and methods

Recently, Li and Hinestroza (the authors) initiated lateral force microscopy research work on the lubrication phenomena for three polymers commonly spun into fibers by the textile industry: cellulose, polyethylene, and polypropylene. The purpose of this work was aimed at studying boundary lubrication behavior. Model surfaces of these polymers were created using traditional spin-coating techniques [3, 15, 49–51]. These specimens were immersed in a liquid cell and studied using LFM. The LFM method allowed the manipulation of a micro-lubrication system at a molecular scale due to the micro dimensions of the contact between the tip and a surface. The lubricants chosen in this particular study were Pluronic products manufactured by BASF and one commercial product (LN336-100I) manufactured by Goulston Technologies Inc.

Pluronic lubricants, commonly used in textile topical treatment formulations, are composed of ABA triblock copolymers consisting of polyoxyethylene (PEO) as the A block and polyoxypropylene (PPO) as the B block. It has been reported that these molecules can absorb on a hydrophobic surface with the more hydrophobic PPO block anchoring on the surface and the PEO block extending into the surfactant solution as tail [52–54]. The commercial lubricant LN336-100I has been reported to be a mixture of standard polypropylene finishes and silicone surfactants although the exact chemical nature of LN336-100I remains proprietary. This mixture is water-dispersible and can be applied at concentrations below and above the critical micelle concentration (cmc) of 0.01%.

12.5.2 Self-assembled molecular structures and micellization

When a tip in lateral force microscopy is sliding on a surface, lateral force profiles can be measured both in air (no lubricant applied) and in solution. Figure 12.13 shows lateral force profiles for a cellulose surface imaged in air and while immersed in Pluronic lubricant P65 solutions. During these experiments, the lubricant was dissolved in ethanol aqueous solutions at various levels of ethanol concentration (22%, 38%, 52%, 66%, and 87%). It is observed that the friction forces measured in air are significantly larger than those in P65 solutions, confirming the lubrication characteristics of P65. However, the force profiles in the five solutions are undistinguishable, making the effect of ethanol concentration unimportant.

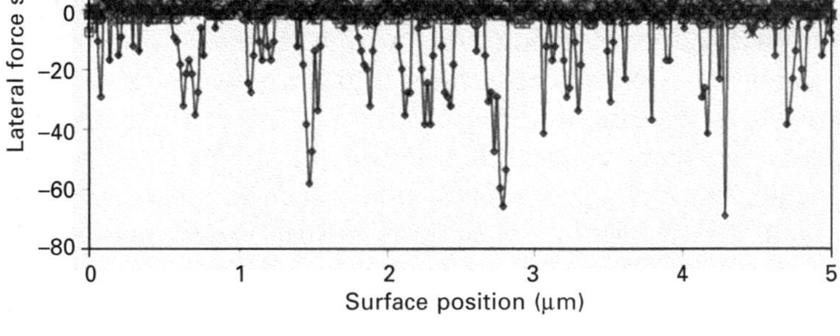

12.13 One line scanning profiles for the cellulose coated surface both in the Pluronic lubricant P65 solutions and in air. P65-1~P65-5 represent a series of ethanol solutions with the increase of ethanol concentration. The lateral forces in air are substantially larger than those in lubricant solutions. The force profiles in the five solutions are similar, without difference in amplitude.

Figure 12.14 displays a line scan across a polyethylene surface tested both in air and in the commercial lubricant solution. A static friction force is seen in Fig. 12.14. It is known that when a tip starts to move forward on a sample's surface, there is a static friction force on the tip [15, 55]. The static friction force was evident in the force profile imaged in air during the forward and the backward scanning. However, when the tip was sliding on a surface coated with the lubricant solution, the high force corresponding to the static friction disappeared and it was replaced by a regularly oscillating force curve. As expected, and seen in Fig. 12.15, the amplitude of force curve in the solution was smaller than that in the air. The distance between two neighboring peaks in solution was regular and approximately 4.7 nm. These regular peaks possibly indicate that lubricant molecules were aligned along PE surface creating regular oscillation in the force profiles. As stated in the previous section, it is possible that these lubricant molecules align themselves in a preferred direction because of secondary molecular interactions in the interface and that these assemblies could influence the friction behavior.

The LFM technique not only reveals the magnitude of the friction forces but also the molecular structure on the surface. As seen in Fig. 12.16, self-assembled monolayer (SAM) of the commercial lubricant LN336-100I can be observed on the polyethylene surface. The lubricant layer was not conformal but formed round-shape domains that packed with an average radius of 8~10 μm. These domains were found to be evenly distributed on the polyethylene surface. This behavior shows that the interaction between the lubricant monolayer and the polymer surface channels the lubricant molecules into the

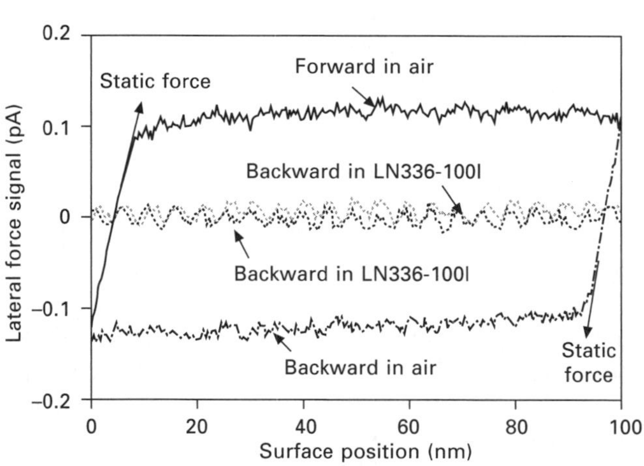

12.14 One line scanning profiles (backward and forward) of a PE surface imaged in air and while immersed in commercial lubricant LN336-100I. The lateral forces from the sample immersed in lubricant solution were smaller than those tested in air.

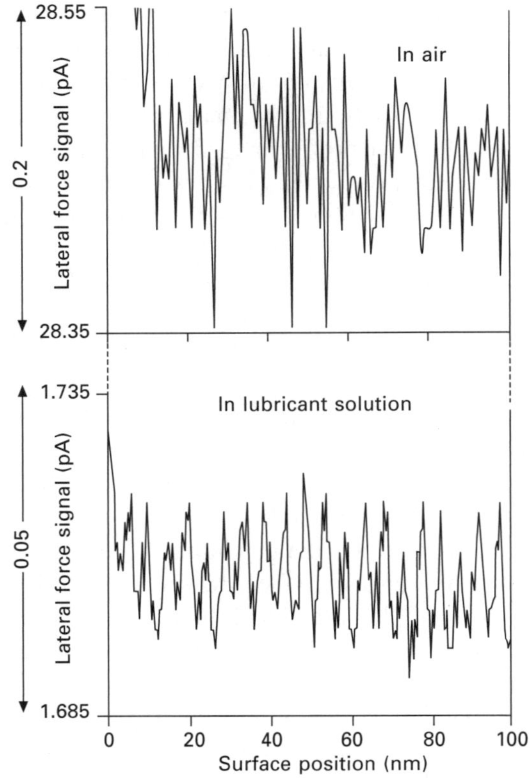

12.15 Magnified views of the lateral force profiles on the PE surface both in air and in lubricant solution. The fluctuation of the curve in air is larger and more random than that in lubricant solution. The distance between two neighboring peaks on the force curve in lubricant solution is approximately constant.

regular domains through molecular aligning/ordering. [56, 57]. The micelle forming characteristics of pluronic surfactants in the aqueous solutions have attracted much attention academically and commercially. Freeze-fracture transmission electron microscopy is conventionally employed to visualize this micellization behavior. However, AFM scanning using a liquid cell is equally capable of discovering these micelles with almost molecular resolution but significant lower sample-preparation work. Fig. 12.17 illustrates distinguishably ribbon-like and rod-like features of Pluronic surfactants on cellulose surface. It is hypothesized that these self-assembled molecular structures were formed by interaction between the lubricant molecules (P65) with the cellulose surface.

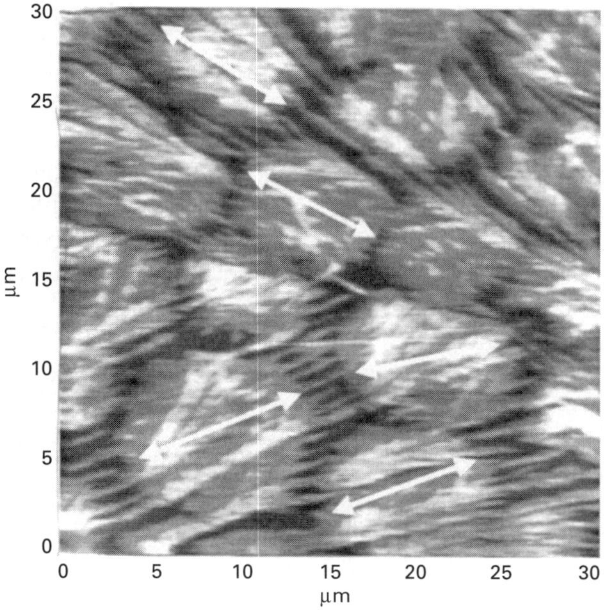

12.16 Lateral force image of a self-assembled monolayer (SAM) on a PE coated surface. The average radius of the domains are in the range of 8~10 μm.

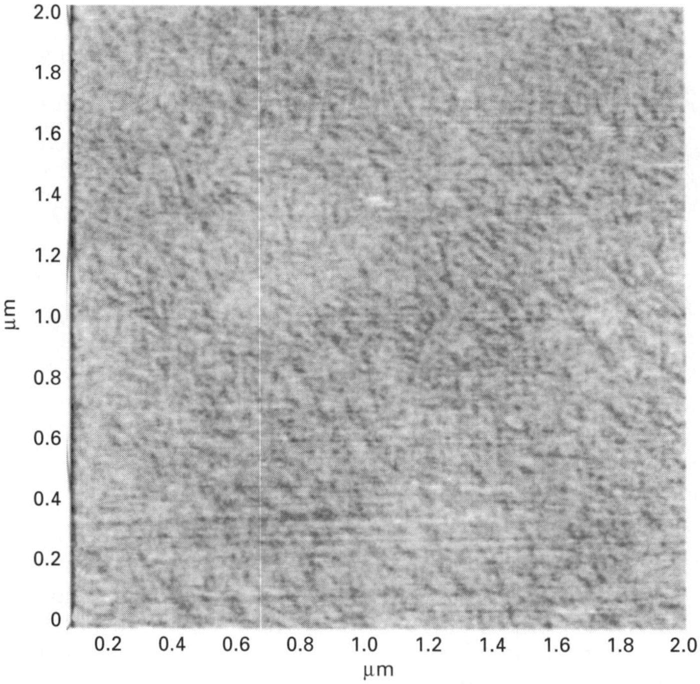

12.17 Lateral force image of a self-assembled monolayer (SAM) on a cellulose coated surface. Ribbon-like structures are seen on the surfaces.

12.5.3 Hypothesis of molecular self-assembled structures controlling friction behavior

Figure 12.18 illustrates the current hypothesis of Li and Hinestroza [58] that particular conformations of lubricants determine friction behavior. In this case, LFM images of cellulose, polyethylene, and polypropylene, surfaces, coated with a commercial lubricant, are presented. The second level of Fig. 12.18 illustrates the conceived conformations of molecular assembled structures and the third level (bottom) shows the friction force curves proposed according to topographical images.

While the samples were coated with the same lubricant, the LFM images obtained are clearly different, highlighting the effect of the three polymer surfaces on the lubricant molecules. On the cellulose surface, the 30 μm image may be recognized as a porous-like surface, but the 3 μm image reveals local orientation formed by the lubricant molecules. Lubricant molecules appear to randomly lay on the surface without a preferred orientation. The molecular chains may be aligned only in very small local areas, giving short range ordered structures. This arrangement leads to a friction force curve that oscillates randomly.

The molecular conformations of the lubricant molecules on the polyethylene surface indicate the creation of large domains. The friction force therefore

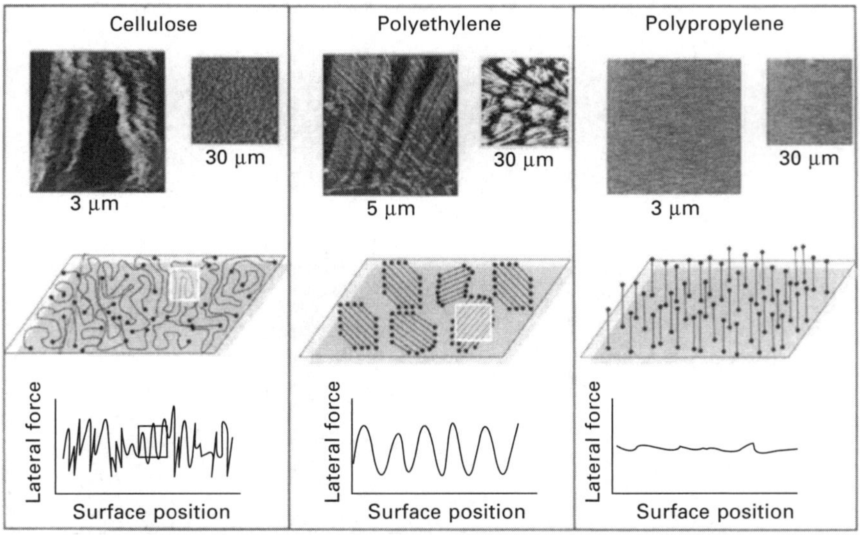

12.18 Molecular assembly conformations of the commercial lubricant on cellulose, polyethylene, and polypropylene films. (Top) images: topography; (middle) conformations: self-assembled molecular structures; (bottom) diagrams: friction force curves for a line scanning.

has a regular amplitude of oscillation for each domains formed on the surface. For the case in which the polypropylene surface is used, the surface appears uniform and featureless. A standing-up conformation of the molecules is proposed for these lubricant chains, which may be supported vertically on the surface. A thick lubricant layer in term of the chain length may be formed resulting in a relatively flat friction force curve.

These LFM profiles are evidences that the chemistry of polymers and lubricants play a significant role in the lubrication phenomena and it should be explored in depth further.

12.5.4 Friction coefficients

In a topographical view of a polyethylene surface coated with a commercial lubricant (Fig. 12.19), a bump in the middle of the specimen can be detected. The friction forces on the bump are lower than those on the valley of the samples shown in the lateral force image. This particular bump is created by self-assembled lubricant molecules hence exhibiting low friction.

Another interesting lubrication behavior probed via LFM is illustrated in Fig. 12.20. It is noted that with the increase in normal forces the friction

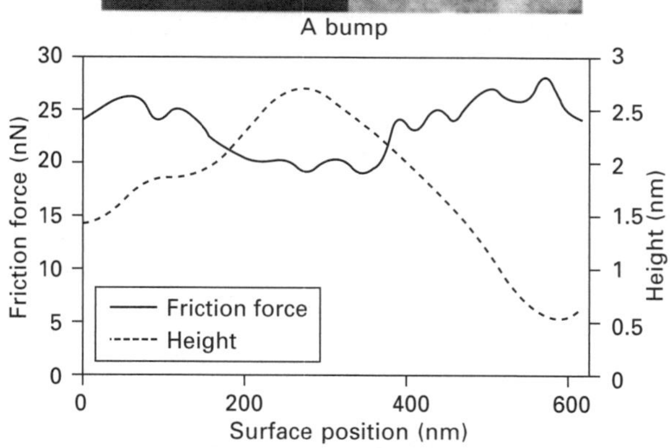

12.19 A bump on a polyethylene film (seen in the topographical image) corresponds to lower friction force than the forces represented by the surrounding low areas (seen in lateral force image). The heights (dashed line) and friction forces (solid line) on one cross scanning line (dash line on the images) also show opposite trends.

force increases. That is to say: with an increase in pressure, the lubricant layers become thinner and, as expected, leads to an increase in friction. This phenomenon is observed in the cases of both polyethylene and polypropylene, although the nature of the effects may be somewhat different. The four lubricants used in this experiment were copolymers of polyethylene oxide (EO) and polypropylene oxide (PO).

As an example, EO(133)PO(50)EO(133) molecules (Fig. 12.20) contains 50 repeat units of PO in the middle and 133 repeat units of EO on both ends.

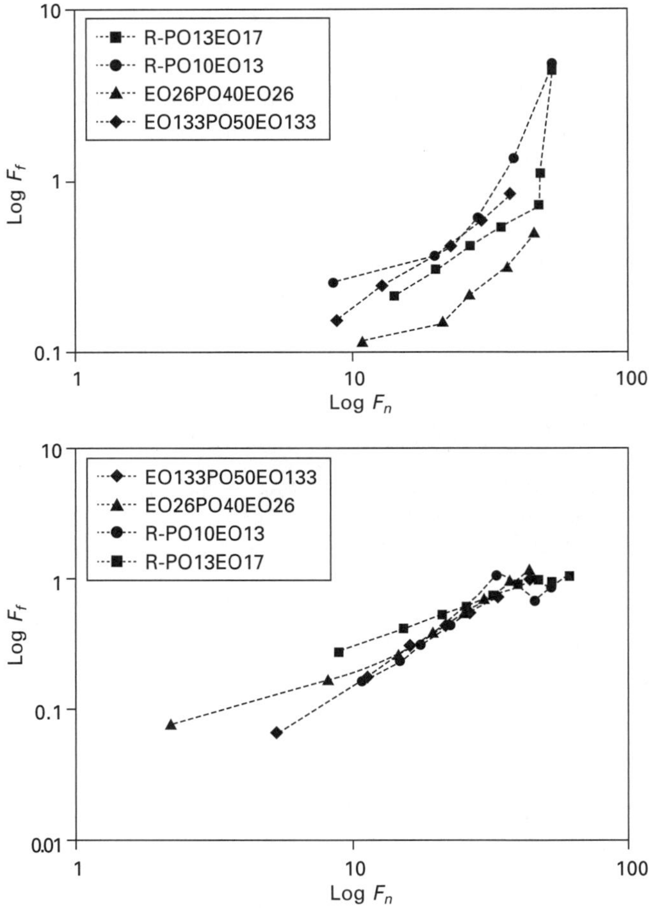

12.20 The relationship of friction force (F_f) and normal force (F_n) on polyethylene (top) and polypropylene (bottom) films using four types of lubricants: (a) EO133PO50EO133, (b) EO26PO40EO26, (c) R-PO10EO13, and (d) R-PO13EO17 (EO: polyethylene oxide; PO: polypropylene oxide; R: alkyl groups). The increase in friction as a function of normal force is valid for both polyethylene and polypropylene surfaces.

R-PO(10)EO(13) molecules contain another R alkyl groups other than the PO and EO groups. With the EO groups exposed and the PO shielded in all cases, the results show that interaction forces between the polymer and the finishes is sensitive to the chemical change in the lubricants in the case of polyethylene (PE), but not in the case of polypropylene (PP). This is more clearly shown in Fig. 12.21, which gives a comparison of the effects of finishes on the two films. In contrast to PE on which the finish constitution showed a significant effect, on PP all finishes gave approximately the same friction values.

12.6 Summary

In boundary lubrication, lubricants might form self-assembled molecular structures on polymer surfaces due to molecular interactions with the surface of the fiber. These interactions appear to determine the alignment modes of the molecules of a lubricant and hence its lubrication efficiency. Lateral force microscopy can be effectively used not only to measure the friction forces in the interfaces but also to probe on a polymer surface the nature of the self-assemblage of these lubricant molecules. These assemblies are expected to control the friction behavior of the lubricated surfaces. It is foreseen that understanding of the molecular structure conformations of the lubricant's

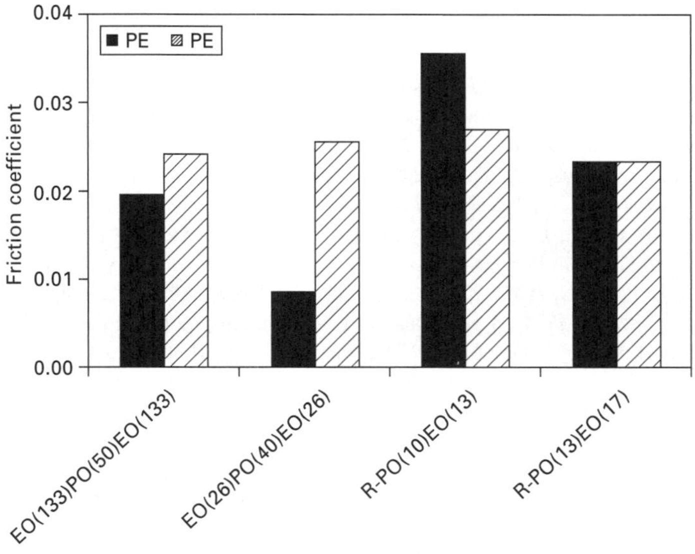

12.21 The coefficients of friction (COF) provided by different lubricants on polyethylene and polypropylene films. The COFs for polypropylene films did not change much with lubricant composition.

chemical compounds on a fiber surface will serve as a fundamental guide for the formulation of improved finishes and lubricant additives.

12.7 Glossary

F_f	Friction force
F_l	Lateral force
F_n	Normal force
AFM	Atomic force microscopy
DHDAA	Dihexadecyl Dimethyl Ammonium Acetate
FFM	Friction Force Microscopy
FOTS	Perfluorooctyltrichlorosilane
JKR	Johnson-Kendal-Roberts Model
LFM	Lateral force microscopy
OTS	Octadecyltrichlorosilane
PDMS	Polydimethylsiloxane
PEO	Polyethylene Oxide (Polyoxyethylene)
PPO	Polypropylene Oxide (Polyoxypropylene)
SAM	Self-Assembled Monolayer
TGS	Triglycine Sulfate

12.8 References

1. Binning G, Quate C F and Gerber Ch, 'Atomic force microscope', *Phys. Rev. Lett.*, 1986, **56**(9), 930.
2. Mate C M, McClelland G M, Erlandsson R and Chiang S, 'Atomic-scale friction of a tungsten tip on a graphite surface', *Phys. Rev. Lett.*, 1987, **59**, 1942.
3. Wang A, Jiang L, Mao G and Liu Y, 'Direct force measurement of Silicone- and Hydrocarbon-based ABA Triblock surfactants in Alcoholic media by Atomic Force Microscopy', *J. Coll. Inter. Sci.*, 2002, **256**, 331.
4. Heinz W F, Hoh J H, 'Spatially resolved force spectroscopy of biological surfaces using the Atomic Force Microscope', *Trends in Biotechnology*, 1999, **17**, 143.
5. Israelachvili J N, *Intermolecular and surface forces,* Academic press, 1992.
6. Radmacher M, Fritz M and Hansma P K, 'Imaging soft samples with the Atomic Force Microscope: Gelatin in Water and Propanol', *Biophys. J.*, 1995, **69**(1), 264.
7. Rief M, Oesterhelt F, Heymann B and Gaub H E, 'Single molecule force spectroscopy on polysaccharides by Atomic Force Microscopy', *Science*, 1997, **275**(5304), 1295.
8. Lee G U, Chrisey L A and Colton R J, 'Direct measurement of the forces between complementary strands of DNA', *Science*, 1994, **266**(5186), 771.
9. Allen S, Chen X, Davies J, Davies M C, Dawkes A C, Edwards J C, Roberts C J, Sefton J, Tendler S J B and Williams P M, 'Detection of antigen-antibody binding events with the Atomic Force Microscope', *Biochemistry*, 1997, **36**(24), 7457.
10. Dammer U, Hegner M, Anselmetti D, Wagner P, Dreier M, Huber W and Guntherodt H-J, 'Specific antigen/antibody interactions measured by Force Microscopy', *Biophys. J.*, 1996, **70**(5), 2437.

11. Florin E-L, Moy V T, and Gaub H E, 'Adhesion forces between individual ligand-receptor pairs', *Science*, 1994, **264**(5157), 415.
12. Butt H-J, 'Measuring electrostatic, van der Waals, and hydration forces in electrolyte solutions with an Atomic Force Microscope', *Biophys. J.*, 1991, **60**(6), 1438.
13. Butt H-J, 'Measuring local surface charge densities in electrolyte solutions with a Scanning Force Microscope', *Biophys. J.*, 1992, **63**(2), 578.
14. Ogletree D F, Carpick R W and Salmeron M, 'Calibration of frictional forces in Atomic Force Microscopy', *Rev. Sci. Instrum.*, 1996, **67**(9), 3298.
15. Liu Y, Wu T and Evans D F, 'Lateral force microscopy study on the shear properties of self-assembled monolayers of Dialkylammonium surfactant on Mica', *Langmuir*, 1994, **10**(7), 2241.
16. Liu Y, Evans D F, Song Q and Grainger D W, 'Structure and frictional properties of self-assembled surfactant monolayers', *Langmuir*, 1996, **12**(5), 1235.
17. Gnecco E, Bennewitz R, Pfeiffer O, Socoliuc A and Meyer E, 'Friction and wear on the atomic scale' in Bhushan B, Ed., *Nanotribology and nanomechanics: An introduction,* Springer, New York, 2005.
18. Cain R G, Biggs S and Page N W, 'Force calibration in lateral Force Microscopy', *J Coll. Inter. Sci.*, 2000, **227**(1), 55.
19. Barrena E, Kopta S, Ogletree D F, Charych D H and Salmeron M, 'Relationship between friction and molecular structure: Alkylsilane lubricant films under pressure', *Phys. Rev. Lett.*, 1999, **82**(14), 2880.
20. Lio A, Charyc D H and Salmeron M, 'Comparative Atomic Force Microscopy study of the chain length dependence of frictional properties of Alkanethiols on gold and Alkysilanes on Mica', *J. Phys. Chem. B.*, 1997, **101**(19), 3800.
21. Devaprakasam D, Khatri O P, Shankar N and Biswas S K, 'Boundary lubrication additives for Aluminum: A journey from nano to macro tribology', *Tribol. Intern.*, 2005, **38**(11–12), 1022.
22. Carpick R W, Sasaki D Y and Burns A R, 'Large friction anisotropy of a Polydiacetylene monolayer', *Tribol. Lett.*, 2000, **7**(2–3), 79.
23. Bluhm H, Schwarz U D, Meyer K P and Wiesendanger R, 'Anisotropy of sliding friction on the Triglycine sulfate (010) surface', *Appl. Phys. A*, 1995, **61**(5), 525.
24. Hirano M, Shinjo K, Kaneko R and Murata Y, 'Anisotropy of frictional forces in Muscovite Mica', *Phys. Rev. Lett.*, 1991, **67**(19), 2642.
25. Hirano M, Shinjo K, Kaneko R and Murate Y, 'Observation of superlubricity by Scanning Tunneling Microscopy', *Phys. Rev. Lett.*, 1997, **78**(8), 1448.
26. Overney R M, Takano H, Fujihira M, Paulus W and Ringsdorf H, 'Anisotropy in friction and molecular stick-slip motion', *Phys. Rev. Lett.*, 1997, **72**(22), 3546.
27. Takano H and Fujihira M, 'Study of molecular scale friction on Stearic acid crystals by Friction Force Microscopy', *J. Vac. Sci. Technol. B*, 1997, **14**(2), 1272.
28. Overney R M, Meyer E, Frommer J, Güntherodt H-J, Fujihira M, Takano H and Gotoh Y, 'Force microscopy study of friction and elastic compliance of phase-separated organic thin films', *Langmuir*, 1994, **10**(4), 1281.
29. Burns A R, Houston J E, Carpick R W and Michalske T A, 'Friction and molecular deformation in the tensile regime', *Phys. Rev. Lett.*, 1999, **82**(6), 1181.
30. Burns A R, Houston J E, Carpick R W and Michalske T A, 'Molecular level friction as revealed with a novel scanning probe', *Langmuir*, 1999, **15**(8), 2922.
31. Dubois L H and Nuzzo R G, 'Synthesis, structure, and properties of model organic surfaces', *Annu. Rev. Phys. Chem.*, 1992, **43**, 437.
32. Dowson D, *History of tribology*, Longman: London and New York, 1979.

33. Derjaguin B, *Research in surface forces,* Consultants Bureau: New York, 2, 312, 1966.
34. Rabinowicz E, *Friction and wear of materials*, John Wiley: New York and London, 1965.
35. Homola A M, Israelachvili J N, McGuiggan P M, Gee M L, 'Fundamental experimental studies in tribology: The transition from "Interfacial" friction of undamaged molecularly smooth surfaces to "Normal" friction with wear', *Wear*, 1990, **136**(1), 65.
36. Yoshizawa H, Chen Y and Israelachvili J, 'Fundamental mechanisms of interfacial friction: 1. Relation between adhesion and friction', *J. Phys. Chem.*, 1993, **97**(16), 4128.
37. Zeng H, Maeda N, Chen N, Tirrell M and Israelachvili J, 'Adhesion and friction of Polystyrene surfaces around T_g', *Macromolecules*, 2006, **39**(6), 2350.
38. Adam N K and Parkburst A K G, 'Solubility of some Paraffin-chain salts', *Trans. Faraday Soc.*, 1946, **42**, 523.
39. Chen Y L, Helm C and Isrealachvili J N, 'Measurements of the elastic properties of surfactant and lipid monolayers', *Langmuir*, 1991, **7**(11), 2694.
40. Maugis D, 'Subcritical crack growth, surface energy, fracture toughness, stick-slip and embrittlement', *J. Mater. Sci.*, 1985, **20**(9), 3041.
41. Marti A, Hähner G and Spencer N D, 'Sensitivity of frictional forces to pH on a nanometer scale: A lateral force microscopy study', *Langmuir*, 1995, **11**(12), 4632.
42. Johnson K L, Kendall K and Roberts A D, 'Surface energy and the contact of elastic solid', *Proc. R. Soc. London A*, 1971, **324**(1558), 301.
43. Kendall K, 'Inadequacy of Coulomb's friction law for particle assemblies', *Nature* (London), 1986, **319**(6050), 203.
44. Moreau J P and Bailey A V, 'Erucoyl Glycolates as lubricants for cotton yarns', *J. Coat. Fab.*, 1988, **18**(Oct.), 85.
45. Moreau J P, Bailey A V, and DeLucca A J, *United States Patent 5015419*, 1989.
46. Sun C, Baird M, Anderson H. A and Brydon D L, 'Separation of broadly distributed nonylphenol Ethoxylates and determination of Ethylene Oxide oligomers in textile lubricants and emulsions by high-performance liquid chromatography', *J. Chroma. A*, 1996, **731**(1–2), 161–169.
47. Childers J T and Fleming I D, *European Patent 261415*, 1988.
48. Rodenberg D G, *European Patent 0796908*, 1997.
49. Helman J S, Baltensperger W and Holyt J A, 'Simple model for dry friction', *Phys. Rev. B.*, 1994, **49**(6), 3831.
50. Bennewitz R, Gyalog T, Guggisberg M, Bammerlin M, Meyer E and Güntherodt H-J, 'Atomic-scale stick-slip processes on Cu(111)', *Phys. Rev. B*, 1999, **60**(16), R11301.
51. Zwörner O, Hölscher H, Schwarz U D and Wiesendanger R, 'The velocity dependence of friction forces in point-contact friction', *Appl. Phys. A*, 1998, **66** (Supplement 1, Scanning Tunneling Microscopy/Spectroscopy and Related Techniques), S263.
52. Eskilsson K, Ninham B W, Tiberg F and Yaminsky V V, 'Effects of adsorption of low-molecular-weight triblock copolymers on interactions between hydrophobic surfaces in water', *Langmuir*, 1999, **15**(9), 3242.
53. Eskilsson K, Grant L M, Hansson P and Tiberg F, 'Self-aggregation of triblock copolymers at the solid silica-water interface', *Langmuir*, 1999, **15**(15), 5150.
54. Karin S, Claesson P M, Malmsten M, Linse P and Booth C, 'Properties of Poly(ethylene oxide)-Poly(butylene oxide) diblock copolymers at the interface between hydrophobic surfaces and water', *J. Phys. Chem. B*, 1997, **101**(21), 4238.

55. Liu Y and Evans D F, 'Structure and frictional properties of self-assembled surfactant monolayers', *Langmuir*, 1996, **12**(5), 1235.
56. Ruths M, 'Boundary friction of aromatic self-assembled monolayers: Comparison of system with one or both sliding surfaces covered with a Thiol monolayer', *Langmuir*, 2003, **19**(17), 6788.
57. Ruths M, Alcantar N A and Israelachvili J N, 'Boundary friction of aromatic silane self-assembled monolayers measured with the Surface and Friction Force Microscopy', *J. Phys. Chem. B*, 2003, **107**(40), 11149.
58. Li Y, Hinestroza J P and Rojas O J, 'Lateral Force Microscopy on thin films of cellulose', *Abstracts of Papers, 233rd ACS National Meeting*, Chicago, IL, US, March 25–29, 2007, Cell-032.

Index